S0-BQI-293

CELL ADHESION AND MOTILITY

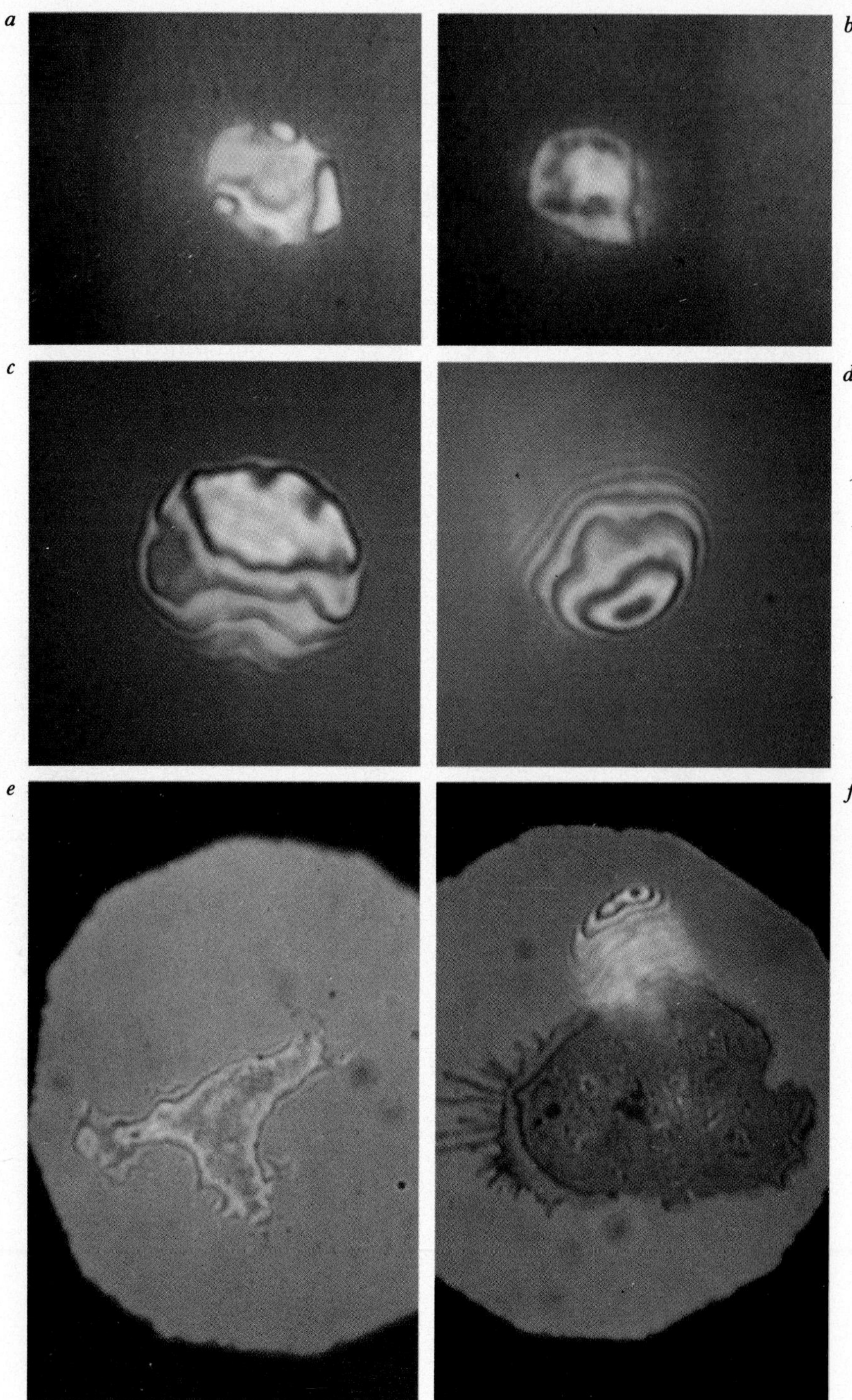

For explanation see p. 14.

THE THIRD SYMPOSIUM OF

THE BRITISH SOCIETY FOR CELL BIOLOGY

CELL ADHESION AND MOTILITY

EDITED BY

A. S. G. CURTIS

Professor of Cell Biology, University of Glasgow

J. D. PITTS

Reader in Biochemistry, University of Glasgow

CAMBRIDGE UNIVERSITY PRESS

CAMBRIDGE

LONDON · NEW YORK · NEW ROCHELLE

MELBOURNE · SYDNEY

Published by the Press Syndicate of the University of Cambridge
The Pitt Building, Trumpington Street, Cambridge CB2 1RP
32 East 57th Street, New York, NY 10022, USA
296 Beaconsfield Parade, Middle Park, Melbourne 3206, Australia

First published 1980

Printed in Great Britain at the
University Press, Cambridge

Library of Congress cataloguing in publication data
Main entry under title:
Cell adhesion and motility.
(The Symposium of the British Society for Cell Biology; 3)
Symposium organized by the British Society for Cell Biology.
Includes bibliographical references and index.
1. Cell adhesion – Congresses. 2. Cells – Motility – Congresses. I. Curtis, A. S. G. II. Pitts, J. D. III. British Society for Cell Biology. IV. Series: British Society for Cell Biology. Symposium – British Society for Cell Biology; 3. [DNLM: 1. Cell adhesion – Congresses. 2. Cell movement – Congresses. W3 BR459 no. 3 1979/QH604.2 C391 1979]
QH623.C44 574.8′75 79-53315

ISBN 0 521 22936 7

CONTENTS

PREFACE

Interest in cell adhesion and motility sprang initially from studies of developmental biology. In the course of embryogenesis many cells migrate, passively or actively, individually or cooperatively, and these migrations play a major role in the establishment of cell position which ultimately defines the anatomy and structure of the organism. Even in the seventeenth century there was some premonition that cell adhesion was an important feature of this process (Robinet, Jean Baptiste R., 1735–1820, *De la nature*, Amsterdam: E. van Harrevelt (1761):

Vous pourrez aussi vous représenter le germe sous la forme d'une éponge comprimée, dont par conséquent les cellules sont assaissées les unes sur les autres, étant vuides due fluide qui doit les tenir gonflées. Cette comparaison est d'autant plus naturelle que l'Anatomie a démontré que les solides du corps sont ou un tissu cellulaire, ou des paquets de fibres et de fibrilles cresses.

Les germes adherent entr'eux dans la semence: on y en a vu plusieurs groupes. Tant qu'ils sont ainsi réunis sous la forme d'un ver spermatique, ils n'ont pas la liberté de s'éntendre, et ainsi ils y restent toujours dans leur état de germe. La faculté de se dilater leur est ôtée tant par leur adhérence réciproque, que par la manière dont ils adherent les uns aux autres. La force d'adhérence est très-grande:. . .)

William Roux, often regarded as the father of embryology, initiated studies on dissociation and reaggregation of embryonic cells (not long after Schiefferdecker (in 1886) first obtained cell suspensions by trypsinising tissues) which are still continuing in many laboratories today.

By the 1950s it was also realised that adhesion and cell movement played important roles in the related phenomena of regeneration and metastasis and it is interesting to note that two earlier books on cell adhesion, both published in 1967 (Curtis, 1967; Weiss, 1967), examined the field in different ways, one reflecting the earlier interests of the developmental biologist and the other from the viewpoint of the oncologist. During the past twelve years a sizable congregation representing other biological interests has entered the study of adhesion or motility. Chief amongst these studies have been the following.

1. The investigation of the adhesion of blood platelets.
2. The normal behaviour and pathology of lymphocyte and leukocyte circulations which involved adhesion and motility leading the cells to, and perhaps away, from specific sites.
3. The investigation of cell adhesion in cell tissue culture, which has proved of interest both for other aspects of cell culture and for the practical advantages that such systems possess.
4. The adhesion of bacteria to one another, to eukaryote cells and to other surfaces such as ships' plates; a field of academic, pathological and economic interest.
5. The adhesions involved in mating-type reactions in micro-organisms.
6. The adhesive processes that occur between different species, for example of sponge, which may represent cell–cell recognition systems.
7. The investigation of adhesion processes in slime moulds, which provide apparently elegant models.
8. Biophysical studies treating cell adhesion as a branch of surface science.
9. Work on the interaction of the cell cytoskeleton with the points of adhesive contact.

We have attempted to strike a balance between these varied interests in our selection of authors for this volume. At first glance the titles may appear diverse, but the reader will soon realise that the different approaches have much in common and that in their own ways they all contribute to a better understanding of the basic mechanisms and functions of cell adhesion and motility and to the methods of investigation that may be used. Biophysical approaches marry with biochemical ones.

It would be pompous to suggest that this book is written at a time when light is dawning on a field where little understanding has come from much observation and considerable effort down the years. However, we feel that the present concentration of effort, summarised in part by this volume, must soon show at least in which direction the dawn lies.

We should like to thank Drs John Edwards, John Lackie and Geoffrey Moores who gave much help and wise advice in the planning as well as in the administration of the meeting. Our special thanks are due to Mrs U. Miller who provided her secretarial expertise both for the meeting and for the preparation of typescripts.

July 1979

ADAM CURTIS
JOHN PITTS

REFERENCES

CURTIS, A. S. G. (1967). *The Cell Surface: Its Molecular Role in Morphogenesis.* Logos Press.

WEISS, L. (1967). *The Cell Periphery, Metastasis, and other Contact Phenomena.* North-Holland.

Long-range forces and adhesion: an analysis of cell-substratum studies

DAVID GINGELL AND SHEILA VINCE

Department of Biology as Applied to Medicine, The Middlesex Hospital Medical School, London W1P 6DB, UK

1. INTRODUCTION

Trying to understand cell adhesion places one, to some extent, in the position of a blind man examining an elephant: what is detected is crucially dependent on where the subject is touched; but, with perseverance, something will almost certainly be found which fulfils expectations. Often the points which have been perceived appear to be an ill fit and have uncertain functional relationships. In discussing adhesion, we shall emphasise limited elements of the anatomy of the process, taking care not to fall into the trap of equating them with the whole. The major experimental and theoretical dichotomy is to be found between chemical and physical aspects of adhesion. Our object is to critically assess evidence concerning the operation of some physical forces in adhesion under experimental conditions, and then comment on their significance in physiological processes.

Most cells will stick to most surfaces, whether the surfaces are those of other cells or merely inert materials. *Amoeba proteus* in distilled water can attach strongly to plastic; washed red blood cells in physiological saline stick to glass (George, Weed & Reed, 1971); tissue cells in culture adhere to polystyrene or glass in the presence or absence of added serum proteins (Rappaport, Poole & Rappaport, 1960; Garvin, 1961; Taylor, 1961; Harris, 1973; Rabinovitch & de Stefano, 1973; Grinnell, 1976*a*). Fibroblasts adhere to, but apparently do not spread on, paraffin wax or uncharged polystyrene (Maroudas, 1973, 1975*a*, 1977; Martin & Rubin, 1974). An apparent exception to this generalisation is that the upper surfaces of attached epithelial cells in culture are reported to be non-adhesive to other cells (Middleton, 1973; Vasiliev *et al.*, 1975). Also, the mutual adhesiveness of red blood cells in saline solutions lacking fibrinogen is too small to be measured viscometrically (Chien, 1975).

The rather general ability of cells to adhere to a wide variety of materials suggests that this process is relatively non-specific. The high degree of specificity which has been demonstrated in mating type interactions between yeasts (Crandall, 1977) and *Chlamydomonas* (Wiese & Wiese, 1978) has so far proved to be exceptional. Most cell–cell interactions which have been studied show merely graded selectivity (see review by Steinberg, 1978). Particularly striking are retinal–tectal cell interactions, which show very weak adhesive preferences in culture (Roth, 1973). This implies that their interactions in development are not governed by adhesive differences, in contrast to the concepts of Gustafson & Wolpert (1967) derived from visual impressions of the behaviour of mesodermal cells in sea urchin gastrulation. Among sponge species, only a limited number have been claimed to exhibit interspecific sorting out, and even in these the degree of specificity has been hotly contested (Moscona, 1968; Curtis & van de Vyver, 1971; Humphreys, Humphreys & Sano, 1977). However, it is not disputed that interspecific cell adhesions occur during the putative sorting-out process.

We conclude, therefore, that whatever the mechanism of cell–cell adhesion, it is likely to be a common property of many cell types. This in itself does not exclude the possibility that the mechanism is specific in a stereochemical sense, but, taken with the fact that cells can adhere to glass and plastics in the absence of proteins, it suggests that the mechanism is both widespread and non-specific. Nevertheless, it is likely that evolution has produced certain specialised adhesive recognition mechanisms which complement a basic generalised adhesive propensity of little or no specificity. For example, in the case of the cellular slime mould *Dictyostelium discoideum*, of the two types of membrane site (A and B) governing cell–cell adhesion, only the B site is apparently inactivated by a low molecular weight component (Garrod, Swan, Nicol & Forman, 1978). This complements the work of Beug *et al.* (1970), who demonstrated distinct antigenic specificities of sites A and B. Here we apparently have a system where chemically specific adhesiveness is the predominant mechanism of cell–cell adhesion. It cannot, however, be excluded at present that highly specific control mechanisms operate in such cases to modulate a non-specific adhesive process.

Because of inadequacies in many of the arguments which have been presented for chemically specific adhesion, and because of the clear ability of cells to adhere to diverse surfaces, we have been encouraged to look for a mechanism responsible for generalised adhesiveness. While all chemical mechanisms are ultimately physical, it is operationally convenient to distinguish between chemical and physical mechanisms. We shall be concerned with electrostatic forces, which are repulsive between similarly charged surfaces, and van der Waals (electrodynamic) forces, which cause attraction between similar bodies.

Long-range forces of attraction were originally detected by Derjaguin, Abrikosova & Lifshitz (1956) and the many subsequent measurements have been reviewed elsewhere (Israelachvili, 1974; Israelachvili & Adams, 1978; Parsegian, 1975; Shih & Parsegian, 1975). In one system involving biological materials, lecithin multilayers in water, there is a clear indication of long-range attractive forces. This attraction is sufficiently strong to create a lattice of lecithin lamellae alternating with water, having an equilibrium bilayer separation of about 2.8 nm. (Small, 1967; Luzzati, 1968; Le Neveu, Rand, Parsegian & Gingell, 1977). Such forces, capable of operating over relatively large distances (tens of nanometres) in physiological media, were first postulated to be important in adhesion by Bangham & Pethica (1960) and in a series of papers by Curtis (1960; review, 1973). Additional short-range forces, such as steric repulsion, hydration forces, hydrogen-bonding etc., may also come into play at very small separations; in this regime the distinction between 'chemical' and 'physical' forces becomes semantic. Antigen–antibody interaction, for example, ultimately depends on electrostatic and electrodynamic forces acting over small distances, complemented by hydrogen-bonding; the overall interaction energy depends on optimal juxtaposition of interacting atoms, which is dictated by steric factors.

It has sometimes been supposed that chemical and physical modes of adhesion could be distinguished if specificity of adhesion could be demonstrated, on the supposition that chemical bonding could confer a degree of specificity on adhesion which would not be the case with physical forces. As discussed above, however, the very high degree of specificity which characterises some enzyme–substrate interactions has rarely been demonstrated in cell adhesion; the degree of adhesive specificity exhibited by most cells is much weaker. It is important to realize that weak specificity is a natural feature of electrodynamic interactions (Jehle, 1969; Parsegian & Gingell, 1973), so that if a

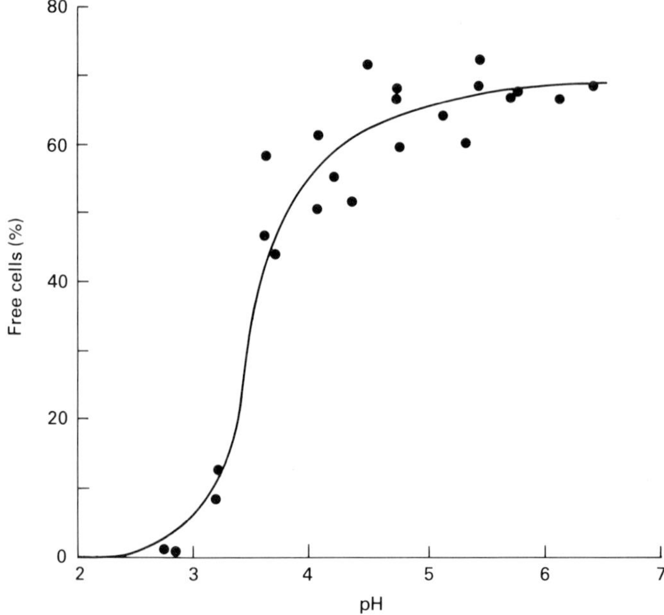

Fig. 1. Aldehyde-treated human red blood cells allowed to aggregate for 3 h in an oscillating shaker in a medium of constant ionic strength (0.150) consisting of NaCl buffered with 2 mM maleic acid/NaOH at 6.9 > pH > 4.8 or 2 mM formic acid/NaOH at 2.8 < pH < 4.8. Only *c.* 75 % of cells remain single above pH 6 due to cell–glass adhesion. Each point represents one experiment.

graded kind of specificity were to be unambiguously demonstrated it would not necessarily help us to decide between chemical and physic modes of adhesion.

2. A QUEST FOR PHYSICAL FORCES

(i) *Attraction and repulsion*

Over the past few years we have selected two simple systems for physical analysis, taking extreme care to control the plethora of variables which exist under physiological conditions. This has enabled us to create conditions rigorous enough to distinguish between physics and biochemistry. We have avoided the pitfalls of undefined interfaces and of unwittingly varying two parameters simultaneously: for example, in altering surface charge density to investigate electrostatics, the chemical composition (which determines the electrodynamic force) should ideally remain constant. The main body of our

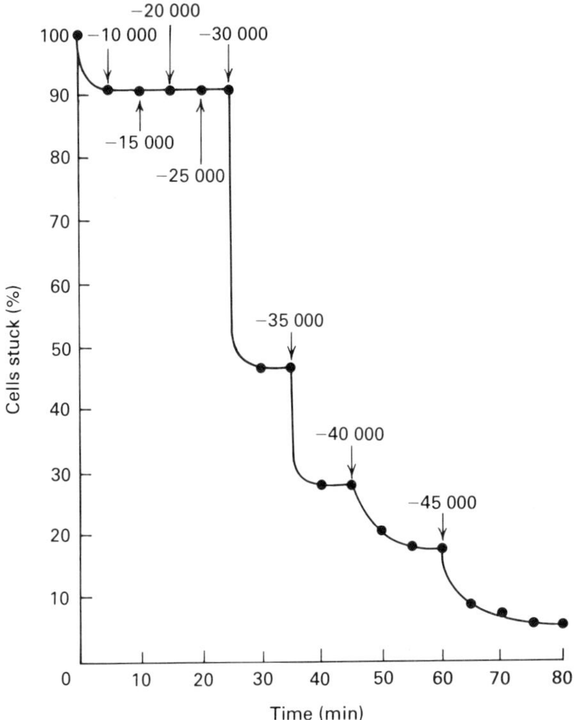

Fig. 2. Time-course of the response of aldehyde-fixed red blood cells adherent to a polarised lead electrode in 1.2 mM NaF to stepwise increases in electrode negative charge density (in esu cm^{-2}). For comparison, the charge density on a red cell is *c*. 5000 esu cm^{-2}. (From Gingell & Fornés, 1976.)

work has involved red blood cells, though recently we have also employed preaggregation amoebae of *Dictyostelium discoideum*. Red blood cells are convenient in requiring no disaggregating procedures and having relatively simple shapes, factors which can be very important in contact studies. Fixation in glutaraldehyde was carried out in the majority of our studies to prevent adventitious contamination of clean surfaces by protein, but control experiments with live cells were carried out where practicable. Since glass does not have a particularly well-defined surface, its surface state being sensitive to its past history including the degree of hydration, most of the cell-surface studies were carried out with oil/water or metal/water interfaces. Before describing this work, we shall consider a very simple experiment on cell–cell adhesion which demonstrates electrostatic repulsion.

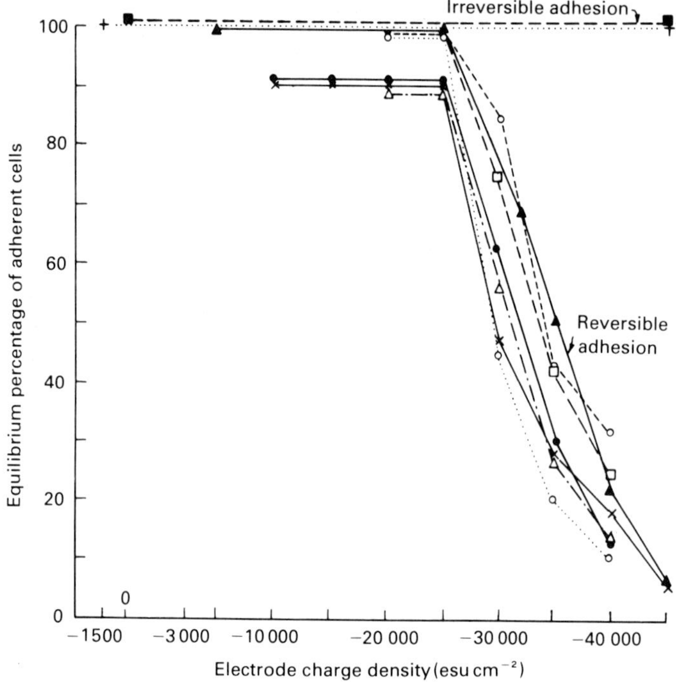

Fig. 3. Adhesion of red blood cells to a polarised lead electrode as a function of electrode surface charge density. Per cent adherent cells are equilibrium values taken from plateau regions of curves like those in Fig. 2. Different symbols represent separate experiments. (From Gingell & Fornés, 1976.)

Fig. 1 shows the proportion of glutaraldehyde-treated human red blood cells which remain single after 3 h oscillatory shaking in 150 mM saline at a series of pH values (Gingell & Todd, unpublished results). As the pH falls at constant ionic strength, the surface sialic acid carboxyls associate, the electrostatic repulsion falls and the cells aggregate. The reduction in single-cell number is not due exclusively to the progressive adhesion of cells to the glass walls of the vessel: counting cells on glass and removing aliquots to a haemocytometer shows increasing cell clustering as the pH falls. This simple approach suggested that electrostatic events to which the cells are apparently sensitive could be monitored more precisely using defined surfaces. One such surface is very pure lead, which has the overriding advantage that it is polarisable: when connected to a source of voltage it stores electrons at its aqueous interface. Since there is no chemical reaction and these electrons to not discharge into the solution, the

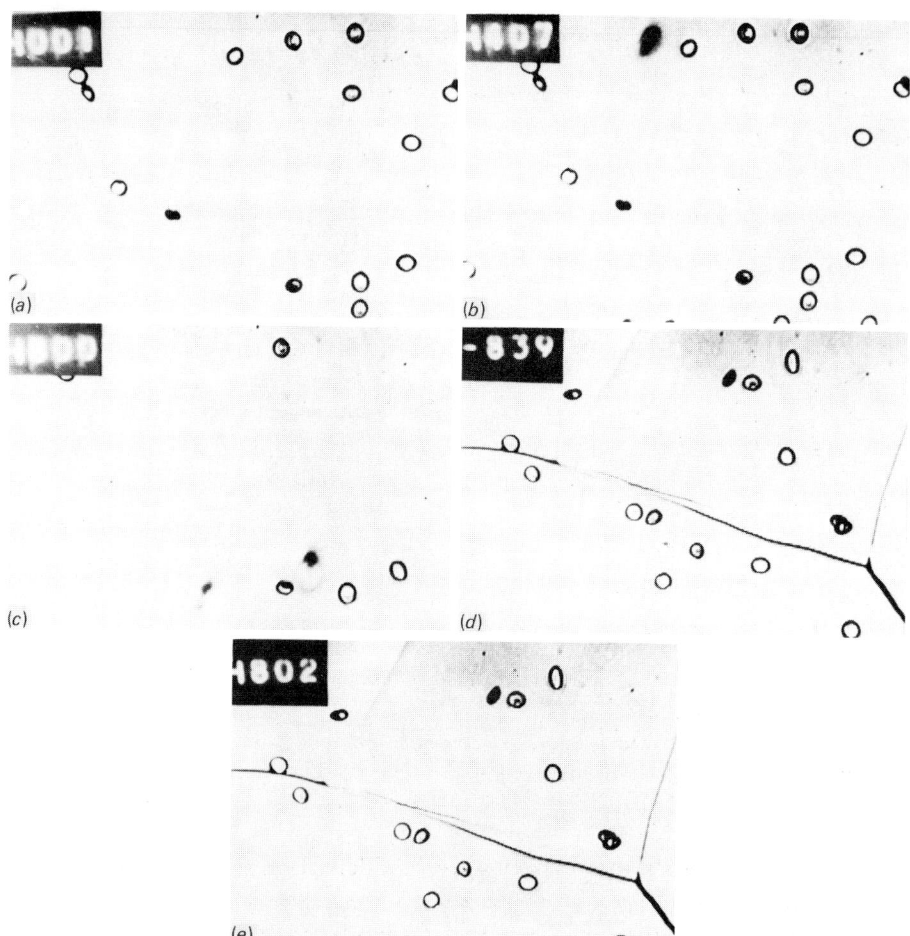

Fig. 4. Photographs taken from a videotape recording of red cells being released from adhesion to a polarised lead electrode in 1.2 mM NaF as electrode charge density increases. (*a*) Shows cells which have initially settled on the electrode at -3000 esu cm^{-2}; (*b*) shows the same cells after increasing the negative charge density to -25000 esu cm^{-2}, no cells have detached in the field but one is seen drifting across out of focus; (*c*) shows the same field at -40000 esu cm^{-2}, most cells have detached or are detaching; compare Fig. 3 (reversible-adhesion regime).

(*d*, *e*) Red cells which have initially settled onto a polarised lead electrode at the potential of zero charge (zero charge density in Fig. 3). On increasing the polarisation to -40000 esu cm^{-2} none fell off. Compare Fig. 3 (irreversible-adhesion regime). Three crystals of lead are seen in this field.

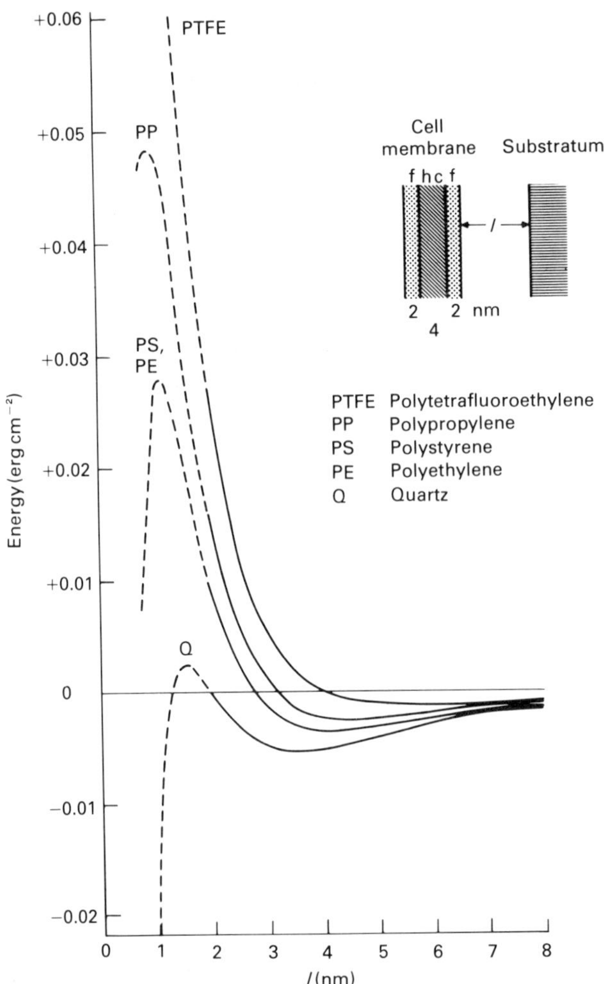

Fig. 5. A family of curves showing summed electrostatic repulsion and electrodynamic attraction for a cell membrane interacting with several substrata in physiological saline. Inset: hc = hydrocarbon, f = glycoprotein 'fuzz'. Curves are dotted at $l < 2$ nm to indicate the region in which the methods of calculation may be unreliable. The secondary minimum is the local energy-well occurring near 4 nm. At smaller separation a repulsive maximum and then attractive minimum is predicted for each curve. All substrata are modelled as having equal charge densities. Note that PTFE, to which cells are poorly adhesive, is predicted to have a low attraction, whereas quartz, to which cells adhere strongly, is highly attractive. (From Parsegian & Gingell, 1973.)

metal develops a surface charge which increases in density as the applied voltage is raised. In our experiment (Gingell & Fornés, 1975, 1976) the electrode was placed in deoxygenated electrolyte in a chamber and positioned a few millimetres below a coverslip window for microscopic examination, using epiillumination optics. Cells were introduced above the electrode and allowed to sediment onto it at a preset electrode surface charge. The electrode chamber and microscope were then rotated through 180° and cell detachment was recorded. Fig. 2 shows the fractions of cells adherent in 1.2 mM NaF at a series of electrode surface charge densities: after equilibrium at one charge, increasing the charge resulted in further cell detachment. This did not occur, however, when the initial electrode charge was near zero or slightly positive, as shown in Fig. 3, which gives the accumulated data from several experiments. Fig. 4*a–c* shows stages in the electrostatic removal of cells from the electrode. Cells initially adhered at a surface charge of -4500 esu cm^{-2} (Fig. 4*a*); by -25000 esu cm^{-2} there was no change (Fig. 4*b*); at -40000 esu cm^{-2} most of the cells had fallen off (Fig. 4*c*) (average cell surface charge density is *c*. 5000 esu cm^{-2}). Fig. 4*d–e* shows an experiment where cells were initially stuck at near zero charge: even at -40000 esu cm^{-2} none fell off. Since the very small current is insufficient to cause cell detachment by electrophoresis, this experiment shows that increasing the charge density (and hence the electrostatic repulsion) prevents cell adhesion and causes detachment of already adherent cells. It is a useful feature of this system that there is an in-built contamination monitor: the differential capacitance of the electrode, from which its charge is calculated, is sensitive to electrode contamination, and has often been used to measure adsorption. We can therefore be confident that the red blood cells are not interacting with a monolayer of adsorbed protein on the electrode.

The duplex adhesive behaviour shown in this experiment is reminiscent of secondary-minimum adhesion, a concept which was first transferred from colloid chemistry to cell adhesion by Bangham & Pethica (1960) and Curtis (1960). This states that long-range van der Waals attraction and long-range electrostatic repulsion, which obey different force laws, can balance at a finite separation (secondary minimum) and at the limit of close approach ('molecular contact') which is a more energetically stable position. This concept was developed by Parsegian & Gingell (1973), from whose paper the curves in Fig. 5 are taken. This illustrates the interaction of a cell with

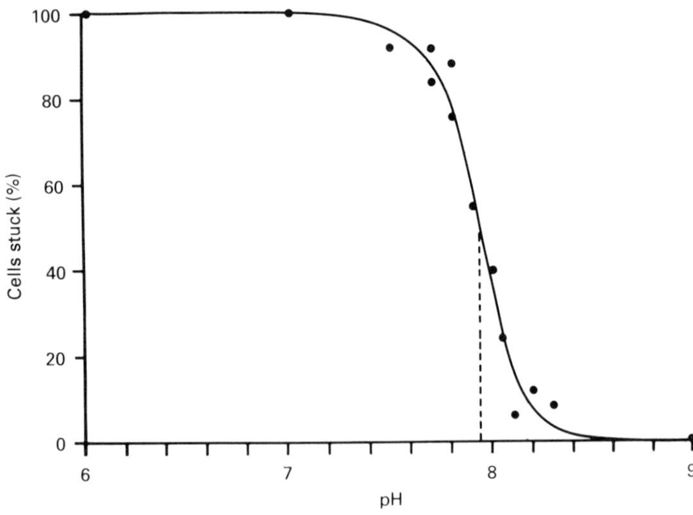

Fig. 6. Adhesion of fixed red cells to a hexadecane/145 mM NaCl interface. Behenic acid dissolved in the oil ionises to give a negative surface charge at the interface as the pH rises. No buffer is present. (From Gingell & Todd, 1975.)

several substrates in physiological saline; in 1 mM NaF the secondary-minimum balance point would be at a much larger separation. It can be seen that increasing the electrostatic repulsion decreases the depth of the secondary-minimum energy 'well' and cells are eventually obliged to detach from the surface. Although the computational formalism does not allow us to make precise estimates of the energy at molecular contact, cells stuck in this mode would be more stably adherent and less vulnerable to subsequent electrostatic eviction. Thus our experimental results correspond qualitatively with the prediction for long-range force theory. It has recently been possible to extract quantitative information from these results; this will be considered with other numerical results at the end of this section.

The second system which has been employed is the hexadecane/saline interface. This is technically simpler than the metal electrode; it can be prepared clean and is readily checked for contamination by surface tension measurements. Observation of red cells at the liquid interface is possible using a chamber wherein the interface is formed between toroidal cheeks (Gingell & Todd, 1975). By their adjustment, the interface can be rendered completely flat. Cells at the interface were observed using a long working-distance 20 × objective. We were able to modify the electrostatic repulsion

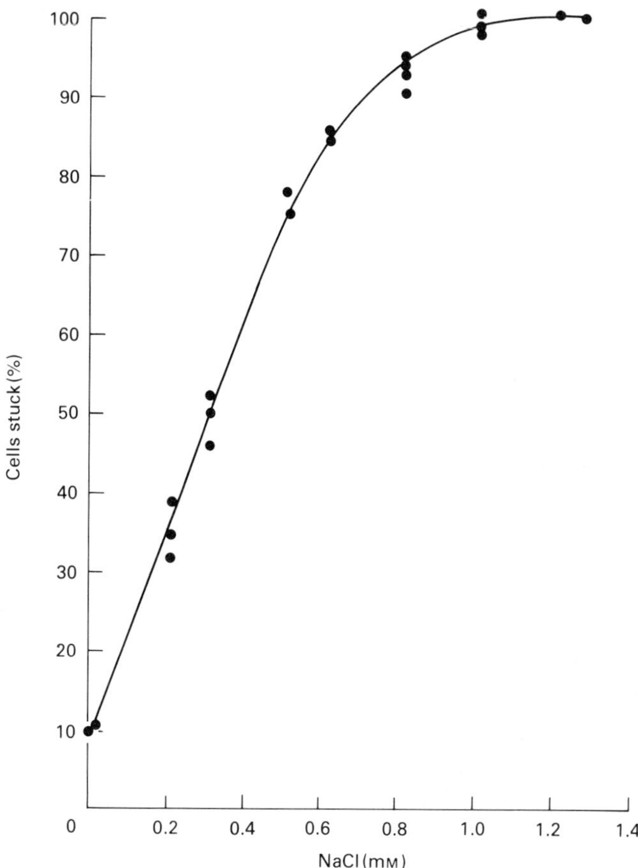

Fig. 7. Adhesion of fixed red cells to a plain hexadecane/NaCl interface as a function of NaCl concentration. No buffer is present: pH 5.6. (From Gingell, Todd & Parsegian, 1977.)

experienced by the red cells by introducing a detergent (behenic acid, C_{22}) into the oil: this orients and ionises at the interface according to the pH. At low pH (Fig. 6) the detergent is not ionised, repulsion is minimal (but not zero; Parsegian & Gingell, 1972) and all cells adhere to the interface. However, repeating the experiment at a series of increasing pH values showed that cells are less able to adhere as repulsion increases. Loss of adhesiveness is apparently not due to the reduced surface tension which accompanies increasing detergent ionisation since a cationic detergent, C_{18} bromide, provides an interface of very low surface tension and positive charge to which all cells adhere.

To test whether the oil/water interface had become contaminated with small amounts of protein or other surfactant after being exposed to high concentrations of fixed cells, we measured the surface tension by the hanging-drop method. A clean oil/water interface without added detergent gave a value of 53 dyn cm^{-1} at 20 °C. When the measurement was repeated using a drop of concentrated cell suspension in 145 mM NaCl at pH 5.4 hanging in hexadecane, the surface tension was found to be unaltered. However, when unfixed cells were used, the surface tension fell very sharply. This important test shows that aldehyde-fixed red cells do not measurably contaminate the interface, even in conditions where they adhere strongly to it (Gingell & Todd, 1979*a*).

These observations at the oil/water interface suggested an even simpler and potentially more powerful approach: we measured cell adhesion to a plain hexadecane/saline interface at fixed pH (5.4) as a function of salt concentration (Gingell, Todd & Parsegian, 1977; Gingell & Todd, 1979*a*). The results of this experiment are shown in Fig. 7. It can be seen that as the salt concentration fell below 1.0 mM NaCl, cells were progressively less able to adhere. Using a flow-through system it was possible to dilute the medium bathing adherent cells: 80–90 % of those initially attached at 0.4 mM could be removed by dilution, but the percentage was reduced above 10 mM. This result is similar to the behaviour seen on the metal electrode, where cells which initially stuck under conditions of very low electrostatic repulsion were irreversibly adherent, but those which stuck under more repulsive conditions were found to be reversibly adherent.

Using electrophoretic data for red blood cells and an emulsion of oil droplets in saline over a range of 0.1 to 20 mM NaCl, it is possible to calculate the repulsive force exerted on a red blood cell approaching an oil/water interface at the 'critical' concentration of 0.3 mM, the concentration at which half the cells fell off after inversion. The computational procedures (Parsegian & Gingell, 1979) involved several steps. The red blood cell was approximated by a torus which was imagined to be opened out into a cylinder, and the high electric potential near the cell surface was treated exactly but was matched up to an approximate form at greater distances where the potential is small. This made it possible to calculate the repulsion as a function of separation by the technique of Brenner & Parsegian (1974). At this stage we can perform an instructive thought-experiment: suppose there is no long-range attraction between red blood cell and hydrocarbon; now imagine a cell settling under gravity onto an oil/water interface; it will settle until the gravitational force exerted on it equals the electrostatic repulsive force, and the cell will balance under the opposing influence of the two forces. Having calculated the repulsive force as a function of separation, we can calculate a distance where the repulsive force is equal to the gravitational force on a cell. This turns out to be *c.* 150 nm. From experiment, half the cells adhere at 0.3 mM when the system is inverted, where both electrostatic repulsion and gravity act to remove them. It follows that there must be an attractive force which can reach out at least to *c.* 150 nm to hold cells in place. Whether the force is sufficiently strong to overcome the electrostatic repulsion and pull cells into molecular contact with the interface, or whether a force-balance situation exists with a finite water gap between cell and surface, can also be deduced if an assumption is made regarding the mathematical form of the attractive force. Assuming it to be electrodynamic with an energy obeying an inverse square law, and further assuming that 5 kT is required to give a stable adhesion, we find a balance region near 150 nm and an attractive force constant of $A = 8 \times 10^{-14}$ erg. The width of the predicted energy well is such that the component of thermal energy of adherent cells in a direction perpendicular to the interface should produce excursions of several tens of nanometers.

A parallel calculation for the metal electrode experiment gives a similar balance with a force constant $A = 4 \times 10^{-14}$ erg. These force constants are reasonably consistent with experimental values obtained in aqueous systems: Haydon & Taylor (1968) found $A = 5.6 \times 10^{-14}$ erg for the electrodynamic force compressing a lipid bilayer in water, and a value $A = 2\text{–}3 \times 10^{-14}$ erg can be calculated from the results of studies on lecithin liquid crystals (Le Neveu *et al.*, 1977) by equating attraction with the repulsive pressure between the lamellae at equilibrium separation. These experimental values are in the same range as those calculated on the basis of the macroscopic theory of electrodynamic forces (Gingell & Parsegian, 1972).

The experimental results which we have described show unambiguously that in dilute media electrostatic repulsion can completely prevent adhesion. However, recent calculations indicate that in physiological saline, even high negative charge densities may be unable to reduce adhesive energies to the point where thermal energy alone can break adhesions if long-range attraction with a coefficient

$A = 6 \times 10^{-14}$ erg is present. Thus we are uncertain at present how to interpret the complete loss of adhesion of red cells from the behenic acid–oil/water interface at maximal ionisation of behenic acid (Parsegian & Gingell, 1979). In contrast, loss of adhesion from a plain oil/water interface at low ionic strength is predicted and has allowed us to estimate the size of the attractive force. It may be that rapid solubilisation of ionised behenic acid drives cells from the interface, or possibly adventitious vibrations aid detachment of weak adhesions, since no precautions were taken to avoid vibrations. It would also be possible to rationalise this result if the attractive force coefficient is smaller than our present estimates.

(ii) *Optical measurement of cell–substratum separation*

The reasonable quantitative correspondence with theory which we have described does not constitute proof that long-range forces are acting in our experiments. To this end we have made optical measurements of the separation between cells and substrata using interference reflection microscopy (Gingell & Todd, 1979*b*). This

Fig. 8. (*a*) White light interference-reflection image of an aldehyde-fixed red blood cell adherent to a hexadecane/0.4 mM NaCl interface. Illuminating numerical aperture (INA) = 0.69. Photographed from a video recording. Low INA was used in photography to increase contrast.

(*b*) As in (*a*), except that the concentration of NaCl has been raised to 70 mM. The cell shows dark grey/black zero-order regions of closest approach to the interface. Photographed from a video recording.

(*c*) A red blood cell adherent to a coverslip in 2 mM NaCl at INA = 0.69. Aided by the *en-face* orientation of this cell, which results in a small cytoplasmic thickness, *c*. 1 μm, low INA has resulted in an interference pattern from the far side of the cell. This is superimposed on the pale yellow closest-contact region derived from the near face/glass gap.

(*d*) Zero-order black contact between a red blood cell and a glass coverslip in 150 mM NaCl. INA = 0.69.

(*e*) White light interference-reflection image of *Dictyostelium discoideum* amoeba adherent to HF-cleaned glass in distilled water. INA = 0.69. Photomicrograph.

(*f*) As in (*e*), except that the amoeba is adherent to poly-L-lysine-coated glass in 20 mM NaCl. A red cell is shown in the same field for comparison. Note the track of very thin cytoplasmic extensions which appear black due to interference across their thickness.

Note: The greater contrast obtained with red cells compared with *Dictyostelium* is largely due to the much higher cytoplasmic refractive index of the former. In the case of red blood cells on oil, the images are less sharp than on glass due to slight motion of the liquid/liquid interface as well as a small loss of quality inherent in videotape recording.

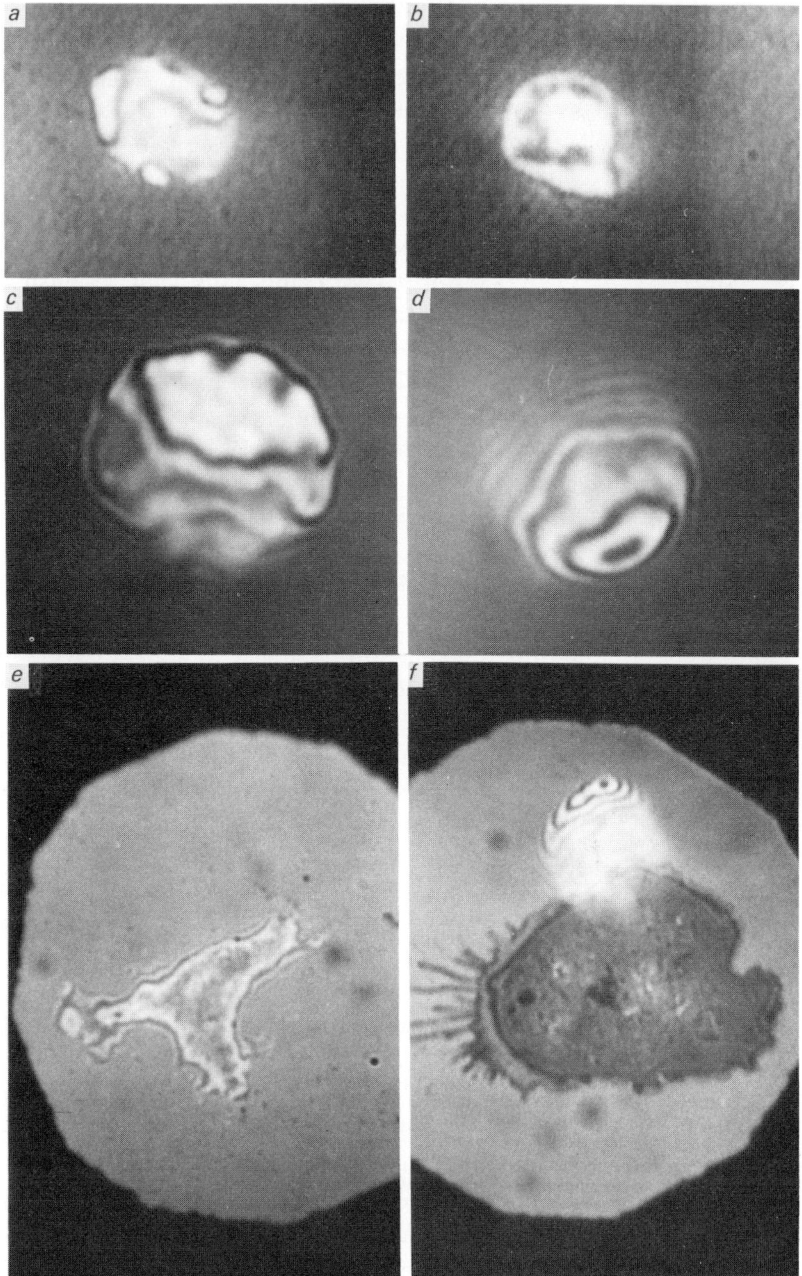

For legend see opposite

technique was pioneered in biology by Curtis (1964), who used it to estimate the separation between fibroblasts and a glass surface under a variety of experimental conditions. It has subsequently become a fashionable tool in cell biology (Abercrombie & Dunn, 1975; Izzard & Lochner, 1976; Heath & Dunn, 1978; Preston & King, 1978), largely because it is simple to use. Image interpretation, however, is less straightforward, as discussed below. In white light, red blood cells attached to a hexadecane/saline interface at 0.4 mM Na Cl show pale yellow regions of closest approach (Fig. 8*a*, Frontispiece *a*) which can sometimes be seen to oscillate in colour from yellow to reddish brown. As the salt concentration is continuously increased, the closest-contact zones change to silver, then pale grey and eventually become black at *c*. 70 mM NaCl (Fig. 8*b*, Frontispiece *b*). These observations, made at an illuminating numerical aperture (INA) of 0.91 but illustrated by photographs taken at INA = 0.69 for enhanced contrast, indicate that in 0.4 mM NaCl cells are separated by a distance of *c*. 100 nm and can oscillate by Brownian motion between smaller and greater separations. The zero-order black contact in 72 mM NaCl probably indicates a separation of $\leqslant$ 10 nm. Corresponding results were obtained for red cells interacting with a glass coverslip in 2 mM NaCl. These cells characteristically show a pale yellowish closest-contact region. The cell illustrated in Fig. 8*c* (Frontispiece *c*) shows a large pale yellow area, in which faint bands resulting from interference from the far side of the cell can be seen due to the use of a low INA value (0.69). Figure 8*d* (Frontispiece *d*) shows zero-order black contact between a red cell and glass in *c*. 150 mM NaCl.

There are two closely interwoven problems associated with distance assessment by this technique. First, as the angle of the illuminating cone of light is increased (high INA) mutual interference between wavefronts inclined at different angles becomes significant. This at first sight suggests the use of a narrow pencil of illuminating light (low INA) but it raises the second problem, that as the INA decreases the image includes more information from the far side of the cell. Both these problems were discussed by Izzard & Lochner (1976). By measuring the intensity of monochromatic light (546 nm) and applying a new theory of microscope interference (Gingell & Todd, 1979*c*), we have been able to make quantitative estimates of cell–substratum separations. Working at INA = 1.18, we obtain a value of *c*. 120 nm for the cell/water/oil aqueous separation in 0.4 mM NaCl. At this INA it is possible to show theoretically that if the cell

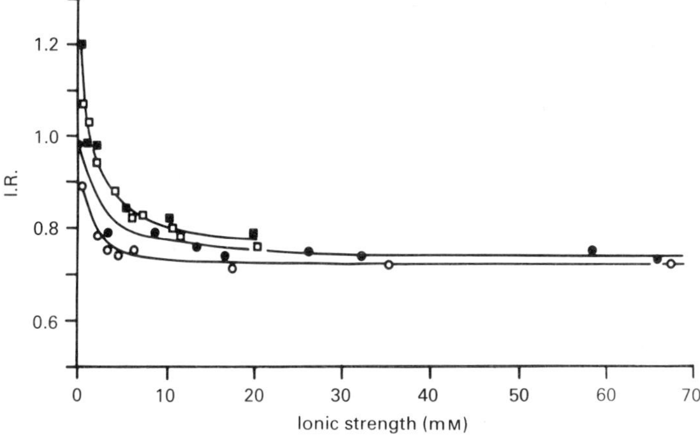

Fig. 9. Relative irradiance of amoebae of *Dictyostelium discoideum* adherent to glass in salt solutions. Ionic strength is indicated on the abscissa. K^+ and Na^+ follow identical curves. (Irradiance ratio, IR, is defined as the monochromatic irradiance of the cell divided by that of the glass/water background.) Cell–substratum separations can be computed from IR values. Symbols: □, Na^+; ■, K^+; ○, Ca^{2+}; ●, Mg^{2+}.

thickness exceeds *c.* 1.0 μm the image intensity is relatively independent of cytoplasmic thickness. Because of the extreme technical difficulties associated with aligning the photometer pinhole with the 0·05 μm^2 area of the cell being measured as it dances by Brownian motion on the interface, this experimental result is very approximate. The white-light observations and photometric measurements taken together show that there is a gap of *c.* 100 nm between a cell and the oil/water interface in 0.4 mM NaCl. This compares only approximately with the *c.* 200 nm predicted from long-range force theory, but it does support the interpretation that such forces are responsible for holding cells at the interface. One possible source of discrepancy in the measured and calculated values lies in the surface potential of the red cell estimated from electrophoretic data; it is known that the zeta potential does not always provide a reliable estimate of surface charge and potential (but see Donath, 1978) and the problem can become severe at the high potentials which exist in dilute salt solutions (MacDonald & Bangham, 1972; Carroll & Haydon, 1975).

Our quantitative interference-reflection method has also been applied to *D. discoideum* amoebae moving on glass. Irradiances were measured in monochromatic light (546 nm) at INA = 1.18 in increasing concentrations of Na^+, K^+, Ca^{2+}, Mg^{2+} as chlorides, ranging from 0.1 to 20 mM. Plotting the irradiance ratio (IR) against ionic strength (Fig. 9) it can be seen that IR decreased as the salt concentration rose, K^+ and Na^+ behaving identically. Since a decrease in IR in this region indicates a decrease in cell–substratum separations, it can be seen that cells in Mg^{2+} and Ca^{2+} are closer to the interface than in an equal ionic strength of Na^+ or K^+. As the ionic strength increases, cell bilayer-glass separations tend asymptotically to minimal values of a few tens of nm, the limiting separation being different for monovalent and divalent cations. This fact, together with the discrepancy between the Mg^{2+} and Ca^{2+} curves, indicates ion adsorption as well as Debye screening. In 20 mM NaCl on poly-L-lysine-coated glass, a similar minimal separation occurs, (Fig. 8*f*, Frontispiece *f*) whereas in distilled water on clean glass separation is maximal, (Fig. 8*e*, Frontispiece *e*). Estimates of separation depend on the optical properties of the surface glycoprotein, as yet unknown. Preston & King (1978) have obtained similar qualitative results for *Naegleria* amoebae exposed to salts. It has been found by many workers that cell–glass adhesion is increased by divalent cations (Armstrong, 1966; Collins, 1966; Armstrong & Jones, 1968; Garvin, 1968; Takeichi & Okada, 1972). This is particularly marked in cases where serum protein is present. The presence of some macromolecules in an optical gap between *Dictyostelium* and glass is likely since there appears to be a limiting separation which cannot be reduced even by high concentrations of divalent cations. Le Neveu *et al.* (1977) have reported a strong repulsive component due to water structure at separations between lecithin multilayers less than 3 nm wide. This implies a repulsive component which is immune to electrostatic screening at elevated salt concentrations. This component is of uncertain origin, but it may be steric repulsion, an entropic effect of extended polymer chains. Weakly charged or uncharged polymer chains may be responsible for the salt-insensitive repulsion seen in *D. discoideum*. Although this phenomenon is well documented in polymer chemistry (Hesselink, Vrij & Overbeek, 1971; Doroszkowski & Lambourne, 1973; Smitham, Evans & Napper, 1975) and has been considered as a candidate for intercellular repulsion (Maroudas, 1975*a*), there is as yet little evidence for a biological role.

3. OTHER EVIDENCE RELATED TO LONG-RANGE FORCES

(i) *Modification of substrata: electrostatics*

Correlations between substratum surface charge and adhesions are remarkably rare in the literature and are frequently unconvincing. One of the first correlations came from the work of Dan (1936) who observed that adhesiveness of marine eggs to a glass plate decreased with the concentration of salt in the bathing medium. This was interpreted in terms of progressively increasing electrostatic repulsion as the salt solution was diluted. More recently, Rappaport *et al.* (1960) and Rappaport & Bishop (1960) obtained evidence of a contrary effect in the spreading and growth of a variety of cell types in protein-free media on glasses having a range of negative charge densities. Growth was promoted as negative charge density increased, as determined by cationic crystal violet binding. Interpretation of these carefully conducted experiments is impeded by several difficulties. First, as Taylor (1961) pointed out, spreading and adhesion are not the same thing; spreading requires metabolism (Wolpert, MacPherson & Todd, 1969; Michaelis & Dalgano, 1971), whereas adhesion can proceed without it. Also, while spreading requires adhesion, the converse is not so. As Maroudas has emphasised, cells which spread into stellate forms stress the substratum in a direction parallel to the surface. Consequently, the inability to spread even in metabolically favourable situations may not stem from the inability to adhere, since the ability to resist such parallel stresses is not a necessary consequence of strong adhesive forces in a perpendicular sense. The second difficulty with this type of experiment is that glasses of different chemical compositions would be expected to have different electrodynamic properties as well as different electrostatics. Third, when cells are left for perhaps an hour or more in contact with a substrate they have the opportunity of coating the interface with macromolecular materials which they have synthesised (Poste *et al.*, 1973). It is also necessary to distinguish carefully between long- and short-term adhesive mechanisms, as is the case with cells which develop specialised junctions, such as hemi-desmosomes. The fourth problem is that dye binding was not shown to be stoichiometrically related to the surface density of negative charge on glass.

Maroudas (1975*a*, 1977) has conducted an interesting analysis of the interaction of BHK cells in simple saline solutions with a series of polystyrene surfaces carrying different amounts of negative charge, as determined by crystal violet binding. He reported that cells spread better on polystyrene as its charge density increased, but beyond a charge which corresponded with sulphonation of all superficial styrene groups, cell spreading decreased. Some of the problems in interpreting the work of Rappaport *et al.*, occur here also. Cell adhesion was not measured; to do this, some distracting force must be applied. The author makes it clear that his spreading criterion omits non-spreading cells which might nevertheless be adherent to the surfaces. Dye binding was not proven to be stoichiometrically related to the density of negative charge at the surface. It can be calculated that at the highest densities reached charges must be distributed in depth in the surface suggesting that at lower charge densities the same thing may happen. It is also possible that sulphonation of the surface modifies the attractive forces. Despite these difficulties the results merit attention in so far as they do not lend themselves to a simple electrostatic interpretation.

Maroudas also reported that cells were not able to spread on deionised wax in the presence or absence of serum protein. This would appear to directly contradict our results which show that red blood cells can adhere to liquid hydrocarbon, both in 'molecular contact' and when separated by *c.* 100 nm. The discrepancy may be due to different experimental criteria: fixed red blood cells do not spread, yet spreading was used as the criterion for adhesion in Maroudas' experiments. Cells may adhere to solid hydrocarbon but possibly find it hard to spread, obtaining insufficient tangential grip. In the presence of serum protein, however, Margolis, Dyatlovitskaya & Bergelson (1978) using films of phospholipids and glycerides on glass made the very interesting observation that fibroblast adhesion and spreading is reduced on films whose phase-transition temperature is below 37 °C but takes place on films which are in a gel state at this temperature. It appears that cells obtain insufficient reaction to tensile forces to spread on liquid or liquid crystalline layers, and this, taken with Maroudas' results, suggests that spreading on gelled lipids is distinct from that on solid hydrocarbon. Facilitated spreading on the former may indicate increased adhesion due to electrostatic interaction with phospholipid head-groups. Even if similar interactions occurred on liquid crystalline phospholipids, no tangential reaction forces could be generated and no spreading could occur.

(ii) *Modification of substrata: long-range attraction*

Rosenberg (1962) observed the rate of spreading of conjunctiva cells onto multilayers of fatty acids deposited on various solid substrates. His results showed that cell spreading was sensitive to the nature of the substrate, even when the multilayers were built up to thicknesses of *c.* 140 nm, and suggested that one interpretation was the action of a long-range force between the substrate and the cells through the intervening lipid. His alternative explanation was that a short-range order effect is transmitted from each molecular layer to the next, so that the outermost layer reflects the properties of the substratum on which the films were formed. There have been suggestions of a progressive electrical polarisation between multilayers and a metal base which builds up as the number of layers increases (see Langmuir, 1938, for references and discussion). This interesting observation, which also included cyclic changes in contact angle with layer number, was, however, only reported for X-layers, whereas under the conditions of deposition in Rosenberg's experiment it is almost certain that Y-layers would have formed. X-layers are those in which the molecules in each successive layer of a multilayer have the same orientation, whilst Y-layers are those in which each successive layer has the opposite orientation from the preceding one. It is possible that 'skeletonisation' of multilayers, involving localised loss of patches of lipid, may expose metal, glass or plastic substrates to different degrees. Exposure may also depend on multilayer thickness, since skeletonisation is known to increase with the number of deposited layers. Cells may in some cases interact with these regions on the coated surfaces. A further factor may be inversion of the outermost layer of the monolayer which can occur under certain circumstances (Langmuir, 1938) and may be induced by cell contact to a degree dependent on multilayer thickness. Despite these objections, Rosenberg's innovative experiment has never received a satisfactory explanation and has not, to our knowledge, been satisfactorily re-investigated.

Folkman & Moscona (1978) have described a particularly striking relationship between cell spreading, macromolecular metabolism and the thickness of poly(HEMA) films deposited on polystyrene. Their data show that as film thickness increased from *c.* 100 nm to *c.* 3 μm adherent cells became less flattened and made smaller apparent areas of contact. Since it is improbable that long-range attractive forces can act over distances considerably in excess of 100 nm, the explanation is unlikely to be a variation in film surface energy with thickness. Cell processes may readily penetrate thinner films, pulling cells onto the substratum, whereas they may be unable to penetrate thick polymer films.

(iii) *Modification of cells : electrostatics*

Attempts to test the importance of electrostatic repulsion in both intercellular and cell-to-substratum adhesion by enzymic modification of the charged groups of the cell periphery face one of the difficulties already referred to. That is, since the attractive force depends on the chemical nature of the interacting surfaces, such procedures may affect attractive forces as well as repulsive forces. This criticism applies to trypsin/neuraminidase treatment which has been reported to increase (Vickers & Edwards, 1972) and decrease (Kemp, 1970) adhesions, as well as to chemical procedures intended to increase the charge density (Kemp & Jones, 1970). Another way of modifying electrostatics with minimal perturbation of attractive forces is to insert charged molecules in the cell surface.

To this end, Owen, Clifford & Marson (1978) have investigated the effects of detergents on cell aggregation and assessed surface-charge changes by cell electrophoresis. Although anionic detergent increases the negative surface charge and decreases aggregation, cationic detergent decreases aggregation with hardly any detectable change in cell-surface charge. These results cannot be interpreted in straightforward electrostatic terms. It is likely that chemical modifications of cell surfaces cause changes in membrane fluidity and redistributions of peripheral macromolecules, with consequent effects on charge distributions. Attachment of cytoplasmic skeletal elements to the surface membrane may also be modified, resulting in altered ability to protrude microvilli, which may be important in forming initial adhesions. The general difficulties inherent in interpreting these types of experiment are well discussed by Kemp, Lloyd & Cook (1978).

Only in cases where the cell surface is not directly modified have readily interpretable results been obtained. Born & Garrod (1968) showed that the aggregation rate of *Dictyostelium* amoebae was increased by divalent cations in the medium. This is qualitatively in accord with the notion that divalent cations both reduce the Debye screening by an ionic strength effect, and also adsorb to cells, reducing the negative surface charge (Armstrong, 1966; Collins, 1966; Armstrong & Jones, 1968; Garvin, 1968; Takeichi & Okada, 1972).

(iv) *Long-range attraction between cells*

Curtis (1969, 1970) and Curtis & Hocking (1970) have employed an ingenious analysis of aggregation rates from which they were able to interpret the aggregation of mammalian tissue cells in terms of their mutual long-range attractive forces. Cells suspended in a shear viscometer will collide because of the velocity gradient in the medium. The fraction of collisions that result in a stable association is expected to be increased by a long-range attraction which pulls approaching cells together and then prevents them from moving apart. Because of the rapidity of approach and recession relative to the time required for metabolic change, this assay detects short-term stickiness of cells. An unresolved problem in the analysis of these results lies in the simplifying assumption that repulsive forces can be disregarded. Even if this omission were remedied, the method as a basis for the quantitative estimation of long-range forces is very indirect and model dependent. Although the interpretation of collision efficiencies in terms of long-range forces seems plausible, the results appear to be consistent with both long- and short-range attraction.

A new approach to adhesion has been employed by Curtis and his colleagues (Curtis, Campbell & Shaw, 1975; Curtis, Chandler & Picton, 1975; Curtis, Shaw & Spires, 1975; Schaeffer & Curtis, 1977). The lipid composition of fibroblasts and neural retina cell membranes was altered by a metabolic technique. Increased incorporation of saturated hydrocarbon chains increased cell–cell and cell–polystyrene adhesiveness in the presence of serum proteins. Conversely, it was found that an increase in the proportion of unsaturated lipids decreased adhesion. While it is possible that electrodynamic forces are sufficiently sensitive to the degree of hydrocarbon saturation and chain length, this is not the only possible or most likely interpretation. It is well known that lipid fluidity is significantly affected by the degree of saturation. This raises the possibility that changes in adhesiveness may be due to modified lateral mobility of macromolecules. Similar results have been reported by Hoover, Lynch &

Karnovsky (1977). They found that long-chain unsaturated fatty acids caused a striking decrease in the adhesion of cells to cell monolayers in serum-free media. There seems no doubt that these findings are of considerable interest but the nature of the mechanism is far from clear. Ueda, Ito, Okada & Ohnishi (1976) found that BHK-21 cell adhesion undergoes a marked increase near 10 °C, where ESR spectroscopy suggests a phase transition of membrane lipid. This implies that cell adhesion increases as the membrane lipid fluidity increases, whereas the results of Curtis's group suggests the converse.

4. ARE LONG-RANGE FORCES IMPORTANT IN ADHESION?

(i) *Interaction in dilute media*

The majority of experiments we have performed were carried out in dilute salt solutions. Only under these conditions can cell-substratum separations greater than the probable extent of cell-surface macromolecules be obtained. Low ionic strength is therefore indispensable for the detection of long-range attractive and repulsive forces. It will be necessary, however, to distinguish very carefully between the results obtained at low ionic strength and those anticipated in physiological media, where electrostatic repulsion is far smaller and interaction distances probably do not exceed macromolecular dimensions.

We have shown that cells can adhere to surfaces of glass, metal and liquid hydrocarbon in the absence of proteins or other macromolecules in the medium at low and high ionic strength. Attachment to hexadecane shows that adhesion can occur without any chemical interactions, since the oil is a pure inert hydrocarbon. We have been able to measure the large electrostatic force required to prevent red blood cells from sticking to clean metallic and hydrocarbon surfaces, from which we have deduced the size of the attractive force. This force behaves qualitatively like an electrodynamic attraction, and the experimental estimates of its coefficient are in reasonable agreement with theoretical predictions. We have also shown that cells in dilute salt solutions can adhere with an optical gap of *c*. 100 nm separating them from the substratum, and that this decreases to what is probably molecular contact as the salt concentration rises. This is in accord with an electrostatic mechanism. It has further been demonstrated that cells adherent to substrata but separated from them by an optical gap are reversibly adherent, whereas those exhibiting no demonstrable gap are more strongly stuck and cannot be removed by subsequent exposure to maximal electrostatic repulsion. These features strongly suggest that a long-range force balance is in operation. Reservations to this conclusion are the lack of demonstrable lateral sliding under appropriate conditions on metal and the technical difficulty of attempting to detect sliding on the oil/water interface. It remains possible that a few extremely long flexible macromolecular links could attach cells to substrata over separations *c.* 100 nm and thus inhibit sliding.

(ii) *Predicted properties of secondary minima in physiological saline*

We shall briefly review the size of predicted secondary minima in physiological saline before discussing the physical limitations of the secondary minimum concept in physiological situations.

Calculations (Parsegian & Gingell, 1973) show that the depth of this minimum, viewed as a mechanical energy holding cells together, is very small. For example, if one imagines two flat bodies sitting in a minimum of 5×10^{-3} to 5×10^{-4} erg cm^{-2} interacting over square faces of $(10\ \mu\text{m})^2$, the force required to slide one face parallel to the other is a negligible 5×10^{-7} to 5×10^{-6} dyn. The work of peeling apart two surfaces in planar apposition is of similar magnitude. The calculated theoretical force for perpendicularly separating two flat bodies in secondary-minimum adhesion is around 10^{-3} to 10^{-4} dyn cm^{-2}. Although this is a million times greater than the gravitational force on a cell attached to a rigid surface over an area of $(10\ \mu\text{m})^2$, it is probably about the same size as the forces which cells can exert (Kamiya, 1964; James & Taylor, 1969) and may be compared with the force of contraction of an actomyosin filament, 3×10^{-5} dyn (Wolpert, 1965), which implies that secondary-minimum adhesions could be readily broken.

The measured forces for separating cells (see Curtis, 1967) span a range from 10 to 10^{-6} dyn cm^{-2}, there being much uncertainty attached to the area of adhesive contact and the contribution of specialised junctions. Since this generous range easily encompasses the range of theoretical force estimates, the data are of little use in helping us to reject any theory of cell adhesion.

Compared with Brownian energy of a particle ($kT \simeq 4 \times 10^{-14}$ erg) the theoretical binding energy conferred by a secondary minimum is quite large: for a contact area of 1 μm^2 the energy depth is 5×10^{-4} erg cm^{-2} or about 100 $kT\, \mu m^{-2}$. While this would be a totally insignificant barrier to a motile cell, it might conceivably have a statistical influence on the distribution of passive cells.

The energy of adhesion can also be compared with cell-surface energy in order to answer the question whether the energy required for cell-surface deformation, which occurs on adhesive contact, could be provided by adhesion. To take an extreme example, could adhesive energy drive the shape change from a spherical cell to a completely flattened form without help from metabolic energy? If measured cell-surface energies (0.01–0.1 erg cm^{-2}) are a reliable determination of the work required to cause a shape change (regardless of whether a real increase in surface area or an unfolding of existing surface occurs), the answer is almost certainly 'no' for secondary-minimum adhesion and a qualified 'yes' for primary adhesion.

Since major cell shape changes involved in spreading and locomotion require metabolism, it is instructive to realise how small surface energies and interaction energies are compared with the metabolic energy which may be available to the locomotor machinery.* Transfer of phosphate from ATP yields about 10 kcal mol^{-1} or 17 kT per molecule ($RT = 0.6$ kcal mol^{-1}). An interaction of 100 kT is thus equivalent to the energy available from negligibly few biochemical reactions; alternatively, it could be provided by about 200 hydrogen bonds between cell surfaces.

The foregoing considerations make it likely that even if secondary minimum adhesions occur naturally they would be weak in comparison to metabolic energy and the forces of cellular locomotion.

(iii) *Limitations on the applicability of the long-range force-balance concept in physiological saline*

Consideration of the calculations in section (ii) must be tempered by the fact that there are severe restrictions on the legitimacy of the force-balance concept in physiological conditions, as opposed to the dilute salt condition used in most of our experiments. The problem is that a Debye length of 0.8 nm in 145 mM NaCl implies a secondary-

* We thank Dr V. A. Parsegian for this idea.

minimum stabilisation, using plausible values of the electrodynamic force coefficient, at only 5–7.5nm between the interacting surfaces. This separation is probably about the same as the distance of projection of cell-surface macromolecules, although reliable experimental data do not exist. The form of the glycophorin molecule, for example, suggests that this may be a minimal value. It follows that it is not realistic to discuss interaction between molecularly rough surfaces at such short distances as if they are smooth planes; the details of surface mosaicism also become of paramount importance. A further but closely related restriction concerns the range of validity of the physical technique used in calculating long-range forces. Electrostatic methods assuming evenly smeared-out charge are only legitimate where the interacting surfaces are separated by distances exceeding the separation of charges or charge clusters on the individual surfaces. The macroscopic electrodynamic approach for calculating attractive forces is limited to separations greater than the minimum dimensions over which the dielectric properties of matter can be considered homogeneous. These two limitations imply almost total uncertainty regarding predictions of interaction energies at separation less than about 2 or 3 nm. (The limit is not well defined.) It follows that for molecularly rough interdigitating cell surfaces in physiological saline, calculations may be based on an insecure foundation.

(iv) *Molecular contact*

As shown above, even if secondary minima exist under physiological conditions they are weak compared with metabolic energies and locomotive pressures. This suggests that closer (molecular) contacts may occur. Under these conditions, hydrogen-bonds might be additional determinants in cell adhesion and spreading. An appropriate conceptual framework for discussing both long- and short-range forces operating synergistically at the limit of close approach may be that of surface energy (Baier, Shafrin & Zisman, 1968; Baier & Zisman, 1975), although no satisfactory theory of surface energy exists at present. Long-range force theory predicts that a significant repulsive energy barrier exists before molecular contact can be achieved. However, the limitation of the physical techniques discussed above weakens this conclusion; the predicted barrier may be altered but it is at present impossible to estimate its size with confidence. Only improved theoretical methods coupled with a detailed description of the cell surface would make this possible. It may also be necessary to take into account the increased repulsive force measured at separations of < 3 nm between lipid lamellae (Le Neveu *et al.*, 1977) which has been tentatively interpreted in terms of water structure. The presence of this additional repulsion may help to prevent membrane fusion at cell–cell contact regions.

Specialised surface projections such as microvilli may penetrate repulsive electrostatic barriers (Weiss, 1964) by virtue of low radius of curvature (Pethica, 1961) or they may possess special surface properties which facilitate molecular contact. Candidates for such surface effects are locally reduced negative charge density or possibly a bridging molecule having the properties of minimal electrostatic and steric repulsion. While claims relating to distinct surface properties of microvillus membranes have been made (Weiss & Subjek, 1974) the situation is not yet resolved. Cell 'feet', first detected interferometrically in fibroblasts by Lochner & Izzard (1973) and confirmed by others (Abercrombie & Dunn, 1975; Lloyd, Smith, Woods & Rees, 1977; Badley *et al.*, 1978; Heath & Dunn, 1978) may represent sites at which molecular contact with the substratum occurs. Although the interference microscope is probably incapable of distinguishing a molecular contact from one of 5 nm (uncertainty in refractive indices of interacting media introduces imprecision of at least this much) and little is known at present about the chemistry of the outer face of the cell surface membrane at feet, it is possible that their adhesiveness may stem from lowered negativity. This appears to be the case at gap junctions, where carbohydrates, which carry most known cell surface charge, have not been detected (see Gilula, 1978). Cell feet may be important in obtaining contacts on protein-coated substrates, perhaps by penetrating the coat, and with this in mind, it would be interesting to re-examine the results of Harris (1973), Letourneau (1975) and Shay, Porter & Krueger (1977) on the locomotion of cells on grid-patterned heterogeneous substrata. Feet are not, however, limited to cells which live in physiological saline: they have been described for *Naegleria* locomoting on glass in dilute saline by Preston & King (1978). *Dictyostelium* amoebae appear to differ in this respect; they can adhere and move on glass without forming characteristic interferometrically distinctive localised areas of close contact.

(v) *Modification of cell–substratum adhesion by macromolecules*

Exposure of high-energy surfaces in biological situations would cause cells to stick strongly to them by physical forces. In physiological situations, interactions are reduced by the adsorption of serum proteins. As discussed earlier, tissue cells characteristically require divalent cations in order to stick to serum-protein-coated glass but not to clean glass – the protein coat reduces adhesion. We have observed Brownian motion of red cells attached to protein-coated glass, showing a first-order grey rolling contact zone by white light interferometry. Without protein the contact is completely static and black, indicating closer and more energetic apposition of membrane to glass (Gingell & Todd, 1979*b*). Although adsorbed proteins reduce adhesiveness, they do not appear to completely mask the physical properties of the underlying substrata, as is clear from the important studies of Harris (1973), which showed that cells in serum can detect and respond adhesively to the difference between glass, metal and plastic. Since physical adhesion to some substrata is strong, protein adsorption may be biologically useful in preventing indiscriminate adhesions of this type developing. In certain circumstances, the reduced attraction exerted by protein-coated surfaces may require a compensatory event so that stronger adhesion can occur. The fibrous protein fibronectin (reviewed by Yamada & Olden, 1978) is capable of greatly increasing cell spreading on tissue-culture dishes by adsorbing to the substratum, and of providing regions of increased intercellular attachment (Grinnell, 1976*b*; Mauntner & Hynes, 1977). The molecule can attach both to cells and culture dishes, and appears to link the two. Cells which lack ricin receptors cannot bind fibronectin and do not enjoy increased adhesion in its presence (Pena & Hughes, 1978). A similar mediating molecule appears to be necessary for cell adhesion to collagen (Klebe, 1974; Pearlstein, 1978).

5. CONCLUSIONS

Where does long-range force theory stand in relation to cell adhesion? While it may, with some justification, be called the only quantitative theory in colloid science (Israelachvili & Adams, 1978), it cannot claim equal success in its power to explain cell contact behaviour. Despite many attempts to correlate cell adhesion with electrostatic properties of either the cell surface or the substratum, we feel that only under non-physiological conditions have convincing results been obtained. These results have shown that in certain situations cell contact can be dependent on electrostatics. Evidence consistent with rather weak electrodynamic attraction at long range has also been found. The major part of the difficulty of accounting for cell adhesion in physical terms almost certainly lies in the extreme complexity of the cell surface and, in particular, its specialised contact regions. This is not merely a smoke screen to retreat behind in the face of unresolved difficulties: it is inevitable that the charged components of the surface which generate electrostatic fields play some part in contact processes; likewise it is certain that electrodynamic forces act between cell surfaces as they come into apposition. What still remains much less certain is the significance of both these forces in adhesion. At secondary-minimum distances the stabilisation energy from these forces is apparently small compared with locomotor forces and energy available from metabolism. We feel, however, that the more energetically significant adhesive events may well occur at the limit of close approach ('molecular contact'), where the methods of long-range force calculations become unreliable. These very close contact regions may coincide with the interferometrically observable cell feet. If it is correct that only a tiny and possibly highly specialised proportion of the cell surface is involved in the most energetic adhesions, it is immediately clear why correlations between gross surface properties and adhesiveness are so hard to demonstrate. Furthermore, if the most adhesive contacts are molecular, it would not be surprising to find correlations between surface energy and adhesion, as van Oss and coworkers have described for bacterial phagocytosis (van Oss, Gillman & Neumann, 1975). This is not to say that long-range forces are not operating, but that they are acting at short distances where they are much stronger, perhaps together with hydrogen bonds, where the language of surface energy may be more appropriate. A further difficulty in any attempt at mathematical modelling of cell interactions in physiological saline is that the distances over which intermolecular forces act are apparently of the same order as the topographic irregularity of the cell surface. While these difficulties will not discourage serious attempts to quantify the physics of cell interactions, they should interject a note of caution into over-sanguine assessments of experimental results. At the same time the complexity of the cell periphery in relation to making and breaking adhesions is such that a complete description is unlikely to be found within the confines of one theoretical methodology.

Acknowledgements

S.V. is in receipt of a training award from the Medical Research Council. D.G. wishes to thank the Science Research Council for continued support. We would like to thank Mr L. Ginsberg for critically reading the manuscript.

REFERENCES

Abercrombie, M. & Dunn, G. A. (1975). Adhesions of fibroblasts to substratum during contact inhibition observed by interference reflection microscopy. *Experimental Cell Research*, **92**, 57–62.

Armstrong, P. B. (1966). On the role of metal cations in cellular adhesion: effect on cell surface charge. *Journal of Experimental Zoology*, **163**, 99–109.

Armstrong, P. B. & Jones, D. P. (1968). On the role of metal cations in cellular adhesion: cation specificity. *Journal of Experimental Zoology*, **167**, 275–82.

Badley, R. A., Lloyd, C. W., Woods, A., Carruthers, L., Allcock, C. & Rees, D. A. (1978). Mechanisms of cellular adhesion. III. Preparation and preliminary characterisation of adhesions. *Experimental Cell Research*, **117**, 231–44.

Baier, R. E., Shafrin, E. G. & Zisman, W. A. (1968). Adhesion: mechanisms that assist or impede it. *Science, New York*, **162**, 1360–8.

Baier, R. E. & Zisman, W. A. (1975). Wetting properties of collagen and gelatin surfaces. In *Applied Chemistry at Protein Interfaces*, ed. R. E. Baier, Advances in Chemistry Series, vol. 145, pp. 155–74. Washington, D.C.: American Chemical Society.

Bangham, A. D. & Pethica, B. (1960). The adhesiveness of cells and the nature of the groups at their surfaces. *Proceedings of the Royal Physical Society, Edinburgh*, **28**, 43.

Beug, H., Gerisch, G., Kempff, S., Riedel, N. & Cremer, G. (1970). Specific inhibition of cell contact formation in *Dictyostelium* by univalent antibodies. *Experimental Cell Research*, **63**, 147–58.

Born, G. V. R. & Garrod, D. (1968). Photometric demonstration of aggregation of slime mould cells showing effects of temperature and ionic strength. *Nature, London*, **220**, 616–18.

Brenner, S. & Parsegian, V. A. (1974). A physical method for deriving the electrostatic interaction between rod-like polyions at all mutual angles. *Biophysical Journal*, **14**, 327–34.

Carroll, B. J. & Haydon, D. A. (1975). Electrokinetic and surface potentials at liquid interfaces. *Journal of the Chemical Society, Faraday Transactions 1*, **71**, 361–77.

CHIEN, S. (1975). Biophysical behaviour of red cells in suspension. In *The Red Blood Cell*, 2nd edn, vol. 2, ed. C. W. Bishop & D. M. Surgenor, chapter 26. London, New York: Academic Press.

COLLINS, M. (1966). Electrokinetic properties of dissociated chick embryo cells: I and II. *Journal of Experimental Zoology*, **163**, 23–47.

CRANDALL, M. (1977). Mating-type interactions in micro-organisms. In *Receptors and Recognition*, Series A, vol. 3, ed. P. Cuatrecasas & M. F. Greaves, pp. 45–100. London: Chapman & Hall.

CURTIS, A. S. G. (1960). Cell contacts: some physical considerations. *American Naturalist*, **94**, 37–56.

CURTIS, A. S. G. (1964). The mechanism of adhesion of cells to glass. A study of interference reflection microscopy. *Journal of Cell Biology*, **20**, 199–215.

CURTIS, A. S. G. (1967). The cell surface: its molecular role in morphogenesis. London: Logos, Academic Press.

CURTIS, A. S. G. (1969). The measurement of cell adhesiveness by an absolute method. *Journal of Embryology and Experimental Morphology*, **22**, 305–25.

CURTIS, A. S. G. (1970). On the occurrence of specific adhesion between cells. *Journal of Embryology and Experimental Morphology*, **23**, 253–72.

CURTIS, A. S. G. (1973). Cell adhesion. *Progress in Biophysics and Molecular Biology*, **27**, 317–86.

CURTIS, A. S. G., CAMPBELL, J. & SHAW, F. M. (1975). Cell surface lipids and adhesion. I. The effects of lysophosphatidyl compounds, phospholipase A_2 and aggregation-inhibiting protein. *Journal of Cell Science*, **18**, 347–56.

CURTIS, A. S. G., CHANDLER, C. & PICTON, N. (1975). Cell surface lipids and adhesion. III. The effects on cell adhesion of changes in plasmalemmal lipids. *Journal of Cell Science*, **18**, 375–84.

CURTIS, A. S. G. & HOCKING, L. M. (1970). Collision efficiency of equal spherical particles in a shear flow. *Transactions of the Faraday Society*, **66**, 1381–90.

CURTIS, A. S. G. & VAN DE VYVER, G. (1971). The control of cell adhesion in a morphogenetic system. *Journal of Embryology and Experimental Morphology*, **26**, 295–312.

CURTIS, A. S. G., SHAW, F. M. & SPIRES, V. M. C. (1975). Cell surface lipids and adhesion. II. The turnover of lipid components of the plasmalemma in relation to cell adhesion. *Journal of Cell Science*, **18**, 357–73.

DAN, K. (1936). Electrokinetic studies of marine ova II. *Physiological Zoology*, **9**, 43–57.

DERJAGUIN, B. V., ABRIKOSOVA, I. I. & LIFSHITZ, E. M. (1956). Direct measurement of molecular attraction between solids separated by a narrow gap. *Quarterly Reviews (London)*, **10**, 295–329.

DONATH, E. (1978). Electrophoretical investigation of cell surface properties – application of theoretical considerations to neuraminidase-treated human red blood cells. *Studia Biophysica*, **74**, 19–20.

DOROSZKOWSKI, A. & LAMBOURNE, R. (1973). The measurement of the dependence of the strength of steric barriers on their solvent environment. *Journal of Colloid and Interfacial Science*, **43**, 97–104.

FOLKMAN, J. & MOSCONA, A. (1978). Role of cell shape in growth control. *Nature, London*, **273**, 345–9.

GARROD, D. R., SWAN, A. P., NICOL, A. & FORMAN, D. (1978). Cellular recognition in slime mould development. In *Cell–Cell Recognition*, ed. A. S. G. Curtis, 32nd Symposium of the Society for Experimental Biology, pp. 173–202. Cambridge University Press.

GARVIN, J. E. (1961). Factors affecting the adhesiveness of human leucocytes and platelets *in vitro*. *Journal of Experimental Medicine*, **114**, 51–73.

GARVIN, J. E. (1968). Effects of divalent cations on adhesiveness of rat polymorphonuclear neutrophils *in vitro*. *Journal of Cellular Physiology*, **72**, 197–212.

GEORGE, J. N., WEED, R. I. & REED, C. F. (1971). Adhesion of human erythrocytes to glass: the nature of the interaction and the effects of serum and plasma. *Journal of Cellular Physiology*, **77**, 51–60.

GILULA, N. B. (1978). Structure of intercellular junctions. In *Receptors and Recognition*. Series B, vol. 2, ed. P. Cuatrecasas & M. F. Greaves, pp. 1–22. London: Chapman & Hall.

GINGELL, D. & FORNÉS, J. A. (1975). Demonstration of intermolecular forces in cell adhesion using a new electrochemical technique. *Nature, London*, **256**, 210–11.

GINGELL, D. & FORNÉS, J. A. (1976). Interaction of red blood cells with a polarized electrode. Evidence of long-range intermolecular forces. *Biophysical Journal*, **16**, 1131–53.

GINGELL, D. & PARSEGIAN, V. A. (1972). Computation of van der Waals interactions in aqueous systems using reflectivity data. *Journal of Theoretical Biology*, **36**, 41–52.

GINGELL, D. & TODD, I. (1975). Adhesion of red blood cells to charged interfaces between immiscible liquids. A new method. *Journal of Cell Science*, **18**, 227–39.

GINGELL, D. & TODD, I. (1979*a*). Red blood cell adhesion. Electrostatic repulsion and long-range attraction. *Journal of Cell Science* (in press).

GINGELL, D. & TODD, I. (1979*b*). Interferometric examination of the adhesion of red blood cells to hydrocarbon and glass. *Journal of Cell Science* (in press).

GINGELL, D. & TODD, I. (1979*c*). Interference reflection microscopy: a quantitative theory for image interpretation and its application to cell-substratum separation measurement. *Biophysical Journal* **26**, 507–26.

GINGELL, D., TODD, I. & PARSEGIAN, V. A. (1977). Long-range attraction between red cells and a hydrocarbon surface. *Nature, London*, **268**, 767–9.

GRINNELL, F. (1976*a*). The serum dependence of baby hamster kidney cell attachment to a substratum. *Experimental Cell Research*, **97**, 265–74.

GRINNELL, F. (1976*b*). Cell spreading factor. Occurrence and specificity of action. *Experimental Cell Research*, **102**, 51–62.

GUSTAFSON, T. & WOLPERT, L. (1967). Cellular contact and movement in sea urchin morphogenesis. *Biological Reviews*, **42**, 442–98.

HARRIS, A. (1973). Behaviour of cultured cells on substrata of variable adhesiveness. *Experimental Cell Research*, **77**, 285–97.

HAYDON, D. A. & TAYLOR, J. L. (1968). Contact angles for thin lipid films and the determination of London-van der Waals forces. *Nature, London*, **217**, 739–40.

HEATH, J. P. & DUNN, G. A. (1978). Cell to substratum contacts of chick fibroblasts and their relation to the microfilament system. A correlated interference-reflexion and high-voltage electron-microscope study. *Journal of Cell Science*, **29**, 197–212.

HESSELINK, F. Th., VRIJ, A. & OVERBEEK, J. Th. G. (1971). On the theory of the stabilization of dispersions by adsorbed macromolecules. II. Interaction between two particles. *Journal of Physical Chemistry*, **75**, 2094–2103.

HOOVER, R. L., LYNCH, R. D. & KARNOVSKY, M. J. (1977). Decrease in adhesion of cells cultured in polyunsaturated fatty acids. *Cell*, **12**, 295–300.

HUMPHREYS, S., HUMPHREYS, T. & SANO, J. (1977). Organization and polysaccharides of sponge aggregation factor. *Journal of Supramolecular Structure*, **7**, 339–51.

ISRAELACHVILI, J. N. (1974). Van der Waals forces in biological systems. *Quarterly Reviews of Biophysics*, **6**, 341–87.

ISRAELACHVILI, J. N. & ADAMS, G. E. (1978). Measurement of forces between two mica surfaces in aqueous electrolyte solutions in the range 0–100 nm. *Journal of the Chemical Society, Faraday Transactions 1*, **74**, 975–1001.

IZZARD, C. S. & LOCHNER, L. R. (1976). Cell-to-substrate contacts in living fibroblasts: an interference reflexion study with an evaluation of the technique. *Journal of Cell Science*, **21**, 129–59.

JAMES, D. W. & TAYLOR, J. F. (1969). The stress developed by sheets of chick fibroblasts *in vitro*. *Experimental Cell Research*, **54**, 107–10.

JEHLE, H. (1969). Charge fluctuation forces in biological systems. *Annals of the New York Academy of Sciences*, **158**, 240–55.

KAMIYA, N. (1964). The motive force of endoplasmic streaming in the amoeba. In *Primitive Motile Systems in Cell Biology*, ed. R. A. Allen & N. Kamiya, pp. 237–75. London, New York: Academic Press.

KEMP, R. B. (1970). The effect of neuraminidase (3:2:1:18) on the aggregation of cells dissociated from embryonic chick muscle tissue. *Journal of Cell Science*, **6**, 751–66.

KEMP, R. B. & JONES, B. M. (1970). Aggregation and electrophoretic mobility studies on dissociated cells. *Experimental Cell Research*, **63**, 293–300.

KEMP, R. B., LLOYD, C. W. & COOK, G. M. W. (1978). Glycoproteins in cell adhesion. *Progress in Surface and Membrane Science*, **7**, 271–318.

KLEBE, R. J. (1974). Isolation of a collagen-dependent cell attachment factor. *Nature, London*, **250**, 248–51.

LANGMUIR, I. (1938). Overturning and anchoring of monolayers. *Science, New York*, **87**, 493–500.

LE NEVEU, D. M., RAND, R. P., PARSEGIAN, V. A. & GINGELL, D. (1977). Measurement and modification of forces between lecithin bilayers. *Biophysical Journal*, **18**, 209–30.

LETOURNEAU, P. (1975). Cell-to-substratum adhesion and guidance of axonal elongation. *Developmental Biology*, **44**, 92–102.

LLOYD, C. W., SMITH, C. E., WOODS, A. & REES, D. A. (1977). Mechanisms of cellular adhesion. II. The interplay between adhesion, the cytoskeleton and morphology in substrate-attached cells. *Experimental Cell Research*, **110**, 427–37.

LOCHNER, L. & IZZARD, C. S. (1973). Dynamic aspects of cell–substrate contact in fibroblast motility. *Journal of Cell Biology*, **59**, abstr. 199*a*.

LUZZATI, V. (1968). X-ray diffraction studies of lipid–water systems. In *Biological*

Membranes, ed. D. Chapman, pp. 71–123. London, New York: Academic Press.

MacDonald, R. C. & Bangham, A. D. (1972). Comparison of double layer potentials in lipid monolayers and lipid bilayer membranes. *Journal of Membrane Biology*, **7**, 29–53.

Malmberg, C. G. & Maryott, A. A. (1950). Dielectric constants of aqueous solutions of dextrose and sucrose. *Journal of Research of the National Bureau of Standards*, **45**, 299.

Margolis, L. B., Dyatlovitskaya, E. V. & Bergelson, L. D. (1978). Cell–lipid interaction. Cell attachment to lipid substrates. *Experimental Cell Research*, **111**, 454–7.

Maroudas, N. G. (1973). Chemical and mechanical requirements for fibroblast adhesion. *Nature, London*, **244**, 353–4.

Maroudas, N. G. (1975*a*). Adhesion and spreading of cells on charged surfaces. *Journal of Theoretical Biology*, **49**, 417–24.

Maroudas, N. G. (1975*b*). Polymer exclusion, cell adhesion and membrane fusion. *Nature, London*, **254**, 695–6.

Maroudas, N. G. (1977). Sulphonated polystyrene as an optimal substratum for the adhesion and spreading of mesenchymal cells in monovalent and divalent saline solutions. *Journal of Cellular Physiology*, **90**, 511–20.

Martin, G. R. & Rubin, H. (1974). Effects of cell adhesion to the substratum on the growth of chick embryo fibroblasts. *Experimental Cell Research*, **85**, 319–33.

Mauntner, V. & Hynes, R. O. (1977). Surface distribution of LETS protein in relation to the cytoskeleton of normal and transformed cells. *Journal of Cell Biology*, **75**, 743–68.

Michaelis, F. B. & Dalgano, L. (1971). Biochemical aspects of the attachment of a pig-kidney monolayer cell line to glass surfaces. *Experimental Cell Research*, **65**, 43–8.

Middleton, C. A. (1973). The control of epithelial cell locomotion in tissue culture. *Ciba Foundation Symposium*, **14**, 251–70.

Moscona, A. A. (1968). Cell aggregation properties of specific cell ligands and their role in the formation of multicellular systems. *Developmental Biology*, **18**, 250–77.

Owen, E., Clifford, J. & Marson, A. (1978). The effects of surfactants on cell aggregation. *Journal of Cell Science*, **32**, 363–76.

Parsegian, V. A. (1975). Long range van der Waals forces. In *Physical Chemistry: Enriching Topics from Colloid and Surface Science*, ed. H. van Olphen & K. J. Mysels. La Jolla: Theorex.

Parsegian, V. A. & Gingell, D. (1972). On the electrostatic interaction across a salt solution between two bodies bearing unequal charges. *Biophysical Journal*, **12**, 1192–1204.

Parsegian, V. A. & Gingell, D. (1973). A physical force model of biological membrane interaction. In *Recent Advances in Adhesion*, ed. L. H. Lee, pp. 153–92. New York: Gordon & Breach.

Parsegian, V. A. & Gingell, D. (1979). Determination of the repulsive force sufficient to prevent adhesion of red cells to planar surfaces. (In preparation.)

Pearlstein, E. (1978). Substrate activation of cell adhesion factor as a prerequisite for cell attachment. *International Journal of Cancer*, **22**, 32–5.

Pena, S. D. J. & Hughes, R. C. (1978). Fibronectin–plasma membrane interactions in the adhesion of hamster fibroblasts. *Nature, London*, **276**, 80–3.

Pethica, B. A. (1961). The physical chemistry of cell adhesion. *Experimental Cell Research*, Suppl. **8**, 123–40.

Poste, G., Greenham, L. W., Mallucci, L., Reeve, P. & Alexander, D. J. (1973). The study of cellular 'microexudates' by ellipsometry and their relationship to the cell coat. *Experimental Cell Research*, **78**, 303–13.

Preston, T. M. & King, C. A. (1978). Cell–substrate associations during the amoeboid locomotion of *Naegleria*. *Journal of General Microbiology*, **104**, 347–51.

Rabinovitch, M. & de Stefano, M. J. (1973). Manganese stimulates adhesion and spreading of mouse sarcoma I ascites cells. *Journal of Cell Biology*, **59**, 165–76.

Rappaport, C. & Bishop, C. B. (1960). Improved method for treating glass to produces surfaces suitable for the growth of certain mammalian cells in synthetic medium. *Experimental Cell Research*, **20**, 580–4.

Rappaport, C., Poole, J. P. & Rappaport, H. P. (1960). Studies on properties of surfaces required for growth of mammalian cells in synthetic medium. *Experimental Cell Research*, **20**, 465–510.

Rosenberg, M. D. (1962). Long-range interactions between cell and substratum. *Proceedings of the National Academy of Sciences, USA*, **48**, 1342–9.

Roth, S. (1973). A molecular model for cell interactions. *Quarterly Reviews of Biology*, **48**, 541–63.

Schaeffer, B. E. & Curtis, A. S. G. (1977). Effects on cell adhesion and membrane fluidity of changes in plasmalemmal lipids in mouse L929 cells. *Journal of Cell Science*, **26**, 47–55.

Shay, J. W., Porter, K. R. & Krueger, T. C. (1977). Motile behaviour and topography of whole and enucleate mammalian cells on modified substrates. *Experimental Cell Research*, **105**, 1–8.

Shih, A. & Parsegian, V. A. (1975). Van der Waals forces between heavy alkali atoms and gold surfaces: comparison of measured and predicted values. *Physical Reviews*, **A12**, 835–41.

Small, D. M. (1967). Phase equilibria and structure of dry and hydrated egg lecithin. *Journal of Lipid Research*, **8**, 551–7.

Smitham, J. B., Evans, R. & Napper, D. H. (1975). Analytical theories of the steric stabilization of colloidal dispersions. *Journal of the Chemical Society, Faraday Transactions 1*, **71**, 285–97.

Steinberg, M. (1978). Cell–cell recognition in multicellular assembly: levels of specificity. In *Cell–Cell Recognition*, 32nd Symposium of the Society for Experimental Biology, ed. A. S. G. Curtis, pp. 25–49. Cambridge University Press.

Takeichi, M. & Okada, T. S. (1972). Roles of magnesium and calcium ions in cell-to-substrate adhesion. *Experimental Cell Research*, **74**, 51–60.

Taylor, A. C. (1961). Attachment and spreading of cells in culture. *Experimental Cell Research*, Suppl. **8**, 154–73.

Ueda, M. J., Ito, T., Okada, T. S. & Ohnishi, S.-I. (1976). A correlation between membrane fluidity and the critical temperature for cell adhesion. *Journal of Cell Biology*, **71**, 670–4.

van Oss, C. J., Gillman, C. F. & Neumann, A. W. (1975). *Phagocytic Engulfment and Cell Adhesiveness as Cellular Surface Phenomena*. New York: Marcel Dekker.

Vasiliev, J. M., Gelfand, I. M., Domnina, L. V., Zacharova, O. S. & Ljubimov, A. V. (1975). Contact inhibition of phagocytosis in epithelial sheets. Alterations of cell surface properties induced by cell-cell contacts. *Proceedings of the National Academy of Sciences, USA*, **72**, 719–22.

Vicker, M. G. & Edwards, J. G. (1972). The effect of neuraminidase on the aggregation of BHK21 cells and BHK21 cells transformed by polyoma virus. *Journal of Cell Science*, **10**, 759–68.

Weiss, L. (1964). Cellular locomotive pressure in relation to initial cell contacts. *Journal of Theoretical Biology*, **6**, 275–81.

Weiss, L. & Subjeck, J. R. (1974). The densities of colloidal iron hydroxide particles bound to microvilli and the spaces between them: studies on glutaraldehyde-fixed Ehrlich ascites tumour cells. *Journal of Cell Science*, **14**, 215–23.

Wiese, L. & Wiese, W. (1978). Sex cell contact in *Chlamydomonas*, a model for cell recognition. In *Cell–Cell Recognition*, 32nd Symposium of the Society for Experimental Biology, ed. A. S. G. Curtis, pp. 83–103. Cambridge University Press.

Wolpert, L. (1965). Cytoplasmic streaming and amoeboid movement. In *Function and Structure of Micro-organisms*, 15th Symposium of the Society for General Microbiology, ed. M. R. Pollock & M. H. Richmond, pp. 270–93. Cambridge University Press.

Wolpert, L., MacPherson, I. & Todd, I. (1969). Cell spreading and cell movement: an active or a passive process? *Nature, London*, **223**, 512–13.

Yamada, K. M. & Olden, K. (1978). Fibronectins – adhesive glycoproteins of cell surface and blood. *Nature, London*, **275**, 179–84.

A physical theory of cell–cell and cell–substratum interactions

KRZYSZTOF DOŁOWY

Medical Centre of Postgraduate Education, Department of Biophysics and Biomathematics, Marymoncka 99, 01-813 Warszawa, Poland

GLOSSARY

A	surface area
C	electric capacitance
c	salt concentration
d	distance or thickness
d_m	thickness of the cell membrane
d^*	effective thickness of the cell membrane
E	energy
E_A	attractive van der Waals-type energy per unit area
E_R	electrostatic repulsive energy per unit area
ΔE_γ	surface energy change per unit area
F	Faraday constant
H	Hamaker constant
k	Boltzmann constant
N_L	number defined by (20)
N_S	number defined by (21)
Q	electric charge
q	elementary charge
R	gas constant
r	radius
T	temperature
U	electrophoretic mobility
V	electric potential at the lipid plane
ΔV	potential change
V_L	electric potential originated in lipid molecules
V_S	electric potential originated in sialic acid groups
V^*	mean molar volume of adsorbed ions
x_0	distance defined by (6)
Γ_{max}	concentration per unit area of sites able to adsorb an ion
γ	interfacial tension
γ_e	electric term of the interfacial tension of the liquids
γ_0	interfacial tension of uncharged liquid
ϵ	dielectric constant

ϵ_0	8.85×10^{-12} F m^{-1}
ζ	electrokinetic potential
η	viscosity
κ	Debye–Hückel parameter defined by (13)
κ_{ion}	Debye–Hückel parameter for a different ionic strength
κ_{m}	Debye–Hückel parameter of the cell membrane
σ	surface charge density
σ_{L}	surface charge density originated in lipid molecules
σ_{S}	surface charge density originated in sialic acid groups
σ_0	surface charge density at the slip plane
σ_1	surface charge density of the Stern layer
σ_2	surface charge density of the outer diffuse layer
σ_3	surface charge density of the inner diffuse layer
ϕ	membrane potential
ϕ_{sub}	electric potential at the substratum
ψ	electric potential at the surface of an interacting cell
ψ_d	electric potential at a distance d from the cell surface
ψ_0	electric potential at the surface of non-interacting cell

Subscripts i and j refer to the first and second cell respectively.

INTRODUCTION

Cell biology is facing today a situation which faced seventeenth- and eighteenth-century physics. There are hundreds of facts loosely related to one another. Cell adhesion problems as well as the whole of cell biology are waiting for a general theory of cell behaviour which will arrange these facts in sequence. This does not mean that there are no theories of cell adhesion. In fact there are at least two groups of theory (consult, for example, Curtis 1967, 1973; Weiss & Harlos 1972; Bell, 1978): physical (DLVO theory) and chemical (antigen–antibody type or ligand–receptor/mutual receptor interaction theory). However, these theories do not provide us with the explanations of all adhesion-related phenomena such as cell aggregation, morphogenesis and cell locomotion. We can hope that in the future one of these theories will expand and finally reach the level of a general theory of cell behaviour. It is, however, difficult to decide at the moment which of the two theories is more promising, or if we should rather look for a completely new theory. Thus, theories of cell adhesion have not yet acquired a firm basis.

From one view-point, DLVO theory, as a theory deduced from physics, has to be correct. But I am not very much impressed by this theory since it has difficulty in explaining features of selectivity in cell adhesion. This may be due to the fact that DLVO theory deals with lyophobic colloid particles and not with real cells. Can we improve DLVO theory? On the other hand the major opposition to the DLVO theory comes from the metaphysical conviction, shared by many biologists, that living matter can not be described by purely physical argument.

The alternative to the DLVO theory is the chemical theory of cell adhesion. The chemical theory is apparently based on the well known specific interactions of antigen and antibody, enzyme and substrate, lectin and receptor etc., and was constructed especially to explain the selectivity of cell adhesion. It has, however, a very limited experimental basis. If we accept the hypothesis that both donor and acceptor groups are present on one cell surface then we should frequently observe, at least *in vitro*, self-adhesion, i.e. one part of a cell would adhere to the other part of the same cell. If, on the other hand, there are only antigens on the surface and the cells are bonded together by means of divalent antibodies, then it should be relatively easy to prove that such direct bonds between cells do in fact exist. But even for the simplest case of red blood cells agglutinated by antibodies, the existence of such direct bonds has not been wholly confirmed experimentally and is accepted as a premise of the cell agglutination theory.

Thus, the physical theory needs some theoretical improvement while the chemical theory requires some experimental verification before either reaches the status of the theory of cell adhesion. It would be very desirable if the theory of cell adhesion could also explain selectivity of adhesion, while answering the question why there are few different patterns of cell sorting-out, and account for features of locomotion and patterning of cell position during morphogenesis.

THE LONG-RANGE PHYSICAL FORCES – THE DLVO THEORY

DLVO theory

At the beginning of the twentieth century, the methods of colloid chemistry were applied to the study of cells *in vitro*. Soon it became apparent that all the cells studied had net negative charge at the surface. From the physical view-point the cell with electric charge at the surface and nearly spherical shape (in suspension) resembles a lyophobic colloid particle. Thus, it was not surprising that the theory of lyophobic colloid interactions (DLVO theory) (Derjagin &

Landau, 1941; Verwey & Overbeek, 1948) was applied to study cell interactions (Curtis, 1960, 1967; Weiss & Harlos, 1972). Two forces are considered by the DLVO theory: the electrostatic-like-charge repulsive force and the attractive van der Waals-type force. While the attractive force dominates at larger separations of interacting surfaces, the repulsive force is stronger for smaller distances. Thus, as predicted by the DLVO theory, cells might adhere to each other with energies equal to 30–100 kT without forming a direct contact between interacting surfaces.

The geometry of interacting bodies

Before we go into detailed description of the two forces of the DLVO theory, we should decide which geometry, sphere–sphere or slab–slab, is more suitable for studying cell interaction. The first is strongly favoured in colloid chemistry because of the minute dimensions and perfect shape of colloid particles. Cells, however, are at least two orders of magnitude larger than colloid particles and they are not solid spheres, but elastic bodies. It is known from experience that cells adhere to one another with large flat surfaces (at least it appears so in the light microscope). Thus, it seems to be reasonable to use slab–slab geometry to study cell behaviour. The most useful notation is in energy of interaction per unit area, which makes it possible to calculate the exact energy of interaction without knowing the real area of cell contact. Use of that notation is necessary because, as was shown in interference reflection (Curtis, 1964), and electron microscopy, the cells adhere to each other or to the substratum in tiny specialised areas as well as at whole adjacent surfaces.

Force–energy dependence

Some confusion between the terms 'force' and 'energy' should also be resolved before any further discussion of DLVO theory. While the term 'attractive force' is fully justified, the term 'attractive energy' needs some additional explanation. Energy itself is a property of a physical system and cannot be considered as attractive or repulsive. However, if one compares energies for two systems which differ only in distance between interacting surfaces then it might be found that the energy change will favour attraction or repulsion of surfaces. Thus, the term 'attractive energy' means when two surfaces approach energy is produced. (Consult a textbook or Gingell (1971) for more detail on the subject.)

The van der Waals-type force

Of the two forces considered by the DLVO theory, the nature of one – like-charge repulsion force – seems to be obvious. There is, however, a lot of confusion among biologists on the nature of van der Waals-type force. If we take a closer look at the structure of a molecule then this mysterious force becomes more understandable. It is not surprising that ions of opposite charge attract. It is also not surprising that ion–dipole and dipole–dipole structures attract. One should keep in mind, however, that the electrons of a neutral molecule form a fluctuating cloud. This fluctuating electron cloud may form an instantaneous dipole or an induced dipole when the molecule is in an electric field. Thus, neutral molecules attract by means of ion-induced dipole, dipole-induced dipole and instantaneous-dipole-induced dipole structures. The attractive van der Waals force is electromagnetic in its origin and should not be considered strange.

There are two ways of describing the van der Waals force. The first uses H, the Hamaker constant, as a mathematical convenience to represent all the interactions between all the molecules of adjacent surfaces by a single form. The second way of calculating the van der Waals force is based on the fact that all the electron fluctuations of the molecules influence the electromagnetic field and thus can be detected by spectroscopy. From spectroscopic data one can calculate the van der Waals force. It is, of course, in theory possible to use a microscopic quantum-mechanical approach, but this is not feasible for complex systems. More information about the second method can be found in review articles (Parsegian, 1975; Nir, 1977).

The van der Waals energy per unit area of two interacting flat surfaces is given by the following equation (using Hamaker notation)

$$E_{\mathrm{A}} = -H/12\pi d^2 \qquad (1)$$

The like-charge-repulsion force

The other energy term of the DLVO theory gives rise to some theoretical problems. The assumptions that were used to simplify the solution of the case of interaction of two colloid surfaces cannot be used in the case of cell interaction. Classical DLVO theory considers the interaction of like surfaces in which electric potential is constant and of low value. In fact we are more interested in the interaction of unlike surfaces, e.g. cell and substratum. The cells are not bodies of 'constant potential' but 'constant charge'. This means that the potential at the surface does not originate from ion adsorption but from dissociation of surface groups. (It has been shown (Heard & Seaman, 1960) that substitution of one anion by another does not change the electrical properties of a cell membrane. The results of chemical modification of the cell surface (Gittens & James, 1963; Haydon & Seaman, 1967) lead to the conclusion that the cell-surface charge is caused exclusively by dissociated chemical groups. There are some indications that calcium cation is bound specifically to the cell surface (Collins, 1966; Seaman, Vassar & Kendall, 1969), however selectivity of that reaction is very low.) Thus, during a cell interaction the potential at the cell surface increases, instead of remaining constant, because of ion desorption. If we consider the interaction of approaching surfaces i and j then the potential at each surface would rise from ψ_0 to:

$$\psi_i = \psi_{0i} + \psi_{dj} \tag{2}$$
$$\psi_j = \psi_{0j} + \psi_{di} \tag{3}$$

These equations are accurate for large cell separations. Finally the potential at the cell surface is quite high and the simplified equation of the potential drop outside the cell membrane

$$\psi_d = \psi_0 \exp[-\kappa d] \tag{4}$$

cannot be used. Instead we have to use the non-linear Poisson–Boltzmann equation (Mysels, 1975):

$$\psi_d = (2kT/q)\ln\{\exp[\kappa(x_0+d)]+1)/(\exp[\kappa(x_0+d)]-1\} \tag{5}$$

where:

$$x_0 = (1/\kappa)\ln\{(\exp[\psi_0 q/2kT]+1)/(\exp[\psi_0 q/2kT]-1)\} \tag{6}$$

Now we are able to write an equation for the repulsive energy per unit area of the cell-surface interaction:

$$E_\mathrm{R} = -(\epsilon\epsilon_0\kappa/2)\{(\psi_{0i}^2-\psi_i^2)+(\psi_{0j}^2-\psi_j^2)\} \tag{7}$$

The weak points of the DLVO theory

Both forces of the DLVO theory are additive. Thus, the energy of interaction of unlike surfaces, E_{ij}, lies between the two interaction energies of like surfaces, E_{ii} and E_{jj}. This feature of the total energy of interaction greatly reduces the number of cell sorting-out patterns to be expected and makes it difficult to explain the selectivity of cell adhesion. There have been some attempts to add one more physical factor to the DLVO theory to overcome these weaknesses, e.g. Good (1972) on the area of cell contact. The new models have little or no experimental basis. As we shall see in the next section, classical DLVO treatment lacks an energy term which originates in a difference between a lyophobic colloid particle and a real cell.

THE SURFACE ENERGY CHANGE – THE MISSING ENERGY TERM OF THE DLVO THEORY

To find the missing energy term influencing the total energy of cells interaction (see also Dołowy, 1975; Dołowy & Holly, 1978) let us examine a system which is far from being living matter – a mercury drop. It has been known for about a hundred years that the surface tension of mercury drop decreases rapidly when it is electrically charged. The nature of that phenomenon is obvious; while the surface-tension force tends to decrease surface area, the repulsion between like charges present at the surface tends to increase the area. We can express that relation as follows:

$$\gamma = \gamma_0 - \gamma_e \tag{8}$$

where:

$$\gamma_e = V^2C/2A = Q^2/2CA = \sigma^2 A/2C \tag{9}$$

Equations (8) and (9) are valid not only for a mercury drop in a vacuum but for any charged liquid drop in any environment. Thus instead of the term 'surface tension' which refer to the liquid–vacuum interface, the term 'interfacial tension' should be used.

The interfacial tension of a liquid drop is decreased by electrons (as in case of mercury) as well as by charged molecules emerging at the surface of a liquid. This phenomenon has fascinated scientists for the last hundred years, ever since it was found that a lipid drop in the basic watery solution perform amoeboid-like movement, chemotaxis, pinocytosis and phagocytosis. This living-matter-like behaviour of the lipid drop is caused by hydrolysis of neutral fats to

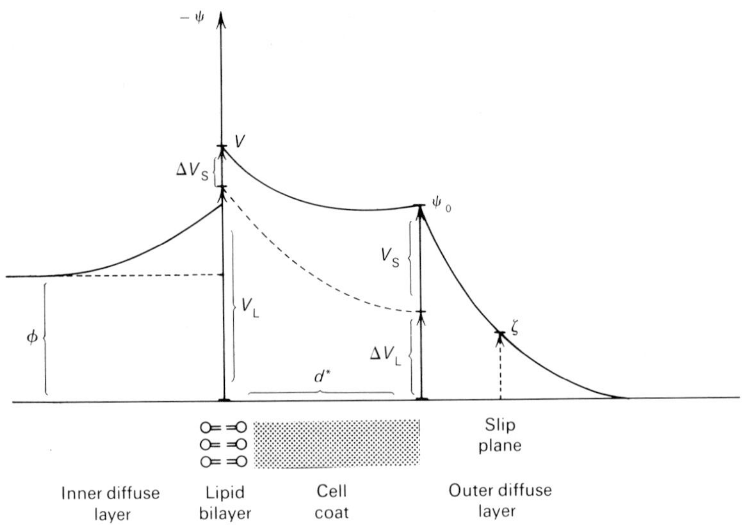

Fig. 1. Potential distribution in cell membrane.

negatively charged fatty acid molecules when in contact with a base in solution. The negatively charged fatty acids emerging at the lipid drop surface cause a local decrease in interfacial tension and thus the amoeboid-like movement of a lipid drop.

It is very difficult to find why the surface-tension theory was abandoned in biology. The reasons were metascientific rather than scientific, because the pretext of a too-low value of the cell membrane surface tension cannot be treated seriously. Just as 1 mV of membrane potential implies a potential gradient of 1000 V cm^{-1}, 0.1 dyn cm^{-1} of interfacial tension change implies 10 atm cell-membrane-expanding pressure. Even if the arguments against surface-tension theories had some power before the cell membrane was discovered (Danielli & Davson, 1935) and its low interfacial tension measured (Danielli & Harvey, 1935) they can no longer be treated seriously after the fluidity of cell membrane has been shown (Curtis, 1961; Frye & Edidin, 1970; Singer & Nicolson, 1972).

To appreciate the importance of interfacial tension changes in cell interaction, let us consider the structure of the cell membrane. The cell membrane (Fig. 1) consists of a fluid lipid bilayer covered completely or partially with fuzz – the glycoprotein coat. A large number of the lipid molecules (about 20 % for red blood cells) have net negative charge, i.e. phosphatidic acid, phosphatidyl serine and phosphatidyl inositol. At the surface of the fuzz, negatively charged molecules of sialic acid can be found. Thus, the cell membrane is not a single-layer structure covering the cytoplasm but a double-layer system consisting of a fluid, negatively charged lipid layer covered by a negatively charged coat layer. This complex structure of the cell membrane could be a cause of the variety of cell phenomena. This would not be the case, however, if the classic 'constant potential' DLVO theory would be applicable to cells. But as it was stated in the previous section, the cell is not a body of 'constant potential' (the potential does not originate in anion adsorption) but 'constant charge' (the cell's surface is charged by dissociation of chemical groups). This means that during cell interaction both electric potentials, ψ_0 and V, increase, leading in consequence (see (8) and (9)) to a decrease in interfacial tension. We can describe a consequent attractive energy term ΔE_γ produced by the change of the potential at the lipid layer as

$$\Delta E_\gamma = \{V_i^2 - (V_i + \Delta V_i)^2\}/2C_i + \{V_j^2 - (V_j + \Delta V_j)^2\}/2C_j \quad (10)$$

Once again let us trace the origin of the additional energy term. The electrostatic part of DLVO theory calculates the interaction of the outer (fuzz) surface of the cell membranes and is not interested in the structure of the cell membrane itself on the potential drop within the cell coat. But the cell membrane consists of two concentric surfaces. the outer (fuzz) and the inner (lipid) surface. While the change of the electric potential at the outer layer is considered in DLVO theory, there is also a change of a potential at the inner layer. This inner layer happens to be fluid. Thus, the change in the potential causes a change in interfacial tension of the fluid (analogous to the mercury drop) and in consequence produces an additional energy term in the cell interaction.

THE STERN-LIKE ELECTROCHEMICAL MODEL OF THE CELL MEMBRANE

As discussed in the previous section, the cell membrane consist of a negatively charged lipid layer covered with a glycoprotein coat. The negatively charged groups of sialic acid are found on the coat surface. To establish the dependence between the potential at the lipid plane, V, and the potential at the cell surface, ψ_0, one should know the electrochemical properties of those layers. For the sake of simplicity, most authors treated the cell membrane as a charged plane or, if more complex models were considered, the cell coat was assumed to have properties identical with the solution. The electrochemical structure of the cell coast itself was not considered, except for the factor representing that part of the cell coat volume not accessible to ions. Those models are, however, completely unrealistic if one compares them with the facts established by surface science.

First of all, water cannot be treated as amorphous medium, it is a highly structured crystal-like substance. Near organic molecules this liquid crystal becomes unstable, because water molecules are attracted to hydrophilic, and repelled from hydrophobic, groups present at the surface of an organic molecule. That effect is present even 0.4–0.6 nm from the surface (Kavanau, 1964; Drost-Hansen, 1971). Thus, it is most unlikely that normal water is present within the cell coat. It should rather be expected that the locations of water molecules are 'restricted' by the structure of the coat.

The second reason why the cell coat cannot be treated as part of the bulk solution is that the coat surface area is very large compared to cell coat volume. This, contrary to the regular plane–solution interface, ions should be expected to be adsorbed within the coat. Ion adsorption, together with water structuring, makes the cell coat resemble, from the electrochemical point of view, a Stern-like layer (see, for example, Overbeek, 1952; Kortüm, 1966).

Adopting the Stern-like model of the cell coat, one is able to find the dependence between V and ψ_0. For the sake of simplicity let us consider the dependence between V_L, potential originated in charged lipid molecules, and ΔV_L, the part of the potential V_L present at the outer cell surface. The problem is also simplified by the assumption that the potential at the outer surface of the cell coat is equal to the membrane potential ϕ (see Fig. 1). Even if the last assumption is not fulfilled the error made for the very thin or thick coats is negligible. (Full discussion in Appendix 1.)

Let us denote the surface charge density within the coat as σ_1 and in the diffuse layer outside the cell coat as σ_2. The dependence between V_L and ΔV_L is given by the following equation:

$$\Delta V_L/V_L = \sigma_2/(\sigma_1+\sigma_2) = 1/(1+\sigma_1/\sigma_2) \tag{11}$$

For the relatively high concentration of uni-univalent electrolyte and lack of specific adsorption, the σ_1/σ_2 ratio is equal to

$$\sigma_1/\sigma_2 = \frac{-2Fc\Gamma_{max}V^*(\exp[F\Delta V_L/RT]-\exp[-F\Delta V_L/RT])}{-(2\epsilon\epsilon_0RTc)^{\frac{1}{2}}(exp[F\Delta V_L/2RT]-\exp[-F\Delta V_L/2RT])}$$
$$\cong F(2c/\epsilon\epsilon_0RT)^{\frac{1}{2}}\Gamma_{max}V^* \qquad (12)$$

If one remembers that Debye–Hückel parameter, κ, is equal to

$$\kappa = F(2c/\epsilon\epsilon_0RT)^{\frac{1}{2}} \qquad (13)$$

and we denote

$$\Gamma_{max}V^* = d^* \qquad (14)$$

where d^* is the equivalent thickness of the cell coat ($\Gamma_{max}V^*$ is given in length units), then we can write the ratio between ΔV_L and V_L as

$$\Delta V_L/V_L = 1/(1+\kappa d^*) \qquad (15)$$

To avoid confusion about the nature of d^*, equivalent thickness, one can also define d^* (Dołowy & Holly, 1978) as:

$$d^* = d_m\kappa_m/\kappa \qquad (16)$$

where d_m is a real thickness and κ_m a real Debye–Hückel parameter of the cell membrane. Unfortunately, it is not possible to measure the value of κ_m directly. Thus, instead of $\kappa_m d_m$ one can use the κd^* value.

Finally one can find the value of the potentials ψ_0 and V (see also Fig. 1)

$$\psi_0 = V_L/(1+\kappa d^*)+V_S \qquad (17)$$

$$V = V_S/(1+\kappa d^*)+V_L \qquad (18)$$

It is worth mentioning that the electric capacitance of the Stern layer is known to be constant and is only slightly dependent on ionic strength of the medium (see, for example, Overbeek, 1952; Kortüm, 1966). The usefulness of this model is now limited only by the ability to measure cell membrane parameters V_L, V_S, and d^*.

THE CALCULATION OF CELL MEMBRANE PARAMETERS

To calculate the cell membrane parameters V,ψ_0 and d^*, one needs to use biochemical and electrophoretical data as well as microscopical data.

In the first step, the electric potential at the slip plane, ζ, and the corresponding surface charge density, σ_0, should be calculated from the electrophoretic mobility, U, using the cell radius value, r, and the Smoluchowski equation (see, for example, Overbeek, 1952)

Table 1. *Comparison of experimental data for erythrocyte electrophoretic mobility for various ionic strengths* with theoretical data of the surface-charged-sphere model*† ($\psi_0 = -80$ mV); *and the two-layer model*‡

Ionic strength (M)	Charged-sphere model	Experimental data	Two-layer model
0.145	1.08	1.08	1.08
0.029	2.55	2.20	2.24
0.0145	3.13	2.78	2.81
0.00725	3.62	3.50	3.40
0.00436	3.91	3.81	3.84
0.00290	4.11	4.15	4.19
0.00218	4.23	4.45	4.43

* At pH 7.4 (Heard & Seaman, 1960).
† The best fit is $\psi_0 = -80$ mV; $d_0 = 1.25$ nm.
‡ $V = -84$ mV; $\psi_0 = -49$ mV; $d^* = 5.2$ nm; $d_0 = 0.79$ nm.

$$U = \zeta\epsilon\epsilon_0/\eta = \sigma_0/\eta\kappa \tag{19}$$

Next, if both the concentration of negatively charged lipid molecules per unit area, σ_L, and the concentration of sialic acid groups per unit area, σ_S, are known from biochemical data then one can compare them with the σ_0 value:

$$N_L = \sigma_L/\sigma_0 = V_L/\zeta \tag{20}$$

$$N_S = \sigma_S/\sigma_0 = V_S/\zeta \tag{21}$$

and calculate the ψ_0 value:

$$\psi_0 = \{N_L/(\kappa_{ion}d^*+1)+N_S\}\zeta \tag{22}$$

Finally, making use of (5) and (6) we can calculate the electric potential at the slip plane, ψ_d. The slip plane is assumed to be located at the distance d_0 from the sialic acid plane.

The measured cell electrophoretic mobility is proportional to the ψ_d potential. The ψ_d potential is the function of V_L, V_S, d^*, d_0 and κ_{ion}. The only operation which remains is to compare the cell electrophoretic mobilities measured for different ionic strengths of medium and calculated for the adopted parameter values and make a computer 'guess' as to which parameter values gives the best fit of theoretical predictions with experimental data. Using this proce-

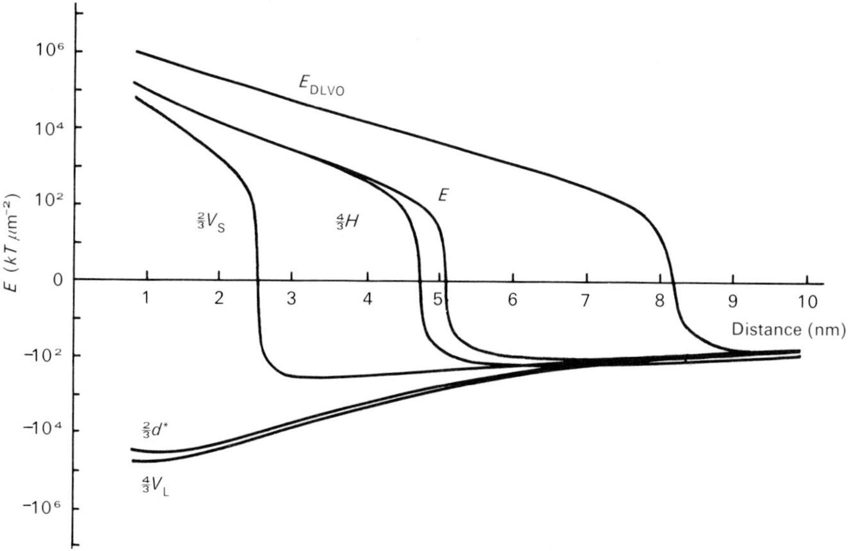

Fig. 2. Effect of cell membrane electric parameter changes on total energy of interaction. $V_L = -75$ mV, $V_s = -24$ mV, $d^* = 0.9$ nm, $H = 1.00 \times 10^{-14}$ erg.

dure parameters were calculated for red blood cells. The comparison between theoretically predicted and experimental data of electrophoretic mobility is given in Table 1. More detailed information on the calculation of the cell membrane parameters will be given elsewhere (Dołowy & Godlewski, in preparation).

It is worth mentioning that the electric potential at the cell membrane calculated from biochemical data and the more realistic cell membrane model is many times higher than would be expected from electrophoretic data. One should also remember that the cell coat is not of uniform thickness. In fact, adhesion may take place at the location of relatively thin coat (Pegrum & Maroudas, 1975). Thus, the value d^* used to calculate energy of adhesion should be adjusted according to electron microscopy data of the coat thickness in the adhesion locations.

CELL MEMBRANE PARAMETERS AND THE TOTAL ENERGY OF INTERACTION

In the previous sections it was shown that the energy of cell interactions depends on four parameters: V_L, the electric potential originating in the number of negatively charged lipid molecules; V_S, the electric potential originating in the number of the sialic acid groups; d^*, the equivalent thickness of the cell membrane which is a function of the number of ions adsorbed in the cell coat, i.e. which depends on structure, hydrophilicity, density and thickness of the cell coat; and H, the Hamaker constant. The changes in the cell membrane parameters will influence the total energy of cell interaction. However, the effect of V_L and d^* changes will affect the total energy of interaction much more significantly than the changes in V_S and H values, as is shown at Fig. 2.

Figure 2 shows also the characteristic difference between the DLVO theory and this new theory which also considers the additional surface-energy term: the parameters V_L and d^*, which play no role in the DLVO theory, in some cases dominate the cell energy-balance. Interacting cells might come to a very close contact to one another when the energy of this interaction per unit area exceeds $10^5\ kT\ \mu m^{-2}$, i.e. equivalent to thousands of chemical bonds. Thus, the additional energy coming from the direct chemical bonds between the cells (if they exist at all) might be safely neglected in the case of close-contact adhesion.

The distance–energy dependence shown in Fig. 2 has important consequences for geometry of cell adhesion. It is obvious that to maintain stable adhesion the energy should be significantly higher than the energy of thermal motion. Thus, the minimum adhesive energy is expected to be 10–20 kT per cell. A rough estimate of the adhesive energy predicted by the DLVO theory of $100\ kT\ \mu m^{-2}$ means that two surfaces have to adhere to each other over considerably large areas of about 0.1 to 1.0 μm^2 to maintain the stable contact in the secondary minimum (5–10 nm apart). If the cell membrane parameters allow the interacting surfaces to come into a very close contact (1–2 nm) as predicted by theory, 10^{-4}–$10^{-3}\ \mu m^2$ will suffice to maintain adhesion. Thus, adhesion might take place on relatively large surfaces at the secondary minimum distance, or the cells might form a very 'tight' contact for which even a very small area, like a microvillus, would suffice to maintain a stable contact.

CELL–SUBSTRATUM INTERACTION

The cell–substratum interaction is one of the most widely investigated phenomena of cell biology. The results of these studies are, however, very inconsistent. Several factors have been found which play a role

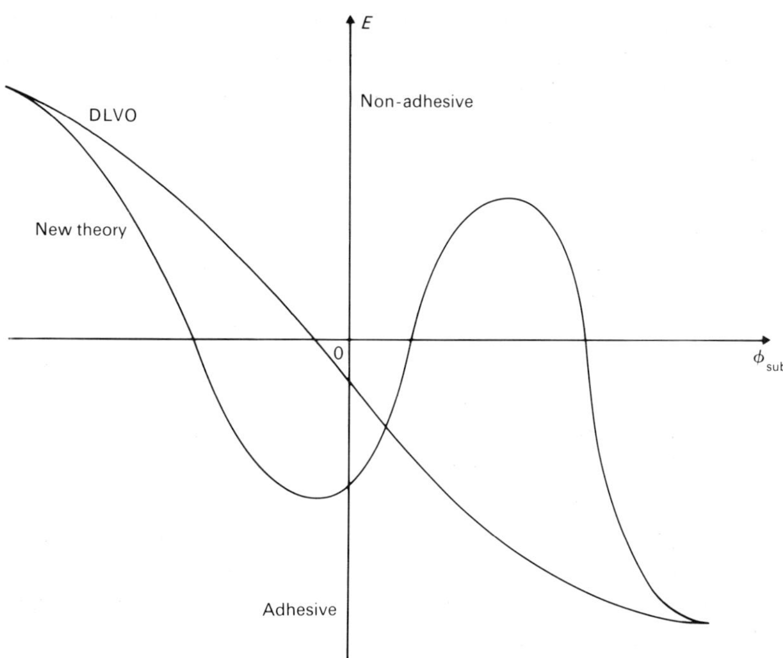

Fig. 3. Energy of adhesion as a function of potential at the substratum surface ϕ_{sub} in DLVO theory and in the new theory.

in the cell–substratum interaction (see for review Curtis, 1967, 1973; Grinnell, 1978) among which there are: the potential at the substratum surface, ϕ_{sub}; substratum structure and roughness; the concentration of proteins and surface-active agents in the medium; ion composition of the medium; specific chemical substances which promote adhesion; state of the cell membrane; temperature etc. Not pretending to explain all the factors important in cell–substratum adhesion let us examine the consequences of the new theory of cell interactions presented in this paper.

Figure 3 shows the difference between the cell–substratum interaction predicted by the DLVO theory and by the new theory. For extreme potential values and for large d^* values the difference between the two theories become negligible. Thus, experiments on the adhesion of erythrocytes (Gingell & Todd, 1975; Gingell & Fornes, 1976) ($d^* = 5.2$ nm) to charged surfaces would be expected to follow DLVO theory as well as the new theory pattern of cell–substratum interaction. On the other hand, the adhesive properties of mesenchymal cells (Maroudas, 1977) and fibroblasts (adhesion of fibroblasts to charged zinc surfaces (Dołowy, Moran & Holly, in preparation)) can be only described by the new theory.

The complex function of adhesive energy versus electric potential at the substratum surface presented at Fig. 3, with four consecutive zones of non-adhesive–adhesive–non-adhesive–adhesive interaction, assures us that the control of surface potential at substratum is crucial for any cell–substratum interaction experiment. The only experiment with a good control of the ψ_S potential was that carried out by Gingell & Fornes (1976), in which the potentiostatic technique was used with a lead substratum. In cell–substratum interaction experiments one should be especially aware of the problems of using a glass substratum, which has highly heterogenous surface, and of having high concentration of proteins or surface-active agents, which can deposit at the substratum. Any protein deposit decreases the absolute value of the surface potential (simply by pushing the adhesion plane off the substratum surface), and this might transfer the cell–substratum interaction from the adhesive to the non-adhesive zone or vice versa. Thus, experimenters should use a uniform-surface substratum and have the potential at the substratum constantly controlled by means of potentiostatic, electrokinetic or any other accurate technique. If these conditions cannot be fulfilled one should not expect consistent results in cell–substratum interactions.

CELL ADHESION, SORTING-OUT AND MORPHOGENESIS

Cell adhesion and sorting-out are related phenomena. It is obvious that adhesion between two cells can take place only if the energy of adhesion is negative, $E_{ii} < 0$. On the basis of the second rule of thermodynamics it is also possible to predict if the mixture of i and j cells would sort out or not. In the case when $E_{ii} + E_{jj} < 2E_{ij}$, cells would not sort out, while in the case when $E_{ii} + E_{jj} > 2E_{ij}$, the energy change would favour the process of sorting-out. The credit should be given to Steinberg (1970, 1976) for pointing out this obvious thermodynamic relation to biologists, even if his theory of cell sorting-out has some shortcomings (Harris, 1976; Curtis, 1978*a*, *b*). Apparently, however, there are many biologists who believe that the sorting-out process might be explained by very strong, stable, specific bond formation between two cells and that these bonds might be

Table 2. *Cell–cell interaction energies, cell sorting-out and cell positioning*

Interaction energies	Mixture of i and j cells	Layer of i cells covered with j cell layers
$E_{ij} < E_{ii} < E_{jj} < 0$	No sorting-out	Both cell types penetrate the opposite layer
$E_{ii} < E_{ij} < E_{jj} < 0$[a]	No sorting out	j cells penetrate i layer
$E_{ij} < E_{ii} < 0 < E_{jj}$	No sorting out (j cells do not adhere to themselves)	j cells penetrate i layer
$E_{ii} < E_{ij} < E_{jj} < 0$[b]	Aggregate of i cells within j aggregate	Stable
$E_{ii} < E_{jj} < E_{ij} < 0$	i and j aggregates in contact	Stable
$E_{ii} < E_{jj} < 0 < E_{ij}$	Two separate aggregates	Separation of the layers
$E_{ii} < E_{ij} < 0 < E_{jj}$	A few j cells within i cells aggregate, rest in suspension	j cells layer detaches cells form a suspension
$E_{ii} < 0 < E_{ij}, E_{jj}$	Aggregate of i cells and suspension of j cells	j cell layer detaches and cells form a suspension
$E_{ij} < 0 < E_{ii}, E_{jj}$	Adhesion only between unlike cells	Cells in suspension; small aggregates of unlike cells form
$0 < E_{ii}, E_{ij}, E_{jj}$	All cells in suspension	All cells in suspension

[a] $E_{ii}+E_{jj} < 2E_{ij}$. [b] $E_{ii}+E_{jj} > 2E_{ij}$.

broken 'at the cell's wish' by means of biochemical processes. This mechanism, though not impossible, just cannot explain sorting-out. The stable 'unbreakable' bonds between cells should have an energy greatly exceeding 100 kT. On the other hand, the energy needed to tear off the small vesicle from the cell membrane ($E = A\gamma$) will be of the order of 30–80 kT (maybe even lower). Thus, the cell could not distinguish between two energies higher than the 'tearing' value, and cells which could adhere to more than one cell type would not undergo the sorting-out process. Taking the adhesive energy values into consideration one might distinguish ten different patterns of sorting-out (Table 2). But the question remains if all those patterns of sorting-out might be explained by physics of cell–cell interactions.

It was shown in previous sections that cell adhesion occurs either on large surfaces, in the secondary minimum (5–10 nm apart) or on a specialised area with very close contact between the interacting surfaces; microscopists might call them zonulae occludents or tight junction. The adhesion in the secondary minimum is controlled by ψ_0 and H, and is not very selective. Unlike the secondary-minimum adhesion, the close-contact adhesion is controlled by two other cell-membrane parameters, V and d^*, and is extremely sensitive and selective.

It is easy to show that for appropriate cell membrane parameters the total energy of cell–cell interactions might fulfil all of the ten theoretically possible patterns. (Table 2) of cells sorting out. Among those ten patterns there are five (rows 1, 3, 5, 6 and 9) which could not be explained on the basis of DLVO theory alone.

While the parameters which play a role in a DLVO theory are not very effectively controlled by the cell, the number of negatively charged lipid molecules (V_L) and the structure of the coat which determines the number of adsorbed ions per unit area of the coat (d^*) are strictly dependent on cell metabolism. Moreover, it is sufficient to change the cell-membrane composition in a very small area to change the cell adhesion pattern completely. Thus, very small changes in cell metabolism occurring during morphogenesis might be responsible for dramatic changes in cell–cell adhesion pattern.

On the basis of the cell-membrane model it is also possible to characterise the properties of the area (I will call it a protrusion) which form the close-contact adhesion with neighbouring cell. The tip of protrusion should have a high density of negatively charged lipid molecules and a relatively thin coat which lead to a high potential value, V. On the other hand, the high value of electric potential at the lipid plane means that the interfacial tension of the protrusion is low (see (8) and (9)). Thus, I think, we can make a reconstruction of the close-contact adhesion formation mechanism.

At some location on the cell membrane a large number of negatively charged lipid molecules appear, probably due to integration with the cell membrane of a highly negatively charged microvesicle. The cell membrane at the location, due to its low interfacial tension, starts to expand in the form of a protrusion. The protrusion, because of its high electric potential at the lipid plane, is highly adhesive and might form the close-contact adhesion with neighbouring cell if the adjacent cell membrane parameters allow such adhesion to take place. However, the close-contact adhesion would not last for a very long time. After the negatively charged lipid molecules disperse in the surrounding lipid bilayer, the protrusion loses its adhesive properties and its low interfacial tension value, which finally leads to the breaking of cell contact and withdrawal of the protrusion. Thus, one should expect cell adhesion to be a highly dynamic process, with expanding protrusions forming and breaking contact with adjacent cell membranes. In the cases when the cell membrane parameters would not favour close-contact adhesion, the expanding protrusion would merely slide at the surface of the neighbouring cell at the secondary minimum distance. The expansion of the protrusion, which pulls the cytoplasm, would also favour the arrangement of microfilaments parallel to the direction of expansion (see Wessells, Spooner & Luduena, 1973; Goldman, Schloss & Starger, 1976).

The parallel arrangement of the microfilaments, on the other hand, would restrict the access of active microvesicles to the side of the protrusion and would promote its expansion forward in the same direction.

It seems to me that the above-described mechanism of close-adhesion formation, and the passive but important role of microfilaments in it, might also be responsible for amoeboid-type locomotion. I have also proposed a possible mechanism of active microvesicle formation, but to avoid confusion of that completely hypothetical mechanism with this theory of cell adhesion I have transferred the discussion to Appendix 2.

APPENDIX 1

The distribution of electric potential within the cell membrane

The simplified solution of the electric potential distribution within the cell membrane was presented on pp. 43–4. The assumption that the electric potential at the inner surface of the lipid bilayer is equal to the membrane potential value (which excluded the possibility of inner diffuse double layer formation) is now to be suspended. As was discussed on pp. 47–9, and shown at Fig. 1, there is a huge Stern-like layer in which the ions are adsorbed. The remaining charge is compensated by two double layers, one outside the cell (with a potential drop (ΔV_L to 0) and charge density σ_2) and one inside the cell (with a potential drop (ΔV_L to ϕ) and charge density σ_3). The potential at the outer and inner diffuse layers must be equal (ΔV_L) to fulfil the condition of electric neutrality of the system. Thus, we can now write analogously to (11) and (12):

$$\Delta V_L/(V_L - \Delta V_L) = (\sigma_2 + \sigma_3)/\sigma_1 \tag{1A}$$

As was shown by the equations (12) to (14)

$$\sigma_1/\sigma_2 = \kappa d^* \tag{2A}$$

and analogously:

$$\sigma_3 = -F(\Delta V_L - \phi)(2\epsilon\epsilon_0 c/RT)^{\frac{1}{2}} \tag{3A}$$

$$\sigma_1/\sigma_3 = [\Delta V_L/(\Delta V_L - \phi)]\kappa d^* \tag{4A}$$

From equations (1A) to (4A) we can obtain:

$$\Delta V_L^2(\kappa d^* + 2) - \Delta V_L(2V_L + \phi) + V_L\phi = 0 \tag{5A}$$

and the solution of that equation will be

$$\Delta V_L = \{2V_L + \phi + [(2V_L + \phi)^2 - 4V_L\phi(\kappa d^* + 2)]^{\frac{1}{2}}\}/2(\kappa d^* + 2) \tag{6A}$$

The main complications are, however, still to come. Previously, to obtain the values of V and ψ_0 we just added ΔV_S and V_S values to V_L and ΔV_L respectively. This is not the case now, because the increase in the potential at lipid plane ($V_L + \Delta V_S$) changes the difference between V and membrane potential, ϕ. The proper solution of the problem runs as follows: from (6A) one can obtain the value of

$$N = \Delta V_L/V_L = \Delta V_S/V_S \tag{7A}$$

Now we can obtain the potential at the lipid plane

$$V = V_L + \Delta V_S \tag{8A}$$

substituting the potential V_L by the potential V', one can calculate $\Delta V'$ from (6A), and using $\Delta V'$ obtain a value for N'

$$N' = \Delta V'/V^1 = \Delta V'_L/V_L = \Delta V'_S/V_S \tag{9A}$$

Then analogously to (8A) we can write

$$V'' = V_L + \Delta V'_S \tag{10A}$$

and keep repeating this procedure until the differences between $N^{(n-1)'}$ and $N^{(n)'}$ becomes negligible, then we can finally calculate the values of the potentials at the cell membrane:

$$V = V_L + \Delta V_S{}^{(n)'} \tag{11A}$$

$$V_{inn} = \Delta V^{(n)'} \tag{12A}$$

$$\psi_0 = V_S + \Delta V_L{}^{(n)'} \tag{13A}$$

It is worth mentioning the characteristic property of the membrane potential. By the way of solving equation (5A) the following condition appears:

$$\phi^2 - \phi\, 4V(\kappa d^* + 1) + 4V^2 \geqslant 0 \qquad (14A)$$

which leads to the two solutions

$$\phi_+ \geqslant 2V\{(\kappa d^* + 1) + [\kappa d^*(\kappa d^* + 2)]^{\frac{1}{2}}\} \qquad (15A)$$

or

$$\phi_- \leqslant 2V\{(\kappa d^* + 1) - [\kappa d^*(\kappa d^* + 2)]^{\frac{1}{2}}\} \qquad (16A)$$

The meaning of these solutions is that the membrane potential cannot assume all values, because there is a restricted area, $\phi_- < \phi < \phi_+$, for the membrane potential values. For the normal cell membrane, the typical values of the membrane potentials will be probably either $\phi < -600$ mV or $\phi > -30$ mV, but for the weakly charged cell membrane as in case of the axon (Segal, 1968) these values will be $\phi < -60$ mV or $\phi > -3$ mV. The obvious importance of this solution with comparison to the observed membrane potential changes of the nerve cell need not be stressed.

APPENDIX 2

The mechanism of active microvesicle formation and the mechanism of cell membrane protrusion expansion and adhesion

The consecutive stages of this mechanism are shown in Fig. 4. The course of active microvesicle production and the turnover of negatively charged lipids could run as follows:

At some location the cell coat starts to become more ion-permeable. At this location, cations penetrate to the negatively charged lipid layer, which becomes partially neutralised as a result. Partial charge neutralisation results in an influx of negatively charged lipid molecules into that location (Fig. 4*a*). This occurs because of the tendency towards the formation of a uniformly charged lipid layer.

The process continues, leading to a concentration, in that area, of a large number of negatively charged lipid molecules that are partially neutralised by cations. The partially neutralised outer lipid monolayer has a higher interface tension than the inner lipid monolayer. As a result of these differences in interfacial tension, the cell membrane becomes concave (*b*), which leads to the detachment of this part of the cell membrane in the form of microvesicles. The pinocytotic

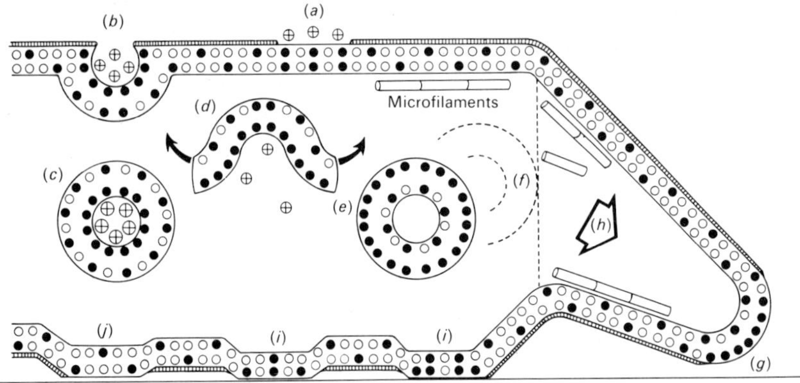

Fig. 4. The dynamic mechanism of active microvesicle formation cell membrane expansion and adhesion (see text for details). ●, negatively charged lipids; ○, neutral lipids; ⊕, cations. (Figure not to scale.)

microvesicle is composed of a large number of negatively charged lipids with cations closed inside (*c*).

Due to the high membrane potential (deficiency of cations inside the cell) the pinocytotic microvesicles would have a tendency to open up and discard the accumulated cations (*d*). The membrane potential is the energy source for that step – the only endothermic step of the whole process.

After the cations are removed from a pinocytotic microvesicle, its inner lipid monolayer is no longer partially neutralised, so its interfacial tension decreases to a value lower than that of the outer lipid monolayer. As a result of the decrease in interfacial tension, the pinocytotic microvesicle closes again, this time inside out (*e*).

Active microvesicles (*e*) integrate with the cell membrane at random due to chaotic motion, which causes a local decrease in surface tension (*f*). The decrease in surface tension produces violent expansion of the cell membrane (*g*). The cell membrane expansion pulls the cytoplasm in the direction of the membrane expansion (*h*). The cytoplasm flow is expected to be highly turbulent, which causes mixing of microvesicle-rich with microvesicle-deficient cytoplasm. The cytoplasmic flow transports other microvesicles in the direction of expansion, these then integrate with the cell membrane and microfilaments which form a parallel arrangement, restricting the active microvesicle access to the sides of the protrusion.

The expanding cell-front is rich in negatively charged lipid molecules and thus it is very adhesive and easily forms an adhesion with the substratum (*g*). During cell movement, the former front adhesion-site might be found further and further away from the cell front (*i*). During that time negatively charged lipids disappears in the surrounding lipid bilayer and finally the site becomes non-adhesive (*j*).

REFERENCES

Bell, G. I. (1978). Specific adhesion of cells to cells. A theoretical framework for adhesion mediated by reversible bonds between cell surface molecules. *Science, New York*, **200**, 618–27.

Collins, M. (1966). Electrokinetic properties of dissociated chick embryo cells. I and II. *Journal of Experimental Zoology*, **163**, 23–47.

Curtis, A. S. G. (1960). Cell contacts: some physical considerations *American Naturalist*, **94**, 37–56.

Curtis, A. S. G. (1961). Timing mechanisms in the specific adhesion of cells. *Experimental Cell Research*, Supplement **8**, 107–22.

Curtis, A. S. G. (1964). The adhesion of cells to glass: a study by interference reflection microscopy. *Journal of Cell Biology*, **19**, 199–215.

Curtis, A. S. G. (1967). *The Cell Surface: Its Molecular Role in Morphogenesis*. London: Logos Press.

Curtis, A. S. G. (1973). Cell adhesion. In *Progress in Biophysics and Molecular Biology*, vol. 27, ed. A. J. V. Butler & D. Noble, pp. 315–86. Oxford: Pergamon Press.

Curtis, A. S. G. (1978*a*). Cell positioning. In *Receptors and Recognition*, Series B, vol. 4, ed. P. Cuatrecasas & M. F. Greaves, pp. 157–95. London: Chapman & Hall.

Curtis, A. S. G. (1978*b*). Cell–cell recognition: positioning and patterning systems. In *Cell–Cell Recognition*, ed. A. S. G. Curtis, 32nd Symposium of the Society for Experimental Biology, pp. 51–82. Cambridge University Press.

Danielli, J. F. & Davson, H. (1935). A contribution to the theory of permeability of thin films. *Journal of Cellular and Comparative Physiology*, **5**, 495–508.

Danielli, J. F. & Harvey, E. N. (1935). The tension at the surface of mackerel egg oil, with remarks on the nature of the cell surface. *Journal of Cellular and Comparative Physiology*, **5**, 483–94.

Derjagin, B. V. & Landau, L. D. (1941). Theory of the stability of strongly charged lyophobic sols and of the adhesion of strongly charged particles in solutions of electrolytes. *Acta physicochimica USSR*, **14**, 633–62.

Dołowy, K. (1975). Uniform hypothesis of cell behaviour – movement, contact inhibition of movement, adhesion, chemotaxis, phagocytosis, pinocytosis, division, contact inhibition of division, fusion. *Journal of Theoretical Biology*, **52**, 83–97.

Dołowy, K. & Holly, F. J. (1978). Contribution of interfacial tension changes during cellular interaction to the energy balance. *Journal of Theoretical Biology*, **75**, 373–80.

DROST-HANSEN, W. (1971). Structure and properties of water at biological interfaces. In *Chemistry of the Cell Interface*, vol. B, ed. H. D. Brown, pp. 1–184. London, New York: Academic Press.

FRYE, L. D. & EDIDIN, M. (1970). The rapid intermixing of cell surface antigens after formation of mouse-human heterokaryons. *Journal of Cell Science*, **7**, 319–36.

GINGELL, D. (1967). Membrane surface potential in relation to a possible mechanism for intercellular interactions and cellular responses, a physical basis. *Journal of Theoretical Biology*, **17**, 451–82.

GINGELL, D. (1971). Computed force and energy of membrane interaction. *Journal of Theoretical Biology*, **30**, 121–49.

GINGELL, D. & FORNES, J. A. (1976). Interaction of red blood cells with a polarized electrode: evidence of long-range intermolecular forces. *Biophysical Journal*, **16**, 1131–53.

GINGELL, D. & TODD, I. (1975). Adhesion of red blood cells to charged interfaces between immiscible liquids. A new method. *Journal of Cell Science*, **18**, 227–39.

GITTENS, G. J. & JAMES, A. M. (1963). some physical investigations of the behaviour of bacterial surfaces VI. Chemical modification of surface components. *Biochimica et Biophysica Acta*, **66**, 237–49.

GOLDMAN, R. D., SCHLOSS, J. A. & STARGER, J. M. (1976). Organizational changes of Actinlike Microfilaments during animal cell movement. In *Cell Motility*, ed. R. Goldman, T. Pollard & J. Rosenbaum, pp. 217–45. Cold Spring Harbor Laboratory.

GOOD, R. J. (1972). Theory of the adhesion of cells and the spontaneous sorting-out of mixed cell aggregates. *Journal of Theoretical Biology*, **37**, 413–32.

GRINNELL, F. (1978). Cellular adhesiveness and extracellular substrata. *International Review of Cytology*, **53**, 65–144.

HARRIS, A. K. (1976). Is cell sorting caused by differences in the work of intercellular adhesion? A critique of the Steinberg hypothesis. *Journal of Theoretical Biology*, **61**, 267–85.

HAYDON, D. A. & SEAMAN, G. V. F. (1967). Electrokinetic studies on the ultrastructure of the human erythrocyte. I. Electrophoresis of high ionic strength – the cell as a polyanion. *Archives of Biochemistry and Biophysics*, **122**, 126–36.

HEARD, D. & SEAMAN, G. V. F. (1960). The influence of pH and ionic strength on the electrokinetic stability of the human erythrocyte membrane. *Journal of General Physiology*, **43**, 635–54.

KAVANAU, J. L. (1964). *Water and Solute–Water Interactions*. San Francisco: Holden-Day.

KORTÜM, G. (1966). *Lehrbuch der Elektrochemie*. Weinheim: Verlag Chemie.

MAROUDAS, N. G. (1977). Sulphonated polystyrene as an optimal substratum for the adhesion and spreading of mesenchymal cells in monovalent and divalent saline solutions. *Journal of Cellular Physiology*, **90**, 511–20.

MYSELS, K. J. (1975). The direct measurement of $1/\kappa$ the Debye length. In *Physical Chemistry : Enriching Topics from Colloid and Surface Science*, ed. N. van Olphen & K. J. Mysels, pp. 73–85. La Jolla: Theorex.

NIR, S. (1977). Van der Waals interactions between surfaces of biological interest. *Progress in Surface Science*, **8**, 1–58.

OVERBEEK, J. Th.G. (1952). Electrochemistry of the double layer *and* Electrokinetic phenomena. In *Colloid Science*, vol. 1, ed. H. R. Kruyt, pp. 115–93 and 194–244. Amsterdam: Elsevier.

PARSEGIAN, V. A. (1975). Long range Van der Waals forces. In *Physical Chemistry: Enriching Topics from Colloid and Surface Science*, ed. N. Van Olphen & K. J. Mysels, pp. 27–72. La Jolla: Theorex.

PEGRUM, S. M. & MAROUDAS, N. G. (1975). Early events in fibroblast adhesion to glass. *Experimental Cell Research*, **96**, 416–22.

SEAMAN, G. V. F., VASSAR, P. S. & KENDALL, M. J. (1969). Electrophoretic studies on human polymorphonuclear leukocytes and erythrocytes: the binding of calcium ions within the peripheral regions. *Archives of Biochemistry and Biophysics*, **135**, 356–62.

SEGAL, J. R. (1968). Surface charge of giant axons of squid and lobster. *Biophysical Journal*, **8**, 470–89.

STEINBERG, M. S. (1970). Does differential adhesion govern self-assembly processes in histogenesis? Equilibrium configurations and the emergence of a hierarchy among populations of embryonic cells. *Journal of Experimental Zoology*, **173**, 395–434.

STEINBERG, M. S. (1976). Adhesion-guided multicellular assembly: a commentary upon the postulates, real and imagined of the differential adhesion hypothesis, with special attention to computer simulations of cell sorting. *Journal of Theoretical Biology*, **55**, 431–43.

VERWEY, E. J. W. & OVERBEEK, J. Th.G. (1948). *Theory of the Stability of Lyophobic Colloids*. Amsterdam: Elsevier.

WEISS, L. & HARLOS, J. P. (1972). Short-term interactions between cell surfaces. *Progress in Surface Science*, **1**, 355–405.

WESSELLS, N. K., SPOONER, B. S. & LUDUENA, M. A. (1973). Surface movements, microfilaments and cell locomotion. In *Locomotion at Tissue Cells*, ed. M. Abercrombie, pp. 53–77. Amsterdam: Elsevier.

The hydrophobic effect and charge effects in the adhesion of enterobacteria to animal cell surfaces and the influences of antibodies of different immunoglobulin classes

L. EDEBO, E. KIHLSTRÖM, K.-E. MAGNUSSON AND O. STENDAHL

Department of Medical Microbiology, University of Linköping, S-581 85 Linköping, Sweden

There are a few now-classical observations by Wood and coworkers (see Davis *et al.*, 1973) on the interaction between polymorphonuclear leukocytes (PMNL) and bacteria on a microscopic slide under the microscope. Virulent bacteria such as pneumococci, streptococci, klebsiellae and others, which are provided with capsules, can escape phagocytosis on smooth surfaces like glass, cellophane, albumin and paraffin by simply gliding away, but become phagocytosed on various inert rough surfaces such as moistened filter paper, cloth and fibre-glass. However avirulent bacteria, which lack capsules, are fairly rapidly ingested. Because of the important role of surfaces this sort of phagocytosis, for which antibody is not required, was called 'surface phagocytosis'. It was also stated that, unlike phagocytosis aided by opsonins, surface phagocytosis does not take place when the cells are floating free in a fluid medium. Furthermore, observations in this system and elsewhere have indicated that the capsules are of great importance in the resistance of bacteria to phagocytosis. The capsules consist largely of water, usually bound as a gel to neutral or acidic homo- or heteropolysaccharides (Wilkinson, 1958; Dudman, 1977).

The fact that several micro-organisms may be phagocytosed by PMNL in the absence of antibody indicated that PMNL do not require very special microbial surface structures in order to be able to phagocytose them. However, complement and other substances present in normal serum might have mediated the attachment to the

PMNL in the early experiments. The capacity of primitive protozoa, such as free-living amoebae, to feed on a variety of micro-organisms and other materials and even avidly phagocytose polystyrene particles (Weisman & Korn, 1967) indicates, however, that phagocytosis is a very primitive and general process. Therefore, we have not dealt so much with the question of which structures of microbes and phagocytes that promote adhesion and engulfment; rather we have focused onto the question: What mechanisms prevent adhesion and hinder engulfment? Thus, in the following presentation we will discuss models that have been developed to study the interaction of enterobacteria and other model particles with animal cells and other model surfaces. Recent reviews on more or less specific interactions between bacteria and different animal cell surfaces have been published by Costerton, Ingram & Cheng (1974), Jones, (1977) and Smith (1977) and are given at this meeting by others. One fundamental argument in this type of approach is the selectivity of the microbe for certain host cells. This interaction is based on specific structures on the surface of the microbe and complementary structures on the host cell similar to the fit between antigen and antibody. An alternative explanation, which has long been forwarded, particularly in connection with phagocytosis by PMNL, is that the physicochemical surface properties of the outermost interface of bacteria and other particles, as well as of phagocytic cells, determine the outcome of the interaction. The main physicochemical properties that may be involved are hydrophobicity and charge (Fenn, 1923; Mudd, McCutcheon & Lucké, 1934; van Oss, 1978).

In order to study if general mechanisms may govern the interaction between bacteria and animal cells we have chosen to study *Salmonella typhimurium*. *S. typhimurium* is extremely virulent for mice, producing a lethal, typhoid-like disease. It is also pathogenic for a wide variety of other animals, such as man and several other mammals, birds and reptiles. Various kinds of infection, usually of enteric origin, are produced, most often as a consequence of oral contamination. However, other routes of infection exist: the respiratory tract and the conjunctiva, the urinary tract, the meninges, the joints and the reticuloendothelial system may also become infected (Wilson & Miles, 1975). These facts indicate that general properties rather than very specific interactions are important in the pathogenicity of *S. typhimurium*.

Like pneumococci, streptococci and klebsiellae, *S. typhimurium*

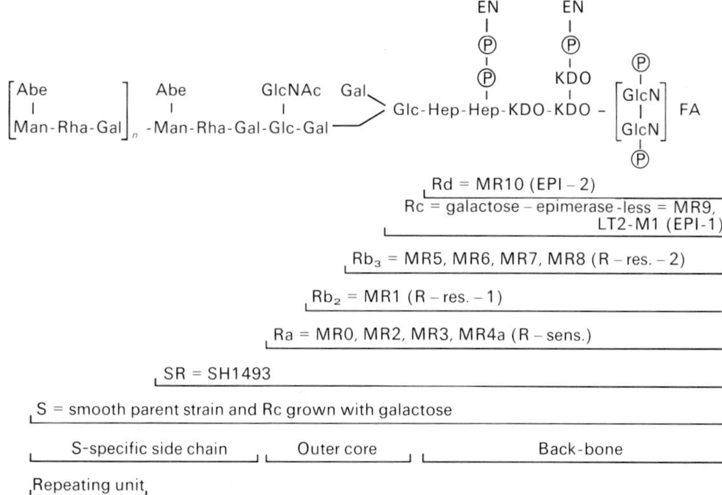

Fig. 1. Proposed structure of the LPS of *S. typhimurium*. The lower part of the figure shows the relation between structure and chemotype (S, SR, Ra–Rd), phage pattern (smooth, rough-sens. etc.) and strains used in this study. Abbreviations: Abe, abequose; Man, D-mannose; Rha, L-rhamnose; Gal, D-galactose; Glc, D-glucose; GlcNAc, *N*-acetyl-D-glucosamine; Hep, L-glycero-D-manno-heptose; P, phosphate; EN, ethanolamine; KDO, 2-keto-3-deoxyoctonate; GlcN, D-glucosamine; FA, fatty acid.

escapes phagocytosis by PMNL. *S. typhimurium*, however, has no genuine capsule but a thinner and more firmly bound microcapsule which consists of the polysaccharide portion of the cell-surface lipopolysaccharide (LPS). Furthermore, *S. typhimurium* bacteria have the advantage of being easily grown in a simple mineral–glucose medium and usually existing as suspensions of single cells, which are useful model particles. A large number of mutants of *S. typhimurium* defective in the synthesis of the LPS are also being analysed with respect to genetics, LPS structure and synthesis, protein and phospholipid composition of the cell surface, phage pattern and virulence. Most of our experiments have been performed with *S. typhimurium* 395 MS and its R-mutants MR0–MR10 (Fig. 1), well described by Holme, Lindberg and coworkers (Holme, Lindberg, Garegg & Onn, 1968; Lindberg & Holme, 1968; Lindberg, 1977).

The virulence for mice after intraperitoneal injection of *S. typhimurium* 395 MS and the R-mutants MR0–MR10 was dependent on the length of basal core oligosaccharide and the S-antigen repeating unit carbohydrate, such that their presence reduced the lethal dose

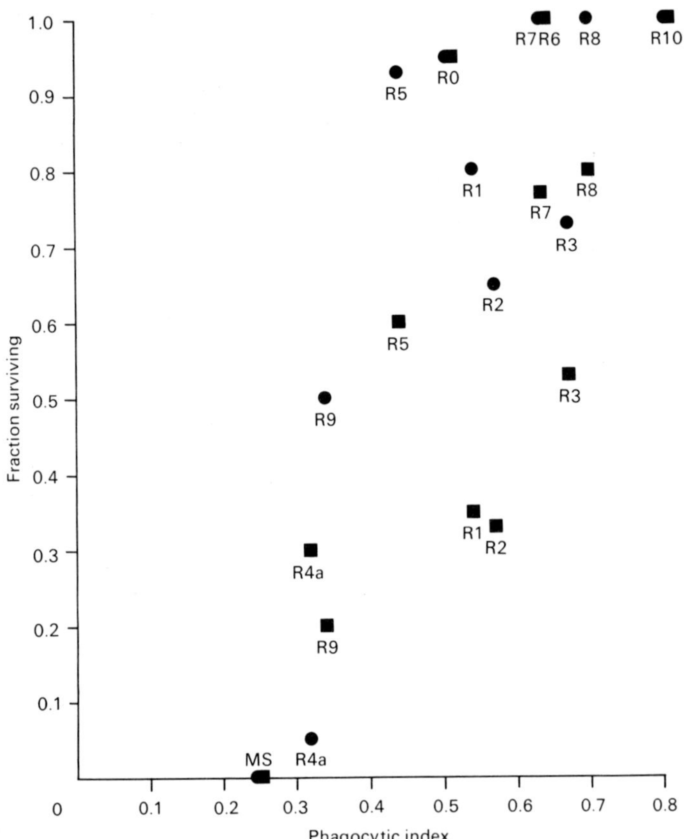

Fig. 2. Relationship between virulence and phagocytosis of *S. typhimurium* 395 MS and its mutants R0–R10. The surviving fraction after i.p. injection into mice of 10^6 bacteria suspended in saline (■) or 100 bacteria suspended in 5% gastric mucin (●) was plotted against the phagocytic index with rabbit peritoneal PMNL in 0.1% albumin.

of organisms (Edebo & Normann, 1970). Some of the mutants have an incomplete defect, leakiness, leading to synthesis of the S-antigen in reduced quantities. Such leakiness enhanced the virulence. Including 5% hog gastric mucin in the suspension medium reduced the lethal dose for most of the mutants to less than 10^{-4}. The phagocytosis by rabbit peritoneal PMNL of the above bacteria was tested with bacteria killed by heat at 56 °C for 1 h (Stendahl & Edebo, 1972). The PMNL were attached to cellulose filters and the bacteria labelled, one strain with ^{51}Cr another with ^{125}I, and both added to one batch of

PMNL. By always including *S. typhimurium* MR10 as one in the pair, a reference was obtained to correct for different qualities of different batches of PMNL. Phagocytosis was tested in preimmune rabbit serum and in precolostral calf serum as well as in a medium containing only salts, glucose and serum albumin (Fig. 2). In all media, MR10 bacteria, which possess the shortest LPS chain, were most liable to phagocytosis and MS bacteria were most resistant. Phagocytosis was generally more rapid for R-mutants with phage patterns typical of shorter LPS chain lengths, but, as in the case of virulence, the presence of S-specific repeating unit on the bacterial surface overshadowed the effect of chemotype. One fundamental result of these experiments was that also in the absence of antibody and complement, the PMNL were capable to discriminate between the different mutants. Humoral factors of the host immune system should, therefore, not be necessary as mediators of the interaction between the bacteria and the PMNL. Since several different bacteria with different surface structures were able to interact with the PMNL, specific receptors on the PMNL for each bacterial strain were not likely. Furthermore, thermodynamic reasoning predicts that spontaneous association occurs between a bacterium (B) and a phagocyte (P), when the overall change of the free energy $\Delta F_{\text{net}} < 0$. For the process of engulfment

$$\Delta F_{\text{net}} = \gamma_{\text{PB}} - \gamma_{\text{BW}}$$

where γ_{BW} is the interfacial tension between the bacterium and water (W) and γ_{PB} is the interfacial tension between the phagocyte and bacterium. Values of γ_{PB} and γ_{BW} can be obtained by contact-angle (θ) measurements, saying that $\Delta F_{\text{net}} < 0$ when $\theta_{\text{B}} > \theta_{\text{P}}$, i.e. a situation favouring engulfment (van Oss, Gillman & Neumann, 1975; van Oss, 1978). Thus, general physicochemical surface properties of the bacteria rather than specific receptors for them on the PMNL seem to determine the outcome of the interaction.

PHYSICOCHEMICAL PROPERTIES OF S. TYPHIMURIUM BACTERIA AND LPS

Partition in aqueous polymer biphasic systems

Some methods are available to study electrical surface properties of particles (Sherbet, 1978), whereas very few ways exist to study surface hydrophobicity and interfacial tension. Mudd used two-phase

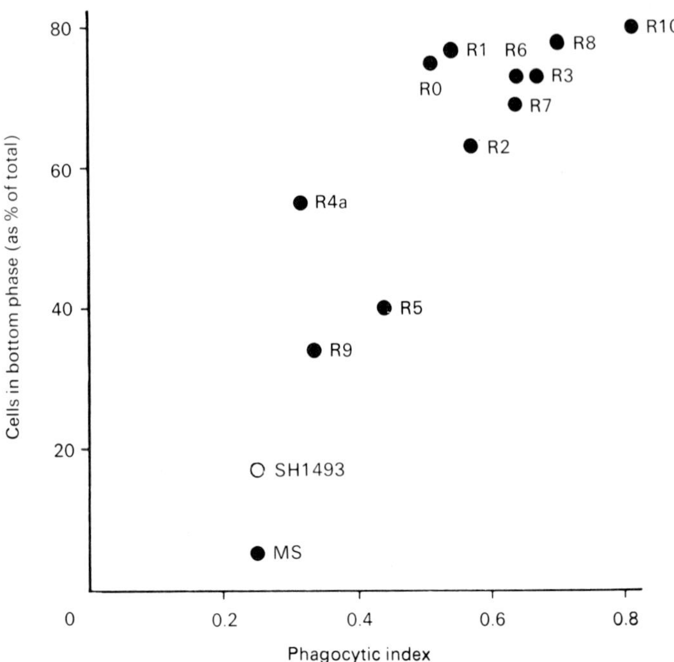

Fig. 3. Relation between accumulation in bottom phase (dextran-rich) and phagocytic index for *S. typhimurium* 395 MS, the mutants R0–R10 derived from it, and the SR-mutant *S. typhimurium* SH1493.

systems made up from mixtures of immiscible liquids, usually oil and water, and showed the relationship between hydrophobicity and phagocytosis (Mudd *et al.*, 1934). The resolving power in fractionation studies is much greater for aqueous polymer two-phase systems as devised by Albertsson (1971), in which subtle changes in surface properties can be detected as a change in surface affinity (Walter, 1977). In a system containing 6.2 % (w/w) dextran-500 (Pharmacia, Uppsala, Sweden) and 4.4 % (w/w) polyethyleneglycol-6000 (PEG), *S. typhimurium* MS accumulated into the PEG-rich top phase, whereas the R-mutants derived from it showed preference for the dextran-rich bottom phase (Stendahl, Magnusson, Tagesson, Cunningham & Edebo, 1973; Stendahl, Tagesson & Edebo, 1973). When the uridine diphosphate–galactose-4-epimeraseless mutants MR9 and LT2-M1 were grown without galactose in the medium, the bacteria partitioned like the other R-mutants, whereas growth in the presence of galactose, which allows the synthesis of the complete S-type LPS, changed the partition properties to that of S bacteria

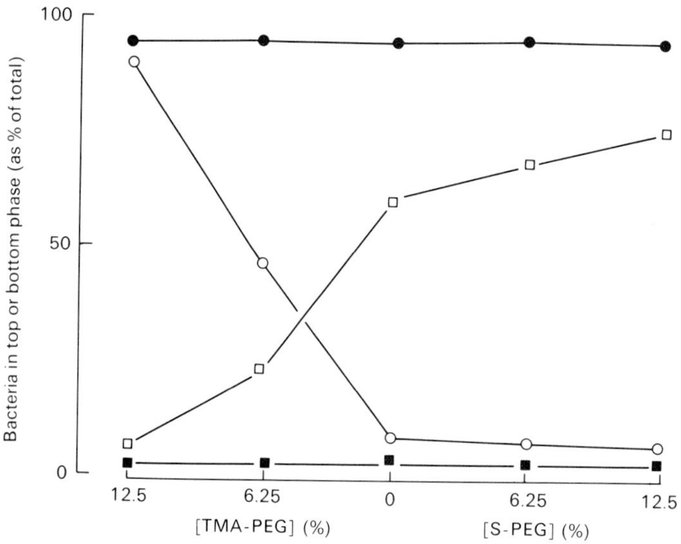

Fig. 4. Partition of bacteria, *Salmonella typhimurium* 395 MR10 (open symbols) and MS (filled symbols), in the presence of different concentrations of TMA-PEG and S-PEG. ○, top phase, □, bottom phase. The cells were labelled with ^{51}Cr. (*X*-PEG concentrations are expressed as a percentage of *X*-PEG+PEG.)

(Fig. 3). Since this phenotypic difference was a consequence of the synthesis of the S-specific polysaccharide side chain, the presence of the surface polysaccharide seems to strongly influence the physico-chemical properties of the bacteria.

When certain alkali halides are added to the two-phase system they become unequally distributed between the two phases and affect the partition by electrostatic interaction. In this way it was shown that MS bacteria and MS LPS were virtually uncharged, whereas MR10 bacteria and MR10 LPS showed a negative net surface charge (Magnusson *et al.*, 1977). Corresponding results were obtained with free electrophoresis and with the use of positively charged tri(methyl)amino-PEG (TMA-PEG) or negatively charged sulphonyl PEG (S-PEG) in the two-phase system (Fig. 4). The charged PEG attracts particles with opposite charges and repels those with the same charges in relation to the PEG-rich phase (Stendahl, Edebo, Tagesson, Magnusson & Hjertén, 1977).

Similarly, when PEG or dextran esterified with fatty acids is included in the phase system, the partition of particles is influenced by hydrophobic interaction between the fatty acid moiety and the particles, and the tendency of the PEG and dextran moiety, respec-

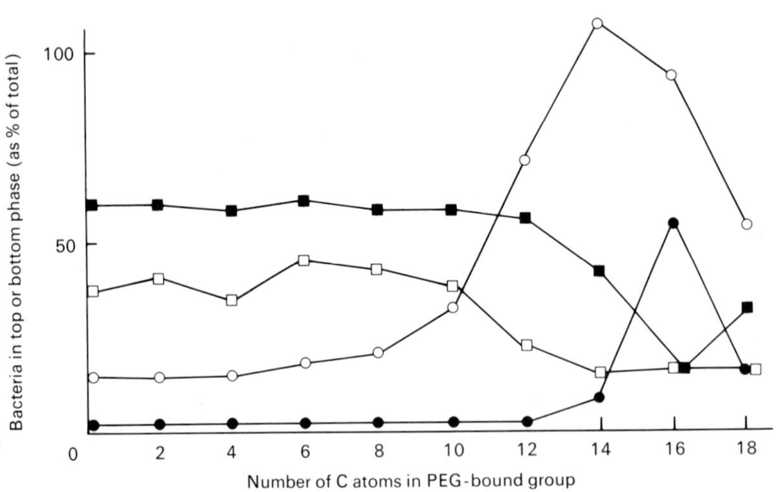

Fig. 5. Effect of the number of carbon (C) atoms in the hydrophobic PEG-bound group (*X*) on partition of *S. typhimurium* 395 MR10 (●, top phase; ■, bottom phase; *X*-PEG/(*X*-PEG+PEG) $\simeq$ 2.2%), and of *S. minnesota* R595 (○, top phase; □, bottom phase; *X*-PEG/(*X*-PEG+PEG) $\simeq$ 1.1%).

tively, each to dissolve in its alike polymer phase. The partition of MS bacteria and of MS LPS was not influenced by the hydrophobic polymers. In contrast, the partition of MR10 bacteria and of MR10 LPS increased towards the phase containing the hydrophobic polymer (Magnusson *et al.*, 1977). Using PEG-ligands of different chain lengths, i.e. the PEG esters of acetic (C_2) to stearic (C_{18}) acid, with various R-bacteria increased the partition towards the PEG-rich phase only provided a minimum length of the aliphatic chain was used. With *S. minnesota* R595 (chemotype Re) at least 8–10 carbon atoms and with *S. typhimurium* 395 MR10 (chemotype Rd) at least 12–14 carbon atoms in the aliphatic chain covalently linked to PEG were needed to increase the partitioning towards the PEG rich phase (Fig. 5). The influence on the partitioning by equal concentrations of esterified PEG increased with the number of carbon atoms in the aliphatic chain to a maximum with PEG palmitate (C_{16}; Magnusson & Johnansson, 1977*a*). Lower concentrations of hydrophobic PEG were required to change the partition of the chemotypes Rd and Re than to change that of the less defective mutants.

Hydrophobic interaction chromatography

Chromatography of suspensions of the *Salmonella* bacterium on hydrophobic columns of Phenyl–Sepharose or Octyl–Sepharose (Pharmacia, Uppsala, Sweden) showed retardation and Sticking of the R-bacteria to the columns, confirming their hydrophobic character (Magnusson, Stendahl, Stjernström & Edebo, 1979; Kihlström & Magnusson, to be published). By contrast, S-bacteria were recovered almost entirely in the void volume (Stjernström, Magnusson, Stendahl & Tagesson, 1977; Kihlström & Magnusson, to be published). These general features were observed also for R-mutants of *S. minnesota* S99 (Kihlström & Magnusson, to be published).

Ion-exchange chromatography

When suspensions of *S. typhimurium* and *S. minnesota* bacteria were applied to columns of DEAE–Sephacel or DEAE–Sepharose (Pharmacia, Uppsala, Sweden), the S-strains showed little interaction with the gel. By contrast, the R-mutants from either bacterium were retarded on the columns and were not desorbed even by buffers of high ionic strength. These results confirm the low tendency to electrostatic interaction of S-bacteria and the negative surface charge of R-bacteria described above. They also agree with earlier measurements of electrophoretic mobilities of S and R variants of *Escherichia coli* and *Salmonella typhi* which have shown considerably higher mobilities for the R-bacteria (Sherbet, 1978).

Contact-angle measurements

The *S. typhimurium* mutants described above were spread onto a flat, solid surface, dried, and the contact angle of a drop of saline measured (Cunningham, Söderström, Gillman & van Oss, 1975). The angle for *S. typhimurium* 395 MS was 17°, whereas for MR10 the contact angle was 22°. These data fit well with the hydrophobicity and liability to phagocytosis of the R-mutants and hydrophilicity and phagocytosis resistance of the S-bacteria. It should be mentioned, however, that the contact angle is a measure of the overall interfacial tension, whereas the cellular structure governing partition and hydrophobic interaction chromatography may be hydrophobic patches in a non-homogeneous surface.

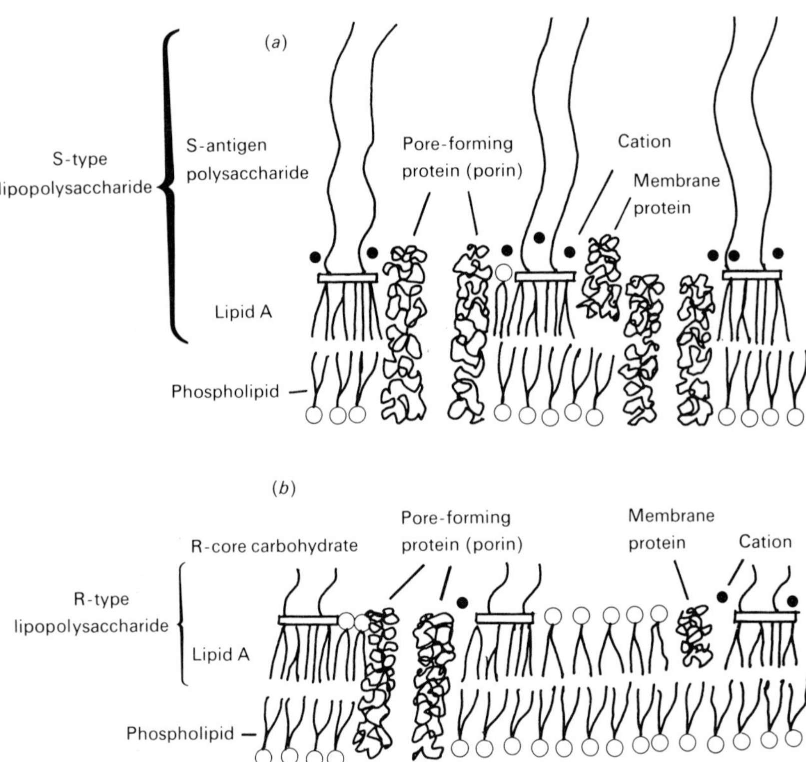

Fig. 6. Outer membrane of *S. typhimurium*. (*a*) S-type; (*b*) R-type.

Structure and properties of the surface of S. typhimurium

Information gathered by the use of various methods to analyse the cell envelope of *S. typhimurium* indicates that differences exist between S- and R-bacteria that not only concern the LPS. Thus, in R-mutants the quantity of outer membrane proteins and LPS is reduced, whereas the phospholipid is increased (Ames, Spudich & Nikaido, 1974; Koplow & Goldfine, 1974; Bayer, Koplow & Goldfine, 1975; Parton 1975; Nikaido, personal communication). This change seems to be particularly pronounced for the more defective mutants, particularly the Rd and Re chemotypes (Fig. 6). Thus, not only is

---

Fig. 7. Thin layer counter-current distribution (CCD) (58 transfers) of a UDP-gal-4-epimeraseless mutant (LT2-M1) grown in medium with D-galactose for different times; (*a*) = 0 min, (*b*) = 10 min, (*c*) = 40 min and (*d*) = 180 min. The cells were labelled with ^{51}Cr or ^{125}I.

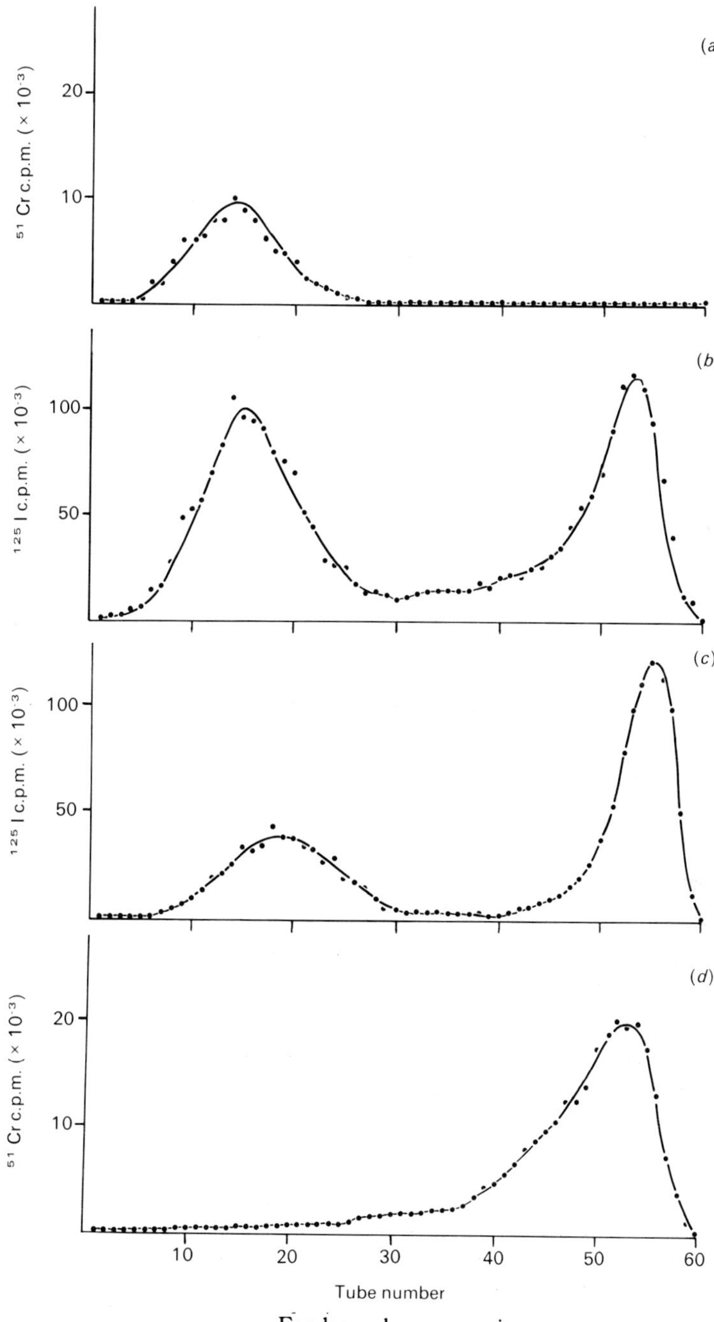

For legend see opposite

the accessibility of the outer membrane proteins, lipid A and LPS core enhanced in R-bacteria as a consequence of the absence of polysaccharide side chains in the LPS and evidenced by the sensitivity to rough specific phages with outer membrane protein receptors (Lindberg, 1977), the presence of larger quantities of phospholipids in the outer membrane of R-mutants should also facilitate hydrophobic interaction. Which of the mechanisms is most important for hydrophobic interaction is not known. However, the preconditions for aliphatic hydrocarbon chains and other hydrophobic structures to interact with the phospholipid of R-mutants ought to be promoted by both mechanisms. Since the partition and phagocytosis of the UDP–galactose-4-epimeraseless mutant *S. typhimurium* LT2-M1 was changed within 10 min after addition of galactose (Fig. 7; Stendahl, Tagesson & Edebo, 1973) we think that the polysaccharide side-chains play a great role in preventing the access to other constituents of the outer membrane.

To achieve a measurement *in situ* of the relative quantities of the elements carbon, oxygen and nitrogen, which might reflect the relative occurrence of carbohydrate and proteinaceous material in the outer membranes, *S. typhimurium* 395 MS and MR10 bacteria and isolated LPS were studied by X-ray photoelectron spectroscopy (ESCA, Electron Spectroscopy for Chemical Analysis; Siegbahn *et al.*, 1967). ESCA is a surface-sensitive method, since electrons analysed originate from a surface layer not thicker than 10 nm in biological material. When the peak areas for the different elements had been corrected for background and sensitivity of detection, the relative occurrence of carbon and oxygen in isolated LPS, as measured with the ESCA technique, correlated well with chemical analysis of the LPS (Magnusson & Johansson, 1977*b*). It was, however, recognised that more nitrogen was present in the outer membrane of the MR10 mutant than of MS. If nitrogen reflects the presence of proteinaceous material, these results are in conflict with those obtained by chemical analysis or freeze-etching of isolated outer membranes, stating that with a reduction of LPS, proteins are also lost (Ames, *et al.*, 1974; Koplow & Goldfine, 1974; Bayer *et al.*, 1975; Irvin, Chatterjee, Sanderson & Costerton, 1975).

INTERACTION OF *S. TYPHIMURIUM* S- AND R-BACTERIA WITH ANIMAL CELLS

Phagocytosis by polymorphonuclear leukocytes in vitro

As mentioned earlier, the phagocytosis *in vitro* by PMNL of *S. typhimurium* was much greater for the R-mutants, especially for the mutants more defective in the LPS and with partition properties more different from that of MS bacteria (Fig. 2, Stendahl, Tagesson & Edebo, 1973). Bacteria killed at 56 °C for 1 h were phagocytosed in the absence of known opsonins. This emphasises the aspecific character of the process, since we consider it most unlikely that PMNL are provided with receptors specialised for the attachment of a vast number of different bacteria. Within this series of mutants there are substantial differences between the surfaces of individual mutants related to the chemotype. This is apparent from the phage pattern (Lindberg & Holme, 1968), the serological specificity (Lindberg & Holme, 1968), and the specificity of the reaction, when different anti-R-mutant antisera after binding to the homologous bacteria or bacteria of the same chemotype activate complement without killing the bacteria (Edebo & Normann, 1971).

Phagocytosis by HeLa cells in vitro

Different to the phagocytosis by PMNL the association to and phagocytosis by HeLa cells *in vitro* was negligible for heat-killed *S. typhimurium*. Neither were bacteria inactivated by ultraviolet irradiation associated to the HeLa cells. Living *S. typhimurium* MR10, however, interacted extensively with the HeLa cells, whereas no difference was observed between living and heat-killed MS bacteria (Kihlström & Edebo, 1976). The differences between MR10 and MS bacteria were obvious after 1 h. After 5 h incubation of MR10 and HeLa cells, 84% of the HeLa cells showed associated bacteria and approximately half of them had more than ten bacteria. Incubation with MS bacteria yielded approximately 20% HeLa cells with associated bacteria, usually only 1–2 bacteria per HeLa cell. By the use of the indirect fluorescent antibody technique and considering that intracellular bacteria become stained only after acetone fixation, as well as by the use of gentamicin to kill only extracellular bacteria, the fraction of internalised bacteria could be determined. Most of the

MS bacteria appeared to be extracellular, whereas several of the MR10 bacteria were intracellular, about 50% with the fluorescence method, lower with the gentamicin method (Kihlström, 1977). The lower capacity for invasion of MS bacteria might seem to be a paradox considering the much higher virulence of MS than that of MR10 (Edebo & Normann, 1970). However, it supports the hypothesis that the wide range of animals susceptible to infection with *S. typhimurium* S-bacteria is less due to a particular aggressiveness toward cells from susceptible hosts and more due to the capacity of the bacteria to evade and survive the host defence (Magnusson *et al.*, 1977). The similarity between PMNL and HeLa cells with respect to preference of prey, and the inhibition of the internalisation process by glycolytic inhibitors such as iodoacetic acid or *N*-ethylmaleimide that reduce cellular ATP levels (Kihlström & Nilsson, 1977), indicate that there are fundamental mechanisms in common. Phagosomes containing MR10 bacteria can also be observed in HeLa cells (Kihlström & Latkovic, 1978). The requirement for living bacteria to infect HeLa cells indicates, however, that factors in addition to surface properties might be needed.

Association of S. typhimurium *with mouse intestinal mucosa* in vitro

A very primitive model was developed to study *Salmonella* strains with different physicochemical surface properties with regard to possible differences in liability to association to mucosal surfaces. Two different bacterial strains killed at 56 °C for 1 h and each of them labelled with either ^{51}Cr or ^{125}I were pumped through a length of mouse small intestine (20 cm taken proximal to the ileo-caecal valve; Perers *et al.*, 1977). The association of the bacteria was dependent on the flow rate such that a slow flow yielded higher association. At a flow rate of 0.1 ml min^{-1} the association of MR10 was 21.8%, whereas at 0.2 ml min^{-1} it was 3.9%. The association of MS was smaller at both rates, 3.9% and 1.0% respectively. Experiments performed at the slower rate required more time, such that the void volume contained the peak concentration of bacteria after about 60 min. At this time microscopical changes like oedema and damage of the mucosal surface lining had started to occur. *Escherichia coli* O14 with two-phase partition similar to MR10 showed a relatively high tendency to association similar to MR10, whereas *E. coli* O111 was

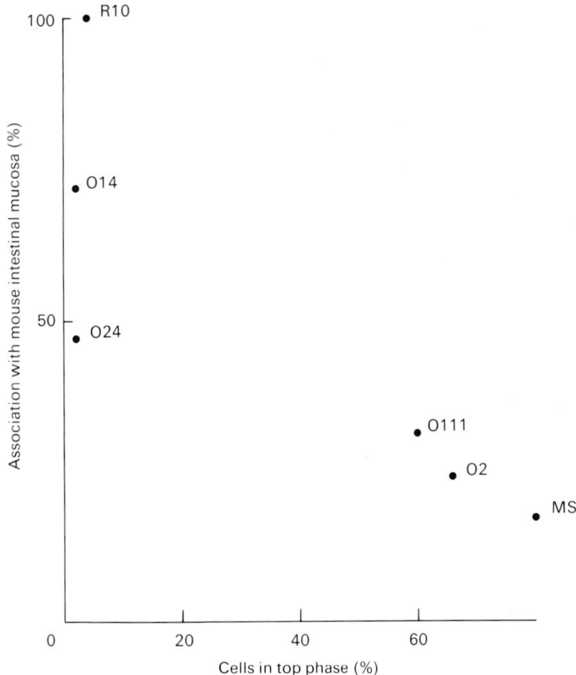

Fig. 8. Relation between accumulation in top phase (PEG-rich) and association with mouse intestinal mucosa for *S. typhimurium* 395 MS and MR10 and for the *E. coli* strains O2, O14, O24 and O111.

more similar to MS in both systems (Fig. 8). The fact that association was favoured in experiments where the mucosal surface had started to degenerate indicates that the intestinal barrier against bacteria and other foreign material may be impaired under certain conditions such as inflammation and disease. Alternatively certain bacterial activities might be operative in the derangement of a healthy mucosa such that the association of bacteria is enhanced.

Endocytosis by non-mammalian phagocytes

There are distinct differences in the foods ingested by the large amoebae which may help to distinguish them from one another (Bovee & Jahn, 1973). Some seem to be carnivores, some herbivores, and others omnivores. Some amoebae show localised areas of food intake. Endocytosis seems to be initiated by the interaction on the mucous coat or glycocalyx with the material to be ingested (Chapman-Andresen, 1973). *Acanthamoeba castellanii* actively phagocytoses

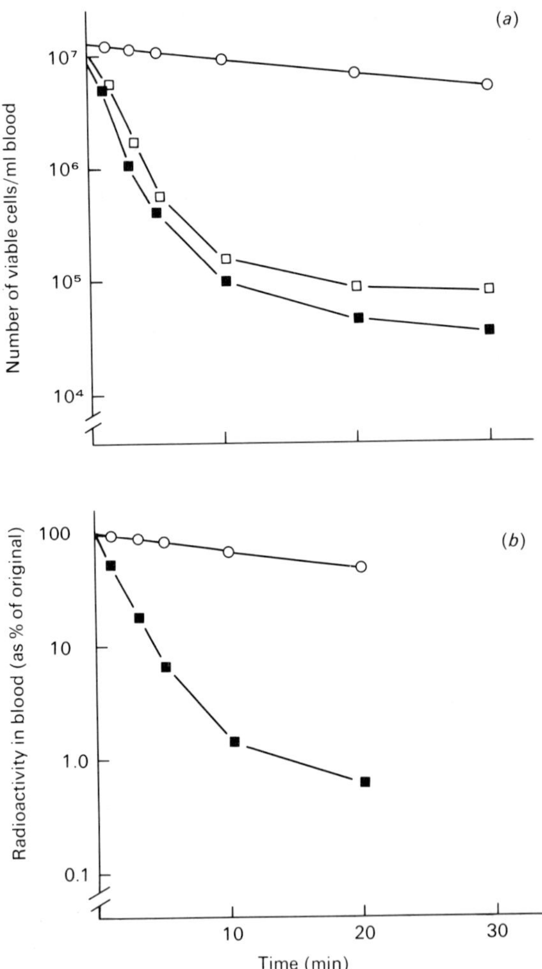

Fig. 9. Clearance in mice, after intravenous injection of 10^7–10^8 viable (*a*) or heat-killed, ^{51}Cr-labelled bacteria (*b*). The plotted lines represent the means of the K-values (phagocytic indices) in five individual mice. The points indicate, when samples were withdrawn. Symbols: ○, *S. typhimurium* 395 MS; ■, MR4a; □, MR10.

polystyrene latex spheres, provided the size of the particle or the number of small particles is large enough (Korn & Weisman, 1967). *A. castellanii* also feeds on a number of different enterobacteria and yeast cells. *S. typhimurium* MR10 seems to promote growth of the amoebae better than MS bacteria (M. Edebo, personal communication). Similarly, *Dictyostelium discoideum* phagocytoses R-bacteria more actively than S-bacteria (Gerisch, Lüderitz & Ruschmann, 1967).

Clearance of S. typhimurium *after intravenous injection into mice*

The clearance of micro-organisms from the blood is mainly a function of the reticuloendothelial system, primarily that of the liver and spleen. The clearance of the R-mutants MR10 and MR4a was very fast; after 5 min more than 90 % was cleared from the circulation (Fig. 9). It was estimated that the liver was capable of efficiently clearing the blood of all the bacteria that were circulating through this organ (Stendahl, 1973). In contrast, MS bacteria were cleared so slowly that approximately 90 min was needed to eliminate 90 % of them from the circulation. Whereas very few R-mutants ended up in the spleen, due to the rapid clearance by the liver, almost as many MS bacteria were trapped by the spleen as by the liver. The different localisation may affect both the pathogenicity and the immunogenicity of particles of different physicochemical properties.

Effects of the properties of the enterobacterial surface on the host–parasite relationship

Whereas the sensitivity to hydrophilic antimicrobial substances is nearly equal for S- and R-bacteria, the sensitivity to hydrophobic substances is much greater for R-bacteria. One such group of hydrophobic substances is the bile salts. Thus, S-bacteria ought to have a selective advantage in the intestinal tract. Our earlier results have indicated that S-bacteria have little tendency to interact with animal cells, so endocytosis of these bacteria by the healthy intestinal mucosa should be quantitatively insignificant. Normally, the immune mechanisms ought to be capable to inactivate the few bacteria that leak through. However, if the immune mechanisms are unable to eliminate even very few bacteria they will multiply and cause disease.

Surface characteristics of Escherichia coli *strains in relation to phagocytosis*

It is widely accepted that different *E. coli* strains show differences in their pathogencity and in their mechanisms of infection. There is a correlation between the above characters and the immunoelectrophoretic patterns of extracts from the *E. coli* strains (Ørskov, Ørskov, Jann & Jann, 1977). These extracts contain mainly K-antigen and O-antigen which are surface constituents of the bacteria. The O-antigen is the LPS and the K-antigen usually consists of acidic polysaccharide. Strains of *E. coli* with different O- and K-antigens were tested in two-phase systems of dextran and PEG. Tri(methyl)-amino-PEG or sulphonyl PEG were added to the system to analyse the surface charge. Liability to phagocytosis was tested with rabbit PMNL *in vitro* (Stendahl, Normann & Edebo, 1979). The bacteria were tested fresh and after heating at 70 °C for 45 min to remove the K-antigen. According to the experimental results the bacteria were arranged into three main groups. These groups agree well with the immunoelectrophoretic patterns, indicating that the K- and O-antigens not only dominate the serological properties but also the physicochemical properties of the surface of *E. coli* bacteria. The K-antigens tested were strongly negatively charged. Two different kinds of O-antigens were observed. One was hydrophilic and uncharged. This kind was called the S-type O-antigen, since it was similar to the O-antigen of *S. typhimurium* S-bacteria. The other kind was hydrophobic and slightly negatively charged and called R-type O-antigen, being similar to that of *S. typhimurium* R-bacteria. Presence of either K-antigen or S-type O-antigen was enough to prevent phagocytosis by PMNL. The first group of *E. coli* bacteria, which possessed both K-antigen and S-type O-antigen has been suggested as the cause of primarily extra-intestinal infections, such as urinary tract infection and sepsis (Ørskov *et al.*, 1977). The second group of strains showed only the S-type O-antigen. This group contains several strains considered pathogenic in infantile diarrhoea, with mechanisms of infection similar to that of salmonella enteritis. Bacteria with K-antigen and R-type O-antigen comprised the third group and are a common cause of dysentery-like infections. A few *Shigella sonnei* strains tested showed similar partition patterns (unpublished). These results indicate that physicochemical properties of the surface of *E. coli* bacteria are important in the host–parasite relationship and seem to be instrumental in the mechanisms of infection. It also seems as if similarities between different genera among enterobacteria with respect to physicochemical surface properties might be at least as important to pathogencity and mechanisms of infection as traditional taxonomic criteria.

EFFECT OF ANTIBODIES OF DIFFERENT IMMUNOGLOBULIN CLASSES ON THE PHYSICOCHEMICAL SURFACE PROPERTIES AND PHAGOCYTOSIS OF SALMONELLA BACTERIA

It has long been known that IgG antibodies increase the phagocytosis and subsequent killing of bacteria by PMNL. In this way bacteria are more efficiently eliminated from those compartments of the body which are normally sterile. When IgG antibodies bind to bacteria *in vitro*, very large quantities usually are required to cause agglutination. With *S. typhimurium* MS, no agglutination was observed when less than 100000 IgG molecules per bacterium were bound (Stendahl, Tagesson, Magnusson & Edebo, 1977). Thus, each antibody is usually bound to only one bacterium. This binding is probably bivalently mediated by the two Fab fragments of the IgG molecule, which gives a much higher binding strength than monovalent binding (Greenbury, Moore & Nunn, 1965). By coupling dinitrophenol (DNP) to bacteriophage ϕX-174 and measuring the phage-neutralising effect by bivalently binding IgG antibodies, an association constant, $K = 1.1 \times 10^{11}$ M^{-1} was calculated. Monovalent Fab showed $K = 6 \times 10^{6}$ M^{-1} (Hornick & Karush, 1971). The double binding between antibody and one antigen particle thus raised the binding constant between antigen and antibody by more than 10^4. This type of antibody binding, which has been called monogamous, is facilitated by the flexibility in the hinge region of the IgG molecule and serves the purpose of extensive binding at low concentration of antibody and orientation of the molecule such that the Fc portion is facing outwards.

The agglutinating capacity of IgM, which is decavalent, is on a weight basis 22-fold that of IgG for *S. typhimurium*. As shown by Feinstein & Munn (1966), IgM may bind monogamously to particles presenting repeating antigenic determinants, for example bacterial flagella. After the initial monogamous binding there is a rearrangement favouring binding to other particles and agglutination. At monogamous binding it is obvious that the non-antigen-binding portion of the antibody, the Fc part, will contribute to the surface of the particle. As a consequence the chemical structure of the surface of the Fc part will determine the surface properties of the antibody-coated particle and its tendency to interaction. The gross chemical

Table 1. *Presence of carbohydrate in antibodies*

Immunoglobulin class	Concentration (%)	Location
IgM	12	Surface
IgG	3	Inside
IgA	7.5	Mainly surface[a]
SIgA	11–15	Surface
IgE	12	Mainly surface[b]

[a] Not in SC-binding domains (Heremans, 1974; Beale & Feinstein, 1976).
[b] Not in Cε4 and carboxy-terminal portion of Cε3 (Beale & Feinstein, 1976) which are supposed to bind to the basophil–mast cell membrane (Bennich & von Bahr-Lindström, 1974).

composition of the different immunoglobulin classes shows the greatest differences with respect to carbohydrate content; IgM, SIgA and IgE show concentrations of 11–15 %, whereas IgG has only 3 % (Table 1). Further, what is known about location of the carbohydrate (Beale & Feinstein, 1976) indicates differences (Fig. 10). In IgG the oligosaccharide is located in segment b4 of Cγ2 in the inside of the molecule, probably having the function of preventing interaction between domains. In IgM there are five carbohydrate moieties on the μ-chain, most of them located at the outside of the molecule. Even the carbohydrate in segment b4 and Cμ3, which is at the corresponding position as that of IgG, may face the outside after monogamous binding of an IgM antibody to antigen in the 'table' form (Feinstein, 1975). The location of carbohydrate in serum IgA is similar to that of IgG, but there are five additional carbohydrate sites in the hinge region (Feinstein, Munn & Richardson, 1971; Beale & Feinstein, 1976). The binding to dimeric IgA of the secretory component (SC), which is extremely rich in carbohydrate (22.8 %; Lindh, 1975*b*), probably leads to superficial location of the carbohydrate moieties of the SC, since the binding requires protein–protein interaction with disulphide bridge formation. Thus, SIgA antibodies monogamously bound to antigen particles ought to render a hydrophilic character to the antibody-coated particles. In IgE, six different oligosaccharide sites are described, most of them probably being on the outside of the Cε1 and Cε2 domains and on the NH_2-terminal portion of the Cε3 domain of the molecule (Beale & Feinstein, 1976).

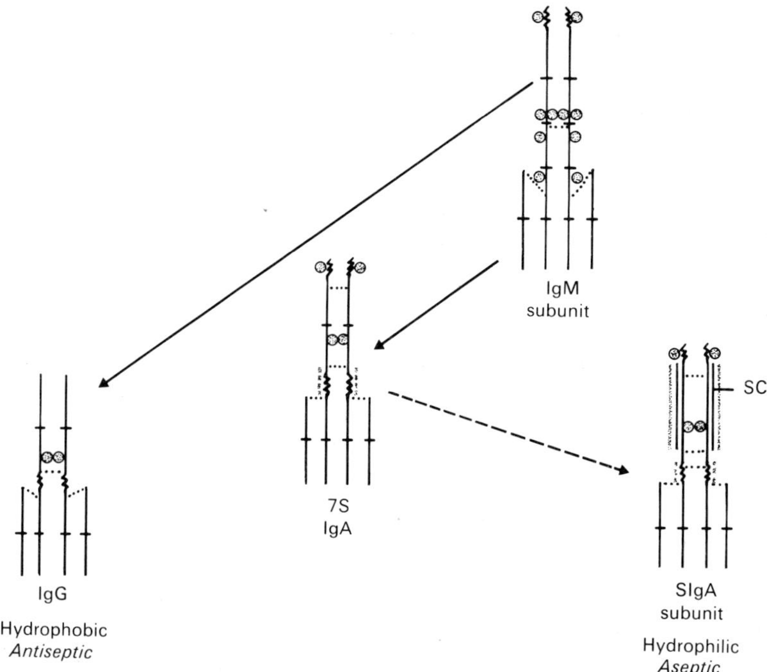

Fig. 10. Development and carbohydrate location of the different immunglobulin classes. Carbohydrate shown as solid balls with the exception of the hinge region of IgA and SIgA where, due to lack of space, five points are drawn. The carbohydrate of the SC has not been localised precisely and is shown as a superficial line.

This would allow interaction with the basophil–mast cell membrane in the terminal Cϵ3–Cϵ4 portion of the molecule (Bennich & von Bahr-Lindström, 1974) without involvement of the carbohydrate.

Binding of IgG antibodies to S. typhimurium *MS*

The binding of IgG antibodies changed the surface properties of *S. typhimurium* MS in several respects, all similar to the differences brought about by an S → R mutation (Table 2). The affinity for the PEG-rich top phase of the dextran–PEG two-phase system was reduced (Stendahl, Tagesson & Edebo, 1974; Stendahl, Tagesson, Magnusson & Edebo, 1977). In hydrophobic interaction chromatography on a column of Octyl–Sepharose, the retardation was increased (Fig. 11; Stjernström *et al.*, 1977), and the contact angle increased (Cunningham *et al.*, 1975). All these results fit with a hydrophobic

Table 2. *Physicochemical and adhesive properties of* S. typhimurium *S- and R-bacteria and SIgA and IgG antibodies*

	Physicochemical property		Relation to mammalian cells	
	Hydrophilic	Hydrophobic	Anti-adhesion	Pro-adhesion
Bacteria	S	R	S	R
Antibody	SIgA	IgG	SIgA	IgG

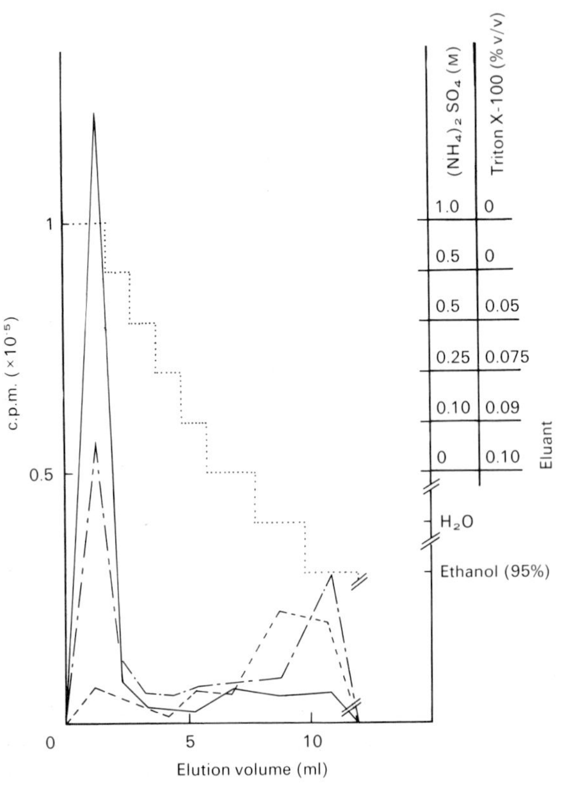

Fig. 11. Elution pattern after hydrophobic interaction chromatography on Octyl–Sepharose of *S. typhimurium* 395 MS (2.1×10^9 ml^{-1}, [^{3}H]leucine labelled) non-sensitised (——) and sensitised with hyperimmune IgG (75 μg ml^{-1} (–·–·–) and 1,200 μg ml^{-3} (–––). The eluants (right, ···) were 0.01 M with respect to phosphate and of pH 6.8, except for ethanol (95 %) and distilled water.

effect by the antibody binding. However, tests to show affinity for palmitoyl PEG have failed.

As is well known from studies with several different bacteria, animal cells and other particles, the IgG antibody binding to *S. typhimurium* MS also opsonised the bacteria for phagocytosis by PMNL *in vitro* and for clearance by the reticuloendothelial system after intravenous injection into mice. Around 8000 molecules per bacterium were required to produce a detectable change of the partition in the two-phase system as well as in phagocytosis by PMNL *in vitro* (Stendahl, Tagesson, Magnusson & Edebo, 1977). Only 2700 molecules per bacterium sufficed to enhance the binding to the Octyl–Sepharose column (Stjernström *et al.*, 1977) and less than 4000 molecules per bacterium were enough to promote clearance in mice (Stendahl, Tagesson, Magnusson & Edebo, 1977). Both the physicochemical effect and the opsonising effect were enhanced by the presence of complement and reduced by digestion of the Fc part with pepsin. Furthermore, artificial liposomal membranes were perturbed by MS bacteria sensitised with antibody IgG, as shown by release of entrapped marker substance to a greater extent than by non-sensitised bacteria (Tagesson, Magnusson & Stendahl, 1977). Thus, the binding of antibody IgG to a hydrophilic bacterium as *S. typhimurium* MS accomplished a physicochemical effect about as measurable as the opsonising effect for PMNL at limiting concentrations of antibody. We visualise this change as a consequence of covering of the hydrophilic polysaccharide side-chains of the bacteria by the more hydrophobic Fc moiety. This aspecific hydrophobic change and the enhanced reactivity for simple liposomal membranes as well as for the slime mould *Dictyostelium discoideum* (Gerisch *et al.*, 1967) indicate that very specific receptors for IgG are not required for opsonin-stimulated membrane interaction and phagocytosis. A thermodynamic calculation of the interfacial free energy needed for engulfment of an IgG molecule by a PMNL, -0.23 erg cm^{-2}, i.e. a positive force, predicts that complexes of three or more IgG molecules are bound to the phagocyte, since then the thermal energy of the complex is lower than the energy of attraction. These findings seemed to fit with experimental data (van Oss, Gillman & Neumann, 1974, 1975). Thus the 'Fc receptor' of PMNL may be a phenomenological way of describing the tendency to hydrophobic interaction brought about by orientation and cooperation of a number of IgG Fc components.

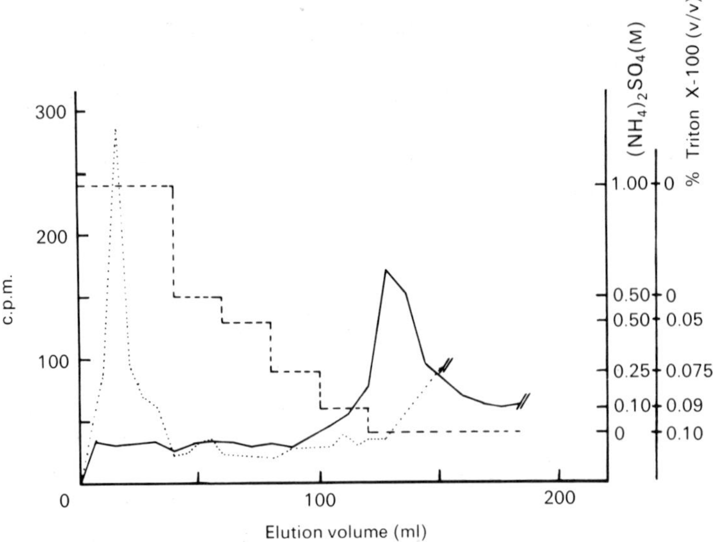

Fig. 12. Elution pattern after hydrophobic interaction chromatography on octyl–Sepharose of *S. typhimurium* 395 MR10 ([^{3}H]leucine labelled) non-sensitised (——) and sensitised with SIgA (280 μg ml^{-1}, · · ·). The eluants (right, – – –) were 0.01 M with respect to phosphate and of pH 6.8, except for ethanol (95 %) and distilled water.

Binding of SIgA antibodies to S. typhimurium *MR10 and* E. coli *086*

Whereas IgG is the predominant immunoglobulin in serum and extracellular fluid, SIgA is the main immunoglobulin of secretions, e.g. colostrum, parotid fluid and saliva, as well as of the mucosal membranes. By means of partition in the dextran–PEG two-phase systems we have shown that the physicochemical effect of the binding of components of human colostrum and of purified colostral antibody SIgA on *E. coli* 086 was opposite to that of IgG (Edebo *et al.*, 1975). Using partition and hydrophobic interaction chromatography it was recently shown that human colostrum as well as colostral SIgA rendered *E. coli* 086 (Magnusson, Stendahl, Stjernström & Edebo, 1978) and *S. typhimurium* MR10 less hydrophobic and less negatively charged (Fig. 12). This treatment also reduced the liability to phagocytosis by PMNL (Table 2; Magnusson, Stendahl, Stjernström & Edebo, 1979).

The function of SIgA is less clear than that of IgG (cf. Heremans, 1974). No opsonising effect has been found (Zipursky, Brown & Bienenstock, 1973). Rather, SIgA may reduce the attachment to the cells lining the mucosal surfaces and the absorption of microbes and non-viable antigens, particularly from the gastrointestinal tract (Hanson & Johansson, 1970; Heremans, 1974; Tomasi, 1976). A propensity for bacteria with bound SIgA antibodies to bind to a mucin-coated Sepharose column supports this mechanism (Magnusson, to be published).

Binding of MOPC315 myeloma protein to bacteria reacted with dinitrophenol (*DNP*)

Since it is hard to produce IgA antibodies by immunisation, the DNP-binding capacity of serum dimeric IgA from mice with the transplantable MOPC315 myeloma was employed. The binding affinity for DNP groups of this myeloma protein is of the same order as that of anti-DNP antibodies, the K_a for ϵ-DNP-L-lysine being 10^7 M^{-1}. The topology of the binding of DNP to the hapten-binding site of the Fab has been extensively investigated. Several hydrophobic amino acid side-chains seem to make up the boundaries of the cavities or depressions into which the DNP group can enter (Potter, 1977). There are side-chain groups that can form hydrogen bonds with the NO_2 groups of DNP.

Coupling of DNP groups to *S. typhimurium* MS made it possible to study the effect of dimeric IgA binding to the modified bacteria (Edebo *et al.*, 1977, 1978). In the ordinary two-phase system of dextran and PEG the partition of DNP–MS was changed by the IgA binding from being mainly in the top phase to the bottom phase. Approximately 2×10^6 IgA molecules per bacterium were estimated to be required to give a detectable effect. This is considerably more than the IgG needed. Presence of trimethylamino-PEG or palmitoyl PEG in the two-phase system shifted the partition towards the top phase, which indicated that hydrophobic moieties and negatively charged groups became exposed at the surface of bacteria that had bound IgA antibody. The hydrophobic character of these bacteria was further substantiated by hydrophobic interaction chromatography on Octyl–Sepharose columns. The IgA opsonised the bacteria for PMNL *in vitro*, whereas the clearance after intravenous injection in the mouse was reduced compared to the DNP–MS. However, the DNP coupling itself enhanced the clearance conspicuously. The preliminary interpretation is that the rapid clearance of DNP–MS in the mouse is mainly facilitated by natural anti-DNP antibodies, the effect of them being counteracted by the presence of IgA bound to the DNP groups. The lower hydrophobic effect and opsonising capacity of IgA compared to IgG fits this interpretation.

Effects of IgM

Our attempts to test the physicochemical effect of IgM antibody binding have failed, partly because of agglutination. In several systems it has been shown, however, that IgM alone does not promote the adherence to PMNL, whereas the additional presence of complement increases the adherence markedly (Scribner & Fahrney, 1976). From experiments with IgG and complement we conclude that complement itself has a hydrophobic effect (Stendahl, Tagesson, Magnusson & Edebo, 1977). After complement activation by IgM antibodies bound to bacterial flagella, a substantial layer of probably complement components covered the flagella (Feinstein & Munn, 1966). Therefore, a hydrophobic effect after complement activation by IgM bound to antigen may be anticipated. Since a monogamous binding of IgM to antigen in the 'table' form exposes carbohydrate, particularly the moieties located at $C\mu 2$ and $C\mu 3$ which will make up the edge of the 'table', a hydrophilic effect is expected. This hypothesis is supported by the fact that IgM itself does not promote the adherence.

Comparison between antibodies of different immunoglobulin classes

There is a drawback to the present experiments in that antibodies with different specificities have been used and tested on particles with different physicochemical properties. Direct comparisons between antibodies of different immunoglobulin classes are, therefore, not possible. However, when IgG and SIgA antibodies were tested on *S. typhimurium* MS, it was observed that antibody SIgA blocked the hydrophobic and opsonising effects of IgG antibodies (Magnusson, Skogh & Stjernström, unpublished).

Aseptic and antiseptic mechanisms in the immune defence

IgG is formed mainly in the spleen and in the lymph nodes and is the main immunoglobulin in serum and extracellularly in tissue. In contrast, IgA is the predominant immunoglobulin which is being

Table 3. *Antibody location and function*

Antibody	Site	Cofactor	Carbohydrate concentration (%)	Physicochemical effect	Phagocyte		Proposed function
					Adhesion	Ingestion	
IgM	Mucous membranes	(SC)			–	–	Aseptic
	Blood	Complement	12 (alone)		+ +	(–)	Antiseptic
IgG	Blood	Complement	(3 alone)	Hydrophobic + + +	+ + +	+ + +	Antiseptic
	Tissues	(Complement)		Hydrophobic + +	+ +	+ +	Antiseptic
IgA	Blood		7.5	Hydrophobic +	+		Antiseptic
	Mucous membranes	SC	11–15	Hydrophilic	–	–	Aseptic
IgE	Mucous membranes	Basophils Mast cells	12				Aseptic

synthesized subepithetially at the mucous membranes. During its passage through glandular cells dimeric IgA becomes conjugated with the secretory component (SC) and then secreted onto the mucous membrane. The bound SC increases resistance against proteolytic enzymes (Heremans, 1974; Lindh, 1975*a*; Tomasi, 1976). Free SC combines as readily with 19S IgM as with dimeric IgA (Brandtzaeg, 1974).

The hydrophobic and opsonising effect of IgG antibodies usually facilitates efficient phagocytic ingestion and destruction of the antigen. This is a kind of antiseptic mechanism that is necessary to maintain the integrity of the body against microbes that have slipped through the barriers of the body (Table 3).

Compared to IgG antibodies, exocrine IgA antibodies against a *S. typhimurium* S-strain (Eddie, Schulkind & Robbins, 1971) and *E. coli* O111 (Knop & Rowley, 1974) showed little or no complement-dependent bactericidal effect or opsonisation after intravenous or intraperitoneal infection into mice. In places like the mucosal surfaces, where a large microbial population is present, the hydrophilic and dysopsonising effect of SIgA seems more fit than an antiseptic mechanism. In these sites the immune mechanisms that serve to maintain the integrity of the body seem to seek to prevent the passage of microbes across the barriers of the body. This is an aseptic mechanism that rather aims to exlude the antigen from the underlying host tissue than to destroy it (Table 3). This hydrophilic and dysopsonising effect of SIgA seems to be supported by an affinity for mucus, as shown by the sticking of SIgA-sensitised bacteria to a mucin–Sepharose column. Both these mechanisms are expected to cooperate to keep the microbes in the lumen.

IgM, which is the most primitive immunoglobulin, may effect either function. In the blood, where abundant complement is available, complement activation should be substantial so that the antiseptic effect dominates. On the mucous membranes, where most complement components are missing and IgM may contain the secretory component, the aseptic effect is more likely and fitting (Table 3). The formation of both IgG and IgA seems to have developed from that of IgM (Hobart, 1975). It is hypothesised that the antibody system of blood and tissues which are to be sterile has become selected due to hydrophobic and antiseptic principles, leading to IgG, whereas hydrophilic and aseptic functions of antibodies have been improved in the development of the immune system of the secretions resulting in SIgA.

There are quantitative differences between the different immunoglobulin classes along the respiratory tract of the dog (Kaltreider & Chan, 1976) and along the genital tract of heifers (Corbeil *et al.*, 1976). Closest to the orifices at the body surface IgA is dominating, whereas in the deepest parts, which ought to be sterile, IgG is the main immunoglobulin. According to our hypothesis (Table 3) the body usually seeks to maintain its integrity by excluding the microbes of the mucosal membranes with the aid of aseptic mechanisms until these mechanisms have failed. This failure shows up as microbes in the deeper parts. Such microbes must be destroyed usually after opsonisation and phagocytosis. The abundant presence of alveolar macrophages are an example of this necessity.

CONCLUSION AND PERSPECTIVES

Our presentation of the mechanisms of adhesion has implicated the hydrophobic effect (Tanford, 1973) in 'non-specifically' promoting the adhesion and subsequent engulfment of R-bacteria by PMNL. Compared to the specificity of an antigen–antibody reaction, the above interaction appears non-specific. However, when the 'specific' binding between the mouse myeloma immunoglobulin MOPC315 and DNP hapten was analysed it was concluded that at the submolecular level in the antigen-binding pocket of the antibody, hydrophobic interaction seemed to play a great role (Potter, 1977). The characterisation of the interaction, 'specific' or 'non-specific', will therefore, ultimately, be quantitative rather than qualitative. An arbitrary limit based on size of ligand and association constant might be used to define the interaction. The association constant will determine the number of ligands necessary to establish adhesion such that interactions with a low association constant require multiple sites of interaction to achieve the same adhesion as a single site with a high association constant. One way to study the above difference is by testing the inhibition of the association by 'non-specific' hydrophobic compounds. A possible example of this approach is given by Freter (1972). The adsorption to slices of rabbit ileum of *Salmonella senftenberg* was in the same range as that of *Vibrio cholerae*. However, the adsorption was strikingly reduced by 0.02% sodium lauryl sulphate only for *S. senftenberg*, whereas L-fucose has been effective in preventing the adsorption of *V. cholerae* (Jones, 1977). These results indicate that the adsorption of *S. senftenberg* to the mucosa was mediated by non-specific hydrophobic interactions, whereas the

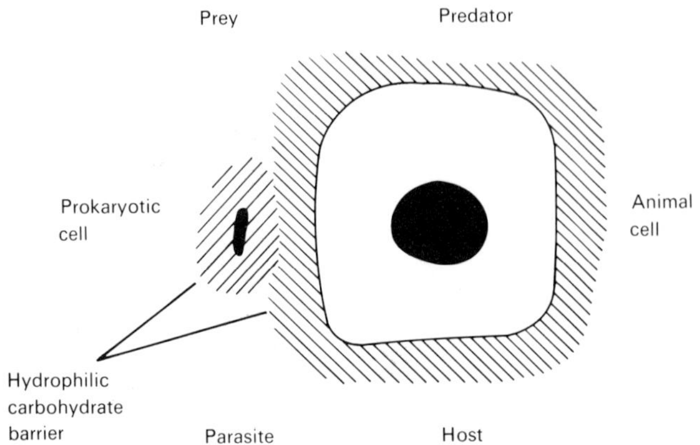

Fig. 13. Surface carbohydrate as barrier between prokaryotic and animal cells.

adsorption of *V. cholerae* might be more specific. The adsorption of each species would, however, be prevented by inert hydrophilic barriers between the bacteria and the mucosal cells.

Surface carbohydrates are common in prokaryotes and are extensively diversified (Sutherland, 1977; Wilkinson, 1977). They generally consist of neutral or acidic polysaccharide chains which show a great capacity to bind water (Wilkinson, 1958; Grasdalen, Svare & Smidsrød, 1974; Dudman, 1977) such that the cells become provided with a hydrophilic surface (Eylar, 1973). The survival value of this kind of surface may be, for example, in protection against desiccation, hydrophobic bactericidal substances and predators (Dudman, 1977). In the same way as the prokaryote prey is protected from animal predators by a hydrophilic coat, animal host cells, especially at the mucosal membranes, seem to be shielded from bacterial parasites by the glycocalyx (Fig. 13; Eylar, 1973; Knop, Ax, Sedlacek & Seiler, 1978). In this way the surface carbohydrates of prokaryotes as well as of animal cells generally seem to form hydrophilic barriers which prevent interference by protecting the victim from the aggressor (Fig. 13).

However, the barrier function of the mammalian mucosal membranes is not perfect. Therefore, the immune system of secretions (Tomasi, 1976) has been developed to support the barrier function of the mucosal membranes. This mechanism also seems to exploit the hydrophilic property of carbohydrates, now that of the secretory antibodies, particularly SIgA. The efficiency of this system seems to be enhanced by an affinity for mucus (Fig. 14).

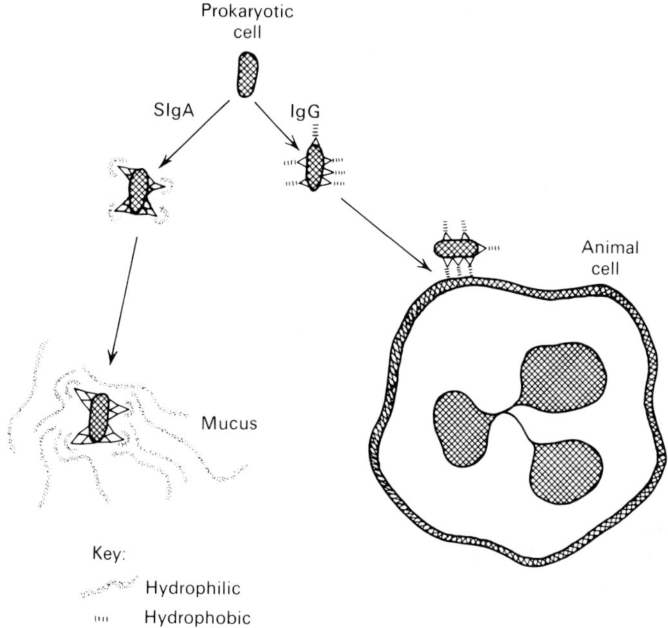

Fig. 14. Aseptic mucophilic effect of SIgA and antiseptic opsonising effect of IgG antibodies.

Bacteria that have passed the mucosal barrier play double roles. On the organism level they are obviously aggressors, whereas on the cellular level they are the victims of the phagocytes of the host. Under these circumstances the immune mechanisms take the side of the host and work to reduce the barrier. The IgG antibodies increase the hydrophobic character of the prey, such that professional phagocytes like PMNL and macrophages become able to engulf it.

In order to maintain the integrity of the body, the hydrophilic and the hydrophobic immune mechanisms need to be topologically separated. The hydrophilic and mucophilic mechanisms should operate mainly on the mucous membranes and the hydrophobic mechanisms in normally sterile tissue. If hydrophilic barrier-enhancing mechanisms prevailed in normally sterile tissue when it becomes infected, adverse effects might arise. An example of this is the great reduction of the lethal dose of R-bacteria (*c.* 10^{-4}) that was a consequence of suspending the bacteria in mucin before intraperitoneal injection (Fig. 2). Another example is the impaired phagocytosis in the presence of SIgA. Other adverse effects might arise if, for example, SIgA antibodies were to operate in the circulation. Then the dysopsonising effect of SIgA might impair antigen clearance by the reticuloendothelial system such that complexes of antigen and antibody might localise in other tissues. The increased serum concentration of SIgA in rheumatoid arthritis, the deposition of IgA in the skin of patients with alimentary intolerance to gluten and dermatitis herpetiformis (Heremans, 1974; Tomasi, 1976) may be consequences of upset topological distinction of the immune system.

REFERENCES

Albertsson, P. Å. (1971). *Partition of Cell Particles and Macromolecules*, 2nd edn. Uppsala: Almqvist & Wiksell; New York: John Wiley & Sons.

Ames, G. F., Spudich, E. N. & Nikaido, H. (1974). Protein composition of the outer membrane of *Salmonella typhimurium*: Effect of polysaccharide mutations. *Journal of Bacteriology*, **117**, 406–16.

Bayer, M. E., Koplow, J. & Goldfine, H. (1975). Alterations in envelope structure of heptose-deficient mutants of *Escherichia coli* as revealed by freeze-etching. *Proceedings of the National Academy of Sciences, USA*, **72**, 5145–9.

Beale, D. & Feinstein, A. (1976). Structure and function of the constant regions of immunoglobulins. *Quarterly Reviews of Biophysics*, **9**, 135–80.

Bennich, H. & von Bahr-Lindström, H. (1974). Structure of immunoglobulin E (IgE). In *Progress in Immunology II*, vol. 1, ed. L. Brent & J. Holborow, pp. 49–58. Amsterdam: North-Holland.

Bovee, E. C. & Jahn, T. L. (1973). Taxonomy and phylogeny. In *The Biology of Amoeba*, ed. K. W. Jeon, pp. 38–82. London, New York: Academic Press.

Brandtzaeg, P. (1974). Characteristics of SC-Ig complexes formed *in vitro*. In *The Immunoglobulin A System*, ed. J. Mestecky & A. R. Lawton, pp. 87–97. New York: Plenum Press.

Chapman-Andresen, C. (1973). Endocytotic processes. In *The Biology of Amoeba*, ed. K. W. Jeon, pp. 319–348. London, New York: Academic Press.

Corbeil, L. B., Hall, C. E., Lein, D., Corbeil, R. R. & Duncan, J. R. (1976). Immunoglobulin classes in genital secretions of mycoplasma-infected and normal heifers. *Infection and Immunity*, **13**, 1595–1600.

Costerton, J. W., Ingram, J. M. & Cheng, K.-J. (1974). Structure and function of the cell envelope of Gram-negative bacteria. *Bacteriological Reviews*, **38**, 87–110.

Cunningham, R. K., Söderström, T. O., Gillman, C. F. & van Oss, C. J. (1975). Phagocytosis as a surface phenomenon. V. Contact angles and phagocytosis of rough and smooth strains of *Salmonella typhimurium*, and the influence of specific antiserum. *Immunological Communications*, **4**, 429–42.

Davis, B. D., Dulbecco, R., Eisen, H. N. Ginsberg, H. S. & Wood, W. B. (1973). *Microbiology*. Hagerstown, Md: Harper & Row.

DUDMAN, W. F. (1977). The role of surface polysaccharides in natural environments. In *Surface Carbohydrates of the Prokaryotic Cell*, ed. I. W. Sutherland, pp. 357–414. London, New York: Academic Press.

EDDIE, D. S., SCHULKIND, M. L. & ROBBINS, J. B. (1971). The isolation and biologic activities of purified secretory IgA and IgG anti-*Salmonella typhimurium* 'O' antibodies from rabbit intestinal fluid and colostrum. *Journal of Immunology*, **106**, 181–90.

EDEBO, L., HED, J., KIHLSTRÖM, E., MAGNUSSON, K.-E., STENDAHL, O. & TAGESSON, C. (1977). Bacterial invasion and antigen elimination as a cell surface interaction phenomenon. Abstracts Symposium on Biology of Connective Tissue, Uppsala. *Uppsala Journal of Medical Science*, **82**, 128.

EDEBO, L., HED, J., MAGNUSSON, K.-E., RICHARDSON, N. & STENDAHL, O. (1978). The relationship of hydrophilic and hydrophobic properties of different immunoglobulin classes to antibody function. *Abstracts 4th European Immunology Meeting, Budapest*, p. 91.

EDEBO, L., LINDSTRÖM, F., SKÖLDSTAM, L., STENDAHL, O. & TAGESSON, C. (1975). On the physical-chemical effect of colostral antibody binding to *Escherichia coli* O86. *Immunological Communications*, **4**, 587–601.

EDEBO, L. & NORMANN, B. (1970). Virulence and immunogenicity of mutant strains of *Salmonella typhimurium*. *Acta Pathologica et Microbiologica Scandinavica*, Section B, **78**, 75–84.

EDEBO, L. & NORMANN, B. (1971). Extinction of the bactericidal effect of normal human serum on rough *Salmonella typhimurium* mutants by rabbit immune serum. 27th Symposium on Enterobacterial Vaccines, Berne, 1969. *Symposia Series in Immunobiological Standardization*, **15**, 41–8.

EYLAR, E. H. (1973). Foreword. In *Surface Carbohydrates of the Eurkaryotic Cell*, ed., G. M. W. Cook & R. W. Stoddart, pp. vii–ix. London, New York: Academic Press.

FEINSTEIN, A. (1975). The three-dimensional structure of immunoglobulins. In *The Immune System*, ed. M. J. Hobart & I. McConnell, pp. 24–41. Oxford: Blackwell Scientific.

FEINSTEIN, A. & MUNN, E. A. (1966). An electron microscopic study of the interaction of macroglobulin (IgM) antibodies with bacterial flagella and of the binding of complement. Proceedings of the Physiological Society, *Journal of Physiology*, **186**, 64–6P.

FEINSTEIN, A., MUNN, E. A. & RICHARDSON, N. E. (1971). The three-dimensional conformation of γM and γA globulin molecules. *Annals of the New York Academy of Sciences*, **190**, 104–21.

FENN, W. O. (1923). The adhesiveness of leucocytes to solid surfaces. *Journal of General Physiology*, **5**, 143–79.

FRETER, R. (1972). Parameters affecting the association of vibrios with the intestinal surface in experimental cholera. *Infection and Immunity*, **6**, 134–41.

GERISCH, G., LÜDERITZ, O. & RUSCHMANN, E. (1967). Antikörper fördern die Phagozytose von Bakterien durch Amöben. *Zeitschrift für Naturforschung*, **226**, 109.

GRASDALEN, H., SVARE, I. & SMIDSRØD, O. (1974). A method of studying the competitive binding of small molecules to macromolecules in the gel state by high resolution ^{1}H NMR spectroscopy. *Acta Chemica Scandinavica*, **28B**, 966–8.

GREENBURY, C. L., MOORE, D. H. & NUNN, L. A. C. (1965). The reaction with red cells of 7S rabbit antibody, its sub-units and their recombinants. *Immunology*, **8**, 420–31.

HANSON, L. Å. & JOHANSSON, B. G. (1970). Immunological studies of milk. In *Milk Proteins, Chemistry and Molecular Biology*, vol. 1, ed. H. A. McKenzie, pp. 45–123. London, New York: Academic Press.

HEREMANS, J. F. (1974). Immunoglobulin A. In *The Antigens*, vol. 2, ed. M. Sela, pp. 365–522. London, New York: Academic Press.

HOBART, M. J. (1975). The evolution and genetics of antibody and complement. In *The Immune System*, ed. M. J. Hobart & I. McConnell, pp. 76–90. Oxford: Blackwell Scientific.

HOLME, T., LINDBERG, A. A., GAREGG, P. J. & ONN, T. (1968). Chemical composition of cell-wall polysaccharide of rough mutants of *Salmonella typhimurium*. *Journal of General Microbiology*, **52**, 45–54.

HORNICK, C. L. & KARUSH, F. (1971). The energetic significance of antibody bivalence. In *Developmental Aspects of Antibody Formation and Structure*, ed. J. Sterzl & I. Riha, pp. 433–43. London, New York: Academic Press.

IRVIN, R. T., CHATTERJEE, A. K., SANDERSON, K. E. & COSTERTON, J. W. (1975). Comparison of the cell envelope structure of a lipopolysaccharide-defective (heptose-deficient) strain and a smooth strain of *Salmonella typhimurium*. *Journal of Bacteriology*, **124,** 930–41.

JONES, G. W. (1977). The attachment of bacteria to the surfaces of animal cells. In *Microbial Interactions*, ed. J. L. Reissig. *Receptors and Recognition*, Series B, vol. 3, pp. 139–76. London: Chapman & Hall.

KALTREIDER, H. B. & CHAN, M. K. L. (1976). The class-specific immunoglobulin composition of fluids obtained from various levels of the canine respiratory tract. *Journal of Immunology*, **116**, 423–9.

KIHLSTRÖM, E. (1977). Infection of HeLa cells with *Salmonella typhimurium* 395 MS and MR10 bacteria. *Infection and Immunity*, **17**, 290–5.

KIHLSTRÖM, E. & EDEBO, L. (1976). Association of viable and inactivated *Salmonella typhimurium* 395 MS and MR10 with HeLa cells. *Infection and Immunity*, **14**, 851–7.

KIHLSTRÖM, E. & LATKOVIC, S. (1978). Ultrastructural studies on the interaction between *Salmonella typhimurium* 395 M and HeLa cells. *Infection and Immunity*, **22**, 804–9.

KIHLSTRÖM, E. & NILSSON, L. (1977). Endocytosis of *Salmonella typhimurium* 395 MS and MR10 by HeLa cells. *Acta Pathologica et Microbiologica Scandinavica*, Section B **85**, 322–8.

KNOP, J., AX, W., SEDLACEK, H. H. & SEILER, F. R. (1978). Effect of *Vibrio cholerae* neuraminidase on the phagocytosis of *E. coli* by macrophages *in vivo* and *in vitro*. *Immunology*, **34**, 555–63.

KNOP, J. G. & ROWLEY, D. (1974). The antibacterial efficiencies of ovine IgA, IgM and IgG. *Journal of Infectious Diseases*, **130**, 368–73.

KOPLOW, J. & GOLDFINE, H. (1974). Alterations in the outer membrane of the cell envelope of heptose-deficient mutants of *Escherichia coli*. *Journal of Bacteriology*, **117**, 527–43.

Korn, E. D. & Weisman, R. A. (1967). Phagocytosis of latex beads by *Acanthamoeba*. II. Electron microscopoic study of the initial events. *Journal of Cell Biology*, **34**, 219–27.

Lindberg, A. A. (1977). Bacterial surface carbohydrates and bacteriophage adsorption. In *Surface Carbohydrates of the Prokaryotic Cell*, ed. I. W. Sutherland, pp. 289–356. London, New York: Academic Press.

Lindberg, A. A. & Holme, T. (1968). Immunochemical studies on cell-wall polysaccharide of rough mutants of *Salmonella typhimurium*. *Journal of General Microbiology*, **52**, 55–65.

Lindh, E. (1975*a*). Increased resistance of immunoglobulin A dimers to proteolytic degradation after binding of secretory component. *Journal of Immunology*, **114**, 284–6.

Lindh, E. (1975*b*). Studies on Human Secretory Immunoglobulins. In Gross Conformation of Secretory IgA and Binding of Secretory component to Immunoglobulins A and M. Thesis. *Acta Universitatis Upsaliensis*. Abstracts of dissertations from the Uppsala Faculty of Medicine.

Magnusson, K.-E. & Johansson, G. (1977*a*). Probing the surface of *Salmonella typhimurium* and *Salmonella minnesota* SR and R bacteria by aqueous biphasic partitioning in systems containing hydrophobic and charged polymers. *FEMS Microbiology Letters*, **2**, 225–8.

Magnusson, K.-E. & Johansson, L. (1977*b*). Envelope composition of *Salmonella typhimurium* 395 MS and 395 MR10 assessed by X-ray photoelectron spectroscopy. *Studia Biophysica, Berlin*, **66**, 145–53.

Magnusson, K.-E., Stendahl, O., Stjernström, I. & Edebo, L. (1978). The effect of colostrum and colostral antibody SIgA on the physico-chemical properties and phagocytosis of *Escherichia coli* O86. *Acta Pathologica et Microbiologica Scandinavica*, Section B, **86**, 113–20.

Magnusson, K.-E., Stendahl, O., Stjernström, I. & Edebo, L. (1979). Reduction of phagocytosis, surface hydrophobicity and charge of *Salmonella typhimurium* 395 MR10 by reaction with secretory IgA (SIgA). *Immunology*, **36**, 439–47.

Magnusson, K.-E., Stendahl, O., Tagesson, C., Edebo, L. & Johansson, G. (1977). The tendency of smooth and rough *Salmonella typhimurium* bacteria and lipopolysaccharide to hydrophobic and ionic interaction, as studied in aqueous polymer two-phase systems. *Acta Pathologica et Microbiologica Scandinavica*, **85**, 212–18.

Mudd, S., McCutcheon, M. & Lucké, B. (1934). Phagocytosis. *Physiological Reviews*, **14**, 210–75.

Ørskov, I., Ørskov, F., Jann, B. & Jann, K. (1977). Serology, chemistry, and genetics of O and K antigens of *Escherichia coli*. *Bacteriological Reviews*, **41**, 667–710.

Parton, R. (1975). Envelope proteins in *Salmonella minnesota* mutants. *Journal of General Microbiology*, **89**, 113–23.

Perers, L., Andåker, L., Edebo, L., Stendahl, O. & Tagesson, C. (1977). Association of some enterobacteria with the intestinal mucosa of mouse in relation to their partition in aqueous polymer two-phase systems. *Acta Pathologica et Microbiologica Scandinavica*, Section B, **85**, 308–16.

Potter, M. (1977). Antigen-binding myeloma proteins of mice. *Advances in Immunology*, **25**, 141–211.

Scribner, D. J. & Fahrney, D. (1976). Neutrophil receptors for IgG and complement: their roles in the attachment and ingestion phases of phagocytosis. *Journal of Immunology*, **116**, 892–7.

Sherbet, G. V. (1978). *The Biophysical Characterization of the Cell Surface.* London, New York: Academic Press.

Siegbahn, K., Nordling, C., Fahlman, A., Nordberg, R., Hamrin, K., Hedman, J., Johansson, G., Bergmark, T., Karlsson, S.-E., Lindgren, I. & Lindberg, B. (1967). *ESCA. Atomic, Molecular and Solid-State Structure Studied by means of Electron Spectroscopy.* Uppsala: Almqvist & Wiksell.

Smith, H. (1977). Microbial surfaces in relation to pathogenicity. *Bacteriological Reviews*, **41**, 475–500.

Stendahl, O. (1973). Role of Cell Surface Properties of *Salmonella typhimurium* in the Interaction with Phagocytic Cells. Thesis, Linköping University. Linköping University Medical Dissertations 10.

Stendahl, O. & Edebo, L. (1972). Phagocytosis of mutants of *Salmonella typhimurium* by rabbit polymorphonuclear cells. *Acta Pathologica et Microbiologica Scandinavica*, Section B, **80**, 481–8.

Stendahl, O., Edebo, L., Tagesson, C., Magnusson, K.-E. & Hjertén, S. (1977). Surface-charge characteristics of smooth and rough *Salmonella typhimurium* bacteria determined by aqueous two-phase partitioning and free-zone electrophoresis. *Acta Pathologica et Microbiologica Scandinavica*, Section B, **85**, 334–40.

Stendahl, O., Magnusson, K.- E., Tagesson, C., Cunningham, R. & Edebo, L. (1973). Characterization of mutants of *Salmonella typhimurium* by countercurrent distribution in aqueous two-polymer phase system. *Infection and Immunity*, **7**, 573–7.

Stendahl, O., Normann, B. & Edebo, L. (1979). Influence of O and K antigens on the surface properties of *Escherichia coli* in relation to phagocytosis. *Acta Pathologica et Microbiologica Scandinavica*, Section B, **87**, 85–91.

Stendahl, O., Tagesson, C. & Edebo, M. (1973). Partition of *Salmonella typhimurium* in a two-polymer aqueous phase system in relation to liability to phagocytosis. *Infection and Immunity*, **8**, 36–41.

Stendahl, O., Tagesson, C. & Edebo, L. (1974). Influence of hyperimmune immunoglobulin G on the physico-chemical properties of the surface of *Salmonella typhimurium* 395 MS in relation to interaction with phagocytic cells. *Infection and Immunity*, **10**, 316–19.

Stendahl, O., Tagesson, C., Magnusson, K.-E. & Edebo, L. (1977). Physicochemical consequences of opsonization of *Salmonella typhimurium* with hyperimmune IgG and complement. *Immunology*, **32**, 11–18.

Stjernström, I., Magnusson, K.-E., Stendahl, O. & Tagesson, C. (1977). Liability to hydrophobic and charge interaction of smooth *Salmonella typhimurium* 395 MS sensitized with anti-MS IgG and complement. *Infection and Immunity*, **18**, 261–5.

Sutherland, I. W. (1977). Bacterial exopolysaccharides – their nature and production. In *Surface Carbohydrates of the Prokaryotic Cell*, ed. I. W. Sutherland, pp. 27–96. London, New York: Academic Press.

TAGESSON, C., MAGNUSSON, K.-E. & STENDAHL, O. (1977). Physico-chemical consequences of opsonization: Perturbation of liposomal membranes by *Salmonella typhimurium* 395 MS opsonized with IgG antibodies. *Journal of Immunology*, **119**, 609–13.

TANFORD, C. (1973). *The Hydrophobic Effect. Formation of Micelles and Biological Membranes*. New York: John Wiley & Sons.

TOMASI, T. B. (1976). *The Immune System of Secretions*. Prentice-Hall Foundations of Immunology Series. Englewood Cliffs, N.J.: Prentice-Hall.

VAN OSS, C. J. (1978). Phagocytosis as a surface phenomenon. *Annual Review of Microbiology*, **32**, 19–39.

VAN OSS, C. J., GILLMAN, C. F. & NEUMANN, A. W. (1974). Phagocytosis as a surface phenomenon. IV. The minimum size and composition of antigen–antibody complexes that can become phagocytized. *Immunological Communications*, **3**, 77–84.

VAN OSS, C. J., GILLMAN, C. F. & NEUMANN, A. W. (1975). *Phagocytic Engulfment and Cell Adhesiveness as Cellular Surface Phenomena*. New York: Marcel Dekker, Inc.

WALTER, H. (1977). Partition of cells in two-polymer aqueous phases: A surface affinity method for cell separation. In *Methods of Cell Separation*, vol. 1, ed. N. Catsimpoolas, pp. 307–54. Amsterdam: North-Holland.

WEISMAN, R. A. & KORN, E. D. (1967). Phagocytosis of latex beads by *Acanthamoeba*. I. Biochemical properties. *Biochemistry*, **6**, 485–97.

WILKINSON, J. F. (1958). The extracellular polysaccharides of bacteria. *Bacteriological Reviews*, **22**, 46–73.

WILKINSON, S. G. (1977). Composition of bacterial lipopolysaccharides. In *Surface Carbohydrates of the Prokaryotic Cell*, ed. I. W. Sutherland, pp. 97–175. London, New York: Academic Press.

WILSON, G. S. & MILES, A. A. (1975). *Topley and Wilson's Principles of Bacteriology, Virology and Immunity*, 6th edn, vol. 1, pp. 939–47. London: Edward Arnold.

ZIPURSKY, A., BROWN, E. J. & BIENENSTOCK, J. (1973). Lack of opsonization potential of 11S human secretory γA. *Proceedings of the Society for Experimental Biology and Medicine*, **142**, 181–4.

The physical chemistry of the adhesion of bacteria and other cells

P. R. RUTTER

Dental School, The London Hospital Medical School, Turner Street, London, E1 2AD, UK

GLOSSARY

A Hamaker constant
C ionic strength of suspending medium
C_0 particle concentration
D particle diffusion coefficient
N Avogadro's constant
R gas constant
T absolute temperature
V_A London-van der Waals energy
V_R electrostatic energy
V_T total interaction energy
a particle radius
h separation between two approaching surfaces
k Boltzmann constant
w angular velocity
ϵ permittivity
κ reciprocal of the Debeye–Hückel screening length
η fluid viscosity
ψ_C collector surface potential
ψ_P particle surface potential
ν kinematic fluid viscosity

INTRODUCTION

Bacteria occur in most, if not all, natural fluid environments on the surface of the planet. Bacteria are perhaps most obvious when their presence leads to the spoilage of food, but they can be collected relatively easily from virtually all natural sources of water and air. Generally speaking, although bacteria exist in bulk fluid environments, they tend to show a preference for the interfaces between bulk media. In this respect bacteria resemble macromolecules, which are often readily adsorbed by solid/liquid, air/liquid or liquid/liquid interfaces.

The apparent affinity of bacteria for interfaces poses a number of questions. The major question is whether bacteria are intrinsically sticky particles or whether bacteria actively produce sticky substances to prevent them being swept away from ecologically desirable locations. These questions are important when considering the apparent specificity with which certain organisms accumulate in particular sites. If the particulate nature of the bacterial cell endows it with a propensity to be collected non-specifically by interfaces, this property will have to be overcome if the organisms are to accumulate at a specific site, otherwise most will be collected long before they reach their desired location. If, on the other hand, bacteria are essentially non-sticky particles whose adhesion is mediated by specific glues, the organisms will remain free to move about in the fluid environment, colliding with surfaces with impunity until a surface is found which is capable of attaching to the glue.

It is perhaps naive to assume that the subject of bacterial adhesion can be considered in terms of general principles. The diversity of bacterial species and of the locations they inhabit, which include wide variations of, for example, temperature and ionic strength, suggest that a number of different mechanisms may be responsible for bacterial adhesion. However, a consideration of the physical chemistry of the deposition and capture of small particles at interfaces, coupled with data from experiments designed to quantify bacterial adhesion, can at least give some insight into the types of interactions that might be utilised by a bacterial cell as it approaches and attaches to a surface. Other authors in this volume also treat this topic. Watt describes the interactions between bacteria and eukaryotes; Edebo *et al.* consider average properties of the bacterial surface and their relationship to the adhesive processes involved in phagocytosis.

In order to simplify discussion, surfaces to which bacteria will adhere will be referred to as collectors unless specified otherwise.

THE DEPOSITION AND CAPTURE OF SMALL PARTICLES

In the absence of external forces, colloidal particles dispersed in a fluid possess a kinetic energy equal to $\frac{3}{2}RT/N$ or $\frac{3}{2}kT$, where k is the Boltzmann constant, R is the gas constant, N is Avogadro's constant and T is the absolute temperature. A consequence of this kinetic energy is that in dilute suspensions a spherical rigid particle will experience a time-dependent displacement from its original position. The relationship between displacement and time was described almost simultaneously by Einstein and von Smoluchowski (Kruyt, 1952) and is given by the following formula:

$$\overline{\Delta x^2} = 2t(RT/N)1/6\pi\eta a$$

in which $\overline{\Delta x^2}$ is the mean of the square of the displacement during time t projected on a chosen x direction, η the viscosity of the suspending fluid and a the radius of the particle, which is assumed to be spherical.

The displacement of particles suspended in a liquid under the constant bombardment of the molecules of the suspending liquid is observed as Brownian motion, and is responsible for the diffusion of particles from regions of high concentration to regions of lower concentration. The number (n) of particles that diffuse across unit area is given by Fick's law (Kruyt, 1952) which states that

$$\frac{\mathrm{d}n}{\mathrm{d}t} = -D\frac{\mathrm{d}c}{\mathrm{d}x}$$

where $\mathrm{d}c/\mathrm{d}x$ is the concentration gradient in the x-direction, D is $kT/6\pi\eta a$ and t is the time.

When a large collecting surface is brought into contact with a suspension, a concentration gradient is set up in the immediate vicinity of the collector. This is because some particles will adhere to the surface, so depleting the layers of suspension adjacent to the collector. Diffusion then causes particles to move into the depleted layers. This produces a layer of liquid next to the collector surface in which the concentration of particles changes rapidly from the concentration in the bulk suspension to a concentration of zero at the collector surface. This layer is referred to as the diffusion boundary layer (Levich, 1962). The rate at which particles are supplied to the surface from the bulk suspension is governed by Fick's law and can therefore be calculated, assuming that the thickness, δ, of the boundary layer and the concentration gradient are known. The simplest system to define is that of a disc-shaped collector rotating in a suspension. The number of particles that impinge on the surface of the disc is given by the Levich (1962) equation of convective diffusion:

$$j = 0.62\, D^{\frac{2}{3}}\nu^{-\frac{1}{6}}w^{\frac{1}{2}}C_0$$

where j is the particle flux, D the diffusion coefficient, ν the kinematic viscosity, w the disc angular velocity and C_0 the particle concentration of the suspension far from the disc surface.

Experiments involving the use of rotating discs to study the deposition of small particles have been carried out by a number of authors (Hull & Kitchener, 1969; Clint, Clint, Corkill & Walker, 1973) and the technique can be readily adapted to study bacterial deposition (Rutter & Abbott, 1978).

If the particles are suspended in a flowing fluid, they may be entrained in the fluid streamlines and be brought into contact with the collector as the fluid moves past it. This is known as interception, and may again be studied using the rotating disc. The rate of particle capture by interception at the surface of a rotating disc is given by (Spielman & Cukor, 1973):

$$j = 0.51\, w^{\frac{3}{2}} \nu^{-\frac{1}{2}} a^2 C_0$$

Calculation shows that in the case of particles of bacterial dimensions, the major process leading to deposition on the surface of the disc is convective diffusion.

In suspensions containing large particles (or aggregates) or particles of considerably greater density than the fluid, for example aerosols, capture by impaction becomes important (Chamberlain, 1967). In addition to impaction, large particles or aggregates will be influenced by gravity and tend to sediment out of suspension. Gravitational sedimentation may therefore be particularly important in localising bacterial aggregates on horizontal surfaces. In systems where interception is also important, the number of collisions between suspended particles and a collector can be obtained by vector addition of the two separate processes (Spielman, 1978).

In addition to being affected by fluid dynamic transport processes, particles can also move along electrical potential gradients, as in particle microelectrophoresis, or along temperature gradients (Chamberlain, 1967). Particles suspended in liquid drops, e.g. raindrops, can also be brought into close proximity to a collector if the liquid evaporates.

The phenomena outlined above will ensure that any particle suspended in a fluid will eventually collide with a surface. The only case where this might not be true is if the particle is able to direct its own movement through the fluid. This is possible for certain motile bacteria which appear to be able to move quickly enough in

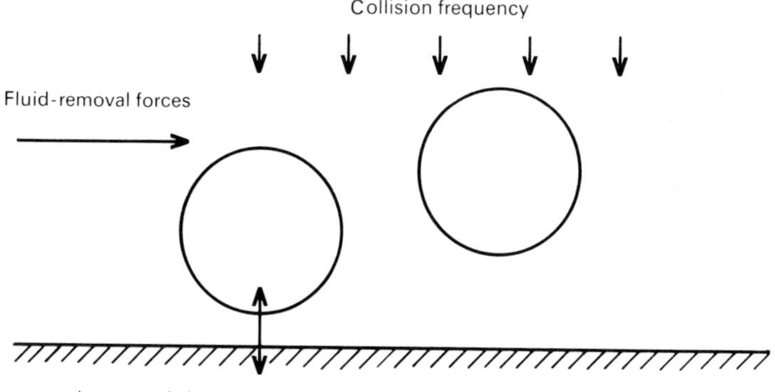

Fig. 1. Collision efficiency. The number of particles that are collected by a surface is determined by the number of collisions, the magnitude of the adhesive interaction and any removal forces acting on the particle.

water to overcome the displacement due to diffusion and sedimentation. A marine pseudomonad, for example, has been shown to move at velocities of 33 μm s^{-1} (Marshall, Stout & Mitchell, 1971). This compares with an effective random-walk displacement of 1 μm s^{-1} due to diffusion. However, its kinetic energy due to Brownian motion is very much greater than that due to motility. This can be illustrated by a simple calculation. The kinetic energy along a given axis due to Brownian motion is equal to $\frac{1}{2}kT$. However, the kinetic energy of a bacterium due to its movement through water at 33 μm s^{-1} under the influence of its flagella is given by $\frac{1}{2}MV^2$ where M is the mass of the bacterial cell and V its velocity. Assuming that the density of a spherical bacterial cell 1 μm in diameter is approximately 1 g cm^{-3} the kinetic energy is equal to:

$$\tfrac{1}{2}[4/3\ (0.5\times10^{-4})^3(33\times10^{-4})^2]$$
$$= 285.3\times10^{-20}\ \text{erg}$$
$$= 0.71\times10^{-4}\ kT$$

Although small particles may be transported to surfaces under the influence of fluids flowing past the collector, the same fluid flow can also facilitate their removal (Dahneke, 1975). Particles greater than 1 μm in diameter can readily be sheared off surfaces under the action of only small velocity gradients. The number of particles that are collected by a surface is therefore determined by the number of collisions, the magnitude of the adhesive interaction between the particle and collector, and the fluid-removal forces acting on the particle (Fig. 1). These considerations can also be applied to bacteria, because although they are complex living entities they are still particulate.

PARTICLE ADHESION

Once a particle has been brought into the close proximity of a surface, particle–surface interactions determine whether or not the particle is captured. These are often termed adhesive interactions. The subject of the adhesion of small particles to surfaces has been reviewed by a number of authors (Zimon & Derjaguin, 1963; Krupp, 1967; Visser, 1976*a*). The physical interactions of particular relevance to bacteria have recently been reviewed by Lips & Jessup (1979).

Perhaps the most familiar treatment of the interactions of small particles at close separations is attributable to Derjaguin & Landau (1941) and Verwey & Overbeek (1948) and is described by the so-called DLVO theory of colloid stability. This states that the total interaction energy, V_T, of two smooth particles is determined solely by the sum of the van der Waals attractive energy and the, usually repulsive, electrostatic energy. This theory can be applied to the case where a particle approaches a flat-plate collector. As the particle approaches the collector it experiences a weak van der Waals attraction induced by the fluctuation dipoles within the molecules of the two approaching surfaces. This attraction increases as the particle moves closer to the collector. In most aqueous systems, however, both the particle and the collector surfaces are negatively charged. This causes a repulsive force to come into effect as the surfaces approach each other, owing to the overlap of the diffuse layers of counterions associated with each charged surface. The magnitude of the repulsion is dependent upon the surface potentials, the ionic strength and dielectric constant of the surrounding medium, and the particle–collector separation.

The variation of the total interaction energy, V_T, as a particle approaches a planar collector can be determined by adding the contributions due to van der Waals (V_A) and electrostatic repulsion (V_R) using the following expression (Visser, 1978):

$$V_T = V_A + V_R$$

where $V_A = (-A/6)[2a(a+h)/h(h+2a) - \ln(h+2a)/h]$

and

$$V_R = \pi \epsilon a\{2\psi_P\psi_C \ln[(\exp(\kappa h)+1)/(\exp(\kappa h)-1)] - (\psi_P{}^2+\psi_C{}^2)\ln[(\exp 2\kappa h)-1)/\exp(2\kappa h)]\}$$

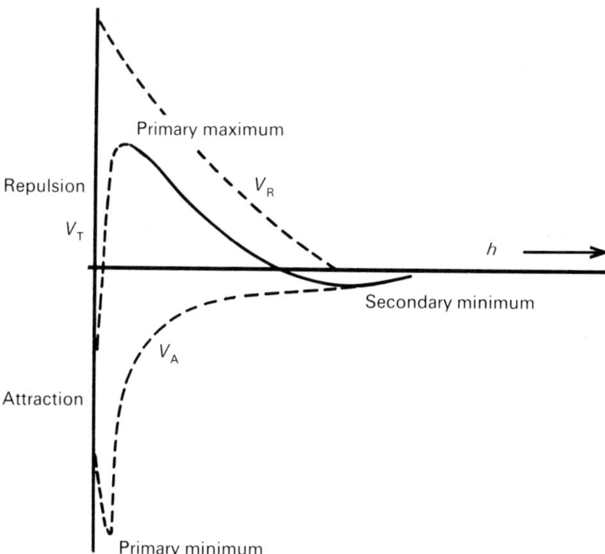

Fig. 2. Diagram to show the variation of the total interaction energy, V_T, with h, the separation distance between a sphere and a plate. The total interaction curve is obtained by the summation of an attraction curve V_A and a repulsion curve V_R.

A is the Hamaker constant, a the particle radius, h the separation between the two approaching surfaces, ϵ the permittivity of the suspending medium, κ is the reciprocal double-layer thickness (which is related to the ionic strength C of the suspension) and ψ_P and ψ_C are the potentials of the two surfaces.

Curves showing the variation of V_T with h (Fig. 2) can show two values of h at which a net attraction occurs. These are referred to as the primary minimum (h very small) and the secondary minimum (h = 5–10 nm). These are separated by a repulsive maximum. Bacteria captured by a surface in a weak secondary minimum are in equilibrium with the remaining bacteria suspended in the bulk phase. The number of captured cells depends upon the number concentration of bacteria in the suspension and the depth of the secondary minimum.

If the bacteria are small and their number concentration in dispersion is low, the depth of the secondary minimum may be insufficient to give significant deposition. However, if the bacteria are relatively large, of the order of 1 μm in diameter, and if the surface charges are low, it is reasonable to expect secondary-minimum interactions to be strong enough to cause attachment.

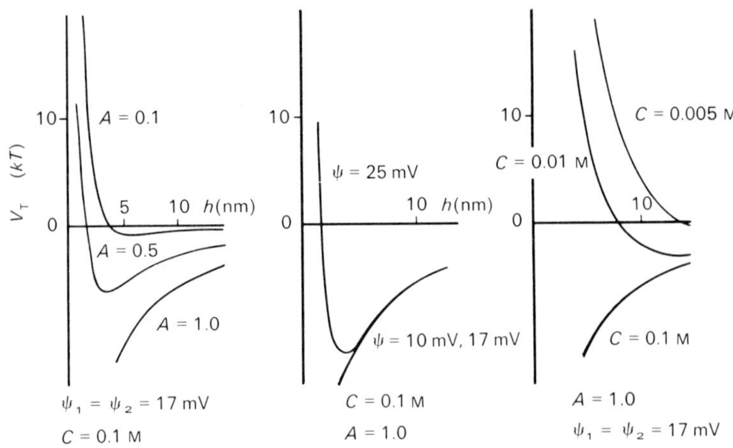

Fig. 3. Diagrams to show the variation of the total interaction energy V_T with h, the separation between a spherical particle (1 μm diameter) and a flat plate. A = Hamaker constant × 10^{13} erg, $\psi_1 = \psi_2$ = surface charge. C = ionic strength of the suspending medium.

Although the DLVO theory is a useful guide in the interpretation of adhesion, it should be used with some caution when considering bacterial systems. There are three main problems: (i) the establishment of the parameters required to calculate V_T, (ii) the depth of the secondary minimum is very sensitive to certain of these parameters, e.g. C, ψ and A, and (iii) at short separations interactions other than those described by the DLVO approach come into play.

The parameters which are required fall into two categories: geometric (a, h) and molecular (C, ϵ, A, ψ_C and ψ_P). In order to define a and h accurately, the geometry of the system must be well-defined, i.e. truly spherical particles and a planar homogenous solid substrate. The greatest problems are associated with assigning values to A, ψ_P and ψ_C. Although A may be calculated with some degree of confidence for non-aqueous homogenous phases, this is not the case for the aqueous heterogenous phases encountered in biological systems. For biological systems, therefore, A is reduced to being an 'adjustable parameter' in DLVO-type calculations. Similarly, the assignment of values to ψ_C and ψ_P involve assumptions. In the absence of more precise information it has been customary to use the zeta potentials (ζ) obtained from micro-electrophoresis experiments for the appropriate surface potentials. The depth of the secondary minimum depends very critically on these parameters. This is illustrated in Fig. 3.

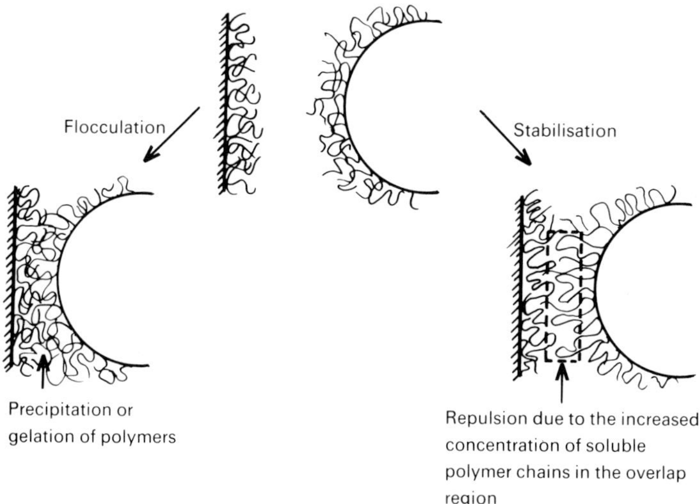

Fig. 4. Polymer-mediated interactions.

In addition to the DLVO interactions, at least two other types of interaction should be considered when discussing the adhesion of small particles. These involve adsorbed layers and surface wettability and relate to interactions at large finite separations and very short range separation respectively.

The presence of solvated layers of macromolecules on both the approaching particle and the collector surfaces can either lead to repulsion or attraction. If the polymers are well solvated the adsorbed layers cause steric repulsion because the approach of the particle and collector surfaces requires the removal of solvent from the polymer layers in the region between. In a good solvent this is clearly thermodynamically unfavourable and leads to an osmotic repulsion (Napper, 1970; Hesselink, Vrij & Overbeek, 1971; Vincent, 1974, Ash & Clayfield, 1976). Greig & Jones (1976) have considered the steric forces due to the overlapping of cell surface glycoproteins using an equation developed by Smitham, Evans & Napper (1975) and have found that the steric forces can be large compared with electrostatic forces.

Under certain conditions, on the other hand, an attractive force can occur, owing to the precipitation of the polymers in the intervening space between the particle and collector (Ash, 1979). Precipitation could be induced by a change of solvent, divalent cations such as calcium, gelation (Morris *et al.*, 1977) or coacervation between different polymers (Feigin & Napper, 1978) (see Fig. 4).

Attraction can also occur if the possibility exists of some polymer molecules having segments attached to both the particle and collector. This phenomenon is known as polymer bridging. The ability of polymers to bridge between approaching surfaces is strikingly demonstrated in the flocculation of inert colloids, where extremely small amounts of polymer can give rise to major effects. Although results obtained by electron microscopy must be treated with caution, Ries & Meyers (1968) have shown the formation of bridges between particles by a high molecular weight cationic polymer. The bridging theory has also received support from Slater & Kitchener (1966) who showed that polymeric flocculants were able to increase the mutual adhesiveness of particles, and by Fleer & Lyklema (1974) who used the concept of bridging to explain the rapid flocculation of polymer-coated silver iodide particles when brought into contact with similar particles that did not have an adsorbed polymer layer.

Polymer-mediated interactions can be of much greater range than van der Waals and electrostatic interactions and very small adsorbed amounts of high molecular weight polymers can induce attachment in deep minima at long range. Under certain circumstances this can occur under conditions of relatively high surface charge and low electrolyte concentration where DLVO theory cannot support the existence of a secondary minimum of sufficient depth.

Interactions at very small separation ($h \rightarrow 0$) can also lead to attraction or repulsion. DLVO theory is unable to predict accurately the depth of primary minimum interactions. The concept of surface wettability, however, has been used by several authors (Green & Halvorson, 1924; van Oss & Gillman, 1972; van Oss, Good & Newman, 1972; Marshall & Cruickshank, 1973) to describe biological interactions that occur at zero separations. This implies that any repulsive interactions between the contacting surfaces have been overcome. If a surface is very hydrophilic it may have stability conferred on it as a result of the organisation of water molecules in the vicinity of the surface. If the collector and bacteria are both hydrophilic one would expect this stabilisation to be large. It follows, however, that strongly hydrophobic contacting interfaces should experience very little solvent-mediated repulsion and on this basis should adhere more strongly (Lips & Jessup, 1979). These generalisations should be used with care, however, since hydrophobic particles can sometimes experience a net van der Waals repulsion under conditions where $A_{\text{collector}} < A_{\text{water}} < A_{\text{particle}}$. This occurs in the system PTFE (teflon)–water–graphite (Visser, 1972).

Another factor to be considered in the case of hydrophobic surfaces is that polymer bridging may be facilitated. Certain proteins, for example, appear to adsorb more readily on hydrophobic surfaces (MacRitchie, 1972; Brash, 1977) than on hydrophilic surfaces.

A consideration of the physical interactions between small particles suspended in fluids and collectors suggests that bacteria will eventually collide with potential collecting surfaces. The resulting interaction depends upon a number of parameters but, in general, particles will tend to adsorb irreversibly in order to reduce the free surface energy of the system unless they are physically prevented from doing so by repulsive interactions. These may be produced by charge repulsion or steric repulsion at separations greater than about 2 nm, or by modifications in water structure at very close separations (Marcelja & Radic, 1976). All these interactions will in turn be dependent upon the ionic strength of the environment, the solvation of the cell-surface polymers and the degree of organisation of solvent molecules on the approaching surfaces.

If an organism is to adhere selectively to a particular collector it must use one or other of these mechanisms to prevent its random deposition on other surfaces and then overcome the repulsion once the appropriate site is located. One way in which a bacterium might achieve this would be to coat itself with a stabilising polymer layer that was only capable of a specific gelation or precipitation reaction with a similar or co-operative polymer on the appropriate collector. This might lead to a long-range attachment which might be subject to reversal by degradation of the bridging polymers. On the other hand, the various parameters might be such that a gradual decrease in separation would occur, resulting in the eventual invasion of the host surface (Ninham & Richmond, 1973). Alternatively, certain bacteria may appear to colonise a surface specifically by attaching non-specifically to a wide variety of surfaces but only thriving on those surfaces to which they are specifically suited.

THE INTERACTION OF BACTERIA AND SURFACES

A number of studies have been carried out in order to quantify and describe the attachment of bacteria to a wide variety of collectors, which range from inert materials such as glass and plastic to complex living organisms and tissue. The influences of complex macromolecules such as immunoglobulins and lectins have also been investigated. In order to try to identify some of the factors important in bacterial adhesion the subject can be arbitrarily divided into three areas. These are (*a*) the interaction of bacteria with clean inert surfaces, (*b*) the influence of cell-surface structures and (*c*) the interactions between macromolecular layers.

The interactions of bacteria and clean inert surfaces

The most popular surface that has been studied as a collector for bacteria is glass. Most experiments involving the attachment of bacteria to glass involve placing a piece of clean glass in a suspension of organisms for a period of time, removing and rinsing the glass, and then counting the attached cells. Although this technique gives an idea of whether the organism will attach or not, it is very difficult to determine the collision efficiencies between different bacteria and glass. The wide variety of organisms taken from different natural environments also makes it difficult to assess the relative importance of the different mechanisms of attachment which may be operating.

Marshall *et al.* (1971), for example, measured the attachment of a marine pseudomonas onto glass and found that the cells attached reversibly. As the ionic strength of the suspending medium increased, the number of cells reversibly attached increased to a maximum of 3×10^3 cm^{-2} at about 0.08 M NaCl. The bacteria also exhibited an irreversible form of attachment which required the synthesis of an extracellular polymer. Meadows (1971) carried out an extensive study of the interaction of a number of motile Gram-negative bacteria to glass under various conditions. He concluded that: (*a*) bacteria readily attach and detach from glass slides and may remain attached from a few seconds to a number of hours and (*b*) bacteria either attach at one end and show slight Brownian movement, or along their length and remain quite immobile. No differences were obtained between fimbriate and non-fimbriate species. When the bacteria were suspended in 0.1 M phosphate buffer at pH 7, the number attached in 60 min was between 10^7 cm^{-2} and 10^9 cm^{-2}.

In another study using marine organisms, Corpe (1974) detected up to 3×10^5 organisms cm^{-2} attached to glass slides that had been immersed in sea water for 12 h. On the other hand, Hendricks (1974) found that certain heterotrophic and enterotrophic organisms obtained from river water would reach a maximum coverage of about 3×10^3 bacteria cm^{-2} after 2 h from a suspension with a bacterial concentration of 10^5 ml^{-1}.

Experiments carried out using oral streptococci show that they have a high affinity for glass. *Streptococcus faecium*, for example, (Orstavick, 1977), will attach to glass to a coverage of 10^6–10^7 bacteria cm^{-2}.

The rate of deposition is very difficult to ascertain from these experiments, since deposition appears to stop before the surface is fully covered. Rutter & Abbott (1978), however, showed that two oral streptococci would deposit irreversibly onto glass at the rate of 2×10^4 bacteria cm^{-2} min^{-1} from a suspension with a cell concentration of 4×10^7 ml^{-1}.

The experiments involving glass suggest that both reversible and irreversible attachment are possible, depending upon the type of bacteria or the conditions of the experiment. Marshall *et al.* (1971) and Rutter & Abbott (1978) both showed an increase in deposition with increasing ionic strength. In one case, however, the attachment was reversible whereas in the other it was not. Marshall *et al.* (1971) proposed that organisms are held close to the glass in a weak secondary minimum which increases in depth as the electrolyte concentration increases. The attractive interaction permits rotation and the occasional break-away of organisms, and also their removal under the shear applied during rinsing. Provided that the organisms are supplied with the correct nutrients, the second stage of attachment then occurs, requiring the synthesis of extracellular polymers able to cross the distance separating the cells from the glass to form anchoring bridges. This mechanism might be usefully extended to cover organisms that already have an extracellular polymer layer and are able to bridge directly to the glass surface. These organisms would not exhibit a reversible phase.

In addition to glass, a number of other substrates have been used in bacterial adhesion studies. Fletcher (1977), for example, has investigated the attachment of a marine organism to polystyrene Petri dishes. In a similar study to that of Orstavick (1977) it was shown that approximately 4×10^7 bacteria cm^{-2} would attach irreversibly to polystyrene in a period of 2 h. This represented a maximum coverage of the surface by bacteria; longer contact with the suspension resulted in very few more attached cells. Fletcher & Loeb (1976) also carried out a number of experiments using substrates with different degrees of hydrophobicity which showed that more bacteria attached to hydrophobic surfaces than to surfaces that were hydrophilic.

The influence of cell-surface and collector-surface hydrophobicity in bacterial accumulation and aggregation has been recognised since

1924 (Green & Halvorson, 1924; Mellon, Hastings & Anastasia, 1924). Mudd & Mudd (1924), for example, used the concept of bacterial surface hydrophobicity to explain why non-acid-fast bacteria were stable at an oil/water interface whereas acid-fast bacteria readily passed into the oil phase. More recently, Baird, Albertsson & Hofsten (1961) showed that the surfaces of different strains of the same type of organism are sufficiently different for the strains to exhibit different partition functions in two-phase polymer systems. Van Oss & Gillman (1972) have also demonstrated that bacteria which have water contact angles greater than that for white blood cells are more readily phagocytosed than bacteria with contact angles less than that of the white blood cells.

Marshall & Cruickshank (1973) have also used the concept of cell-surface hydrophobicity to explain the orientation of certain organisms at air/water, oil/water and solid/water interfaces. Their experiments show a considerable increase in the adsorption of *Hyphomicrobium vulgare* ZV580 to hydrophobic glass and a decrease to glass treated with a non-ionic surfactant (Tween 80). Their results, coupled with transmission electron micrographs showing the separation between the cell wall and the substrate surface and the adsorption of a negatively-charged AgI sol (which suggested that the surface was evenly charged), led them to suggest that one end of the bacterium consisted of an extracellular hydrophobic polymer which preferentially adsorbed at an interface between a hydrophobic and aqueous medium. Norkrans & Sorensson (1976) have also shown that marine organisms will readily accumulate on or in thin lipid microlayers which accumulate at air/water interfaces prepared in experiments *in vitro*.

A considerable amount of research has been carried out concerning the adhesion of oral organisms to various surfaces. In spite of the apparent selectivity with which certain oral bacteria colonise the teeth, Clark, Banman & Gibbons (1978) found that all the oral organisms they tested would adhere to beads of hydroxyapatite (the principal mineral in tooth enamel) confirming earlier work by Orstavick, Kraus & Henshaw (1974). Olsson & Krasse (1976) also found that both *Streptococcus salivarious*, commonly found on the cheeks and tongue, and *Streptococcus sanguis*, usually found on the teeth, would both adhere to glass and tooth surfaces, although *S. anguis* appeared to be both preferentially adsorbed and more resistant to washing. A similar finding was reported by Hillman, van Houte

& Gibbons (1970). Olsson, Glantz & Krasse (1976) followed this study by relating the electrophoretic mobility of oral bacteria with their tendency to adsorb. They interpreted their observations by suggesting that when the electrostatic repulsion forces between the cell and collector surface had been overcome, species- or type-specific properties of the cell surface are of determinant importance for adherence. The physiological condition of the bacterium has also been implicated as a determinant for adherence. Hattori, Hattori & Furusaka (1972) showed that not only would bacteria readily adsorb to anion-exchange resin, but they appear to be able to desorb as they begin to grow actively.

The experiments outlined above enable a number of broad generalisations to be made concerning the adhesion of bacteria to solid/liquid interfaces: bacteria appear to be able to attach to a wide variety of surfaces; in general, these surfaces carry a net negative charge; the tendency for bacteria to adsorb appears to be proportional to their bulk concentration, the contact time, ionic strength, temperature and collector hydrophobicity; bacteria also readily adsorb to positively-charged collectors, such as the edge of clay particles (Marshall, 1969) and in electrostatic filters (Gvozdyak, Chekhovskaya, Grebenyuk & Koshechkina, 1974). These properties suggest that bacteria exhibit the normal properties of colloidal particles and that their interaction with surfaces might be adequately explained by DLVO theory in the same way as, for example, the deposition of polystyrene latex particles to cellophane (Visser, 1976) or polystyrene (Clint *et al.*, 1973).

An important aspect of bacterial adhesion is the fact that bacteria often attach reversibly to surfaces. This suggests that strong primary-minimum interactions are not involved, at least between the dense regions of the cell wall and the collector surface. Secondary-minimum or polymer-bridging interactions may both give rise to reversible attachment depending upon the depth of the minimum or the number and strength of the polymer bridges (see Fig. 5). In the case of secondary-minimum interactions the detachment of a bacterium from a collector would require an external removal force (for example, rinsing), a change in the surface charge densities, or a decrease in the ionic strength of the environment. In the case of polymer bridging, a similar removal force or a degradation of the polymer bridges could result in detachment. A combination of secondary-minimum capture followed by polymer bridging may of course be involved. This in turn

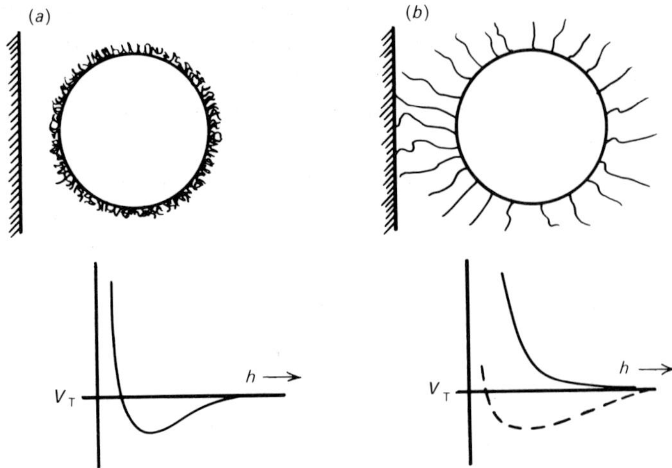

Fig. 5. Polymer bridging. (*a*) Secondary-minimum capture; (*b*) polymer bridging in the absence of a secondary minimum.

may be followed by the synthesis of more bridging polymers to consolidate the cell on the collector. The evidence suggests that all four mechanisms – primary-minimum attachment, secondary-minimum attachment, polymer bridging and polymer consolidation – can occur under certain circumstances. In natural environments, however, it appears that long-range interactions mediated by bridging polymers or other specialised structures may be favoured by the bacteria.

The influence of bacterial surface polymers and structures on bacterial attachment

It has already been suggested that the adsorption of extracellular polymers might cause the attachment of bacteria to surfaces via a polymer-bridging mechanism. Fletcher & Floodgate (1973) and Marshall (1973) have used transmission electron microscopy to demonstrate the presence of extracellular polymers apparently attaching bacteria to surfaces. Corpe (1974) also describes the attachment of marine organisms via extracellular polymers. In an extensive study of the adhesion of *Streptococcus mutans*, Mukase & Slade (1973) suggested that it was the adsorption of an extracellular glucan to the collector surface which caused the bacteria to attach. Similarly

Nalbandian, Freedman, Tanzer & Lovelace (1974) showed that the adhesion of *S. mutans in vitro* was causally and functionally related to the extracellular fibrillar glucan component present on the cell surface. Mutants that do not exhibit this glucan show a dramatic decrease in attachment.

Cell-surface polymers, therefore, appear to play an important part in bacterial attachment. However, certain organisms possess specialised fibres extending from the cell surface which may also be able to 'bridge' the bacterium to a collector surface.

Holdfasts, flagella and pili. A number of appendages to the bacterial cell surface have been identified. The fact that some bacteria seemed to possess tails or flagella was noted in the early 19th century (Hodgekiss, Short & Walker, 1976). These can now be seen with the electron microscope as sinuous protein filaments usually 12–20 nm in diameter and about 6–8 μm in length. Flagella have been implicated in the adhesion of organisms to glass slides by Meadows (1971).

Another major group of surface appendages are termed fimbriae or pili. These are filamentous surface appendages which are quite distinct from flagella. Pili are again composed of protein and can be divided into six groups. The pili that form group 1 vary in length between 0.2 μm and 20 μm and are between 3 nm and 4 nm in diameter. A bacterium can possess between 100 and 300 group 1 pili arranged peritrichiously. Group 2 pili are referred to as the sex pili and groups 3, 4, 5 and 6 can all be differentiated morphologically.

Certain bacteria called Caulobacter possess stalks (Poindexter, 1964). These organisms occur widely in aqueous environments and were found initially attached to microscope slides which had been immersed in a fresh-water lake. It appears that a polysaccharide material is secreted from the distal end of the stalk and facilitates attachment of the bacterium to a wide variety of substrates including glass, collodion, cotton, other bacterial cells and other cells of the same species. The organisms are often found in rosettes containing between 2 and 200 cells, all adhering to a common central point by the distal ends of their stalks. However, the stalks are separated by a common mass of holdfast material 0.3 μm to 0.5 μm in diameter which lies at the centre of the rosette.

An intriguing surface structure is exhibited by certain marine gliding microbes (Ridgeway & Levin, 1973). These have the appearance of minute wine-glasses arranged over the cell surface. It is suggested that these wine-glass subunits may specifically secrete an extracellular slime (probably polysaccharide) which initiates surface attachment.

Whereas the function of stalks and wine-glass structures appears to be to secrete polysaccharide which is then able to bridge the structure to the collector surface as described before, the involvement of pili and flagella in adhesion is by no means clear. These fibres might facilitate adhesion by providing an extension of the cell with a small tip radius which will therefore only suffer a small electrostatic repulsion as it approaches the collector surface (Pethica, 1961). However, unless several pili attach at the same time the adhesion might be very weak. If the adhesion were weak and the repulsive maximum high, the influence of an additional weak secondary-minimum interaction might enable the bacterium to 'walk' about on the collector surface by increasing the number of contacts in advance of the cell and reducing the number behind. De Boer, Golten & Scheffers (1975) suggested that the swarming vibrio *Alginolyticus* possesses a single polar flagellum when living fully in water, but produces a number of undulate flagella 15 nm in diameter arranged peritrichiously when living on a solid substrate. It is suggested that the bacterium would be held by adhesive forces operating at the multiple sites of contact between the substrate and the undulate flagella in such a way as to allow motility and swarming.

The natural function of pili is unclear, although they seem to be more involved in attachment than locomotion (Ottow, 1975). *Sulpholobus*, for example, attaches to sulphur crystals in hot springs (70–75 °C) by means of pili which separate the bacteria from the crystals and permit lateral movement. The same organism, however, will attach irreversibly to glass via its cell wall (Weiss, 1973). There appears to be a clear relationship between the presence of pili and the ability of organisms to attach to red blood cells and cause haemagglutination. The presence of pili does not, however, seem to be closely related to pathogenicity (Ottow, 1975).

It is possible that pili provide a controlled method of attachment whereby the cell utilises the pili to provide bridges to the collector surface. The cell itself is probably normally prevented from strong primary-minimum attachment by the presence of an electrostatic repulsion maximum and possibly by the rigidity of the pili themselves. The strength of attachment could then be controlled by the number of pili in contact with the surface, and in some cases may be increased by the presence of a secondary minimum. Thus, when migration over the collector surface is important, attachment via pili may be the desired mechanism. The interaction between specific macromolecules on the tips or length of the pili and molecules present on the collector surface might provide stronger and more specific attachment. This requires the close-range interaction of macromolecules and is discussed in the next section.

The interaction between macromolecular layers

Transmission electron micrographs of sections through bacterial cells show that bacteria consist essentially of a spherical or tubular envelope enclosing a complex organisation of membranes and particles which make up the contents of the organism. The envelope forms the cell wall, and is an organised layered structure of varying complexity depending upon the type of bacterium. It is the outer layers of the cell wall and any overlaying 'extracellular' polymeric materials that are of major importance in determining the mechanisms by which bacteria adhere. Until relatively recently the cell wall was regarded as a permanent rigid structure; however, the application of continuous-culture techniques to the study of cell wall carbohydrates has demonstrated that cell wall turnover is possible (Ellwood & Tempest, 1972). This has led to the concept of the bacterial cell wall as a dynamic entity with certain cell wall polymers altering or being replaced during growth (Fiedler & Glaser, 1973).

Many genera of bacteria include species capable of producing polysaccharides outside the cell wall. These exopolysaccharides may either take the form of a discrete capsule or an extracellular slime layer which is apparently not covalently bound to the bacterial surface. The exact form of the exopolysaccharide layer probably depends both on the species and on the physiological state of the organisms. An excellent review of the nature and production of bacterial exopolysaccharides has been written by Sutherland (1977). The thickness of the capsule layer may extend 0.1–10 μm beyond the outer layers of the cell wall, although the precise thickness is difficult to assess because of artefacts associated with the preparation of samples for electron microscopy. Roth (1977) has studied the physical structure of bacterial surface carbohydrates and concluded that the physical structure of bacterial slime in various systems appears to be that of a network of interconnected fibres. The slime mat appears to adhere to the surface to which it attaches by the inherent 'stickiness' of the slime. These observations suggest that the slime polymers have a low affinity for the aqueous solvent and precipitate readily on any available surface. The physical structure of the capsule also appears to consist of a collection of fibres. The images of capsules do not, however, give any evidence for interlinking between fibres, except that there is some evidence for the peripheral linking of fibres. This peripheral linking of fibres would fit in well with the idea of a well-defined capsular edge.

It may be suggested therefore, that at any point in time a bacterium should be regarded as a rigid, mainly polysaccharide, envelope (the cell wall) about 1 μm in diameter (if spherical) surrounded in some cases by a radial fibrous polysaccharide layer with a well-defined edge. In addition to this, a slime network of interlinked polysaccharide fibres may also be present.

Bacteria seldom have the opportunity of attaching to clean surfaces. In the majority of environments, potential collectors carry a surface layer of either adsorbed or integrated macromolecules. When, for example, solids are placed in the sea, rivers or the oral cavity they rapidly adsorb macromolecules to form a conditioning layer or film. The thickness of the film appears to depend upon the environment. Ellipsometry carried out by Loeb & Neihof (1975) on marine conditioning films showed a maximum thickness of about 50 nm on a surface immersed in the sea for 6 h. The glycoprotein layer, known as the enamel pellicle, which forms on teeth from saliva, however, can occasionally attain a thickness of 10 μm (Meckel, 1965). The surfaces of eukaryotic cells to which bacteria often attach also possess mucous films or well-organised layers of macromolecules extending into the aqueous environment.

These studies suggest that bacteria approaching surfaces in aqueous environments will first contact a macromolecular layer associated with the collector. Any discussion of bacterial adhesion, therefore, must take into account the potential for interaction between bacterial exopolymers and substrate polymers. The net result of such interactions can either lead to repulsion or adhesion, as discussed on pp. 111–13.

Although little work has been carried out to study the interactions between adsorbed layers of biological macromolecules, it is well known that proteins, lipids and polysaccharides will interact with each other in solution or in air/water films. A number of studies have shown that proteins will readily penetrate lipid films either specifically or non-specifically (Collaccio, 1969). Proteins will also penetrate adsorbed layers of other macromolecules to form mixed monolayers (Fromageot, Groves, Sears & Brown, 1976). The adsorption of dextran to red blood cell surfaces is also a dynamic reversible process in which the adsorbed molecules exchange readily with macromolecules in the bulk or those attached to another cell (Chien, Simchen, Abbott & Jan, 1977).

It is well known that a number of large polymers undergo, in addition to adsorption, immiscible-phase separation in binary mixtures with one polymer predominating in each phase (Dobry & Boyer-Kawenoki, 1947). In principle this is caused by the weak non-specific interactions between the polymer chains. These are generally stronger between like molecules than unlike molecules, so large polymers tend to separate into separate phases (Flory, 1953). According to these considerations, solutions of proteoglycans and glycoproteins should show phase separation (Ogston, 1970). A form of phase separation produces the repulsion caused when two adsorbed polymer layers are brought into proximity. This is a common method of stabilising colloidal suspensions and has been discussed on pp. 111–13.

The reluctance with which macromolecules form multilayers, coupled with the general phenomenon of phase separation, might suggest that bacterial attachment via polymer–polymer interactions would seldom occur. This evidently is not so, since a number of specific interactions have been described. For example, a number of hydrophilic polymers are able to adsorb to bacterial surfaces producing flocculation (Hodge & Metcalf, 1958) presumably by a bridging mechanism. Harris & Mitchell (1973) have reviewed this area extensively. Cationic polymers will also adsorb to bacterial surfaces, producing flocculation probably by producing charge mosaics on the cell surfaces rather than by bridging (Treweek & Morgan, 1977). It has also been shown that bacteria are effectively flocculated by proteins at their isoelectric point. Adsorbed proteins, however, will also substantially reduce the numbers of bacteria that will attach to polystyrene (Fletcher, 1977) and glass (Meadows, 1971).

The above examples probably describe non-specific interactions between macromolecules and bacterial surfaces, leading either to adsorption or to exclusion in the case of the protein layers. However, certain specific macromolecules have been isolated which will flocculate only certain types of bacteria (Gibbons & van Houte, 1973; Levine *et al.*, 1978). In these cases, whilst the macromolecule may have a weak affinity for the bacterial surface because it represents a solid/liquid interface, it is able to bind strongly to molecular species resident in the bacterial cell surface layers that have a suitable spatial geometry or sequence of molecular groups to enable the macromolecule to interact specifically. Such an interaction might be very strong and involve covalent bonding, or somewhat weaker if it arises from hydrogen or hydrophobic interactions. Macromolecules capable of adsorbing to specific types of bacteria could either (i) cause flocculation, (ii) cause bridging to specific surfaces, or (iii) change the surface characteristics of the bacterial surface so as to facilitate a subsequent process such as phagocytosis.

A number of macromolecules with specific affinities for different types of bacteria have been isolated and termed lectins. Lectins are proteins and glycoproteins that possess, principally, the ability to agglutinate red blood cells. They occur widely in nature and have been isolated from a number of plants, certain invertebrates and lower vertebrates such as fish (Sharon & Lis, 1972). The usefulness of lectins lies in the specificity with which they will interact with polysaccharides and glycoproteins. In many cases the interactions of lectins with cells can be inhibited by simple sugars; this has led to the conclusion that lectins bind specifically to the saccharides on the cell surface. The role of lectins in nature is unclear, although it has been suggested that they act as plant antibodies intended to counteract soil bacteria (Sharon & Lis, 1972; Calvert, Lalonde, Bhuvaneswari & Bauer, 1978).

Plant lectins have been shown to flocculate oral organisms (Hamade, Gill & Slade, 1977). However, glycoproteins found in human saliva also have the ability to flocculate certain oral organisms (McBride & Gisslow, 1977). This observation has formed the basis of an extensive literature concerned with the role of salivary aggregation-inducing substances (AIS) in dental plaque formation. The three predominant streptococci in dental plaque, *Streptococcus sanguis*, *Streptococcus mitis* (Kashket & Donaldson, 1972) and *Streptococcus mutans* (Magnusson & Ericsson, 1976), are all aggregated by salivary glycoproteins.

It is also known that the saliva factors inducing aggregation of some strains of *S. mutans*, *S. sanguis* and *S. mitis* will adsorb to hydroxyapatite (Ericsson & Magnusson, 1976). The presence of a high concentration of AIS in saliva, however, was related to a low rate of formation of plaque, the soft sticky bacterial deposit found on teeth (Magnusson, Ericsson & Pruit, 1976). These observations have led to the hypothesis that at low concentrations the AIS may bridge organisms to the enamel surface and also to each other, but at high concentrations both the enamel and the bacterial surfaces are saturated and the amount of bridging is reduced (Magnusson & Ericsson, 1976). Alternatively, the flocculation of organisms in saliva might facilitate their rapid clearance from the mouth under natural shear forces.

The role of salivary secretions, and in particular the adsorbed salivary pellicle which invariably coats the teeth, have been subjected to considerable investigation. Several authors have shown that a number of oral organisms will readily attach to enamel (Orstavick *et al.*, 1974) and hydroxyapatite (Clark *et al.*, 1978). The effect of treating these surfaces with saliva or of suspending the organisms in saliva during the experiment is, however, difficult to determine, owing to the variety of methods used to quantify adherence. Generally speaking, the presence of a saliva film on enamel or hydroxyapatite tends to reduce the tendency of oral organisms to attach, with the exception of *S. mitis*, *Actinomyces viscosus* and possibly *S. sanguis*. In the case of *S. sanguis*, the attachment can readily be blocked by fatty acids and serum proteins (Liljemark, Schaver & Bloomquist, 1978). A number of studies by Orstavick (1978) have led him to suggest that the presence of an adsorbed layer or pellicle on the surface of the teeth may generally render the tooth surface less attractive to organisms. However, if aggregation-inducing substances are included in the pellicle with the correct conformation, the attachment of organisms capable of interacting with the substance will be greatly facilitated.

The specific attachment of certain bacteria to eukaryotic cells, and the involvement of antigens in adhesion, may also be examples of specific macromolecular recognition being written into a generally stabilising layer. The attachment of bacteria to epithelial cells is often characterised by the presence around the bacteria of a 'fuzzy' layer which appears to be in contact with the epithelial cell surface (Barnett, 1973; Brady, Gray & Lava-Garcia, 1975; Brooker & Fuller, 1976). This would suggest that the macromolecular associations which anchor the bacteria to the epithelial cell surface are situated on the outside of the 'fuzzy' layer. Therefore, the material of the fuzzy layer, and not the molecules which link them, provides the

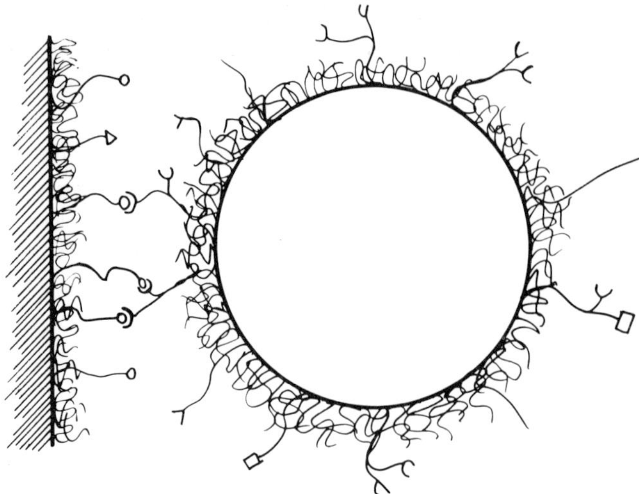

Fig. 6. Specific polymer interactions might cause the bridging of otherwise sterically stabilised bacteria.

bridge. For example, if antibodies to *Escherichia coli* and *Salmonella typhimurium* are labelled with ferritin and brought into contact with the appropriate bacteria, the ferritin particles are seen to be arranged round the outside of the bacterial capsule, sometimes as much as 150 nm from the cell membrane (Shands, 1966). The determination and spatial arrangements of antigenic material in bacterial cell walls has been reviewed by Rogers (1979).

The influence of specific polymer interactions on bacterial attachment may be summarised by considering the approach of two macromolecular layers through an at least partially solvating medium. If the two layers are well solvated, exclusion will generally result and the organism will be unable to approach the collector close enough to utilise the van der Waals forces of attraction. If, however, the macromolecules are able to interact specifically, the organisms will be bridged to the collector (see Fig. 6). Estimates of the strengths of such polymer bridging interactions have been made by Bell (1978). The advantage of this mechanism is that the cell will only strongly attach to macromolecular layers with the correct molecular make-up. The same organism, however, may readily adsorb to surfaces that do not possess adsorbed layers by an entirely non-specific bridging adsorption or van der Waals attraction. *S. salivarius*, for example, shows a reluctance to attach to teeth or saliva-coated hydroxyapatite

(Gibbons & van Houte, 1973; Clark, Banman & Gibbons, 1978). It will readily adsorb to glass and polystyrene, however (Rutter & Abbott, 1978).

Bacteria may also interact with adsorbed layers by means of gelation or incipient flocculation mechanisms. These involve the combination of the surface polymers in such a way as to produce gels or precipitates between the approaching surfaces. The gelation of alginates and pectins in the presence of divalent cations, for example, is well known. Similarly some polysaccharides that are incapable of gelation (at low concentration) can be made to gel by adding a different type of polysaccharide (Dea, McKinnon & Rees, 1972). Xanthan, for example, forms gels in the presence of galactomannans although it does not gel on its own. This interaction may be important in facilitating the attachment of *Xanthamonas campestris* to plants such as cabbages, beans and cotton (Morris, Rees, Thom & Welsh, 1977). Precipitation may be utilised by organisms such as *Streptococcus mutans* which produce copious amounts of insoluble polysaccharide in the presence of sucrose. This mediates its attachment to tooth surfaces (Gibbons & van Houte, 1973). In the absence of sucrose or in the case of mutants that do not produce insoluble glucans (Johnson, Boczola, Schelmeister & Shklair, 1977), attachment is not increased.

It is possible that the frequently observed dependence of bacterial attachment on the presence of calcium is due to the reduction in solubility of cell-surface polymers brought about by calcium neutralisation, rather than the popular concept of calcium bridging often used to explain bacterial attachment (Rolla, 1976).

CONCLUSION

The fact that bacteria are found attached to a wide range of surfaces under conditions as varied as hot sulphur springs, the human gut and the surface films of the sea suggests that bacteria are particularly adhesive particles. Their small size enables them to escape the majority of shear forces that are encountered naturally and their surfaces appear to have evolved in such a way as to enable a certain amount of control over attachment to occur.

Bacteria are negatively charged, in common with the majority of surfaces to which they attach. The range and magnitudes of the electrostatic repulsion exerted by these surface charges as the bacterium approaches a collector surface will depend upon the ionic strength of the medium, the pH and the presence of any potentially adsorbing ions. Under most conditions, however, the repulsion will probably not reduce to zero. The strength of the van der Waals attraction between bacteria and collector surfaces is very difficult to estimate. The values of the Hamaker constants that have been used are at best only estimates, and, since the resultant attractive or repulsive interaction at close range is highly dependent on the value of A, calculations are probably not very helpful at present. It is reasonable to suggest, however, that some residual repulsion does occur between bacteria and collecting surfaces since bacterial cells can usually be removed from surfaces either by sonication or shear without disrupting the cell membrane. It is tempting to suggest in the light of these considerations that a general mechanism for bacterial attachment might be described as follows:

The bacterial cell approaches the collector and is held in a weak DLVO secondary minimum for a short time under the influence of van der Waals forces. Its close approach is prevented by the primary maximum. During this short time-period, cell-surface polymers are either able to adsorb non-specifically to the surface or interact specifically with complementary polymers adsorbed to or forming part of the collector surface. This increases the probability of attachment until enough links or bridges have been formed to attach the cell irreversibly. The presence of a secondary minimum is not an absolute requirement for polymer bridging but its presence would facilitate bridge formation.

Further consolidation might be achieved by the synthesis of insoluble macromolecules that are extruded into the intervening space, or by gelation of the existing polymers, or even precipitation of the polymers by neutralisation with divalent cations.

This general picture is, of course, subject to considerable modification bearing in mind the possible variations in bacterial surface structure and the collector surface. Presumably, the mechanisms that operate in bacterial adhesion may also act in the adhesion of eukaryotes.

I would like to thank Dr B. Vincent for his helpful suggestions during the preparation of this paper.

REFERENCES

Ash, S. G. (1979). Adhesion of micro-organisms in fermentation processes. In *Adhesion of Microorganisms to Surfaces*, ed. D. C. Ellwood, J. Melling & P. R. Rutter, in press. London, New York: Academic Press.

Ash, S. G. & Clayfield, E. J. (1976). Effect of polymers on the stability of colloids: flocculation of polystyrene lattices by polyethers. *Journal of Colloid and Interface Science*, **55**(3), 645–57.

Baird, D., Albertsson, P. A. & Hofsten, B. (1961). Separation of bacteria by counter-current distribution. *Nature, London*, **192**, 236–9.

Barnett, M. L. (1973). Adherence of bacteria to oral epithelium; *in vivo* electron microscope observations. *Journal of Dental Research*, **52**(NS), 1160.

Bell, G. I. (1978). Models for the specific adhesion of cells to cells. *Science, New York*, **200**, 618–27.

Brady, J. M., Gray, W. A. & Lara-Garcia, W. (1975). Localisation of bacteria on the rat tongue with scanning and transmission electron microscopy. *Journal of Dental Research*, **54**(4), 777–82.

Brash, J. L. (1977). Hydrophobic polymer surfaces and their interactions with blood. In *The Behaviour of Blood and its Components at Interfaces*, ed. Leo Vroman & Edward F. Leonard, pp. 356–71. *Annals of the New York Academy of Sciences*, **283**, 1–560.

Brooker, B. E. & Fuller, R. (1976). Demonstration of a carbohydrate layer involved in the attachment of lactobacilli to the chicken crop epithelium. In *Microbial Ultrastructure*, ed. R. Fuller & D. W. Lovelock, Society of Applied Bacteriology, Technical Series 10, pp. 87–100. London, New York: Academic Press.

Calvert, H. E., Lalonde, M., Bhuvaneswari, T. V. & Bauer, W. D. (1978). The role of lectins in plant micro-organism interactions. 4. Ultrastructural localisation of soybean lectin binding sites on *Rhizobium japonicum*. *Canadian Journal of Microbiology*, **24**(7), 785–93.

Chamberlain, A. C. (1967). Deposition of particles to natural surfaces. In *Airborne Microbes*, ed. P. H. Gregory & J. L. Monteith, 17th Symposium of the Society for General Microbiology, pp. 138–64. Cambridge University Press.

Chien, S., Simchon, S., Abbott, R. E. & Jan, K. M. (1977). Surface adsorption of dextrans on human red cell membranes. *Journal of Colloid and Interface Science*, **62**(3), 461–70.

Clark, W. B., Banmann, L. L. & Gibbons, R. J. (1978). Comparative estimates of bacterial affinities and adsorption sites on hydroxyapatite surfaces. *Infection and Immunity*, **19**(3), 846–53.

Clint, G. E., Clint, J. H., Corkill, J. M. & Walker, T. (1973). Deposition of latex particles on to a planar surface. *Journal of Colloid and Interface Science, New York*, **44**(1), 121–32.

Colaccio, G. (1969). Applications of monolayer techniques to biological systems: symptoms of specific lipid-protein interactions. *Journal of Colloid and Interface Science*, **29**, 345–64.

Corpe, W. A. (1974). Microfouling: The role of primary film forming bacteria.

Proceedings of the 3rd International Congress on Marine Corrosion and Fouling, ed. R. F. Acker, B. F. Brown, J. R. de Palma & W. P. Inverson. pp. 598–609. Evanston Illinois: Northwestern University Press.

Dahneke, B. (1975). Kinetic theory of the escape of particles from surfaces. *Journal of Colloid and Interface Science*, **50**, 89–107.

Dea, I. C. M., McKinnon, A. A. & Rees, D. A. (1972). Tertiary and quaternary structure in aqueous polysaccharide systems which model cell wall cohesion. *Journal of Molecular Biology*, **68**, 153–72.

De Boer, W. E., Galten, C. & Scheffers, W. A. (1975). Effects of some physical factors on flagellation and swarming of *Vibrio alginolyticus*. *Netherlands Journal of Sea Research*, **9**(2), 197–213.

Derjaguin. B. V. & Landau, L. (1941). Theory of the stability of strongly charged lypophobic sols and the adhesion of strongly charged particles in solutions of electrolytes. *Acta Physicochemica USSR*, **14**, 633–62.

Dobry, A. & Boyer-Kawenoki, F. (1947). Phase separation in polymer solutions. *Journal of Polymer Science*, **2**(1), 90–100.

Ellwood, D. C. & Tempest, D. W. (1972). Effects of environment on bacterial wall content and composition. *Advances in Microbial Physiology*, **7**, 83–117.

Ericsson, T. & Magnusson, I. (1976). Affinity for hydroxyapatite of salivary substances inducing aggregation of oral streptococci. *Caries Research*, **10**(1), 8–18.

Feigin, R. I. & Napper, D. H. (1978). Heterosteric stabilisation and selective flocculation. *Journal of Colloid and Interface Science*, **67**(1), 127–39.

Fiedler, F. & Glaser, L. (1973). Assembly of bacterial cell walls. *Biochimica et Biophysica Acta*, **300**, 467–85.

Fleer, G. J. & Lyklema, J. (1974). Polymer adsorption and its effect on the stability of hydrophobic colloids. *Journal of Colloid and Interface Science*, **46**(1), 1–12.

Fletcher, M. (1977). The effects of culture concentration and age, time and temperature on bacterial attachment to polystyrene. *Canadian Journal of Microbiology*, **23**(1), 1–6.

Fletcher, M. & Floodgate, G. D. (1973). An electron microscope demonstration of an acidic polysaccharide involved in the adhesion of a marine bacterium to solid surfaces. *Journal of General Microbiology*, **74**, 325–34.

Fletcher, M. & Loeb, G. I. (1976). The influence of substratum surface properties on the attachment of a marine bacterium. In *Colloid and Interface Science*, vol. 3, ed. M. Kerker, pp. 459–69. London, New York: Academic Press.

Flory, P. J. (1953). *Principles of Polymer Chemistry*. Ithaca, New York: Cornell University Press.

Fromageot, H. P. M., Groves, J. N., Sears, A. R. & Brown, J. F. (1976). The interaction of macromolecular solutions with macromolecular monolayers adsorbed on a hydrophobic surface. *Journal of Biomedical Material Research*, **10**, 455–69.

Gibbons, R. J. & van Houte, J. (1973). Formation of dental plaques. *Journal of Periodontology*, **44**(6), 347–60.

Green, R. G. & Halvorson, H. O. (1924). Surface energy as the controlling factor in agglutination and dispersion. *Journal of Infectious Diseases*, **35**, 5–13.

Greig, R. G. & Jones, M. N. (1976). The possible role of steric forces in cellular cohesion. *Journal of Theoretical Biology*, **63**, 405–19.

Gvozdyak, P. I., Chekhovskaya, T. P., Grebenyuk, V. D. & Koshechkina, L. P. (1974). Retention of microorganisms on granular materials in an electric field. *Doklady Academii Nauk SSSR*, **214**(2), 454–5.

Hamade, S., Gill, K. & Slade, H. D. (1977). Binding of Lectins to *Streptococcus mutans* cells and type-specific polysaccharides, and effect on adherence. *Infection and Immunity*, **18**(3), 708–16.

Harris, R. H. & Mitchell, R. (1973). The role of polymers in microbial aggregation. *Annual Review of Microbiology*, **27**, 27–50.

Hattori, R., Hattori, T. & Furusaka, C. (1972). Growth of bacteria on the surface of anion exchange resins. *Journal of General and Applied Microbiology, Tokyo*, **18**(4), 271–84.

Hendricks, C. W. (1974). Sorption of heterotrophic and enteric bacteria to glass surfaces in the continuous culture of river water. *Applied Microbiology*, **28**, 572–8.

Hesselink, F. Th., Vrij, A. L. & Overbeek, J. Th. G. (1971). On the theory of stablisation of dispersions by adsorbed macromolecules. II. Interaction between two flat particles. *Journal of Physical Chemistry*, **75**(14), 2094–2103.

Hill, T. L. (1960). *Introduction to Statistical Mechanics*. London, New York: Addison-Wesley.

Hillman, J. D., van Houte, J. & Gibbons, R. J. (1970). Sorption of bacteria to human enamel powder. *Archives of Oral Biology*, **15**, 899–903.

Hodge, H. M. & Metcalfe, S. N. (1958). Flocculation of bacteria by hydrophilic colloids. *Journal of Bacteriology*, **75**, 258–64.

Hodgkiss, W., Short, J. A. & Walker, P. D. (1976). Bacterial surface structures. In *Microbial Ultrastructure*, ed. R. Fuller & D. W. Lovelock, Society of Applied Bacteriology, Technical Series 10, pp. 49–72. London, New York: Academic Press.

Hull, M. & Kitchener, J. A. Interaction of spherical colloidal particles with planar surfaces. *Transactions of the Faraday Society*, **65**, 3093–3104.

Johnson, M. C., Boczola, J. T., Schelmeister, I. L. & Shklair, I. L. (1977). Biochemical study of the relationship of extracellular glucan to adherence and cariogenicity in *Streptococcus mutans* and an extra-cellular polysaccharide mutant. *Journal of Bacteriology*, **129**(1), 351–7.

Kashket, S. & Donaldson, C. G. (1972). Saliva-induced aggregation of oral streptococci. *Journal of Bacteriology*, **112**(3), 1127–33.

Krupp, H. (1967). Particle adhesion: theory and experiment. *Advances in Colloid and Interface Science*, **1**(2), 111–239.

Kruyt, H. R. (1952). *Colloid Science*, vol. 1. Amsterdam, London, New York: Elsevier.

Levich, V. G. (1962). *Physicochemical Hydrodynamics*. Englewood Cliffs: Prentice-Hall.

Levine, M. J., Herzberg, M. C., Levine, M. S., Ellison, S. A., Stinson, M. N., Li, H. C. & van Dyke, T. (1978). Specificity of salivary-bacterial interactions: role of terminal sialic acid residues in the interaction of salivary glycoproteins with *Streptococcus sanguis* and *mutans*. *Infection and Immunity*, **19**(1), 107–15.

Liljemark, W. F., Schaver, S. V. & Bloomquist, C. G. (1978). Compounds which affect the adherence of *Streptococcus sanguis* and *Streptococcus mutans* to hydroxyapatite. *Journal of Dental Research*, **57**(2), 373–9.

Lips, A. & Jessup, N. (1979). Colloidal aspects of bacterial adhesion. In *Adhesion of Microorganisms to Surfaces*, ed. D. C. Ellwood, J. Melling & P. Rutter, in press. London, New York: Academic Press.

Loeb, G. I. & Neihof, R. A. (1975). Marine conditioning films, In *Applied Chemistry at Protein Interfaces*, ed. R. E. Baier, Advances in Chemistry Series 145, pp. 319–35. Washington: American Chemistry Society.

McBride, B. C. & Gisslow, M. T. (1977). Role of sialic acid in saliva-induced aggregation of *Streptococcus sanguis. Infection and Immunity*, **18**(1), 35–40.

MacRitchie, F. (1972). The adsorption of proteins at the solid/liquid interface. *Journal of Colloid and Interface Science*, **38**(2), 484–8.

Magnusson, I. & Ericsson, T. (1976). Effect of salivary agglutinins on reactions between hydroxy-apatite and a serotype C strain of *Streptococcus mutans. Caries Research*, **10**(4), 273–86.

Magnusson, I., Ericsson, T. & Pruit, K. (1976). Effect of salivary agglutinins on bacterial colonisation of tooth surfaces. *Caries Research*, **10,** 113–22.

Marcelja, S. & Radic, N. (1976). Repulsion of interfaces due to boundary water. *Chemical Physics letters*, **42**, 129–30.

Marshall, K. C. (1969). Orientation of clay particles sorbed onto bacteria possessing different ionogenic surfaces. *Biochimica et Biophysica Acta*, **193,** 472–4.

Marshall, K. C. (1973). Mechanisms of adhesion of marine bacteria to surfaces. In *Proceedings of the 3rd International Congress on Marine Corrosion and Fouling*, ed. R. F. Acker, B. F. Brown, J. R. de Palma & W. P. Iverson, pp. 625–32. Evanston, Illinois: Northwestern University Press.

Marshall, K. C. & Cruickshank, R. H. (1973). Cell surface hydrophobicity and the orientation of certain bacteria at interfaces. *Archiv für Mikrobiologie*, **91**, 29–40.

Marshall, K. C., Stout, R. & Mitchell, R. (1971). Mechanism of the initial events in the sorption of marine bacteria to surfaces. *Journal of General Microbiology*, **68**, 337–48.

Meadows, P. S. (1971). The attachment of bacteria to solid surfaces. *Archiv für Mikrobiologie*, **75**, 374–81.

Meckel, A. (1965). The formation and properties of organic films on teeth. *Archives of Oral Biology*, **10**, 585–97.

Mellon, R. R., Hastings, W. S. & Anastasia, C. (1924). On the nature of the cohesive factor in spontaneous agglutination of bacteria especially considering the interfacial tension. *Journal of Immunology*, **9**, 365–81.

Morris, E. R., Rees, D. A., Thom, D. & Welsh, E. J. (1977). Conformation and intermolecular interactions of carbohydrate chains. *Journal of Supramolecular Structure*, **6**, 259–74.

Morris, E. R., Rees, D. A., Young, G., Walkinshaw, M. D. & Darke, A. (1977). Order–disorder transitions for a bacterial polysaccharide in solution. A role for polysaccharide conformation in recognition between *Xanthamonas pathogens* and its plant host. *Journal of Molecular Biology*, **110**, 1–16.

MUDD, S. & MUDD, E. B. H. (1924). Certain interfacial tension relations and the behaviour of bacteria in films. *Journal of Experimental Medicine*, **40**, 647–60.

MUKASE, H. & SLADE, H. D. (1973). Roles of insoluble dextran levan synthetase enzymes and cell wall polysaccharide antigen in plaque formation. *Infection and Immunity*, **8**, 555–62.

NALBANDIAN, J., FREEDMAN, M. L., TANZER, J. M. & LOVELACE, S. M. (1974). Ultrastructure of mutants of *Streptococcus mutans* with reference to agglutination, adhesion and extracellular polysaccharide. *Infection and Immunity*, **10**(5), 1170–9.

NAPPER, D. H. (1970). Colloid Stability. *Industrial and Engineering Chemistry. Product Research and Development*, **9**(4), 467–77.

NINHAM, B. W. & RICHMOND, P. (1973). Multi-molecular adsorption on cell surfaces under the influence of van der Waals forces. *Journal of the Chemical Society, Faraday Transactions, 2*, **69**, 658–64.

NORKRANS, B. & SORENSSON, F. (1976). On the marine lipid surface microlayer – Bacterial accumulation in model systems. *Botanica marine*, **20**, 473–8.

OGSTON, A. G. (1970). The biological functions of the glycosaminoglycans. In *Chemistry and Molecular Biology of the Intercellular Matrices*, vol. 3, ed. E. A. Balzacs, pp. 1231–40. London, New York: Academic Press.

OLSSON, J., GLANTZ, P. O. & KRASSE, B. (1976). Surface potential and adherence of oral streptococci to solid surfaces. *Scandanavian Journal of Dental Research*, **84**, 240–2.

OLSSON, J. & KRASSE, B. (1976). A method for studying adherence of oral streptococci to solid surfaces – glass, human enamel and whale dentin. *Scandanavian Journal of Dental Research*, **84**, 20–8.

ORSTAVICK, D. (1977). Sorption of *Streptococcus faecium* to glass. *Acta Pathalogica et Microbiologica Scandanavica*, **B85**, 38–46.

ORSTAVICK, D. (1978). The *in vitro* attachment of an oral streptococcus to the acquired tooth enamel pellicle. *Archives of Oral Biology*, **23**(3), 167–73.

ORSTAVICK, D., KRAUS, F. W. & HENSHAW, L. C. (1974). *In vitro* adherence of streptococci to the tooth surface. *Infection and Immunity*, **9**(5), 794–800.

OTTOW, J. C. G. (1975). Ecology, physiology and genetics of fimbriae and pili. *Annual Review of Microbiology*, **29**, 79–108.

PETHICA, B. (1961). The physical chemistry of cell adhesion. *Experimental Cell Research*, Supplement **8**, 123–40.

POINDEXTER, J. S. (1964). Biological properties and classification of the Caulobacter group. *Bacteriological Reviews*, **28**, 231–95.

RIDGEWAY, H. F. & LEVIN, R. A. (1973). Goblet-shaped subunits from the wall of a marine gliding microbe. *Journal of General Microbiology*, **79**, 119–28.

RIES, H. E. & MEYERS, B. L. (1968). Flocculation mechanism: charge neutralisation and bridging. *Science, New York*, **160**, (3835), 1449–50.

ROGERS, H. J. (1979). Adhesion of microorganisms to surfaces: some general considerations of the role of the envelope. In *Adhesion of Microorganisms to Surfaces*, ed. D. C. Ellwood, J. Melling & P. Rutter, in press. London, New York: Academic Press.

ROLLA, G. (1976). Inhibition of adsorption: general considerations. In *Microbial Aspects of Dental Caries: Workshop Proceedings*, vol. 2, ed. H. M. Stiles, pp.

309–24. Information Retrieval Inc. US. *Microbiology Abstracts*, special supplement.

Roth, I. L. (1977). Physical structure of surface carbohydrates. In *Surface Carbohydrates of the Prokaryotic Cell*, ed. I. Sutherland, pp. 5–27. London, New York: Academic Press.

Rutter, P. & Abbott, A. (1978). A study of the interaction between oral streptococci and hard surfaces. *Journal of General Microbiology*, **105**, 219–26.

Shands, J. W. (1966). Localisation of somatic antigen on Gram-negative bacteria using ferritin antibody conjugates. *Annals of the New York Academy of Sciences*, **133**, 292–8.

Sharon, N. & Lis, H. (1972). Lectins: cell-agglutinating and sugar-specific proteins. *Science, New York*, **177**(4053), 949–59.

Slater, R. W. & Kitchener, J. A. (1966). Characteristics of flocculation of mineral suspensions by polymers. *Discussions of the Faraday Society*, **42**, 267–75.

Smitham, J. B., Evans, R. & Napper, D. H. (1975). Analytical theories of the steric stabilisation of colloidal dispersions. *Journal of the Chemical Society, Faraday Transactions, 2*, **71**(2), 285–97.

Spielman, L. A. (1978). Particle capture mechanisms. In *Deposition and Filtration of Particles from Gases in Liquids*, pp. 1–20. London: Society of Chemical Industry.

Spielman, L. A. & Cukor, P. M. (1973). Deposition of non-Brownian particles under colloidal forces. *Journal of Colloid and Interface Science*, **43**(1), 51–65.

Sutherland, I. W. (1977). Bacterial exopolysaccharides: their nature and production. In *Surface Carbohydrates of the Prokaryotic Cell*, ed. I. Sutherland, pp. 27–97. London, New York: Academic Press.

Treweek, G. P. & Morgan, J. J. (1977). Polymer flocculation of bacteria. The mechanism of *E. coli* aggregation by polyethyleneimine. *Journal of Colloid and Interface Science*, **60**(2), 258–73.

van Oss, C. J. & Gillman, C. F. (1972). Phagocytosis as a surface phenomenon: 1. Contact angles and phagocytosis of non-opsonised bacteria. *Journal of the Recticuloendothelial Society*, **12**, 283–92.

van Oss, C. J., Good, R. J. & Neumann, A. W. (1972). The connection of interfacial free energies and surface potentials with phagocytosis and cellular adhesiveness. *Journal of Electroanalytical Chemistry*, **37**, 387–91.

Verwey, E. J. W. & Overbeek, J. Th. G. (1948). *Theory of the Stability of Lyophobic Colloids*. Amsterdam, London, New York: Elsevier.

Vincent, B. (1974). Adsorbed polymers and dispersion stability. *Advances in Colloid and Interface Science*, **2**(3).

Visser, J. (1972). On Hamaker constants: A comparison between Hamaker constants and Lifshitz–van der Waals constants. *Advances in Colloid and Interface Science*, **3**, 331–63.

Visser, J. (1976*a*). Adhesion of Colloidal Particles. In *Surface and Colloid Science*, vol. 8, ed. E. Matijevic, pp. 3–84. New York: Wiley.

Visser, J. (1976*b*). The adhesion of colloidal polystyrene particles to cellophane as a function of pH and ionic strength. *Journal of Colloid and Interface Science*, **55**(3), 664–77.

VISSER, J. (1978). Colloid and other forces in particle adhesion and removal. In *Deposition and Filtration of particles from Gases and Liquids*. London: Society of Chemical Industry.

WEISS, R. L. (1973). Attachment of bacteria to sulphur in extreme environments. *Journal of General Microbiology*, **77**, 501–7.

ZIMON, A. D. & DERJAGUIN, B. V. (1963). Adhesion of particles to a plane surface, adhesion in an aqueous medium. *Colloid Journal of USSR*, **25**(2), 159–64.

Adhesive properties of bacteria virulent to man

P. J. WATT

Microbiology Department, Southampton General Hospital, Southampton University, Tremona Road, Southampton SO9 4XY, UK

INTRODUCTION

The natural habitats of many pathogenic bacteria are the mucosal surfaces of man and other mammals. A critical factor in their adaptation for life in such an hostile environment is the development of a mechanism for anchorage to the mucosal cells, whose surfaces are continually washed by flows of mucus and other secretions.

The mechanism of microbial adhesion to host cell surfaces must, of necessity, be mediated by macromolecules associated with the bacterial outer membrane. The function of this outer membrane in Gram-negative pathogens is complex (DiRienzo, Nakamura & Inouye, 1978), with different components involved in structural integrity, the formation of non-specific diffusion pores, as well as uptake systems for iron, sugar, vitamins, etc. In Gram-positive bacteria the cell wall teichoic and teichuronic acids are important in the assimilation of divalent cations particularly Mg^{2+} (Heckels, Lambert & Baddiley, 1977). Clearly, such molecules will contribute to the physical characteristics of the bacterial surface and, by their effects on charge density, hydrophobicity and polymer interactions, must influence the cohesive properties of any specific mediators of microbial adhesion. In pathogens this situation is further complicated by the need for surface structures to act as a defence against host phagocytic systems and to form a diffusion barrier protecting the vulnerable cytoplasm membrane from the lytic action of antibodies and complement. This induces a conflict of requirements. Bacteria with hydrophobic surfaces readily associate with mammalian cell membranes (van Oss, Gillman & Neuman, 1975; Perers *et al.*, 1977), but bacterial resistance to host defence systems is mediated by hydrophilic, often negatively charged, capsular polymers. Clearly, any understanding of the molecular basis of adhesion requires detailed information on the surface architecture of the pathogen.

The healthy mucosal surfaces of the body teem with commensal bacteria. Thus, the essential difference between a commensal bacterium and a pathogen is not the ability to attach to mucosal surfaces but that an adherent pathogen damages and/or penetrates the membrane of the host cell. Clearly, then, laboratory investigations of these fundamental problems must initially establish the basic pathological features of the natural disease. This review concentrates on selected examples of such mucosal infections. The objective is to illustrate how differing cohesive processes correlate with the distinctive clinical features of the infection. The fate of any pathogens which invade through the body's mucosal surfaces will depend upon their interaction with phagocytes. This important aspect of host–parasite relationships will be discussed in terms of those surface components of pathogens which determine the initial association of the bacterium with the phagocyte.

GONORRHOEA; ADHESION AS A PRECURSOR OF INVASION

The prevalence of gonorrhoea is tribute to the infectiousness of the disease; indeed, some 50% of men having intercourse with an asymptomatic woman carrier develop infection (Holmes, Johnson & Throstle, 1970). Yet relatively few gonococci can be recovered from infectious patients. In men with acute gonorrhoea of around one day's duration, some 10^4 gonococci were recovered by scraping the urethral mucosa (Ward, Watt & Glynn, 1970), whilst in women, vaginal washout techniques recovered some 4×10^2 to 1.8×10^7 gonococci (Lowe & Kraus, 1976). That the infectious dose is small was confirmed by studying human volunteers; the introduction of only 1×10^3 gonococci into the urethra initiated disease (Brinton *et al.*, 1978). The observation that micturition immediately after intercourse may not prevent naturally acquired gonorrhoea (Bernfeld, 1972) suggests that infecting organisms rapidly adhere to the genital mucosa and cannot be flushed from the surface. By day 3 or 4 after exposure, gonococci have penetrated the mucosal surface. In diseased areas there is disorganisation of the epithelial surface, with the widened intercellular spaces packed with polymorphonuclear leukocytes and gonococci (Harkness, 1948). This ability to invade through the mucosal surface, provoking a localised inflammatory response, is characteristic of gonococci, and distinguishes them from closely related commensal Neisseriae, which grow on, but do not damage, mucosal surfaces.

A working hypothesis is that bacterial adherence to the susceptible mucus-secreting columnar epithelia requires high-avidity attachment mechanisms which gonococci, but not the commensals, possess. Certainly, gonococci infect the columnar epithelium in the genital tract at sites like the posterior male urethra and endocervix (Harkness, 1948) which are only transiently colonised with low numbers of commensal organisms (Sparkes, Purrier, Watt & Elstein, 1977). Experimental support for this concept comes from the finding that, unlike gonococci, the commensal *Neisseria subflava* did not attach to the mucus-secreting columnar epithelial surface of human fallopian tube organ cultures (Johnson, Taylor-Robinson & McGee (1977). Essentially, commensal Neisseriae adhere to squamous epithelia, whose surface consists of dead cells forming a passive substrate for bacterial attachment. By contrast, the surface membrane of columnar epithelia is in active movement, necessitating a high-avidity mechanism if the organisms are to adhere. Interestingly, gonococcal attachment to an MRC 5 fibroblast monolayer was markedly enhanced when the epithelial cell surface was immobilised by cytochalasin B (Watt & Ward, 1977), suggesting that cell surface movement may hinder attachment.

Electron microscopic studies of mucosal cells obtained from the urethra of men with early gonorrhoea showed gonococci attached to, and partially embedded in, the surface of epithelial and mucus-secreting cells (Ovčinnikov & Delektorskij, 1971; Ward & Watt, 1972). Studies on experimentally infected human fallopian tube organ cultures show that the initial interaction of gonococci is with microvillous projections from the host cell surface (Ward & Watt, 1975). These microvilli frequently appear twisted towards the invading bacteria and are particularly well developed on the cell surface nearest to the bacteria. This suggests that their formation might be stimulated by the proximity of gonococci; an effect comparable to that reported for the mucosal pathogen *Bordetella pertussis* (De Bault & Yoo, 1974). Some gonococci become enfolded by microvillous processes; resorption of these processes would bring gonococci into contact with the host cell surface (Ward & Watt, 1975). Once the gonococcus is firmly bound to the epithelial surface by high-avidity binding, interiorisation of a proportion of the organisms is inevitable (Watt, Ward, Heckels & Trust, 1978). There is evidence to suggest that the initial discrete contact of surface ligands on particles with receptors on the surface of professional phagocytes does not trigger ingestion (Griffin, Griffin, Leider & Silverstein, 1975). Thus, for attached red cells to be ingested, ligands must be available on the whole circumference of the cell, since proteolytic destruction of ligands not involved in, and protected by, attachment prevented phagocytosis. Similarly, when lymphocytes were capped using anti-membrane antibodies, they attached to macrophages by the anti-globulin cap. Spread of the macrophage membrane was limited to the cap and the lymphocyte could not be ingested (Griffin, Griffin & Silverstein, 1976). Thus, it is reasonable to postulate that the ingestion of gonococci by columnar epithelial cells results from the sequential, circumferential interactions of gonococcal ligands with specific plasma membrane receptors; a 'zipper-mechanism' of phagocytosis. Support for this model comes from the finding that gonococci interiorised by the columnar epithelial cells of human fallopian tube organ cultures lie in tight, membrane-bound vesicles (Ward & Watt, 1975). Moreover, this invasive process was impaired by cytochalasin B (Watt, Ward & Robertson, 1976).

Pilus-mediated adhesion

Pili, protein filaments 7 nm in diameter and extending 2–4 μm from the bacterial surface, are universally present on gonococci when primarily isolated from the patient (Jephcott, Reyn & Birch-Anderson, 1971; Swanson, Kraus & Gotschlich, 1971) After sub-culture, non-piliated variants soon outgrow the piliated gonococci, indicating that growth within the host exerts a selective pressure in favour of the piliated state.

Piliated gonococci rapidly adhere to human cells in tissue culture (Swanson, 1973), to human sperm (James, Knox & Williams, 1976) and to red blood cells (Buchanan & Pearce, 1976). Indeed, they would seem to be more effective at attaching to human vaginal epithelial cells than the normal commensal flora organisms of the vagina (Mårdh & Weström, 1976). The human fallopian tube organ-culture model devised by Taylor-Robinson, Whytock, Green & Carney, (1974) is highly relevant for the study of gonococcal adhesion to and invasion of mucosal surfaces, since infection of the fallopian tube (salpingitis) is a major complication of gonorrhoea. Using this model, we have demonstrated (Watt *et al.*, 1976) that piliated gonococci show four-fold increased adhesion to the mucosal surface than the non-piliated variant of the same strain, and that piliated gonococci appear attached to the surface by pilus bundles (Ward, Watt & Robertson, 1974). Clearly, pilus-mediated adhesion is critical for gonococcal virulence.

Gonococcal pili are composed of multiple units of a single protein termed pilin, which contains some 200 amino acid residues with a molecular weight 19000 ± 2500 dalton in different isolates (Robertson, Vincent & Ward, 1977; Brinton *et al.*, 1978; Buchanan, Pearce & Chen, 1978). The amino acid analyses reported by these groups are remarkably consistant, with a percentage composition of 46% non-polar, 25% acidic and 13% basic, with the remainder being polar, uncharged amino acids. This is a high proportion of hydrophobic amino acids; indeed with the exception of threonine and glutamic acid in positions 2 and 5, the first 24 residues at the N-terminal end are all hydrophobic (Hermodson, Chen & Buchanan, 1978).

Given that 46% of the constituent amino acids of gonococcal pili are non-polar, it is not surprising that pili avidly bind to hydrophobic gels. When ^{125}I-labelled pili were mixed with 1 ml 10% (v/v) phenyl Sepharose, 70% of the pili sedimented with the gel. Since the area of the gonococcal outer membrane and the area of the pili are of comparable magnitude (about 1×10^{-11} m^2), we might expect pili to dominate gonococcal surface hydrophobicity. However, when whole gonococci were interacted with amphipathic gels, pili only conferred a 20% advantage in binding to hexyl-, octyl- and phenyl-substituted Sepharose (these results to be published). This result emphasises the facility with which the surface of the gonococcal outer membrane undertakes hydrophobic binding, and is in marked contrast to the report that the pilus-like K88 antigen was solely responsible for the binding of smooth enteropathogenic *E. coli* to octyl and phenyl Sepharose gels (Smyth *et al.*, 1978).

Adhesion at mucosal surfaces exists in an aqueous environment, so the exclusion of hydrophobic groups from the water lattice will assist the close approximation of hydrophobic surfaces and facilitate non-specific van der Waals binding. Further hydrophobic interaction may assist bonding to specific receptors. Noting that concanavalin A has a hydrophobic region adjacent to the sugar-binding site, Smyth *et al.* (1978) suggest that lectin-like binding of the *E. coli* K88 antigen to mannose residues may be facilitated by the hydrophobic nature of the protein.

Effect of ions on pilus adhesion

Bacteria infecting mucosal surfaces exist in a complex environment, rich in mucins and varying in ionic composition. In particular, gonococci are exposed to extremes of acidity, varying from semen (pH 7.19) and prostatic secretions (pH 6.45) to urine, the pH of which ranges from 4.8 to 8. In women, during the menstrual cycle, the pH of endocervical mucus varies from 5.9 to 7.3; in the ectocervical mucus from 4.0 to 7.4; in the lateral fornix of the vagina from 3.5 to 5.8 and in the vaginal entrance from 3.5 to 5.3 (Kroeks & Kremer, 1977). However, after coitus the ejaculate exerts a powerful buffering effect, perhaps permitting initial attachment at less extreme pH values. Gonococcal adhesion is enhanced by increased acidity. Mårdh & Weström (1976) reported that gonococcal attachment to vaginal epithelial cells was enhanced three-fold at pH 4.5 compared to pH 7.5, while Pearce & Buchanan (1978) have shown comparable increases in the binding of ^{125}I-labelled purified pili to buccal epithelial cells.

Experimental investigation of the role of cations in gonococcal adhesion is fraught with problems. Piliated gonococci are aggregated by moderate concentrations of di- and trivalent cations, and rapidly lyse when deprived of divalent cations. Nevertheless, the fact that the binding of ^{125}I-labelled pili to buccal cells pretreated with 5 mM EDTA was not increased in the presence of 1 mM Ca^{2+}, Mg^{2+} or Zn^{2+} (data to be published), suggests that salt bridging is unimportant for pilus attachment. In contrast, the attachment of piliated gonococci to human sperm was enhanced in the presence of 10 mM iron salts (James *et al.*, 1976). A possible explanation is that Stern layer adsorption of the Fe^{3+} counter-ion enhances adhesion non-specifically by reducing the magnitude and operating distance of the electrostatic repulsive forces (Watt & Ward, 1977). However, Buchanan *et al.* (1978) have suggested a more specific mechanism. These workers showed that 100 μM Fe^{3+} enhanced the binding of ^{125}I-labelled pili to human buccal epithelial cells to 2.5-fold. This enhancement did not occur at pH 4.5 (the average pH of vaginal secretions) and its effect at pH 7.4 was blocked by the prior photo-oxidation of pili with methylene blue, a treatment that selectively destroys histidine residues. The suggestion was that Fe^{3+} binds to the histidine residues on pili when they are in an unprotonated state (pH 7.4), thus bridging between the pili and the buccal epithelial surfaces. There are difficulties in accepting this model of gonococcal adhesion. First, 100 μM $FeCl_3$ at pH 7.4 will form colloidal $Fe(OH)_3$, which can adsorb the pili and epithelial cell surfaces, producing non-specific cross-links. At pH 4.5, $FeCl_3$ is less likely to form $Fe(OH)_3$ and was without effect on pilus binding to cells. Secondly, given that the concentration of Fe^{3+} in serum is about 10^{-14} M (Weinberg, 1978) it is doubtful if bridging through Fe^{3+} plays a significant role in gonococcal attachment in the natural infection.

Electrostatic interaction

The importance of electrostatic repulsive forces in the interaction of gonococci with host cells has been demonstrated (Heckels, Blackett, Everson & Ward, 1976). Free amino groups on the surface of gonococci were blocked using formaldehyde, and the carboxyl groups with 1-ethyl-3-(dimethylaminopropyl) carbodiimide and methylamine. The effect of this treatment on the surface charge of the organisms was determined by measuring the pI by equilibrium isoelectric focusing. The mean number of untreated or chemically modified gonococci firmly adherent to the surface of monolayer cultures of WISH cells was determined microscopically. The results show that blocking amino groups and increasing the surface net negative charge (pI 4.0) reduced binding by 67 %. Blocking carboxyl groups (pI 8.2) effectively reversed the usual negative surface charge (pI 5.6) on the gonococcus and doubled the mean number of gonococci adherent per cell. This increased attachment was not simply due to electrostatic attraction to the now positively charged gonococcal surface, because when both the amino and carboxyl groups on the gonococcus were blocked the enhanced adherence of the gonococcus to the cells was retained. Thus, the critical factor was the removal of the electrostatic barrier to the negatively charged gonococcal surface. An interesting finding was that pili no longer promoted the adhesion of gonococci to human cells when the electrostatic repulsive barrier was reduced by blocking negatively charged groups on the gonococcal surface. Arguably, this effect might have been due to the chemical treatment destroying pilus attachment function. However, in recent experiments in this laboratory (results to be published) in which the attachment of ^{3}H-labelled piliated and non-piliated gonococci to human erythrocytes was quantified, the two-fold enhancement of binding conferred by pili was abolished when acidic groups on the red blood cell were removed by treatment with trypsin or sialidase. The fact that Dextran-80 (molecular length 55 nm) aggregates erythrocytes but Dextran-40 (molecular length 32 nm) cannot, suggest a repulsive force operating up to 20 nm from each red cell surface (Jan & Chien, 1973*a*, *b*). Clearly, the length of gonococcal pili (2 μm) permits interaction with the host cell membrane at distances where electrostatic repulsive forces between the two cell bodies would be non-existent. Further, the absolute energy required for the pilus tip with a radius of 7 nm to penetrate the energy barrier would be small.

Pilus binding to cell-surface carbohydrate receptors

Unlike the mannose-sensitive *Escherichia coli* Type 1 pilus haemagglutination (Salit & Gotschlich, 1977*b*), the binding of piliated gonococci to human buccal cells and gonococcal haemagglutination were resistant to a wide range of simple sugars (Punsalang & Sawyer, 1973). Using ^{3}H-labelled gonococci, we have established that selected representatives of the constituent sugars of surface glycolipids and glycoproteins (including D-galactose, β-methyl-D-galactoside, α-methyl-D-galactoside, lactose (Gal-β-[1 → 4]-Glc), melibiose (Gal-α-[1 → 6]-Glc), mannose, α-methyl-D-mannoside, α-L-fucose, *N*-acetyl-D-galactosamine and *N*-acetyl-D-glucosamine) were without effect on the binding of piliated gonococci to human buccal cells or erythrocytes (results to be published). A second approach has been to block surface carbohydrates by pretreatment of buccal cells with lectins. In these experiments, 50 μl of packed buccal cells were suspended in 15 ml of phosphate-buffered saline containing 0.08 μM lectins specific for: β-D-galactose (Ricin), α-L-fucose (Lotus), α-D-mannose (Con A), *N*-acetyl-D-glucosamine (wheatgerm) and *N*-acetyl-D-galactosamine (Soyabean). When the treated buccal cells were reacted with ^{125}I-labelled pili, the proportion of pili adhering to the cells (40 %) was the same in the untreated and lectin-treated cells. Taken together, the results of these experiments suggest that gonococcal pili do not act as simple lectins binding to single sugars on the host-cell membrane. However, they cannot exclude the possibility that pili bind, perhaps non-specifically, to a sequence of sugars on cell-surface carbohydrates. Evidence supporting this concept comes from the work of Buchanan *et al.* (1978) who showed that the binding of ^{125}I-labelled pili to cells was inhibited by 60% when the buccal cells were treated with an exoglycosidase mixture at 2 mg ml^{-1}. Further, pilus binding to buccal cells was inhibited by 20% in the presence of 1 mg ml^{-1} chitin oligosaccharides ($\beta[1 \rightarrow 4]$-linked polymer of *N*-acetyl-D-glucosamine) and by 20% using 860 μg ml^{-1} of yeast mannan (a branched polymer with a backbone of $\alpha[1 \rightarrow 6]$-linked D-mannose residues and side-chains of two to five mannose units linked $\alpha[1 \rightarrow 2]$ or $\alpha[1 \rightarrow 3]$), but 10 μM α-methyl-D-mannoside was without effect. We can, in part, confirm these findings using whole gonococci. Treatment of buccal cells with 1 mg ml^{-1} exoglycosidases reduced the binding of ^{3}H-labelled piliated gonococci by 27% and 10 mg ml^{-1} enzyme by 78%. In an attempt to confirm the significance of cell-surface carbohydrates we have treated cells by periodate oxidation followed by borohydride reduction of the resulting aldehydes to polyalcohols. The binding of ^{3}H-labelled gonococci to periodate/borohydride-treated buccal cells was unaffected (results to be published).

Candida albicans, with a surface largely composed of mannan, has been used in studies on mannose-sensitive pili (Ofek & Beachey, 1978). We have measured the binding of ^{3}H-labelled pilated gonococci to 10% v/v suspensions of *Candida*; attachment (18.5%) was significant, and although only half the binding seen with buccal and red cells this does not take into account the relative differences in surface area for a given packed cell volume. Although mannans dominate the surface of *Candida*, the underlying chitin is exposed, since yeasts are aggregated by lectins specific for *N*-acetyl-D-glucosamine (G. Bull, personal communication). Thus, there are two polysaccharides accessible for pilus binding.

A further suggestion arising from the purified gonococcal pilus studies of Buchanan *et al.* (1978) is that cell surfaces possess 'ganglioside-like' pilus receptors. The experimental evidence supporting this concept arises from the finding that pili treated with gangliosides (GM_1, GD_{1a}, GD_{1b}, GT) at a concentration of 10–20 μM show impaired binding to buccal cells. At concentrations of 0.3 μM, GD_{1a} and GD_{1b} inhibited attachment by 20% to 30%, but significant binding (40%) persisted even at concentrations of 74 μM. This inhibition may be due to gangliosides binding to specific receptors on pili, but non-specific effects such as a detergent action of ganglioside micelles cannot be excluded. Certainly, we have been unable to demonstrate specific binding of piliated gonococci to gangliosides incorporated into ^{3}H-labelled cholesterol/lecithin liposomes (results to be published).

In summary, the evidence supporting gonococcal pilus adhesion to carbohydrate receptors on the host cell surface is of two types. First, treatment of the cell with a mixture of α-mannosidase, α-L-fucosidase, α-galactosidase, α-*N*-acetyl-galactosaminidase and β-*N*-acetyl-hexosaminidase markedly impairs pilus-mediated adhesion to the cell. Secondly, piliated gonococci adhere to the surface mannans on yeasts and these mannans, together with chitin and gangliosides, inhibit pilus attachment. Although we cannot exclude pilus-mediated binding to a single sugar with unique conformation, this is unlikely because this adhesion is unaffected by high concentrations of constituent sugars of cell-surface carbohydrates or by pretreating cells with lectins specific for these sugars. A simple explanation would be that gonococcal pili bind, perhaps by hydrogen bonding, along the length of the sugar chain to unsubstituted hydroxyl groups. Under these circumstances, specificity would be low and modification of the oligosaccharide sugars to polyalcohols need not inhibit binding. This was observed with periodate/borohydride-treated buccal cells. In support of this concept, we have observed (unpublished) that sugars (tetrasaccharides > trisaccharides > disaccharides > monosaccharides) disperse pilus–pilus aggregates while long-chain carbohydrates, such as dextrans and polysucrose, cross-link pili.

Outer envelope components as adhesins

Pilus-mediated adhesion to experimentally infected human fallopian tube organ cultures is not essential for invasion, since the proportion of attached gonococci penetrating the mucosal surface was 0.85 % for piliated gonococci and 0.67 % for non-piliated gonococci (Watt *et al.*, 1976). Electron micrographs of gonococci sectioned at their point of contact with the mucosal surface, after exposure to colloidal thorium prior to dehydration and embedding, show that particles must have a diameter of less than 13 nm to penetrate the gap between the bacterial outer membrane and the host-cell surface (Ward & Watt, 1975). This gap is consistant with surface–surface interactions at the energy level of secondary-minimum adhesion (Derjaguin & Landau, 1941; Verwey & Overbeek, 1948). The attractive force of secondary-minimum adhesion is small and the association will be redispersible; such weak non-specific adhesion may account for the ability of gonococci to attach to the mucosal surface of animals, for example, guinea-pigs, naturally resistant to gonorrhoea. Thus, Tebbutt, Veale, Hutchison & Smith (1976) were able to quantify the numbers of gonococci attached to the mucosal surface of the guinea-pig's uterus, cervix and urethra by applying a shearing force and counting the numbers of gonococci detached. Such a system in no way precludes the possibility of specific adhesion to receptors on human mucosal surfaces.

A major structure invariably expressed at the gonococcal surface is lipopolysaccharide (LPS). O-antigen chains have not been satisfactorily demonstrated on gonococcal LPS; the core sugars consist of heptose linked to two residues of glucose, two or three residues of galactose and two residues of *N*-acetyl-D-glucosamine with galactose and *N*-acetyl-D-glucosamine as the terminal, immunodominant sugars (Perry *et al.*, 1978). These external sugars could serve as bridging molecules linking to the specific surface glycosyltransferases postulated by Roseman (1970) or to other surface proteins (Yamada, Yamada & Pastan, 1975). Such a mechanism may account for the ability of the macrophage membrane to bind to a wide range of micro-organisms (Freimer *et al.*, 1978). However, the binding of non-piliated gonococci to human buccal epithelial cells was not inhibited by 0.4 M galactose or by 0.4 M glucosamine. Further evidence against the specific attachment of gonococci to host cell surfaces via LPS terminal sugar groups comes from our finding that purified core polysaccharide coupled via KDO to ^{125}I-labelled methylalbumen failed to adhere to buccal cells or to red blood cells (results to be published). These findings suggest that the attachment of non-piliated gonococci is mediated by surface proteins.

The outer membrane of gonococci always contains a major protein, termed Protein I. This protein is responsible for serotype specificity, varies in molecular weight from 32000 to 38000 dalton and is expressed at the gonococcal surface, as shown by intense labelling with ^{125}I using the lactoperoxidase system (Johnston, Holmes & Gotschlich, 1976; Heckels, 1978). Recent studies on gonococci with differing colonial opacities demonstrated that a second protein, Protein II, was present in comparable amounts to Protein I (Swanson, 1978; Lambden & Heckels, 1979). Protein II cannot be considered a single entity, since selected opacity variants of strain P9 possess distinct proteins termed II*, IIa*, IIb*, IIc*, and IId* with molecular weights ranging from 27500 to 29000 dalton (Lambden,

Table 1. *Attachment of opacity variants of non-piliated gonococcus P9 to buccal epithelial cells, erythrocytes and hydrophobic phenyl Sepharose*

	Additional surface protein					
	Nil	II*	IIa*IId*	IIa*	IIb*	II*IIc*
Buccal epithelial cells[a]	12.0	24.0	27.3	26.1	17.7	20.6
Erythrocytes[a]	24.5	19.0	15.1	15.5	11.0	11.4
Phenyl Sepharose[a]	8.9	3.9	8.2	8.5	8.6	8.1

[a] Average values from three separate experiments.

Attachment was expressed as the percentage of ^{3}H-labelled gonococci sedimenting with the buccal cells, erythrocytes or gel.

Heckels, James & Watt, 1979). Data abstracted from this paper are summarised in Table 1.

The variant with a surface composed only of LPS plus Protein I bound least well to human buccal epithelial cells. Preliminary studies of this attachment (results to be published) show that adhesion was unaffected by the addition of the constituent sugars of host-cell glycolipids and glycoproteins (β-methyl-D-galactoside, α-methyl-D-galactoside, α-methyl-D-mannoside, α-L-fucose, *N*-acetyl-D-galactosamine and *N*-acetyl-D-glucosamine) or by the addition of GM_1, GD_{1a} or GT_1 gangliosides. Further treatment of the host cell with sialidase plus mixed exoglycosidases did not inhibit binding of the variant lacking both pili and Protein II. A possible explanation was that the binding was non-specific and it was of interest that this variant shows significant attachment to hydrophobic phenyl Sepharose gel (Table 1). The finding that the additional presence in the outer membrane of any member of the Protein II family significantly increased attachment to buccal cells without enhancing hydrophobic binding suggested that a specific mechanism of adhesion might be involved. Indeed, one variant containing Protein II* showed a marked reduction in hydrophobic interactions but a two-fold greater attachment to buccal cells when compared to isolates lacking Protein II (Table 1). In contrast to these results, Protein II impairs the binding of non-piliated gonococci to red cells. A possible explanation is that red cells lack the receptor for Protein II mediated adhesion; certainly erythrocytes lack HLA antigens.

Further work to establish the molecular basis for Protein II mediated attachment to mucosal cells is urgently required since Protein II must be considered a candidate immunogen for a gonococcal vaccine. This claim is strengthened by the fact that Protein II confers significant resistance to destruction of gonococci by antibody plus complement (Lambden *et al.*, 1979).

Summary

The attachment of gonococci to mucosal surfaces is a complex, multifactorial process involving both pili and outer-membrane adhesins. Experiments designed to elucidate the molecular basis of this adhesion must be interpreted with caution. First, effective blockade of a specific membrane receptor may be masked by non-specific binding to the cell, for example by hydrophobic interactions. Secondly, the available tools are crude, e.g. lectins may block specific sugar residues on cell surfaces but in so doing modify surface charge density, hydrophobicity, membrane fluidity, induce steric hindrance to adjacent groups and affect host cell physiological processes. Nevertheless, it is clear that pili penetrate the electrostatic barrier between the host-cell surface and the gonococcus initiating adhesion. This adhesion may involve hydrophobic interaction and a relatively non-specific binding to sugar chains on membrane glycolipids and glycoproteins. Outer-membrane proteins also function as adhesins, but the mechanism has not been investigated. The importance of advancing our understanding of the mechanism of gonococcal adhesion lies in the field of vaccine development. The promise of this approach is shown by the work of Brinton *et al.* (1978) who immunised four volunteers with purified pili and found that the dose of gonococci needed to produce disease was increased a thousand-fold.

CHOLERA; THE ROLE OF BACTERIAL MOTILITY IN COHESION

Cholera is an acute infectious disease caused by *Vibrio cholerae* and characterised by diarrhoea, vomiting and dehydration which are extremely severe in typical cases. The characteristic short, curved flagellate bacilli are present in enormous numbers throughout the small and large intestine. Diarrhoea is caused by a potent exotoxin which binds specifically to GM_1 ganglioside receptors on the mucosal cells of the small bowel, activating adenylate cyclase. The result is an increase in intracellular levels of cyclic adenosine monophosphate (cAMP) with hypersecretion of salts into the intestinal lumen. This aspect of cholera is the subject of a detailed recent review (Richards & Douglas, 1978).

The pathogenesis of cholera is undergoing intensive research utilising the susceptibility of infant animals to oral challenge with virulent *V. cholerae*. Using the suckling mouse model, Guentzel & Berry (1975) compared the virulence of flagellate and non-flagellate toxogenic isolates of *V. cholerae* and found that loss of motility resulted in 100- to 1000-fold increase in the LD_{50}. These findings have been confirmed by treating vibrios with the mutagen nitrosoguanidine and screening the treated clones for mouse virulence. Lack of flagella was the sole defect in some virulence-deficient mutants (Baselski, Upchurch & Parker, 1978). Interestingly, when small numbers (10^6) of non-motile toxogenic *V. cholerae* were injected into the lumen of isolated segments of rabbit intestine their growth rate was equal to that of the motile wild form, but fluid accumulation was markedly impaired (Yancy, Willis & Berry, 1978). Presumably, then, toxin excreted into the lumen is ineffective, perhaps neutralised by binding to GM_1 present on cell debris, and vibrio motility permits efficient toxin delivery to the mucosal surface. Motility is essential for penetrating the mucous blanket coating the brush border; the vibrios migrate along tracts that parallel the lines of strain within the stretched mucus (Jones, Abrams & Freter, 1976). Some twelve hours post infection of infant mice, vibrios have penetrated into the intervillous spaces and deep into the mucosal crypts (Guentzel, Field, Eubanks & Berry, 1977). Specific adhesion of vibrios to the tips of the microvilli on the outer surfaces of the brush border was dependent upon the presence of Ca^{2+}. Spontaneous elution of vibrios was observed at 37 °C, perhaps because the 'adhesin' had a high rate of turnover. The possibility that flagella function as 'adhesins' was suggested by the finding that the adsorption of ^{14}C-labelled non-motile vibrios to mouse intestinal tissue was reduced to less than 20 % of the control value of the motile wild-type (Guentzel & Berry, 1975). Comparable results were obtained for the binding of ^{35}S-labelled vibrio to the adult rabbit gut (Yancy *et al.*, 1978). The enhanced adhesion of motile vibrios was not simply the result of an ability to overcome energy barriers and make contact with the mucosal surface, since the attachment of non-motile variants was not increased by violent agitation or by centrifuging the bacteria onto the mucosal surface (Jones & Freter, 1976). These workers also investigated the possibility that the 'adhesion' of motile vibrios functioned as a lectin binding to specific sugar residues on the intestinal glycocalyx. Of the possible sugars, galactose, *N*-acetylneuraminic acid, *N*-acetylglucosamine and *N*-acetylgalactosamine were ineffective, but L-fucose at a concentration of 100 μg ml^{-1} inhibited vibrio binding by 60% and D-mannose at 100 μg ml^{-1} by 20%. Fucosides such as *p*-nitrophenyl-L-fucoside and methyl-L-fucoside were effective at lower concentrations. However, it is unlikely that attachment was mediated solely by an adhesion binding specifically to complementary L-fucose and D-mannose residues because increasing the sugar concentration to 10 mg ml^{-1} caused no further inhibition of binding.

These data do not establish whether the flagella function as adhesins or if cohesion is mediated by a factor intimately associated with the flagella. Electron microscopic examination of the mucosal surface in experimental infections show that the majority of bacteria align horizontally with the epithelial surface, but some appeared end-on, with their flagella extending into the lumen (Nelson, Clements & Finkelstein, 1976). These findings suggest that *V. cholerae* can adhere via its surface coat. However, attempts to block attachment using antibodies against the surface coat of non-motile vibrios were ineffective, while crude flagella vaccines were highly protective (Eubanks, Guentzel & Berry, 1977). When this flagellar preparation was treated with 0.5% sodium deoxycholate to remove vesicular material associated with the flagella, the purified flagella retained little of the protective effect of the crude vaccine. A possible explanation was that the flagellar sheath is the true adhesin.

The importance of these studies lies in the possibility of developing a cholera vaccine with the dual effect of neutralising the enterotoxin and preventing the attachment of vibrios to the gut surface. Clearly, the adhesin must be isolated in a pure, immunogenic form and the molecular basis of its cohesive properties established.

URINARY INFECTIONS: BACTERIAL ADHESION MEDIATED BY PILI

Urinary infection is a major medical problem and the subject of numerous monographs and conference reports (Kincaid-Smith & Farley, 1970; Stamey, 1972). Bacteria may multiply in the bladder urine without evidence of tissue invasion; this is termed asymptomatic bacteriuria. Invasion of the bladder (cystitis) producing frequent, painful micturition is extremely common in women and has a marked tendency to recurrence. The most serious consequence of bacteriuria is infection of the kidney, pyelonephritis. Repeated attacks of pyelonephritis, particularly in young girls, results in progressive destruction of the kidneys. In the great majority of childhood urinary infections, the bacteria isolated from the urine correspond to the dominant *Escherichia coli* strain from the patient's stool. A possible explanation for the fact that asymptomatic bacteriuria rarely progresses to pyelonephritis is that strains of *E. coli* differ in virulence for the urinary tract. Svanborg-Edén and his colleagues (1978) have examined the ability of different *E. coli* isolates to attach to normal human uroepithelial cells. An adhesive strain was defined as one able to bind more than ten bacteria per host cell. The results of this study were clear cut; 68% of *E. coli* isolated from kidney infections and 50% isolated from cases of cystitis were 'adhesive', compared to 22% of isolates from asymptomatic bacteriuria and 10% of normal faecal *E. coli*. Presumably, this ability to adhere to cell surfaces confers two advantages: first in preventing washout by the fast flows of urine, and secondly in permitting invasion of the mucosal surfaces of the bladder and renal pelvis. Such adhesive strains are invariably piliated, and removal of pili by vigorous washing or heating to 56 °C impairs binding to uroepithelial cells (Svanborg-Edén & Hansson, 1978).

Type I *E. coli* pili have been purified and many of their properties established (Salit & Gotschlich, 1977*a*, *b*). The pili are arranged in a peritrichous manner extending 1.5 μm from the surface of *E. coli* and have a diameter of 7 nm. By amino acid analysis the pili have a subunit molecular weight of 17099 dalton and contain 45% non-polar and 20% acidic residues. The binding efficiency of the purified pili was confirmed by demonstrating that as little as 3.3 μg pili could agglutinate red blood cells. To investigate the mechanism of pilus-mediated adhesion to renal tissue a standard cell line (Vero) derived from the kidney of the African green monkey was chosen. Replica techniques were used to examine the interaction between pilus and host cell membrane at high resolution while avoiding the disadvantages of sectioned preparations. The electron micrographs show that pilus contact with the membrane extends over a considerable length of the pilus. The authors conclude that attachment was not affected by a specialised tip to the pilus. This may be an unwarranted assumption; the appearance of pili lying on the membrane may be an artefact resulting from the critical-point drying technique used.

The findings of earlier workers (Old, 1972) that haemagglutination mediated by Type I pilus was susceptible to inhibition by mannosides suggested that pili might bind to mannose residues on renal cell membranes. This was confirmed by demonstrating that adhesion to Vero cells could be inhibited or reversed by α-methyl-D-mannose at 0.5 μM and yeast mannan at 10 μM. Further, blocking mannose residues on the cell membrane by treatment with Concanavalin A or *Lens culinaris* lectins inhibited attachment by over 80%. Treatment of Vero cells with α-mannosidase did not impair pilus binding, although treatment of red blood cells had this effect, causing a two-fold drop in the haemagglutination titre. Presumably, then, *E. coli* Type I pili function as lectins binding to mannose residues. In the urine, where osmolality and ionic strength fluctuate widely, interactions between host and pathogen must be affected. With the exception of pH these factors have not been studied. Pilus binding was maximal between pH 4 and pH 5, near to the pI of pili, where charge interactions would be minimal.

In summary, pilus-mediated adhesion to mannose residues on host-cell surfaces is a major determinant of *E. coli* virulence for the kidney. That antibodies block this attachment (Salit & Gotschlich, 1977*b*) raises the possibility of preparing a vaccine for the prevention of kidney infection. Clearly, we need to know if the mannose-binding site on the pilus is immunodominant and how many immunotypes of pili are important in human urinary tract infections. This analysis may be an over-simplification of the cohesive process since there is some evidence that mannose does not inhibit pilus binding to uroepithelial cells (Svanborg-Edén & Hansson, 1978). Future studies must utilise relevant cells from the human urinary tract.

MYCOPLASMA PNEUMONIA: MODIFICATION OF THE HOST CELL MEMBRANE

Mycoplasmas are the smallest and simplest self-replicating procaryotes. Infection with *M. pneumoniae* is usually limited to the upper respiratory tract, and when pneumonia does occur serious respiratory complications are rare. A characteristic feature of pneumonia is the late appearance of diverse symptoms including muscle and joint pains, nausea, diarrhoea and vomiting, nerve palsies, skin rashes and mouth ulcers. Of particular interest is the development of erythrocyte cold agglutinins which can result in anaemia from the breakdown of red cells in the cooler peripheral vessels of the skin. Cold agglutinins are IgM antibodies specific for the blood group *I* and *i* determinants; *I* and *i* are the precursors of the A, B, H and Lewis blood-group substances. Thus, we have a situation where a localised lung infection produces disease in other tissues and induces antibody formation against a component of self. These manifestations of *M. pneumoniae* infections have been reviewed in detail (Murray, Masur, Senterfit & Roberts, 1975).

Culture of sputum from patients with pneumonia grew out some 10^6 *M. pneumoniae* ml^{-1}. Electron micrographs showed that the organisms were attached to the epithelial cell luminal membranes, to microvilli and to cilia by a differentiated terminal structure (Collier & Clyde, 1974). Greater ultrastructural detail was observed using tracheal organ cultures challenged with virulent *M. pneumoniae* (Wilson & Collier, 1975). The terminal structure was seen to consist of a dense, rod-like core running through a lucent space enclosed by the organism's outer membrane. An extracellular layer, presumed to be mucoprotein on the basis of tannic acid staining, was especially thick in the region of the tip. This layer may form the loose network of fibrils observed to extend between the organism and the host cell surface. The specialist tip may also function to translocate mycoplasmas across the host cell surfaces. *M. pneumoniae* contain an actin-like protein (Neimark, 1977) and glide like myxobacteria on liquid-covered surfaces at speeds of 1–2 $\mu m\ s^{-1}$ (Razin, 1978). The tip is always the leading part, and avirulent *M. pneumoniae* unable to attach to cells also lack motility (Bredt, 1974).

No hard data are available on the chemical nature of the adhesion associated with the tip. The finding that treatment with 0.1 M $NaIO_4$ markedly reduced the binding of ^3H-labelled *M. pneumoniae* to tracheal organ cultures raised the possibility that the adhesin was a glycoprotein (Powell *et al.*, 1976). Although a membrane glycoprotein containing 52% glycine, 20% histidine and 5% carbohydrate has been extracted from *M. pneumoniae* using lithium diiodosalicylate (Kahane & Brunner, 1977) the fact that treatment of the organisms with proteases enhanced the binding of lectins (Kahane & Tully, 1976) suggested that little of the glycoprotein was expressed at the surface. Exposure of *M. pneumoniae* to 25 μg ml^{-1} trypsin selectively removed the adhesion, which reappeared on incubation in growth medium (Hu, Collier & Baseman, 1977). In this study a protein, designated P^1, which was highly susceptible to proteases and labelled with ^{125}I using the lactoperoxidase system, was tentatively identified as the 'adhesin'. No carbohydrate residues were detected in protein P^1.

There is now considerable evidence to suggest that sialic acid residues on human cell surfaces are involved in the adhesion of *M. pneumoniae*. In particular, treatment of tracheal organ cultures with neuraminidase or oxidation of the sialic acid residues with $NaIO_4$ impaired the binding of ^{3}H-labelled *M. pneumoniae* by 64% and 58% respectively (Powell *et al.*, 1976). However, extensive treatment of tracheal epithelial cells with neuraminidase still leaves a residual binding capacity of 40% (Engelhardt & Gabridge, 1977). This is explained if the adhesive processes involve more than one receptor. The alternative suggestion is that neuraminidase is unable to remove all sialic acid units from the receptors. Support for this view comes from studies on *M. gallisepticum* (Razin, 1978). When erythrocytes were exhaustively treated with neuraminidase, 5% of the original sialic acid units remained, and the cells exhibited about 20% of their original mycoplasma-binding capacity.

The adherence of *M. pneumoniae* to tracheal cell surfaces induces profound metabolic changes, resulting in death of the host cell (Hu, Collier & Baseman, 1975). Presumably, this interaction induces or uncovers cryptic *I* and *i* antigens on the host cell surface. The simplest explanation is that *M. pneumoniae* enzymes, lipids or toxic products such as H_2O_2 (Razin, 1978) in some way modify the host cell membrane. Support for this concept comes from the finding that human O red blood cells preincubated with *M. pneumoniae* induced four-fold rises in the titres of cold agglutinins in 32% of rabbits tested (Feizi, Taylor-Robinson, Shields & Carter, 1969). This is comparable to the proportion of human infections which develop cold agglutinins, but in man the antibody titres may be considerably greater. Another suggestion is that the *M. pneumoniae* outer membrane perturbs and subsequently fuses with the host limiting membrane (Gabridge, Barden-Stahl, Polisky & Engelhardt, 1977). These workers found that ^{3}H-labelled *M. pneumoniae* membranes bind to neuraminidase-treated tracheal cells, and saturation of receptor sites for viable mycoplasmas fails to inhibit subsequent membrane adhesion. On the basis of these findings Gabridge *et al.* (1977) postulate that the specialist tip serves to make initial contact, permitting intimate contact between, and ultimate fusion of, the two limiting membranes. The potential biological advantages of such a system are that essential metabolites might leak from the host cell directly to the mycoplasma, and enzymes would have free access to the cell. There is no direct experimental evidence to support this concept for *M. pneumoniae*, although electron micrographs of the interaction of *M. gallisepticum* with erythrocytes suggest fusion may occur (Apostolov & Windsor, 1975). In other systems (Jones & Hirsch, 1971) gaps of 10 nm were apparent between the outer surface of *M. pulmonis* and host cells. Unfortunately, the methods used in these studies are inadequate for the problem; exact alignment of membranes and cell–cell junctions requires the use of a goniometer stage which permits tilting of the specimen.

The close association of *M. pneumoniae* with the host cell membrane seriously impairs protein and nucleic acid synthesis (Hu *et al.*, 1975). Of particular interest were studies utilising [^{14}C]galactose, a sugar which can replace glucose in mammalian cells but not in *M. pneumoniae*. Compared to control organ cultures, infected tracheal rings release a considerably greater proportion of [^{14}C]galactose as CO_2. The implication of this finding was that cells parasitised by *M. pneumoniae* could not utilise available carbon units for the synthesis of macromolecules. Thus, one can speculate that incomplete synthesis of membrane macromolecules, e.g. *I* and *i*, might be the basis for the auto-immune phenomena of *M. pneumoniae* infections. Clearly, this interesting disease warrants considerable attention from membrane biologists.

INTERACTION WITH PHAGOCYTES

The function of phagocytic cells, namely the detection, sequestration, killing and digestion of invading micro-organisms, is determined by the surface properties of the bacterium and the phagocytic membrane. The surface of the invading microbe may be considerably modified by interaction in the host with the IgG antibodies and/or the C3b component of complement. C3b may be generated by antibody–antigen interactions on the bacterial surface or as a result of bacterial components directly activating complement by the alternative pathway (Götze & Müller-Eberhard, 1976). The coated bacterium binds directly to receptors for the Fc segment of Ig and for C3b which are present on the phagocyte membrane. These interactions are the subject of a recent review (Elsbach, 1977). This specific opsonisation cannot account for the ability of phagocytes to engulf bacteria in the absence of serum factors or the uptake of inert particles such as latex beads or effete red blood cells. Presumably, then, bacteria with surfaces of an appropriate physicochemical nature directly adhere to, and are subsequently engulfed by, phagocytes.

In an extensive study, van Oss *et al.* (1975) explored the possibility that phagocytosis occurs primarily because of differences in the surface free energies of phagocyte and particle. These studies are based on the simplistic view that if engulfment of a bacterium is accompanied by a decrease in Helmholtz free energy the process will be favoured. Independent of the shape of bacterium (B) surrounded by the aqueous environment (W) an overall free energy change, ΔF_{net}, exists for the process of engulfing a bacterium so that $\Delta F_{net} = \gamma_{PB} - \gamma_{BW}$, where γ_{PB} is the phagocyte/bacterial interfacial tension and γ_{BW} is the bacterium 'water' interfacial tension. Using a telescope with cross-hairs attached to a goniometer, these authors have measured the experimental advancing contact angles (θ_a) for sessile drops of saline water on lawns of bacteria and monolayers of phagocytes. From these data it proved possible to derive values for γ_{PB}, γ_{BW} and γ_{PW} based on modifications of Young's equation, $\gamma_{SV} - \gamma_{SL} = \gamma_{LV} \cos\theta_e$, where γ_{SV} is solid/vapour interfacial tension, γ_{SL} solid/liquid interfacial tension and θ_e the equilibrium contact angle. Plots of contact angles as a function of phagocytosis show that for avirulent bacteria and intracellular pathogens such as *Brucella*, $\theta = 27°$ (exceeding that of the PMN ($\theta = 18°$)), while the value for capsulated organisms resistant to phagocytosis is lower ($\theta = 17°$). Thus, the experimental data support the principle that where $\theta_{bacterium}$ is greater than $\theta_{phagocyte}$ and ΔF_{net} becomes more negative than the phagocyte/water interfacial tension (-0.20 erg cm^{-2}), phagocytosis is enhanced. However, the view that the relative hydrophobicity of the bacterial and phagocyte membranes is the sole determinant of engulfment must be an oversimplification of a complex problem. First, the supporting experimental evidence must be viewed with suspicion because *Streptococcus pyogenes*, *Staphylococcus aureus* and *Salmonella typhimurium*, organisms normally resistant to phagocytosis, cluster in the high contact angle, easily engulfed group. A possible explanation is the failure to ensure the continued presence of labile virulence factors on the surface of the organisms tested. Further measurement of overall surface hydrophobicity by the contact-angle method ignores the possibility of cell membrane movement, of localised differences in hydrophobicity and, in particular, the role of specific receptors. Presumably, the effect of specific receptors accounts for the finding that the contact angle for *Listeria* ($\theta = 25.1°$) was unaffected by opsonisation with IgG plus complement, although this treatment enhanced phagocytosis. We have used the binding of ^{3}H-labelled gonococci to hydrophobic ligands on phenyl Sepharose gel as a means of comparing the relative hydrophobicity of variants of gonococcus P9 (Lambden *et al.*, 1979). The outer-membrane variants possessing proteins IIa* and IIb* were equally hydrophobic, as shown by binding to phenyl Sepharose, but the IIa* protein conferred a seven-fold advantage in leukocyte association. Presumably, Protein IIa* binds in a specific manner to the leukocyte membrane. Thus, while accepting the evidence for a general correlation between bacterial surface hydrophobicity and the facility of their uptake by phagocytes, specific adhesive mechanisms may be the critical determinants of the initial attachment to the leukocyte membrane. (See, however, the article by Edebo *et al.* in this volume.)

Freimer and his colleagues (1978) have presented data suggesting that the non-specific adhesion of bacteria to macrophages results from the binding of sugar residues on the bacterial surface to 'receptors' on the macrophage membrane. In these studies, some 10^8 to 10^{10} bacteria ml^{-1} were suspended in phosphate-buffered saline containing Ca^{2+} and Mg^{2+} and reacted with peritoneal exudate macrophages at 4 °C for 2 h. A wide range of commensal bacteria and the pathogens *Streptococcus pyogenes* and *Staphylococcus aureus* were found to adhere to the macrophage surface. This attachment was inhibited when 10 mM glucose or galactose was incorporated in the medium. The specificity of sugar inhibition was established using wild-type *Salmonella typhimurium* and a rough variant lacking O-chains. The attachment of the wild-type to the phagocyte surface was blocked using the constituent sugars of the lipopolysaccharide, glucose, galactose, glucosamine and rhamnose. The exception was mannose; this sugar is linked $\alpha 1 \rightarrow 4$ to rhamnose and substituted at position 3, with abequose forming a branch point. The core mutant lacking O-chains and galactose was inhibited only by the remaining sugar, glucose, and by lipid A. The implication of these results was that the macrophage membrane carries 'receptors' able to bind to a wide range of carbohydrate units. Presumably, steric factors prevent adhesion to the mannose residue. The nature of the presumptive receptor was explored using *Corynebacterium parvum* as the test organism (Ögmundsdóttir, Weir & Marmion, 1978). Treatment of the macrophage with trypsin, β-galactosidase or phospholipases A, C and D resulted in a 50 % inhibition of bacterial adhesion. Incubation in tissue-culture medium restored the ability to bind bacteria, presumably as a result of resynthesis of receptor molecules.

An interesting finding was that minimal oxidation with sodium periodate (1 mM $NaIO_4$ for 5 min) reduced binding by 60 %, but attachment was restored to 80 % of the control value by borohydride reduction. The authors suggest that at a concentration of 1 mM periodate has a specific effect, removing the terminal two carbon atoms from sialic acid residues and oxidising the alcohol group on C-7 to an aldehyde. Given this assumption, the restoration of adhesion by borohydride reduction of the aldehyde implies that hydroxyl groups on the macrophage membrane are essential for the attachment of carbohydrate units on the bacterial surface. Presumably, the cohesive mechanism would be through hydrogen-bonding. However, the specificity of the periodate effect is doubtful, since 1 mM periodate cross-links the carbohydrate-free protein spectrin in the red cell membrane (Gahmberg, Virtanen & Wartiovaara, 1978). Two factors should be considered before accepting this proposed mechanism for the ability of macrophages to engulf avirulent bacteria in the absence of specific antibody. First, sugar inhibition of bacterial binding to the macrophage surface may, in part, be non-specific. Thus, glucose at concentrations around 10 mM partially blocks phagocyte adhesiveness to glass (van Oss *et al.*, 1975). Secondly, the role of the proposed adhesive mechanism at physiological temperatures should be explored because bacteria such as *Strep. pneumoniae*, *Staph. aureus* and *S. typhimurium*, which adhere to the macrophage membrane at 4 °C, resist phagocytosis at body temperatures (Cline & Lehrer, 1968; Elsbach, 1977).

Many of the bacteria which invade or accidentally penetrate the tissues possess specific mechanisms for attachment to the body's surfaces. Whether these adhesins are important in the interaction of pathogens with phagocytes has been extensively investigated for the gonococcus. Swanson and his colleagues (Swanson *et al.*, 1974; King & Swanson, 1978) have demonstrated that non-piliated gonococci can exhibit either higher or lower levels of association with human polymorphonuclear leukocytes (PMNL) than piliated variants of the same strain. These findings led to the suggestion that a non-pilus factor, termed leukocyte association factor, was the primary determinant of the interaction of gonococci and PMNL. Recent studies (King & Swanson, 1978) show that leukocyte association factor is a protein of molecular weight 29000 dalton in strain MS111 and 28000 dalton in strain C109. That leukocyte association protein (LAP) is expressed at the gonococcal surface was confirmed by demonstrating that LAP was readily labelled by ^{125}I using the lactoperoxidase system, and was destroyed when the gonococci were exposed to trypsin. If, as reported by these workers, LAP does not facilitate gonococcal attachment to mucosal cells it is difficult to envisage the advantage to the gonococcus of developing a specific mechanism for leukocyte association.

Studies from our group (Lambden *et al.*, 1979) on isogenic mutants of gonococcus strain P9 demonstrated a family of some five distinct surface proteins in the molecular weight range 27500 to 29000 dalton which enhanced the binding of gonococci to buccal epithelial cells. One protein (IIb*) of molecular weight 28000 dalton was associated with decreased ability to bind to leukocytes, whereas all the other variants demonstrated an increased association. Thus, we were not able to identify a single LAP, although it may be relevant that the variant which shows greatest leukocyte association contains a single extra protein (IIa*) with a molecular weight of 28500 dalton, which is close to that reported for the LAP of King & Swanson (1978). Nevertheless, these data (Table 1) support the proposition that the primary function of adhesive protein on the gonococcal surface is attachment to mucosal cells.

Antiphagocytic strategies

Those bacteria which cause the classic, life-threatening pyogenic infections are invariably capsulated. Of the six capsular types of *Haemophilus influenzae* only one, capsular type b, is virulent, causing meningitis and septicaemia. By the age of 1 year children have acquired protective antibodies and infection is rare (Hoeprich, 1977). More successful pathogens such as *Neisseria meningitidis* and *Streptococcus pyogenes* possess a range of antigenically distinct capsules. Given that man is the only host of these pathogens, it is of interest that they have been able to generate antigenic diversity while retaining the biological function of the capsular molecule.

Invasive *N. meningitidis* always possess capsules and are resistant to phagocytosis by human PMNL (Frasch, 1979). Immunisation with pure capsular polysaccharide induces opsonic antibodies which stimulate PMNL uptake and destruction of the meningococci (Roberts, 1970). Meningococcal capsules are acidic polymers of sialic acid (Groups B and C) or 2 acetamido-2-deoxy-D-mannopyranosyl phosphate (Group A); the polymers are rigid rods with chain lengths varying from 76 to 150 sugar units (Liu, Gotschlich, Egan & Robbins, 1977). The antiphagocytic properties of such molecules may be multideterminant. First, the capsule is extremely hydrophilic and secondly, their acidic properties must increase the energy barrier of electrostatic repulsion present between the meningococcus and the phagocyte. Finally, the rigid structure of the polymer could give rise to steric repulsive forces as a consequence of mutually excluded volumes of polymer chains between the interacting bacterium and the phagocyte surface. A possible explanation for the ability to retain function while undertaking antigenic change is that the mannosamine phosphate capsular polymer of Group A, *N. meningitidis*, is the primitive structure and that *N*-acetylmannosamine is the precursor of the polyneuraminic acid capsule of Groups B and C (Liu, Gotschlich, Jonssen & Wysocki, 1971).

In *Streptococcus pyogenes* the biological role of the carbohydrate capsule is taken over by a peripherally exposed protein, the M-antigen. Electron-micrographs show the M-protein as hair-like projections from the streptococcal surface (Swanson, Hsu & Gotschlich, 1969). Chemical studies on the M-protein (Fischetti, Gotschlich, Siviglia & Zabriskie, 1976) indicate that it is composed of three proteins of molecular weights 28000, 31000 and 35000 dalton, which are assembled from a single low molecular weight peptide. On amino-acid analysis, M-protein was found to contain more than 75 % of polar amino acids, with acidic amino acids (glutamic, 20.4 % and aspartic, 14.5 %) the most common. That the M-protein, like the carbohydrate capsule of meningococci, is hydrophilic, carries a predominance of acidic groups and appears to be structurally organised in linear manner suggests a common functional basis for their antiphagocytic properties. Peptide maps of antigenically distinct M-proteins show that only a single peptide was common in the different types. Presumably, then, biological activity is dependent upon the tertiary structure exposing the same functional groups despite sequence variation (Fischetti, 1978).

Those properties of capsules which prevent the association of the pathogen with the phagocyte membrane could interfere with adhesion to mucosal surfaces. This has been demonstrated by Craven & Frasch (1978) who found that eight of twelve piliated, freshly isolated meningococci bound poorly to buccal epithelial cells, while variants with small capsules readily adhered. The ideal of a mechanism which both attaches the pathogen to a mucosal surface and serves as an antiphagocytic factor has been attained by the gonococcus. Several groups (Ofek, Beachey & Bisno, 1974; Dilworth, Hendley & Mandell, 1975) have reported that non-piliated gonococci were readily phagocytosed by PMN[2] but piliated variants resisted ingestion. That pili are antiphagocytic has been confirmed by demonstrating the opsonic properties of specific antipilus serum (Buchanan *et al.*, 1978). The microscopic studies of these workers and of Densen & Mandell (1978) show piliated gonococci adherent to the PMNL surface with their pili radiating across the host cell membrane. The mechanism by which these pili block ingestion is unknown. One possibility is that the mesh of pili adherent to the PMNL surface seriously disrupts membrane motility; certainly, it is difficult to envisage how a 'zipper-mechanism' of phagocytosis could function in this situation. Another suggestion is that adherent pili decrease the fluidity of the host cell membrane and this impairs particle ingestion. Preliminary evidence supporting this concept comes from spin-label studies on red blood cells (Senff, Sawyer & Haak, 1977). The concept of an adhesin which interferes with the motility of host cell membranes warrants further investigation.

CONCLUSIONS

Our current understanding of the adhesive properties of pathogenic bacteria is limited largely to describing at the ultrastructural level the processes involved. Nevertheless, many interesting biological problems have emerged.

Pili and flagella-associated adhesins permit bacteria to attach to mucosal surfaces at distances remote from the bacterial surface. This mechanism enables pathogens to retain polar, antiphagocytic surfaces but still adhere to body surfaces. Where adhesins are integral components of the bacterial surface, attachment to the host-cell membranes will result in the cell engulfing the bacterium by a 'zipper-mechanism'. This is the route of gonococcal invasion of mucosal surfaces. By contrast, in *M. pneumoniae*, where adhesins are localised to a specialised tip, invasion does not occur, the infection being limited to the bronchial mucosal surface. An interesting result of this adhesion is modification of the host cell membrane, inducing auto-immune disease.

Investigations into the molecular basis of these phenomena is fraught with difficulties. Certainly, the tools available for modifying or blocking receptors are crude and induce unwanted side-effects. Laboratory models rarely mimic the natural infection and non-specific attachment, for example by hydrophobic interactions, may obscure critical specific adhesive mechanisms. Finally, essential virulence factors may not be phenotypically expressed on laboratory-grown bacteria, such as the adhesive Protein II on the gonococcal surface. This problem is further complicated by the finding that Protein II exists in four molecular-weight variations in isogenic mutants of a single strain. Perhaps, fortunately, progress towards a new generation of vaccines directed against bacterial adhesins does not demand an understanding of the molecular basis of attachment. Inevitably, research will be concentrated on identifying and purifying adhesins in an immunogenic form and establishing the number of antigenic types.

REFERENCES

Apostolov, K. & Windsor, G. D. (1975). The interaction of *Mycoplasma gallisepticum* with erythrocytes. 1. Morphology. *Microbios*, **13**, 205–15.

Baselski, V. S., Upchurch, S. & Parker, C. D. (1978). Isolation and phenotypic characterisation of virulence-deficient mutants of *Vibrio cholerae*. *Infection and Immunity*, **22**, 181–8.

Bernfeld, W. K. (1972). How infectious is gonorrhoea? *British Medical Journal*, **4**, 173.

Bredt, W. (1974). Structure and motility. *Inserm*, **33**, 47–54.

Brinton, C. C., Bryan, J., Dillon, J.-A., Guerina, N., Jacobson, L. J., Labik, A., Lee, S., Levine, A., Lim, S., McMichael, J., Polen, S., Rogers, K., To, A. C.-C. & To, S. C.-M. (1978). Uses of pili in gonorrhea control: role of bacterial pili in disease, purification and properties of gonococcal pili, and progress in the development of a gonococcal pilus vaccine for gonorrhea. In *Immunobiology of* Neisseria gonorrhoeae, ed. G. F. Brooks, E. C. Gotschlich, K. K. Holmes, W. D. Sawyer & F. E. Young, pp. 155–78. Washington D.C.: The American Society of Microbiology.

Buchanan, T. M. & Pearce, W. A. (1976). Pili as a mediator of the attachment of gonococci to human erythrocytes. *Infection and Immunity*, **13**, 1483–9.

BUCHANAN, T. M., PEARCE, W. A. & CHEN, K. C. S. (1978). Attachment of *Neisseria gonorrhoeae* pili to human cells, and investigations of the chemical nature of the receptor for gonococcal pili. In *Immunobiology of* Neisseria gonorrhoeae, ed. G. F. Brooks, E. C. Gotschlich, K. K. Holmes, W. D. Sawyer & F. E. Young, pp. 242–9. Washington D.C.: The American Society for Microbiology.

CLINE, M. J. & LEHRER, R. I. (1968). Phagocytosis by human monocytes. *Blood*, **32**, 423–35.

COLLIER, A. M. & CLYDE, W. A. (1974). Appearance of *Mycoplasma pneumoniae* in lungs of experimentally infected hamsters and sputum from patients with natural disease. *American Review of Respiratory Disease*, **110**, 765–73.

CRAVEN, D. E. & FRASCH, C. E. (1978). Pili-mediated and non-mediated adherence of *Neisseria meningitidis* and its relation to invasive disease. In *Immunobiology of* Neisseria gonorrhoeae, ed. G. F. Brooks, E. C. Gotschlich, K. K. Holmes, W. D. Sawyer & F. E. Young, pp. 250–2. Washington D.C.: The American Society for Microbiology.

DE BAULT, L. E. & YOO, T. J. (1974). Acute effects of *Bordetella pertussis* vaccine on the surface cytomorphology of cells in culture. *Journal of Cell Biology*, **63**, 79a.

DENSEN, P. & MANDELL, G. L. (1978). Gonococcal interactions with polymorphonuclear neutrophils; importance of the phagosome for bactericidal activity. *Journal of Clinical Investigation*, **62**, 1161–71.

DERJAGUIN, B. V. & LANDAU, L. (1941). Theory of the stability of strongly charged lyophobic sols and of adhesion of strongly charged particles in solution of electrolytes. *Acta Physiochemica USSR*, **14**, 633–62.

DILWORTH, J. A., HENDLEY, J. O. & MANDELL, G. L. (1975). Attachment and ingestion of gonococci by human neutrophils. *Infection and Immunity*, **11**, 512–16.

DIRIENZO, J. M., NAKAMURA, K. & INOUYE, M. (1978). The outer membrane proteins of Gram-negative bacteria: biosynthesis, assembly and functions. *Annual Review of Biochemistry*, **47**, 481–532.

ELSBACH, P. (1977). Cell surface changes in phagocytosis. In *The Synthesis, Assembly and Turnover of Cell Surface Components*, ed. G. Poste & G. L. Nicolson, pp. 361–401. Amsterdam: Elsevier, North-Holland Biomedical Press.

ENGELHARDT, J. A. & GABRIDGE, M. G. (1977). Effect of squamous metaplasia on infection of hamster trachea organ cultures with *Mycoplasma pneumoniae*. *Infection and Immunity*, **15**, 647–55.

EUBANKS, E. R., GUENTZEL, M. N. & BERRY, L. J. (1977). Evaluation of surface components of *Vibrio cholerae* as protective immunogens. *Infection and Immunity*, **15**, 533–8.

FEIZI, T., TAYLOR-ROBINSON, D., SHIELDS, M. D. & CARTER, R. A. (1969). Production of cold agglutinins in rabbits immunised with human erythrocytes treated with *Mycoplasma pneumoniae*. *Nature, London*, **222**, 1253–6.

FISCHETTI, V. A. (1978). Streptococcal M protein extracted by non-ionic detergent. III Correlation between immunological cross-reactions and structural similarities with implications for antiphagocytosis. *Journal of Experimental Medicine*, **147**, 1771–8.

Fischetti, V. A., Gotschlich, E. C., Siviglia, G. & Zabriskie, J. B. (1976). Streptococcal M protein extracted by nonionic detergent I. Properties of the antiphagocytic and type-specific molecules. *Journal of Experimental Medicine*, **144**, 32–53.

Frasch, C. E. (1979). Noncapsular surface antigens of *Neisseria meningitidis*. *Seminars in Infectious Diseases*, **2**, 1–47.

Freimer, N. B., Ögmundsdóttir, H. M., Blackwell, C. C., Sutherland, I. W., Graham, L. & Weir, D. M. (1978). The role of cell wall carbohydrates in binding of microorganisms to mouse peritoneal exudate macrophages. *Acta Pathologica et Microbiological Scandinavica*, Section B, **86**, 53–7.

Gabridge, M. G., Barden-Stahl, Y. D., Polisky, R. B. & Engelhardt, J. A. (1977). Differences in the attachment of *Mycoplasma pneumoniae* cells and membranes to tracheal epithelium. *Infection and Immunity*, **16**, 766–72.

Gahmberg, C. G., Virtanen, I. & Wartiovaara, J. (1978). Cross-linking of erythrocyte membrane proteins by periodate and intramembrane particle distribution. *Biochemical Journal*, **171**, 683–6.

Götze, O. & Müller-Eberhard, H. J. (1976). The alternative pathway of complement activation. *Advances in Immunology*, **24**, 1–35.

Griffin, F. M., Griffin, J. A., Leider, J. E. & Silverstein, S. C. (1975). Studies on the mechanism of phagocytosis. I. Requirements for circumferential attachment of particle-bound ligands to specific receptors on the macrophage plasma membrane. *Journal of Experimental Medicine*, **142**, 1263–82.

Griffin, F. M., Griffin, J. A. & Silverstein, S. C. (1976). Studies on the mechanism of phagocytosis. II. The interaction of macrophages with anti-immunoglobulin IgG-coated bone marrow-derived lymphocytes. *Journal of Experimental Medicine*, **144**, 788–809.

Guentzel, M. N. & Berry, L. J. (1975). Motility as a virulence factor for *Vibrio cholerae*. *Infection and Immunity*, **11**, 890–7.

Guentzel, M. N., Field, L. H., Eubanks, E. R. & Berry, L. J. (1977). Use of fluorescent antibody in studies of immunity to cholera in infant mice. *Infection and Immunity*, **15**, 539–48.

Harkness, A. H. (1948). The pathology of gonorrhoea. *British Journal of Venereal Diseases*, **24**, 137–47.

Heckels, J. E. (1978). The surface properties of *Neisseria gonorrhoeae*: topographical distribution of the outer membrane protein antigens. *Journal of General Microbiology*, **108**, 213–19.

Heckels, J. E., Blackett, B., Everson, J. S. & Ward, M. E. (1976). The influence of surface charge on the attachment of *Neisseria gonorrhoeae* to human cells. *Journal of General Microbiology*, **96**, 359–64.

Heckels, J. E., Lambert, P. A. & Baddiley, J. (1977). Binding of magnesium ions to cell walls of *Bacillus subtilis* W23 containing teichoic acid or teichuronic acid. *Biochemical Journal*, **162**, 359–65.

Hermodson, M. A., Chen, K. C. S. & Buchanan, T. M. (1978). *Neisseria* pili proteins: amino-terminal amino-acid sequences and identification of an unusual amino-acid. *Biochemistry*, **17**, 442–5.

Hoeprich, P. D. (1977). *Infectious Diseases*. Hagerstown, Maryland: Harper & Row.

Holmes, K. K., Johnson, D. W. & Throstle, H. J. (1970). An estimate of the risk of men acquiring gonorrhea by sexual contact with infected females. *American Journal of Epidemiology*, **91**, 170–4.

Hu, P. C., Collier, A. M. & Baseman, J. B. (1975). Alterations in the metabolism of hamster tracheas in organ culture after infection by virulent *Mycoplasma pneumoniae*. *Infection and Immunity*, **11**, 704–10.

Hu, P. C., Collier, A. M. & Baseman, J. B. (1977). Surface parasitism by *Mycoplasma pneumoniae* of respiratory epithelium. *Journal of Experimental Medicine*, **145**, 1328–43.

James, A. N., Knox, J. M. & Williams, R. P. (1976). Attachment of gonococci to sperm. Influence of physical and chemical factors. *British Journal of Venereal Diseases*, **52**, 129–35.

Jan, K. M. & Chien, S. (1973*a*). Influence of the ionic composition of fluid medium on red cell aggregation. *Journal of General Physiology*, **61**, 655–68.

Jan, K. S. & Chien, S. (1973*b*). Role of surface electric charge in red blood cell interactions. *Journal of General Physiology*, **61**, 638–54.

Jephcott, A. E., Reyn, A. & Birch-Anderson, A. (1971). *Neisseria gonorrhoeae*. III. Demonstration of presumed appendages to cells from different colony types. *Acta Pathologica et Microbiologica Scandanavica*, Section B, **79**, 437–9.

Johnson, A. P., Taylor-Robinson, D. & McGee, Z. A. (1977). *Neisseria gonorrhoeae* in fallopian tube organ cultures. In *Gonorrhoea Epidemiology and Pathogenesis*, ed. F. A. Skinner, P. D. Walker & H. Smith. London, New York: Academic Press.

Johnston, K. H., Holmes, K. K. & Gotschlich, E. C. (1976). The serological classification of *Neisseria gonorrhoeae*. I. Isolation of the outer membrane complex responsible for serotype specificity. *Journal of Experimental Medicine*, **143**, 741–58.

Jones, G. W., Abrams, G. D. & Freter, R. (1976). Adhesive properties of *Vibrio cholerae*: adhesion to isolated rabbit brush border membranes and haemagglutinating activity. *Infection and Immunity*, **14**, 232–9.

Jones, G. W. & Freter, R. (1976). Adhesive properties of *Vibrio cholerae*: nature of the interaction with isolated rabbit brush border membranes and human erythrocytes. *Infection and Immunity*, **14**, 240–5.

Jones, T. C. & Hirsch, J. G. (1971). Interaction *in vitro* of *Mycoplasma pulmonis* with mouse peritoneal macrophages and L-cells. *Journal of Experimental Medicine*, **133**, 231–59.

Kahane, I. & Brunner, H. (1977). Isolation of a glycoprotein from *Mycoplasma pneumoniae* membranes. *Infection and Immunity*, **18**, 273–7.

Kahane, I. & Tully, J. G. (1976). Binding of plant lectins to mycoplasma cells and membranes. *Journal of Bacteriology*, **128**, 1–7.

Kincaid-Smith, P. & Farley, K. F. (1970). *Renal Infection and Renal Scarring*. Melbourne: Mercedes.

King, G. J. & Swanson, J. (1978). Studies on gonococcus infection. XV. Identification of surface proteins of *Neisseria gonorrhoeae* correlated with leukocyte association. *Infection and Immunity*, **21**, 575–84.

Kroeks, M. V. A. M. & Kremer, J. (1977). The pH in the lower third of the genital tract. In *The Uterine Cervix in Reproduction*, ed. V. Insler & G. Bettendorf, pp. 109–117. Stuttgart: Thieme.

Lambden, P. R. & Heckels, J. E. (1979). The influence of outer membrane protein composition on the colonial morphology of *Neisseria gonorrhoeae* strain P9. *FEMS Microbiology Letters*, **5**, 263–5.

Lambden, P. R., Heckels, J. E., James, L. T. & Watt, P. J. (1979). Variation in surface protein composition associated with virulence properties in opacity types of *Neisseria gonorrhoeae*. *Journal of General Microbiology*, **114**, 305–12.

Liu, T.-Y., Gotschlich, E. C., Egan, W. & Robbins, J. B. (1977). Sialic acid-containing polysaccharides of *Neisseria meningitidis* and *Escherichia coli* strain Bos-12: structure and immunology. *Journal of Infectious Diseases*, **136**, Supplement S71–7.

Liu, T.-Y., Gotschlich, E. C., Jonssen, E. K. & Wysocki, J. R. (1971). Studies on the meningococcal polysaccharides. I. Composition and chemical properties of the group A polysaccharide. *Journal of Biological Chemistry*, **246**, 2849–58.

Lowe, T. L. & Kraus, S. J. (1976). Quantitation of *Neisseria gonorrhoeae* from women with gonorrhoea. *Journal of Infectious Diseases*, **133**, 621–6.

Mårdh, P.-A. & Weström, L. (1976). Adherence of bacteria to vaginal epithelial cells. *Infection and Immunity*, **13**, 661–6.

Murray, H. W., Masur, H., Senterfit, L. B. & Roberts, R. B. (1975). The protean manifestations of *Mycoplasma pneumoniae* infection in adults. *American Journal of Medicine*, **58**, 229–42.

Neimark, H. C. (1977). Extraction of an actin-like protein from the prokaryote *Mycoplasma pneumoniae*. *Proceedings of the National Academy of Sciences, USA*, **74**, 4041–5.

Nelson, E. T., Clements, J. D. & Finkelstein, R. A. (1976). *Vibrio cholerae* adherence and colonisation in experimental cholera: electron microscopic studies. *Infection and Immunity*, **14**, 527–47.

Ofek, I. & Beachey, E. H. (1978). Mannose binding and epithelial cell adherence of *Escherichia coli*. *Infection and Immunity*, **22**, 247–54.

Ofek, I., Beachey, E. H. & Bisno, A. L. (1974). Resistance of *Neisseria gonorrhoeae* to phagocytosis: relationship to colonial morphology and surface pili. *Journal of Infectious Diseases*, **129**, 310–16.

Ögmundsdóttir, H. M., Weir, D. M. & Marmion, B. P. (1978). Binding of microorganisms to the macrophage plasma membrane; effects of enzymes and periodate. *British Journal of Experimental Pathology*, **59**, 1–7.

Old, D. C. (1972). Inhibition of the interaction between fimbrial haemagglutins and erythrocytes by D-mannose and other carbohydrates. *Journal of General Microbiology*, **71**, 149–57.

Ovčinnikov, N. M. & Delektorskij, V. V. (1971). Electron microscope studies of gonococci in the urethral secretions of patients with gonorrhoea. *British Journal of Venereal Diseases*, **47**, 419–39.

Pearce, W. A. & Buchanan, T. M. (1978). Attachment role of gonococcal pili, optimum conditions and quantitation of adherence of isolated pili to human cells *in vitro*. *Journal of Clinical Investigation*, **61**, 931–43.

Perers, L., Andäker, L., Edebo, L., Stendahl, O. & Tagesson, C. (1977). Association of some enterobacteria with the intestinal mucosa of the mouse in relation to their partition in aqueous polymer two-phase systems. *Acta Pathologica et Microbiologica Scandinavica*, Section B, **85**, 308–16.

PERRY, M. B., DAOUST, V., JOHNSON, K. G., DIENA, B. B. & ASHTON, F. E. (1978). Gonococcal R-type lipopolysaccharides. In *Immunobiology of* Neisseria gonorrhoeae, ed. G. F. Brooks, E. C. Gotschlich, K. K. Holmes, W. D. Sawyer & F. E. Young, pp. 101–7. Washington D.C.: The American Society of Microbiology.

POWELL, D. A., HU, P. C., WILSON, M., COLLIER, A. M. & BASEMAN, J. B. (1976). Attachment of *Mycoplasma pneumoniae* to respiratory epithelium. *Infection and Immunity*, **13**, 959–66.

PUNSALANG, A. P. & SAWYER, W. D. (1973). Role of pili in the virulence of *Neisseria gonorrhoeae. Infection and Immunity*, **8**, 255–63.

RAZIN, S. (1978). The mycoplasmas. *Microbiological Reviews*, **42**, 414–70.

RICHARDS, K. L. & DOUGLAS, S. D. (1978). Pathophysiological effects of *Vibrio cholerae* and enteropathogenic *Escherichia coli* and their exotoxins on eukaryotic cells. *Bacteriology Reviews*, **42**, 592–613.

ROBERTS, R. B. (1970). The relationship between group A and group C meningococcal polysaccharides and serum opsonins in man. *Journal of Experimental Medicine*, **131**, 499–513.

ROBERTSON, J. N., VINCENT, P. & WARD, M. E. (1977). The preparation and properties of gonococcal pili. *Journal of General Microbiology*, **102**, 169–77.

ROSEMAN, S. (1970). The synthesis of complex carbohydrates by multiglycosyltransferase systems and their potential function in intercellular adhesion. *Chemistry and Physics of Lipids*, **5**, 270–97.

SALIT, I. E. & GOTSCHLICH, E. C. (1977*a*). Type I *Escherichia coli* pili, characterisation of binding to monkey kidney cells. *Journal of Experimental Medicine*, **146**, 1182–94.

SALIT, I. E. & GOTSCHLICH, E. C. (1977*b*). Hemagglutination by purified type I *Escherichia coli* pili. *Journal of Experimental Medicine*, **146**, 1169–81.

SENFF, L. M., SAWYER, W. D. & HAAK, R. A. (1977). In *Abstracts of the Annual Meeting of the American Society of Microbiology*. Washington D.C.: American Society of Microbiology Publications Office.

SMYTH, C. J., JONSSON, P., OLSSON, E., SÖNDERLIND, O., ROSENGREN, J., HJERTÉN, S. & WADSTRÖM, T. (1978). Differences in hydrophobic surface characteristics of porcine enteropathogenic *Escherichia coli* with or without K88 antigen as revealed by hydrophobic interaction chromatography. *Infection and Immunity*, **22**, 462–472.

SPARKES, R. A., PURRIER, B. G. A., WATT, P. J. & ELSTEIN, M. (1977). The bacteriology of the cervical canal in relation to the use of an intrauterine contraceptive device. In *The Uterine Cervix in Reproduction*, ed. V. Inster & G. Bettendorf, pp. 271–7. Stuttgart: Thieme.

STAMEY, T. A. (1972). *Urinary Infections*. Baltimore: William & Wilkins.

SVANBORG-EDÉN, C., ERIKSSON, B., HANSON, L. Å., JODAL, U., KAIJSER, B., LIDIN JANSON, G. L., LINBERG, U. & OLLING, S. (1978). Adhesion to normal human uroepithelial cells of *Escherichia coli* from children with various forms of urinary tract infection. *Journal of Pediatrics*, **93**, 398–403.

SVANBORG-EDÉN, C. & HANSSON, H. A. (1978). *Escherichia coli* pili as possible mediators of attachment to human urinary tract epithelial cells. *Infection and Immunity*, **21**, 229–37.

SWANSON, J. (1973). Studies on gonococcus infection. IV. Pili: their role in attachment of gonococci to tissue culture cells. *Journal of Experimental Medicine*, **137**, 571–89.

SWANSON, J. (1978). Studies on gonococcus infection. XIV. Cell wall protein differences among color/opacity colony variants of *Neisseria gonorrhoeae. Infection and Immunity*, **21**, 292–302.

SWANSON, J., HSU, K. C. & GOTSCHLICH, E. C. (1969). Electron microscope studies on streptococci. I. M antigen. *Journal of Experimental Medicine*, **130**, 1063–91.

SWANSON, J., KRAUS, S. J. & GOTSCHLICH, E. C. (1971). Studies on gonococcus infection. I. Pili and zones of adhesion: their relation to gonococcal growth patterns. *Journal of Experimental Medicine*, **134**, 886–906.

SWANSON, J., SPARKS, E., ZELIGS, B., SIAM, M. & PARROTT, C. (1974). Studies on gonococcus infection. V. Observations on *in vitro* interactions of gonococci and human neutrophils. *Infection and Immunity*, **10**, 633–44.

TAYLOR-ROBINSON, D., WHYTOCK, S., GREEN, C. J. & CARNEY, F. E. (1974). Effect of *Neisseria gonorrhoeae* on human and rabbit oviducts. *British Journal of Venereal Diseases*, **50**, 279–88.

TEBBUTT, G. M., VEALE, D. R., HUTCHISON, J. G. P. & SMITH, H. (1976). The adherence of pilate and non-pilate strains of *Neisseria gonorrhoeae* to human and guinea-pig epithelial tissues. *Journal of Medical Microbiology*, **9**, 263–73.

VAN OSS, C. J., GILLMAN, C. F. & NEUMANN, A. W. (1975). *Phagocytic Engulfment and Cell Adhesiveness as Cellular Surface Phenomena*. New York: Marcel Dekker.

VERWEY, E. J. W. & OVERBEEK, J. T. G. (1948). *Theory of the Stability of Lyophobic Colloids*. Amsterdam: Elsevier.

WARD, M. E. & WATT, P. J. (1972). Adherence of *Neisseria gonorrhoeae* to urethral mucosal cells: an electron microscopic study of human gonorrhea. *Journal of Infectious Diseases*, **126**, 601–5.

WARD, M. E. & WATT, P. J. (1975). Studies on the cell biology of gonorrhoea. In *Genital Infections and their Complications*, ed. D. Danielsson, L. Juhlin & P.-A. Mårdh, pp. 229–41. Stockholm: Almqvist & Wiksell.

WARD, M. E., WATT, P. J. & GLYNN, A. A. (1970). Gonococci in urethral exudates possess a virulence factor lost on subculture. *Nature, London*, **227**, 382–4.

WARD, M. E., WATT, P. J. & ROBERTSON, J. R. (1974). The human fallopian tube: a laboratory model for gonococcal infection. *Journal of Infectious Diseases*, **129**, 650–9.

WATT, P. J. & WARD, M. E. (1977). The interaction of gonococci with human epithelial cells. In *The Gonococcus*, ed. R. B. Roberts, pp. 652–5. New York: Wiley.

WATT, P. J., WARD, M. E., HECKELS, J. E. & TRUST, T. J. (1978). Surface properties of *Neisseria gonorrhoeae*: attachment to and invasion of mucosal surfaces. In *Immunobiology of* Neisseria gonorrhoeae, ed. G. F. Brooks, E. C. Gotschlich, K. K. Holmes, W. D. Sawyer & F. E. Young, pp. 253–7. Washington D.C.: The American Society for Microbiology.

WATT, P. J., WARD, M. E. & ROBERTSON, J. N. (1976). Interaction of gonococci with host cells. In *Sexually Transmitted Diseases*, ed. R. D. Caterall & C. S. Nichol, pp. 89–100. London, New York: Academic Press.

WEINBERG, E. D. (1978). Iron and infection. *Microbiological Reviews*, **42**, 45–66.

WILSON, M. H. & COLLIER, A. M. (1975). Ultrastructural study of *Mycoplasma pneumoniae* in organ culture. *Journal of Bacteriology*, **125**, 332–9.

YAMADA, K. M., YAMADA, S. S. & PASTAN, I. (1975). The major cell surface glycoprotein of chick embryo fibroblasts is an agglutinin. *Proceedings of the National Academy of Sciences, USA*, **72**, 3158–62.

YANCY, R. J., WILLIS, D. L. & BERRY, L. J. (1978). Role of motility in experimental cholera in adult rabbits. *Infection and Immunity*, **22**, 387–92.

Short-term and incomplete cell–substrate adhesion

JAN DOROSZEWSKI

Department of Biophysics and Biomathematics, Medical Centre of Postgraduate Education, Warsaw, Poland

Cells flowing in the vascular bed form, together with the walls of the blood or lymph, a coherent physiological system, in which adhesive or adhesive-like interactions play the major role. They operate also when the normal composition and functioning of this system is disturbed, e.g. when foreign cells appear or normal ones undergo a change. In these situations the action of these forces is in the first place defensive or repairing, but it may promote also some processes which are harmful for the organism, such as the dissemination of malignant cells and growth of secondary tumours.

Adhesive interactions of cells under flow conditions are difficult to examine in the natural environment and, therefore, most investigations are based on the use of artifical model systems, in which a microscope flow-chamber constitutes the main element. We have applied this technique in our previous studies on dynamic interactions of the leukaemic L1210 cells (Doroszewski, Skierski & Prządka, 1977*a*; Zachara & Doroszewski, 1978; Doroszewski, Gołąb-Meyer & Guryn, 1979) and L5222 cells (Doroszewski, Skierski & Prządka, 1977*b*). Other authors studied various phenomena connected with adhesivity of erythrocytes (Hochmuth *et al.*, 1972; Mohandas, Hochmuth & Spaeth, 1974), leukocytes (Madras, Morton & Petschek, 1971; Hochmuth *et al.*, 1972) and platelets (Friedman *et al.*, 1970; Friedman & Leonard, 1971; Poliwoda, Hageman & Jacobi, 1971; Turitto, Muggli & Baumgartner, 1977) with the aid of the flow-chamber method.

The arrest of cells on a surface adjacent to the stream in which they flow constitutes the basic and most easily detectable effect of action of adhesive bonds. Various observations and theoretical considerations, however, suggest that total immobilisation of a cell is not the only possible manifestation of adhesive forces and that cell arrest is not always a final effect of adhesion. It is the aim of this work (1) to study these incomplete, as it were, cell–substrate interactions of cells flowing in the medium stream which do not result in the final arrest of a cell but only impair its movement, and (2) to assess the time of cell immobilisation. These processes are connected with several problems, such as the mechanism of cell adhesion under flow conditions, the life cycle of normal and pathological cells in the blood and lymph, the kinetics of transplanted cells, malignant dissemination and metastasis. As the present study is a continuation of our previous investigations (cited above), it concerns the same leukaemic cell strains.

METHODS

Two types of malignant cell strains were used throughout this study: the undifferentiated leukaemia L5222 cells and the lymphatic leukaemia L1210 cells. The cells were grown as peritoneal exudate in BD IX rats (L5222) and in DBA/2 mice (L1210). The animals and cells which were necessary for starting the breeding of the L5222 strain had been kindly supplied by the Department of Cancer Research, University of Zurich, Switzerland. The procedure of preparing the cells for the experiments was similar for both cell types. The cells were collected on the fifth or sixth day after inoculation, washed twice with physiological saline and resuspended in Hanks' medium enriched with 0.5 % lactoalbumin hydrolysate or in Eagle's minimal essential medium (MEM) manufactured by Sera and Vaccines Plant, Lublin, Poland; the media were supplemented with 10 or 20 % calf serum. The final cell concentration was 5×10^6 to 10^7 cells ml^{-1}. The viability of cells was estimated by the trypan blue exclusion test by means of which about 95 % of cells were classed as viable. In control experiments, the locomotive activity of L5222 cells (Haemmerli, Felix & Sträuli, 1976) was checked: the cells always exhibited morphological features typical for active movement ability ('hand-mirror' form). During the experiments the medium temperature was maintained at 37 °C.

The cell suspension was perfused through the microscope flow chamber (for details of its construction see Doroszewski *et al.*, 1977*a*) the dimensions of which were length 30 mm, width 10 mm, height 0.12 mm. The upper and lower walls were made of glass (a coverslip and a microscope slide, respectively). The glass chamber was inserted

Table 1. *Hydrodynamic parameters of the flow chamber*

Symbol of fluid velocity	Flow rate (μl s^{-1})	Mean fluid velocity (μm s^{-1})	Fluid velocity at 5 μm height (μm s^{-1})	Shear rate (s^{-1})
I	0.063	52.9	12.8	2.56
II	0.115	95.9	23.2	4.64
III	0.204	169.7	41.0	8.21
IV	0.360	299.8	72.5	14.50
V	0.638	531.5	128.6	25.72

into a metal support and was connected with the inlet and outlet tubes of 2 mm internal diameter. The cell suspension was driven by a mechanical device (Skierski, Oleśniewicz & Wichniewicz, 1979) which consists of a synchronous electric motor connected through a high-precision gear-wheel system to a pushrod acting on a piston of a syringe containing the cell suspension. The device permits one to perfuse the chamber with any of twenty possible flow rates. In the present work, five fluid velocity levels were used; the flow rates were between 0.063 and 0.638 μl s^{-1} (see Table 1). The functioning of the mechanical driving system was precisely stabilised: the variations of the velocity of the syringe piston did not exceed 1 %.

In the experiments designed for verifying the hydrodynamic parameters, the medium flow was visualised by the suspension of blood platelets (taken from a rat), the velocity of the platelets was measured on a television screen (microscope magnification 200 ×). The measurements of the local linear fluid velocity were performed on different heights above the bottom of the chamber by adjusting the microscope focus-plane and estimating the time needed by platelets to cover a certain distance. The use of latex particles proved much less satisfactory in comparison with platelets. The fluid flow rate was estimated by weighing the fluid perfusing the chamber during a certain time. The near-wall cell velocity was estimated by measuring the passage-time of L1210 cells flowing in the focus-plane, which was so adjusted that cells adherent to the surface were seen as sharp images.

Two methods of recording the phenomena under study were used: (*a*) the velocity of cells during their flow in the chamber, was analysed by microcinematography at 16 frames per second. Optical magnifi-

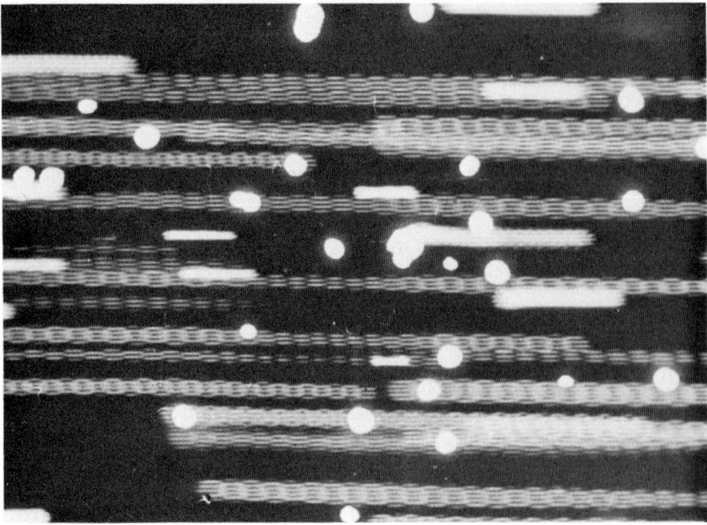

Fig. 1. A fragment of the television monitor screen during an experiment: the adhering cells with sharp contours may be easily differentiated from the moving ones, appearing as blurred shadows. Flow rate 0.638 $\mu l\ s^{-1}$, exposure time 1 s.

cation was 128 ×. The velocity estimation was based on measurements (on a magnifying screen) of the displacements of cell images on the film frames. Cell positions were recorded at 1/4-second intervals, i.e. every four frames. The error of the velocity estimation could not be reduced below ±10%, the main problem being the sharpness of cell images on movie pictures. The experiments were performed at a flow rate of 0.115 $\mu l\ s^{-1}$ (see Table 1). (*b*) In the study of the adhesion time and the cell–surface contact rate, the cells were observed on the screen of a television set (Fig. 1) connected to the microscope and recorded on videotape. In this way, long, continuous observation of the cell suspension flow was possible (up to 40 min with one cassette), while uninterrupted film sequences could not be longer than 30 s. The quality of pictures was poor in comparison with cinematography, but this was not essential for quantitative analysis, which was much easier and less time-consuming. Typical experiments consisted of recording the cell flow for either 5 or 40 min. The first of these two time-periods was sufficient to estimate the distribution of times of adhesions whose duration did not exceed one or two minutes. The time of adhesion of cells which were arrested on the surface for longer periods was assessed during 40 min continuous observation. Short times of adhesion (up to 2–3 min) were measured with a stop-watch, while longer times of adhesion were estimated by counting and identifying the adherent cells in 12.5 min intervals. In some experiments, the recording time was increased to 80 min by replacing a cassette without stopping the cell suspension flow.

During the observation, filming and videotape recording, the microscope objective focus-plane was adjusted at the level of a few micrometres above the chamber bottom; as has been shown in previous work (Zachara & Doroszewski, 1978), practically all leukaemic cells suspended in medium flowing with the velocities used in this study sink to the lower wall of the chamber before appearing in the observation region.

The adhesion efficiency was expressed as a percentage of the total number of cells that entered into the microscope field of vision; this value (N) was calculated according to the formula

$$N = \frac{mvt}{l}$$

where m = mean number of cells on the film frames or television screen, v = fluid velocity, t = time, l = width of the film frame or screen.

The cell adhesion-time distributions were compared by means of the chi-square test based on median values and the differences in the total adhesion rate were examined by Student's t-test.

RESULTS

Flow parameters

Some values characterising the cell suspension flow have been estimated experimentally and compared with theoretical values. The results of measurements of the linear fluid velocity by means of a biological flow-marker (platelets) are represented in Fig. 2. The fluid velocity distribution measured for three different flow rates is approximately parabolic and does not deviate significantly from the theoretical values. The velocity of cells (L1210) flowing in the near-wall region is in the range of values which correspond to those calculated for the distance of 5 to 8 μm above the bottom surface of the chamber (Fig. 3). The angular velocity of L5222 and L1210 cells flowing in conditions as above ranges from 2.8 to 3.3 rad s^{-1} when

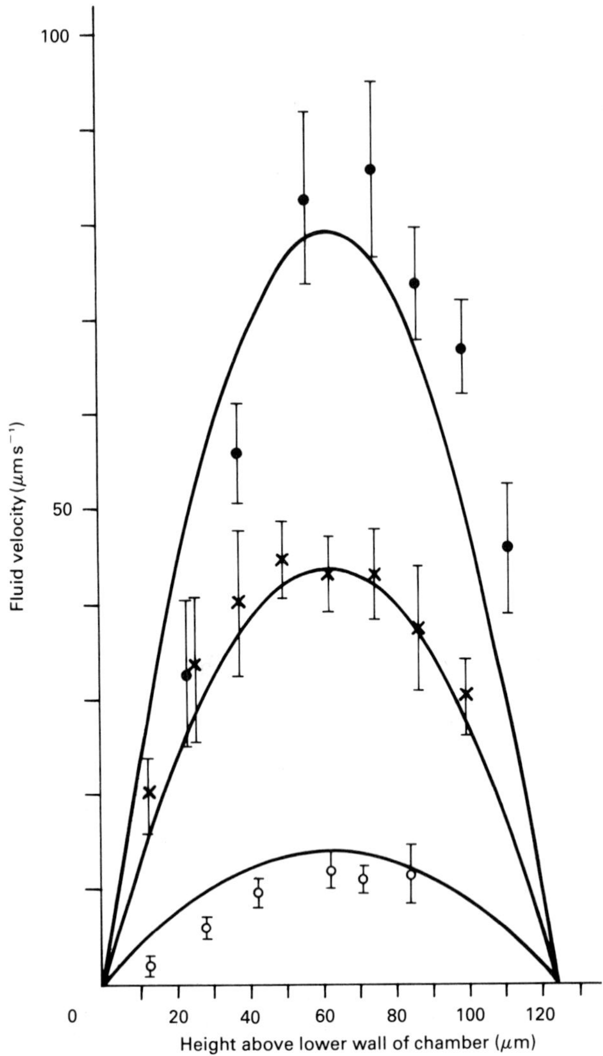

Fig. 2. Velocity distribution of the fluid in the flow chamber in the vertical plane. Measurements were performed for the following flow rates: 0.063 μl s^{-1} (●), 0.035 μl s^{-1} (×), 0.011 μl s^{-1} (○) (bars represent standard deviation), corresponding theoretical velocity-distribution curves are traced with continuous lines.

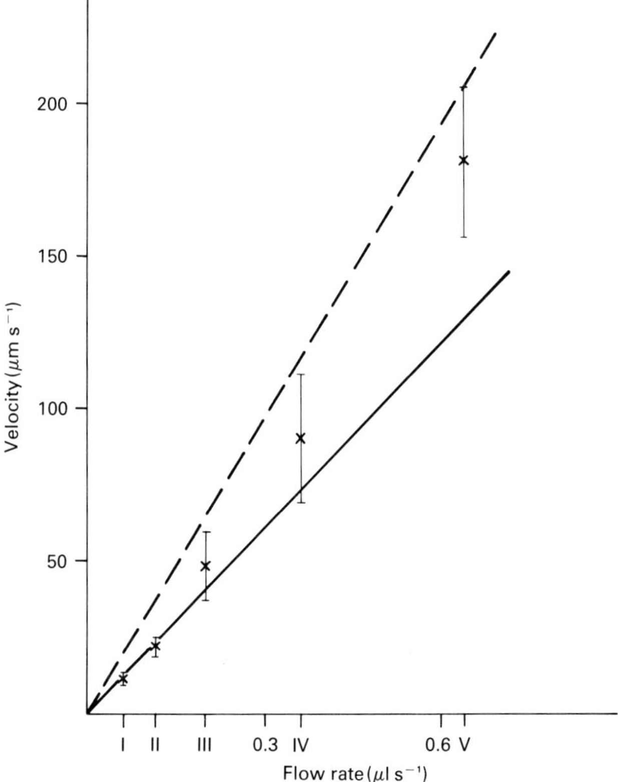

Fig. 3. Velocities of cells (L1210) flowing in the near-wall region. I to V, values used in the experiments (see Table 1). Fluid velocities (theoretical computation) at heights of 8 μm and 5 μm above the chamber bottom are traced by broken and continuous lines respectively.

the shear rate is equal to 5.3 s^{-1} and from 4.7 to 5.4 rad s^{-1} for a shear rate of 9.3 s^{-1} (Duszyk, unpublished data). The cell diameters are: 11.46 ± 1.0 μm and 10.30 ± 1.18 μm for L5222 and L1210 cells respectively.

Cell velocity before adhesion

In the search for cell adhesion records, the film bands corresponding to about 30 min of documented observation were examined (flow rate 0.115 μl s^{-1}). The fraction of cells which were arrested on the surface was 1–2 % of the total number of cells passing through the microscope's field of vision. From the group of adhering cells, those which had detectable movement on the film band for more than 2 s before

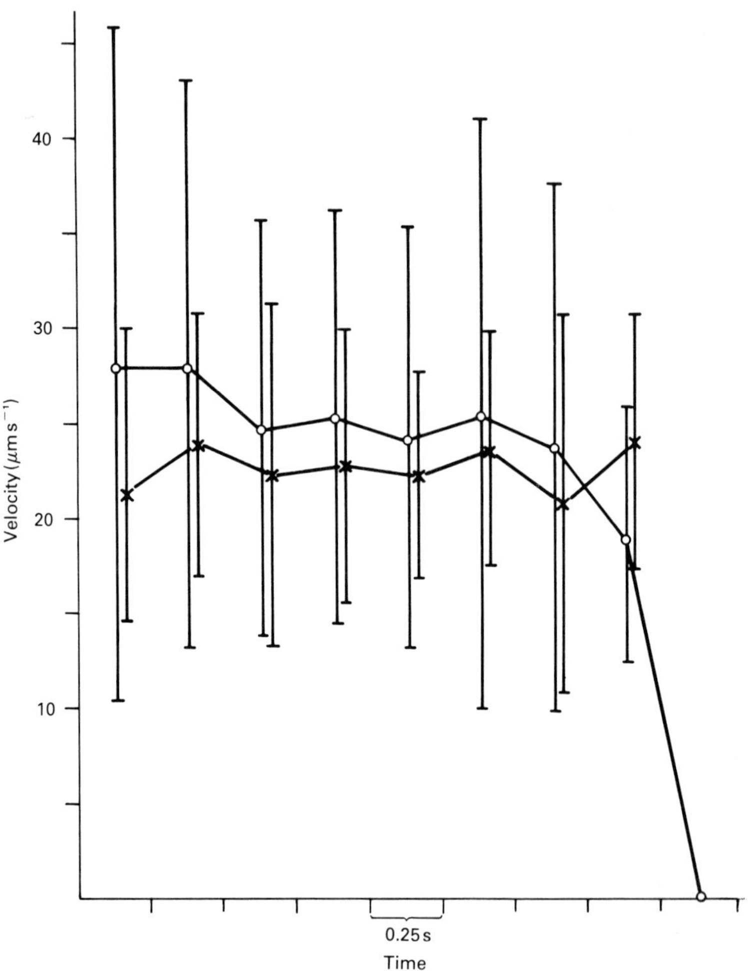

Fig. 4. Mean velocity of L 5222 cells (○) and L 1210 cells (×) during the two-second period preceding the arrest due to adhesion. Bars represent standard deviation.

arrest and which adhered for more than approximately 5 s were chosen for detailed analysis. Twenty-three L5222 cells and seventeen L1210 cells fulfilled these conditions.

Variations of cell velocity. The velocity of cells immediately before adhesion is similar for both cell types (L5222 and L1210 cells (Fig. 4)). In the given experimental conditions, the mean velocity of L5222

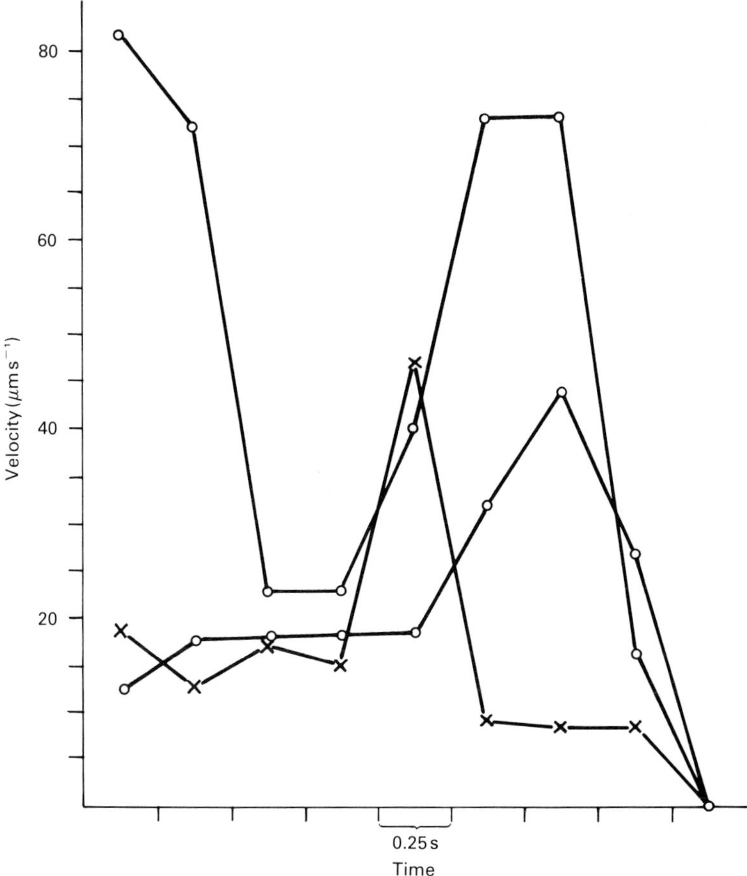

Fig. 5. Examples of abrupt slowing down and momentary acceleration of cells before immobilisation L5222 cells (○) and L1210 cells (×).

cells which are going to adhere to the surface is 25.0 μm s^{-1}, and that of L1210 is 22.3 μm s^{-1} (the fluid velocity at the cell level is 23.2 μm s^{-1}). The average final cell velocity recorded 0.25 s before its drop to zero in the moment of adhesion was 19.0 μm s^{-1} for L5222 cells and 23.9 μm s for L1210 cells. The corresponding values 2 s before adhesion were 27.9 μm s^{-1} and 21.3 μm s^{-1} for L5222 and L1210 respectively. Thus, the average velocity of cells before adhesion does not change in a significant manner until the very last moment (0.25 s or less) of their flow. The dispersion of cell velocities is, however, considerable. The standard deviation is equal to 35–64 % of the mean value, and it is a result of relatively great variability of velocities of

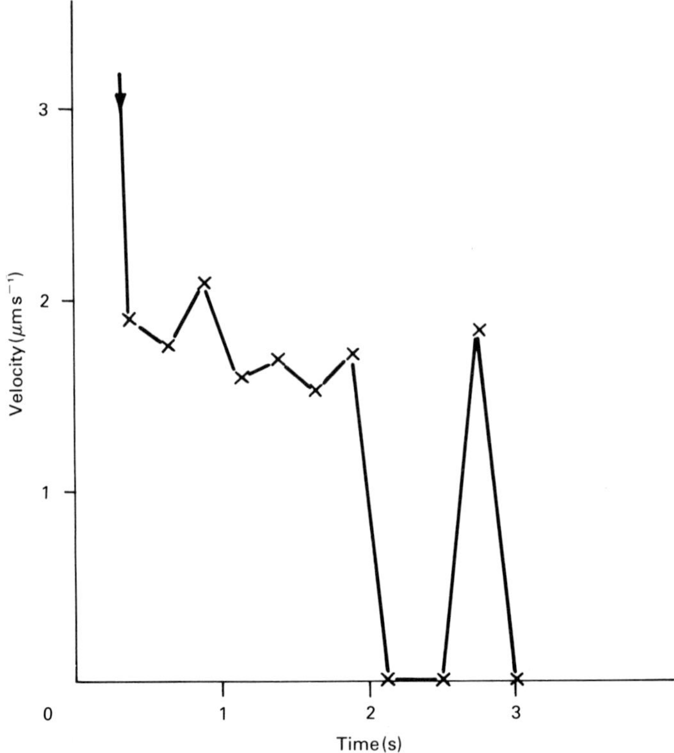

Fig. 6. Very slow movement of a cell (L1210) and final velocity variations preceding the arrest. Fluid velocity at the cell level was 12.7 μm s^{-1}.

individual cells. For example, the slowest cells moved with velocity of 8.8 μm s^{-1} (L5222) and 7.6 μm s^{-1} (L1210) and the maximal observed velocity of an L5222 cell was 49.5 μm s^{-1} and that of an L1210 one was 38.4 μm s^{-1}. Some cells flow during the period of observation with an approximately uniform speed and decelerate rather abruptly only in the last 0.25 s, others begin to slow down earlier and stop more gently. There are also numerous cells which exhibit great speed variations, sometimes abruptly braking, sometimes accelerating suddenly before stopping short (Fig. 5).

Slowed-down steady cell movement. In a certain number of cells a different type of movement may be observed. The cells do not exhibit considerable velocity variations, but during relatively long periods of time flow very slowly (Fig. 6); their speed is only a small fraction of

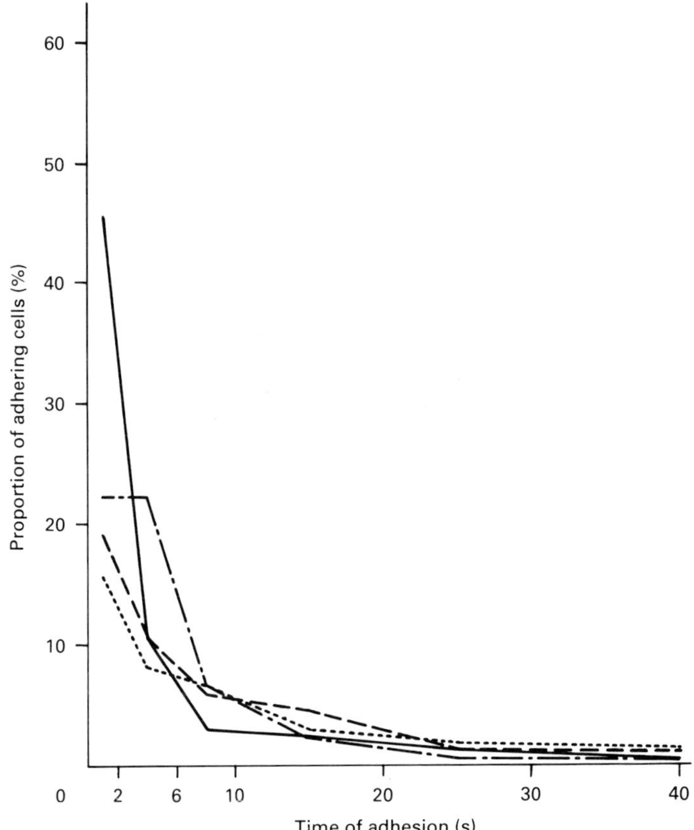

Fig. 7. Distribution of short-lived adhesion times in Hanks' medium +10% serum. L5222 cells: velocity I (–·–·) and V (········), L1210 cells: velocity I (——) and V (– – –).

that of other cells and of the fluid in the near-wall region. It was not possible to estimate exactly the frequency of appearance of such a steadily inhibited movement; the phenomenon is rather rare but unquestionable, and sometimes even very impressive. Mostly, but not always it is followed by an arrest of a cell during the observation period.

Short-lived adhesion

Numerous cells are adherent only for a short moment and then flow further. Such a temporary or short-lived adhesion is a frequent phenomenon with both L5222 cells and L1210 cells. The frame-

Table 2. *Distribution of short-lived cell-adhesion times around the median value (L5222 cells in Hank's medium +10% serum)*

Adhesion time (s)	Number and fraction of adhering cells										Total cell number
	I[a]		II[a]		III[a]		IV[a]		V[a]		
	n	%	*n*	%	*n*	%	*n*	%	*n*	%	
⩽ 9.0[b]	33	73.3	44	62.0	55	46.6	49	40.2	48	43.6	229
> 9.0[b]	12	26.7	27	38.0	63	53.4	73	59.8	62	56.4	237
Total	45	100.0	71	100.0	118	100.0	122	100.0	110	100.0	466
	Group median										
	3.9		5.4		9.8		12.75		13.5		
	Statistical significance										
	–		$P < 0.05$		$P < 0.01$		$P < 0.001$		$P < 0.001$		

[a] Fluid-velocity symbol (see Table 1). [b] Total (intergroup) median.

Table 3. *Distribution of short-lived cell-adhesion times around the median value (L1210 cells in Hank's medium + 10% serum)*

Adhesion time (s)	Number and fraction of adhering cells										Total cell number
	I[a]		II[a]		III[a]		IV[a]		V[a]		
	n	%	*n*	%	*n*	%	*n*	%	*n*	%	
⩽ 7.5[b]	24	72.7	34	63.0	28	36.8	54	50.5	60	49.6	200
> 7.5[b]	9	27.3	20	37.0	48	63.2	53	49.5	61	50.4	191
Total	33	100.0	54	100.0	76	100.0	107	100.0	121	100.0	391
	Group median										
	2.0		3.7		13.35		6.5		8.8		
	Statistical significance										
	–		$P > 0.30$		$P < 0.01$		$P < 0.05$		$P < 0.02$		

[a] Fluid-velocity symbol (see Table 1). [b] Total (intergroup) median.

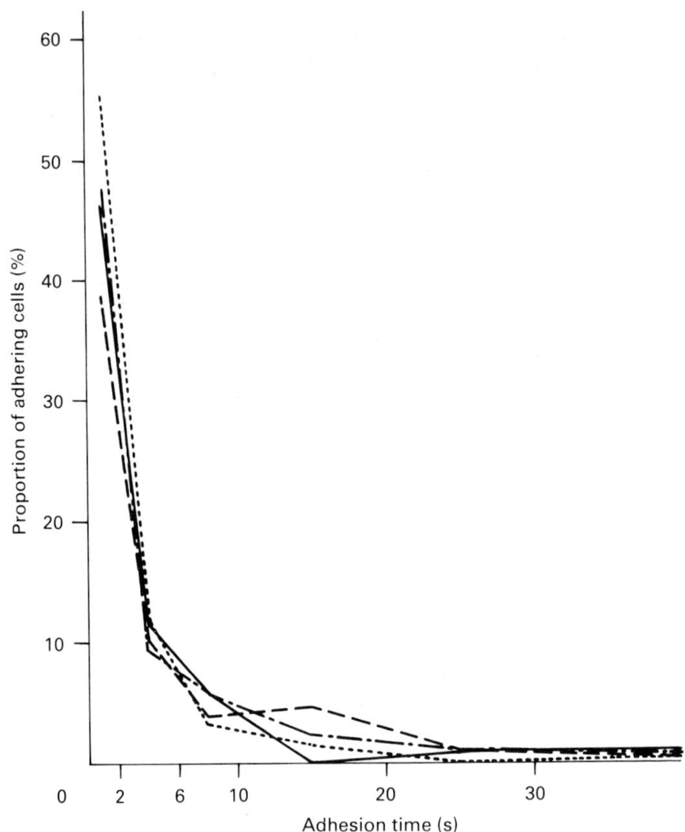

Fig. 8. Distribution of short-lived adhesion times in MEM + 10 % serum. Cell types and velocities are denoted as in Fig. 7.

by-frame film analysis suggests that the time of immobilisation varies from a fraction of a second to several seconds and even more. The possibility of observing the same cell continuously over a longer time is, however, limited by the duration of film sequences. Some cells which adhere during the observation time or are immobile at the beginning of observation do not detach before a given film sequence is finished. It is only by using the video-recording technique that a detailed study of the phenomenon of the short-lived cell adhesion is possible.

The distribution of cell adhesion times, as revealed by the back-play videotape analysis of 5 min sequences, is highly asymmetric, the great majority of cells adhering for periods shorter than 2 s (Fig. 7). The

Table 4. *Distribution of short-lived cell-adhesion times around the median value (L5222 cells in MEM + 10% serum)*

Adhesion time (s)	Number and fraction of adhering cells				Total cell number
	I[a]		V[a]		
	n	%	*n*	%	
⩽ 1.8[b]	25	47.12	26	55.32	51
> 1.8[b]	28	52.83	21	44.68	49
Total	53	100.0	47	100.0	100
	Group median				
	2.3		1.6		
	Statistical significance				
	–		$P > 0.40$		

[a] Fluid-velocity symbol (see Table 1). [b] Total (intergroup) median.

Table 5. *Distribution of short-lived cell-adhesion times around the median value (L1210 cells in MEM + 10% serum)*

Adhesion time (s)	Number and fraction of adhering cells				Total cell number
	I[a]		V[a]		
	n	%	*n*	%	
⩽ 3.6[b]	14	53.85	19	48.72	33
> 3.6[b]	12	46.15	20	51.28	32
Total	26	100.0	39	100.0	65
	Group median				
	2.5		3.7		
	Statistical significance				
	–		$P > 0.50$		

[a] Fluid-velocity symbol (see Table 1). [b] Total (intergroup) median.

adhesion time of the order of 2–10 s is observed in a few per cent of cells, and only an insignificant fraction of cells adheres for several dozen seconds or more. The general shape of the adhesion-time distribution is similar for both types of cells suspended in a given

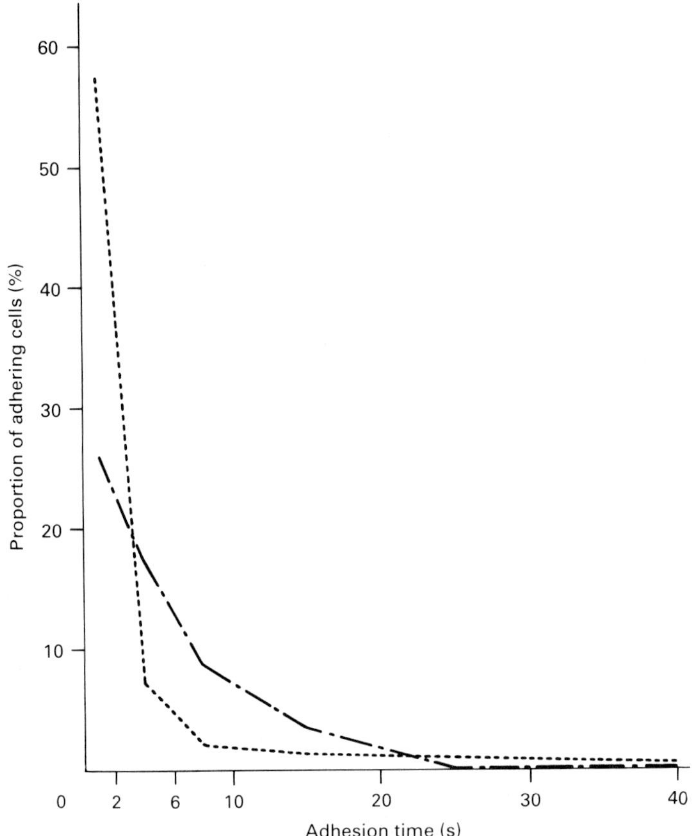

Fig. 9. Distribution of short-lived adhesion times in MEM + 20 % serum. L5222 cells, velocity I (–·–·) and V (········).

medium. The adhesion-time distribution depends, however, to a certain degree on the flow velocity (Tables 2 & 3): the number of cells which adhere for shorter times is greater when the adhesion takes place in slower medium velocity, and the adhesion times tend to increase when the cell suspension flows faster. The statistical analysis reveals in most cases that the differences are significant. In the second medium (MEM) the short-lived adhesion time distribution is comparable to that in Hanks' solution (Fig. 8). When the cells are suspended in the latter medium, however, the differences in the adhesion-time distribution due to the flow velocity are not detectable or statistically not significant (Tables 4 & 5).

The increase in the calf serum concentration, i.e. the supplement

Table 6. *Distribution of short-lived cell-adhesion times around the median value (L5222 cells, MEM + 20 % serum)*

Adhesion time (s)	Number and fraction of adhering cells										Total cell number
	I[a]		II[a]		III[a]		IV[a]		V[a]		
	n	%	*n*	%	*n*	%	*n*	%	*n*	%	
$\leqslant 2.8$[b]	10	43.5	11	44.0	15	57.7	7	26.9	31	66.0	74
> 2.8[b]	13	56.5	14	56.0	11	42.3	19	73.1	16	34.0	73
Total	23	100.0	25	100.0	26	100.0	26	100.0	47	100.0	147
	Group median										
	3.7		3.5		2.4		3.45		1.5		
	Statistical significance										
	–		$P > 0.30$		$P > 0.30$		$P > 0.50$		$P > 0.01$		

[a] Fluid-velocity symbol (see Table 1). [b] Total (intergroup) median.

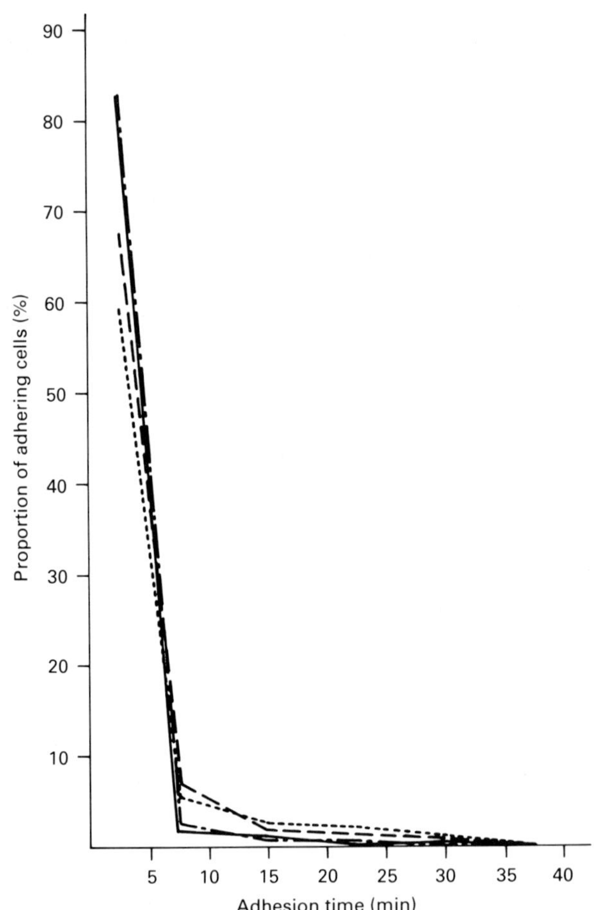

Fig. 10. Distribution of long-lasting adhesion times in MEM + 10 % serum. Cell types and velocities are denoted as in Fig. 7.

of 20 % instead of 10 %, does not influence appreciably the short-lived adhesion-time distribution (Fig. 9 and Table 6), which does not differ significantly in various velocity groups (with one exception, however, where the direction of the velocity effect is rather unexpected).

Adhesion of long duration

During a 5 min recording period not only the cells which stopped and afterwards detached were observed, but also those which either adhered and did not detach or were immobilised at the beginning of observation and moved away before its end. It was evident also from

Table 7. *Distribution of long-lasting cell-adhesion times*

Adhesion time (min)	Number of adhering L5222 cellscells		Number of adhering L1210		Total cell number	%
	I[a]	V[a]	I[a]	V[a]		
< 5.0	218	118	176	262	774	79.38
2.5–12.5	13	21	7	47	88	9.02
10.0–20.0	3	10	3	14	30	3.08
17.5–27.5	2	8	1	8	19	1.95
25.0–35.0	0	4	1	4	9	0.92
Total	251	184	197	343	975	100.0

[a] Fluid-velocity symbol (see Table 1).

the analysis of adhesion-time distribution that there exists a group of cells which adhere for relatively long periods of time. In order to estimate the adhesion time of these cells, the experiments with prolonged recording time (up to 40 and 80 min) were performed. The results of typical experiments of this kind are represented in Fig. 10 and Table 7. A small proportion of the total number of adherent cells are immobilised for time periods of the order of several minutes, and a small fraction of cells adhere for periods longer than half and hour.

Adhesion time of cells adhering under stationary conditions

In order to judge the effect of the conditions in which adhesion takes place (moving or resting medium), experiments were conducted in which the adherent cells contacted the surface while resting on it in an immobile suspension, instead of adhering in flow conditions. It appeared that the fraction of long-lived (adhesion time greater than 40 min) is much greater (Table 8): it amounted to *c*. 90% of the total number of cells which were adherent at the beginning of the action of the medium flow.

Cell detachment

Although this process has not been studied in detail, the observations lead to the conclusion that in numerous cases the detachment of a cell looks like a simple process: at a certain moment it starts to move.

Table 8. *Adhesion times of cells which adhered in stationary medium and were then subjected to medium flow*

	Number and fraction of cells L5222						L1210					
	I[a]			V[a]			I[a]			V[a]		
	n	% of *a*	% of *b*	*n*	% of *a*	% of *b*	*n*	% of *a*	% of *b*	*n*	% of *a*	% of *b*
(*a*) Cells in stationary contact with the surface	116	–	–	65	–	–	110	–	–	67	–	–
(*b*) Cells adhering after 40 min of static surface contact	45	38.8	–	40	61.5	–	58	52.7	–	39	58.2	–
(*c*) Cells from group *b* which have not detached after 40 min of medium flow	42	–	93.3	32	–	80.0	47	–	81.0	31	–	79.5

[a] Fluid-velocity symbol (see Table 1).

The mobilisation of an adherent cell, however, is not always an instantaneous event: frequently it takes some time, during which the cell begins to move at first slowly, then faster and only after such an initial, relatively slow, displacement for a distance of a few cell diameters, does it finally break off from the substrate.

Adhesion efficiency

The total number of adherent cells expressed as a fraction of the number of cells which enter with the flowing medium into the field of vision is either equal to a few per cent or is less than 1 %, depending on the suspension flow velocity. When studied in the slowest flow rate, the adhesion efficiency of L5222 and L1210 cells does not differ significantly, while other differences, i.e. those depending on the flow velocity, type of medium and serum supplement are moderately or highly significant ($P < 0.05$ to $P < 0.001$).

DISCUSSION

The experiments and observations described above were planned to investigate the attachment of cells to a surface from a flowing fluid. This type of adhesion may act on any cell type in the blood or lymph. A clear understanding of the processes involved in these types of attachment should be of value in the investigation of inflammation, metastasis and the normal emigration of leukocytes and lymphocytes into the surrounding solid tissues.

The experimental system used, in which the attachment of cells to the wall of a transparent tube is observed, is, I suggest, a good model for attachment from the blood in certain parts of the vascular bed.

The effect of the following factors upon attachment was studied.

1. Flow rate of the cell suspension
2. Different media with different serum supplement
3. Cell type (L5222 or L1210): L5222 cells are capable of active crawling movement on solid surfaces while L1210 are not.

The experimental system used ensured that flow conditions were identical in all experiments. The cells tended to sediment into a zone 5–8 μm thick above the lower wall of the flow chamber. There were slight indications of a lifting force acting on the cells at the highest flow rates (Saffman, 1965; Blackshear, Bartelt & Forstrom, 1977). Various perturbations of the cell flow were observed, namely.

1. The cells showed turbulent flow patterns, due most probably to temporary bonding of cells to the substrate.

2. Cell velocity lag in a certain fraction of cells. The ratio of cell average velocity to fluid velocity (in the same layer) was much smaller than would be expected on theoretical grounds. (Goldman, Cox & Brenner, 1967; Happel & Brenner, 1973). I suggest that this lag of the cell movement may be due to interactions with the wall material.

3. Arrest of the cells on the walls due to adhesion was a fairly frequent occurrence. Most arrests were short-lived and lasted for seconds or 1–2 min. I am uncertain as to whether the very long arrests (1 h or more) are fundamentally the same in mechanism as the short-term ones. Heterogeneity in the wall surface and elution or addition of adsorbed components to the wall might account for the changes in adhesion.

The general shape of the distribution of the cell adhesion-times suggests, without, however, proving it, that there exists two components of the curve: a rapidly decreasing short-time initial part and a 'tail' extending to relatively very long time intervals. However, as the analytical form of the cell adhesion-time distribution is unknown, it is impossible to know whether the experimental values are to be interpreted as belonging to one or more groups. It may be hoped that a certain light will be thrown on this problem by the mathematical model of the dynamic cell-surface adhesion which is now under preparation (in collaboration with T. Ruijgrok and Z. Gołab-Meyer). It is evident, however, that the conditions in which the adhesion takes place, influence to a considerable degree the subsequent cell-surface interaction time: the fraction of long-lasting adhesion is considerably greater when the cells adhere in an immobile suspension (see Table 8).

There are two processes which may lead to detachment of the adhering cells from the substrate, namely 'pure de-hesion' and disruption of cell membrane (the term 'de-hesion' if the Latin origin is remembered). If the process of dynamic cell adhesion is interpreted in the frame of the DLVO theory (Curtis, 1967; Weiss & Harlos, 1972), it may be supposed that the detaching cells adhere in the secondary energetic minimum. Such an adhesion is reversible, and a cell which receives energy from outside may be released from the surface; in this case the separation would occur at the 'molecular border' between the cell and the surface, without the structure of either cell or substrate being altered. The second possible mechanism is that under the impact of the shearing force, the cell membrane in the region of the adhesion centres is disrupted and the cell breaks off. This process may occur irrespective of the reversibility of stability of the cell–substrate bonds on molecular level. The latter hypotheses seems to be more probable in the light of the suggested role of cell extensions in the adhesive interactions with the substrate and is corroborated by the observation of the detaching process (see Results section and Doroszewski *et al.*, 1977*a*). One can consider only the energetic phenomena on molecular level, but it is much easier to explain the process of slow and gradual tearing away of a cell on the grounds of the cell-membrane disruption hypothesis, which could include the fatigue of bonds under stress.

In the analysis of the short-lived cell adhesion it is necessary to take into account two active biological processes: (1) the generation by the cell of new adhesive centres, e.g. by forming extensions directed towards the substrate, and (2) active 'unfastening' from the substrate. These processes probably play an important role in the cell–substrate adhesion of long duration, perhaps they even constitute the predominant factor in the ability of a cell to resist the flow shearing force during a long period.

The fact that the distribution of adhesion times is similar in different flow conditions suggests that the basic mechanism of the short-lived adhesion does not depend considerably on the value of the shearing force (in the range under examination) and on the medium composition. The only statistically significant difference consists in the relative prevalence of longer adhesion times when the suspension flow rate is greater. The possible explanation is that weak bonds which easily retain a cell for a short time in a slow fluid flow are not sufficiently strong to cause the same effect under the influence of a greater shearing force in the fast flow. The problem, however, is not clear and the above explanation does not account for the fact that the time-distribution differences appear only in experiments performed with Hanks' solution.

In the present study the shearing force operating on the adhering cells lies between 3.17×10^{-12} N and 31.9×10^{-12} N and the torque values range from 6.9×10^{-18} N m to 69×10^{-18} N m. These values are based on the previously published paper (Doroszewski *et al.*, 1979) where the relation between the cell suspension flow velocity and the number of adhering cells was investigated; the results were generally comparable to those obtained in the present study: the

Table 9. *Types of cell–substrate interactions under flow conditions*

	(steady cell flow)	
Without cell arrest	impairment of cell flow	irregular velocity variations
		uniform slowing down
	temporary immobilisation	short-lived
With cell arrest		long-lasting
	(permanent settling)	

adhesion efficiency is significantly smaller when the medium flow rate increases. Several authors have calculated the cell adhesion force (Curtis, 1967; Weiss, 1968, Brooks, Miller, Seaman & Vassar, 1969) and their estimations ranged from 10^{-11} to 10^{-12} N for a secondary energetic minimum. Other values, however, have also been given, e.g. 10^{-9} N as a result of computation based on a flat-plate model (Brooks *et al.*, 1969). Weiss (1968) has estimated that mastocytoma cells were detached from the substrate by forces of 10^{-11} N per cell, but only an insignificant fraction of RPMI No. 41 cells were dislodged under the action of approx. 10^{-9} N per cell. Moreover, it appears that the number of adhering cells depends on the medium composition (Hanks' medium and MEM), although it is impossible to decide which of the medium components are responsible for this effect; the ion concentration, including calcium, is comparable in both media. What is important, however, is that the basic mechanism of the short-lived adhesion under flow conditions does not seem to be directly connected with the processes which are responsible for the global adhesion efficiency, i.e. it operates in a similar way in the situations in which the total number of adhering cells is great or small.

The possible fates of cells flowing in the near-wall fluid region are summarised in Table 9.

The results of this study may be viewed as contributing to the research on the process of malignant dissemination and metastatic growth, and they can also be considered more generally. The validity of the extrapolation of the findings arrived at in an artificial model system over the conditions *in vivo* has obvious limitations; some tentative hypotheses, however, may be advanced. Thus, it seems justifiable to suppose that in the blood and lymph vessels only a few per cent of the circulating cells adhere in one passage to the endothelium and that only an insignificant fraction of the cells which have adhered are able to settle, implant and divide, giving rise to malignant invasion. This last conclusion is based on the presumption that leukaemic cells which adhere to the vascular wall only for a short time are not likely to be able to start to proliferate and infiltrate the surrounding tissue. It appears, therefore, that despite the fact that the total cell adhesion rate under flow conditions is not negligible, the probability of initiating a sequence of events leading to the secondary tumour growth should be greater in the regions of the vascular bed in which the blood or lymph does not move. However, as the magnitude of shearing force in certain limits seems to influence the total cell adhesion rate rather than the cell–substrate contact time, it is probably that the local blood or lymph velocity (if not equal to zero) is not very important from this point of view. It is not clear, however, whether strictly stationary conditions in fact exist somewhere in the vascular system, even in such structure as the liver or lymph node sinuses. The near-wall fluid velocities in the flow chamber used in this study are comparable to those in small pre- and post-capillary vessels, especially in venules.

There are some possible connections between the results of the present work and the role of the lungs in the kinetics of transplanted cells. It has been demonstrated in numerous studies that intravenously injected lymphocytes and L1210 cells (Doroszewski, Kowalczyńska & Zurowski, 1968; Skierski & Doroszewski, 1977 and others), as well as other cell types, are retained in the lungs, but only for a limited period of time, after which they are released into the circulation. Temporary adhesion of these cells in the lung vascular bed would constitute a plausible explanation of this phenomenon; it would require only the assumption of somewhat longer cell arrest times than those which correspond to the short-lived adhesion to the serum-coated glass. The fact that only a small fraction of adhering cells is able to resist the action of a shearing force may be of some importance from the point of view of the mechanisms operating in the passage through the vessel walls of polymorphonuclear leukocytes in the acute inflammatory reaction and of lymphocytes in their recirculation process. From this point of view further investigation embracing not only malignant but also normal cells would be needed.

It seems probable that the adhesive, reversible contacts of cells with the vessel walls, may promote some important processes; (1) the exchange of material between the membrane of the flowing cells and the surface of the endothelial cells, (2) modification of some surface-dependent properties of the flowing cells, and (3) local changes of properties of the endothelial lining. At present the exact significance of these cell–substrate mutual influences could only be the object of more or less justified speculation, and it is only in the light of further study that some of these hypotheses may stand out more clearly.

Acknowledgements

I gratefully acknowledge the valuable aid in statistical analysis of Miss Stanisława Sęk and the excellent technical assistance of Miss Renata Grudniewicz. The work was supported by grant II.1 from the Polish Academy of Sciences.

REFERENCES

Blackshear, P. L., Bartelt, K. W. & Forstrom, R. J. (1977). Fluid dynamic factors affecting particle capture and retention. *Annals of the New York Academy of Sciences*, **283**, 270–9.

Brooks, D. E., Miller, J. S., Seaman, G. V. F. & Vassar, P. S. (1969). Some physico-chemical factors relevant to cellular interaction. *Journal of Cellular Physiology*, **69**, 155–65.

Curtis, A. S. G. (1967). *The Cell Surface: Its Molecular Role in Morphogenesis.* London: Logos Press.

Doroszewski, J., Gołąb-Meyer, Z. & Guryn, W. (1979). Adhesion of cells flowing in the fluid stream: effects of shearing force and cell kinetic energy. *Microvascular Research* (in press).

Doroszewski, J., Kowalczyńska, H. & Zurowski, S. (1968). Kinetics of intravenously transplanted lymphoid cells. *Bulletin de l'Académie Polonaise des Sciences*, **11**, 729–34.

Doroszewski, J., Skierski, J. & Prządka, L. (1977*a*). Interaction of neoplastic cells with glass surface under flow conditions. *Experimental Cell Research*, **104**, 335–43.

Doroszewski, J., Skierski, J. & Przadka, L. (1977*b*). Adhesion of L5222 leukaemia cells under flow conditions. *Leukemia Research*, **2/3**, 207–8.

Friedman, L. I., Liem, H., Grabowski, E. F., Leonard, E. F. & McCord, C. W. (1970). Inconsequentiality of surface properties for initial platelet adhesion. *Transactions of the American Society of Artificial Internal Organs*, **16**, 63–76.

FRIEDMAN, L. J. & LEONARD, E. F. (1971). Platelet adhesion to artificial surfaces: consequences of flow, exposure time, blood condition, and surface nature. *Federation Proceedings*, **30**, 1641–6.

GOLDMAN, A. J., COX, R. G. & BRENNER, H. (1967). Slow viscous motion of a sphere parallel to a plane wall. II. Couette flow. *Chemical Engineering Science*, **22**, 653–60.

HAEMMERLI, G., FELIX, H. & STRÄULI, P. (1976). Motility of L5222 rat leukemic cells. *Virchow's Archiv Abteilung B Cell Pathology*, **20**, 143–54.

HAPPEL, J. & BRENNER, H. (1973). *Low Reynolds Number Hydrodynamics*. Leyden: Noordhoff International Publishing.

HOCHMUTH, R. M., MOHANDAS, N., SPAETH, E. E., WILLIAMSON, J. R. BLACKSHEAR, P. L. Jr. & JOHNSON, D. W. (1972). Surface adhesion, deformation and detachment at low shear of red and white cells. *Transactions of the American Society of Artificial Internal Organs*, **18**, 325–32.

MADRAS, P. N., MORTON, W. A. & PETSCHEK, H. E. (1971). Dynamics of thrombus formation. *Federation Proceedings*, **30**, 1665–76.

MOHANDAS, N., HOCHMUTH, R. M. & SPAETH, E. E. (1974). Adhesion of red cells to foreign surface in the presence of flow. *Journal of Biomedical Material Research*, **8**, 119–36.

POLIWODA, H., HAGEMAN, G. & JACOBI, E. (1971). Velocity dependent interaction between platelets and different surface. In *Theoretical and Clinical Hemorheology*, ed. H. H. Hartet & A. L. Copley, pp. 227–32, Berlin: Springer.

SAFFMAN, P. G. (1965). The lift on a small sphere in a slow shear flow. *Journal of Fluid Mechanics*, **22**, 383–400.

SKIERSKI, J. & DOROSZEWSKI, J. (1977). The role of the lungs and liver in the kinetics of transplanted lymphoid leukemia L1210 cells labelled with radiochromium ^{51}Cr. *Leukemia Research*, **4**, 309–13.

SKIERSKI, J., OLEŚNIEWICZ, M. & WICHNIEWICZ, W. (1979). Mechanical device for stabilized cell suspension perfusion. *Progress of Technology in Medicine* (in Polish) **10**, 241–9.

TURITTO, V. T., MUGGLI, R. & BAUMGARTNER, H. R. (1977). Physical factors influencing platelet deposition on subendothelium: importance of blood shear rate. *Annals of the New York Academy of Sciences*, **283**, 284–92.

WEISS, L. (1968). Studies on cellular adhesion in tissue culture. X. An experimental and theoretical approach to interaction forces between cells and glass. *Experimental Cell Research*, **53**, 603–13.

WEISS, L. & HARLOS, J. P. (1972). Short-term Interactions Between Cell Surfaces. *Progress in Surface Science*, **1**, 355–405.

ZACHARA, J. & DOROSZEWSKI, J. (1978). Analysis of the cell suspension flow in the parallel-plate channel. *Progress in Medical Physics*, **13**, 115–25.

Mechanisms regulating platelet adhesion

J. L. GORDON

ARC Institute of Animal Physiology, Babraham, Cambridge CB2 4AT, UK

1. INTRODUCTION

Platelets, which are the smallest formed elements circulating in the blood, do not normally adhere to any surface but circulate freely within the vascular lumen. Platelet adhesion is, however, important in haemostasis, in vascular disease, and in some aspects of the inflammatory process. Under normal circumstances platelets circulate in a quiescent state but, when activated, platelets can adhere to many surfaces: those that they usually come in contact with are the lining of the blood vessels and the surfaces of other blood cells. These are, therefore, the surfaces of the greatest biological importance in the context of platelet adhesion, but many experiments have been performed to study the interaction of platelets with artificial surfaces. These are also important because vascular prostheses are frequently inserted into damaged blood vessels, and how successful they are depends partly on the degree to which platelets adhere to them.

In practice, although platelets may collide with any other circulating blood cells, the only ones a platelet will stick to are other platelets. There is also specificity with respect to platelet adhesion to the vascular wall: platelets rarely stick to the endothelial cells that line the blood vessels, but if the endothelial layer is damaged platelets will stick avidly to the exposed subendothelial material – in fact, this is the main biological function of blood platelets and it initiates the haemostatic process. Platelet–platelet adhesion is usually called 'aggregation', as it leads to the formation of a platelet clump, or aggregate. Platelet adhesion to the vascular wall and platelet aggregation are intimately interconnected – indeed, haemostasis depends on the first process inducing the second – but they can be studied separately, and it is convenient to discuss some aspects of each process individually. As I indicated above, the main function of blood platelets is to adhere to the damaged surface of a blood vessel wall, then to adhere to each other and form a platelet plug, thus filling any breach in the vessel wall and preventing blood loss. Platelet–vascular adhesion and platelet aggregation can sometimes occur even when there is no complete breach in the vessel wall; in such cases the platelet plug intrudes into the lumen and can grow to occlude it completely, thus stopping blood flow – the process known as thrombosis. The consequences of thrombosis are well known: it is the acute event mainly responsible for strokes, heart attacks, and other manifestations of occlusive vascular disease, which accounts for more than 50 % of the morbidity and mortality in the Western World today. One of the characteristics of a thrombotic event is its suddenness; this, however, is understandable because the biological processes involved are essentially the same as those responsible for haemostasis, and we can easily appreciate the need for blood flow out of a cut vessel to be stopped as soon as possible. In summary, efficient regulation of platelet adhesion is vitally important because blood flow under normal conditions must not be interrupted by platelets sticking to each other or to the vascular lining; however, when a vessel is cut, platelet adhesion and aggregation must occur very rapidly in order to seal the breach.

We can, therefore, ask four interrelated questions.

1. What normally keeps platelets from adhering to the vascular lining or to other blood cells, including platelets?

2. What activates platelets and induces adhesion and aggregation?

3. What regulates the extent and duration of platelet adhesion and aggregation – that is, what switches these processes off?

4. What goes wrong in thrombosis?

I shall address these questions by considering in some detail the molecular mechanisms that regulate the processes of platelet adhesion, aggregation and secretion, drawing comparisons with other cell types where appropriate. Before doing this, however, it is important to set the subject in perspective by giving a brief historical overview of research on platelet functions and by giving an equally brief account of the structure and constituents of platelets.

2. RESEARCH ON PLATELET FUNCTION: AN HISTORICAL OUTLINE

In 1851 the English physician Wharton-Jones discovered that local injury to the small vessels in the web of a frog's foot resulted in 'an agglomeration of colourless corpuscles with a few red ones, held together, apparently, by coagulated fibrin'. This 'agglomeration' adhered to the wall of the vessel and could grow so as to 'more or less completely obstruct it at the place'. What Wharton-Jones had observed was the process of haemostasis (or thrombus formation) in the frog and the 'colourless corpuscles' that adhered to the vessel wall lining and to each other were thrombocytes, the amphibian counterpart of the mammalian blood platelet. It is fortunate that Wharton-Jones chose an amphibian as his experimental animal because although the functions and reactions of thrombocytes are essentially identical to those of platelets (indeed, in earlier literature the term 'thrombocyte' is used to describe the mammalian platelet also) the amphibian thrombocyte is much larger than the mammalian platelet, approximating in size to the red blood cell.

Because the platelets are so much smaller than the other formed elements of the blood, they were not described until much later than the red and white cells, which had been recognised since the very earliest days of microscopy (Leeuwenhoek, cited by Robb-Smith, 1967). However, in 1882 Bizzozero (cited by Robb-Smith, 1967) observed that tiny elements from the circulating mammalian blood adhered to the vessel wall at points of damage and then subsequently, by sticking to each other, built up an aggregated mass attached to the vessel wall and obtruding into the lumen. Similar masses were formed from circulating cells in response to injury in the crustacea, and, since these animals do not possess a mechanism equivalent to that of coagulation and fibrin formation in mammals, this indicated that the processes involved in blood clotting and in the accumulation of platelets or thrombocytes were probably different.

Thus, the fundamental cellular reactions involved in haemostasis were discovered and the mammalian platelet was recognised as an independent entity – although, even in the early years of this century, their existence was not universally accepted: in 1906 Buckmaster concluded that platelets were present in blood samples from healthy subjects only if the blood had been taken with insufficient care. Further research on platelets during the first half of this century soon confirmed that they were not merely a technical or pathological artefact, but much of this work was connected with the contribution that platelets make to the coagulation process, rather than with adhesion or aggregation. Platelets do indeed assist in coagulation, but in the past twenty years or so it has been recognised that platelets exhibit a wide variety of cellular reactions quite unrelated to blood clotting. This has resulted in an enormous upsurge of interest in platelet biology – in 1978 alone there were over a thousand publications on platelets.

Platelets are being increasingly used as model cells to study processes such as prostaglandin metabolism, amine uptake, degranulation, and cell–cell interactions or cell–substrate adhesion. It is with this last aspect that I shall be mainly concerned here, but the capacity of platelets to adhere to other cells or substrates is regulated largely by intracellular mechanisms within the platelets themselves, and therefore it is necessary to consider platelet adhesion in relation to their intracellular structure, constituents and biochemical functions.

3. PLATELET STRUCTURE

Platelets are derived from megakaryocytes in the bone marrow. The mature megakaryocyte transforms from a spherical to a highly irregular shape (Behnke, 1970), its periphery becomes deeply invaginated, and small pieces of cytoplasm are budded off and released as platelets into the circulation. Considering that they are formed by fragmentation of megakaryocyte cytoplasm, the uniformity in platelet size is remarkable – usually 2–3 μm in diameter – and the rate of platelet production is usually fairly constant at about 35000 per microlitre of blood per day ($\mu l^{-1}\ d^{-1}$). The circulating platelet count is approximately 250000 μl^{-1} in the peripheral blood of healthy subjects. The life-span of platelets is around 10 d in man and is terminated either by platelets becoming incorporated into haemostatic plugs or thrombi, or by effete cells being removed by the reticuloendothelial system. The values for platelet size, count, production and turnover can vary substantially in disease states, and can also be altered by changes in diet or by administration of drugs.

The plasma membrane of the platelet is rich in glycoproteins, which play important roles in regulating the response of platelets to many stimuli and in mediating platelet adhesion. Beneath this membrane is a bundle of microtubules (MT in Fig. 1) which form

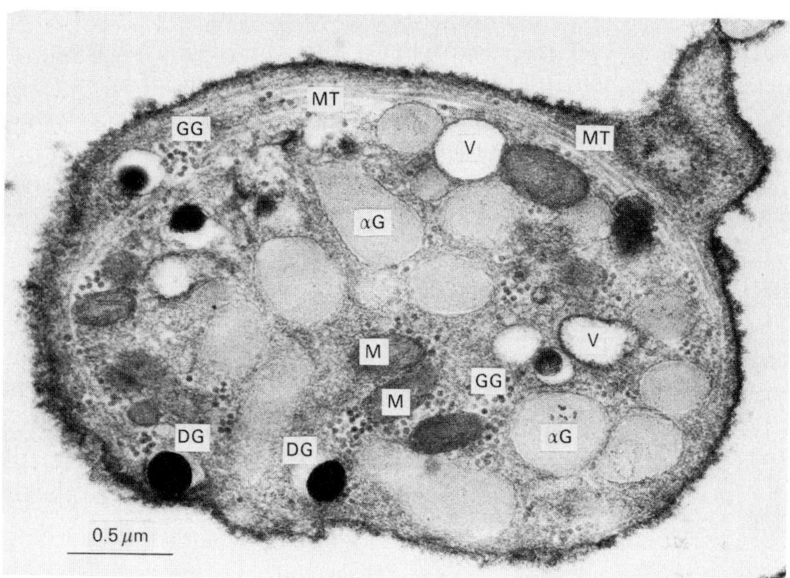

Fig. 1. Platelet isolated from rabbit citrated platelet-rich plasma. Key: αG, alpha granules; DG, dense granules; GG, glycogen granules; M, mitochondria; MT, microtubules; V, vesicles. (From: MacIntyre, Allen, Thomas & Gordon, 1977).

an equatorial ring that helps to maintain circulating platelets in their characteristic discoid shape, apparently acting as a cytoskeleton. Large amounts of the contractile protein actin, and smaller amounts of myosin are also present in platelets. The actin can attach to points on the internal surface of the platelet plasma membrane, and the actin–myosin interaction that occurs during the platelet response to stimuli helps to regulate platelet shape-change, adhesion, aggregation and clot retraction.

Platelets also contain several types of granular organelles, and most of these are shown in Fig. 1. Dense granules contain vasoactive amines (mainly serotonin), adenine nucleotides (chiefly ATP and ADP) and bivalent cations (Ca^{2+}, Mg^{2+}). Alpha granules, so called because they were the first platelet organelles to be observed under the microscope, are a heterogenous population of granules that contain lysosomal enzymes, fibrinogen, and at least three platelet-specific proteins: β-thromboglobulin, platelet factor 4 (an anti-heparin protein), and a protein that powerfully stimulates the proliferation of smooth muscle cells (Ross, Glomset, Kariya & Harker, 1974; Witte *et al.*, 1978). All these granule contents can be secreted when platelets are stimulated.

Glycogen granules are abundant in platelets but mitochondria are sparse with few cristae – consistent with the platelet deriving its metabolic energy mainly from glycolysis rather than oxidative phosphorylation. Finally, the apparently empty vesicles frequently seen in platelets are not usually enclosed vacuoles but rather parts of a complex, surface-connected canalicular system involved in uptake and secretory processes. Intimately associated with this canalicular system (but not evident in Fig. 1) is the network of dense tubules that probably constitutes the site of prostaglandin synthesis (Gerrard, White, Rao & Townsend, 1976). The energy needed for the platelet's response to stimuli is supplied by a metabolic pool of ATP in the cytoplasm, which is in very low equilibrium with the storage pool of ATP in the dense granules. More detailed reviews are given by White (1971) on platelet morphology, Crawford (1976) on platelet contractile proteins and Holmsen, Salganicoff & Fukami (1977) on platelet biochemistry.

4. PLATELET–VASCULAR ADHESION

The importance of platelet-vascular interation in haemostasis and thrombosis has been sufficient justification for many research groups to study this phenomenon but, in addition, its speed and specificity intrigued many of those workers with an interest in cell adhesion *per se*, who realised that whatever mediated the dramatic change from platelets circulating freely in the vascular lumen to the almost instantaneous accumulation of large numbers of platelets at a point of local vascular damage must be well worth investigating. The build-up of such platelet clumps depends partly on the response of platelets to soluble mediators of low molecular weight that can be secreted locally from several cell types (including the platelets themselves) and partly on the platelet's own intracellular regulatory mechanisms. It is more convenient to discuss these aspects of platelet–vascular interaction in the following section, under the heading 'Aggregation', and the theme of this present section will be concerned mainly with the specificity of platelet–vascular adhesion.

As platelets do not normally stick to the endothelial lining, some substance in the vascular wall (presumably revealed when the endothelium was damaged) must be 'recognised' by platelets and cause their adhesion. What this substance was remained a mystery until quite recently, when it was discovered that a connective tissue component was involved; electron microscopic examination revealed that platelets filled gaps between endothelial cells (Majno & Palade, 1961; Tranzer & Baumgartner, 1967) and formed a continuous layer on the subendothelial connective tissue at sites where endothelium had been removed (French, MacFarlane & Sanders, 1964; Baumgartner, Tranzer & Studer, 1967). Several constituents of the subendothelial connective tissue were considered as potential substrates for platelet adhesion, but of the various components tested only fibrillar collagen induced platelet aggregation *in vitro* (Zucker & Borrelli, 1962; Hovig, 1963). This indicated that there was considerable specificity in platelet–collagen adhesion, but the picture became more complicated when morphometric studies of platelet interaction with the subendothelium of vascular preparations, treated to remove selected components, established that the affinity of platelets for different constituents of the subendothelial connective tissue varied considerably (for review see Baumgartner & Muggli, 1976). Also, various treatments alter the interaction of collagen with platelets *in vitro* (Wilner, Nossel & Procupez, 1971; Puett, Wasserman, Ford & Cunningham, 1973). Such studies coincided with the discovery that several chemically distinct subtypes of collagen existed in the vascular wall (Trelstad, 1974) and a logical conclusion from all these findings was that by studying pure preparations of different subtypes, or of chemically modified collagens, the mechanisms through which collagen interacted specifically with platelets might be clarified. It was already evident that when platelets interacted with collagen fibrils either *in vitro* or *in vivo* the platelets were stimulated to secrete constituents which in turn induced the formation of a platelet aggregate, and measurements of platelet secretion or aggregation were much easier to make than morphometric assessment of platelet adhesion. Many research groups therefore adopted these indirect methods to investigate platelet–collagen adhesion and joined the search for a collagen 'receptor' on platelets and for the determinants on collagen that platelets recognised.

Jamieson, Urban & Barber (1971) provided a great stimulus for research on platelet–collagen interaction by advancing the hypothesis that a glucosyl transferase on the platelet surface mediated adhesion to collagen by forming an enzyme–substrate complex with incomplete disaccharide groups on the amino acid backbone of the collagen molecules. This hypothesis was, in effect, a specific case of the general mechanism proposed by Roseman (1970) to explain intercellular adhesion. Other workers obtained evidence that seemed to support this concept of platelet–collagen adhesion (Chesney, Harper & Colman, 1972) but subsequently the weight of evidence against it made this hypothesis untenable (see, for example, Menashi, Harwood & Grant, 1976; Santoro & Cunningham, 1977), despite the impressive rearguard action mounted by its proponents (Jamieson, Smith & Kosow, 1975). An important factor in this debate was the discovery that the quaternary fibrillar structure of collagen regulated its interaction with platelets (Muggli & Baumgartner, 1973; Jaffe & Deykin, 1974), whereas platelet glucosyl transferase accepted only denatured collagen and not fibrillar collagen as a substrate (Menashi *et al.*, 1976).

In experiments using monomeric collagen (that is, individual collagen molecules each composed of three polypeptide chains in a triple helical configuration) where platelet aggregation or secretion was measured, collagen polymerisation is a prerequisite for activity. Thus, in experiments where the physical state of the collagen preparations was not rigorously controlled (as in some of the experiments by Jamieson *et al.*, 1971 and by Chesney *et al.*, 1972) it was erroneously concluded that inhibitors were blocking platelet–collagen interaction when in fact they were interfering with collagen fibrillogenesis. Inhibition was also seen in these indirect systems, of course, if the intracellular processes controlling platelet aggregation and secretion were affected, and it was therefore not surprising that some confusion existed at this time. The importance of fibrillogenesis was exemplified by studies with collagen subtypes which show varying rates of fibril formation: for example, type III is a much more potent stimulant than type I when it is mixed with platelets in monomeric form (Balleisen, Gay, Marx & Kuhn, 1975; Hugues, Herion, Nusgens & Lapiere, 1976), but type III forms fibrils more efficiently *in vitro*, and the potency difference between these two collagens is minimal if preformed fibrils are tested (Barnes, Gordon & MacIntyre, 1976).

However, although it was accepted that collagen had to be in fibrillar form in order to stimulate platelets optimally, the question still remained whether monomeric collagen (or even fragments of individual molecules) could bind specifically to platelets in some circumstances and induce a cellular response. Attempts to address this question experimentally usually took one of three forms. First, collagen monomers or fragments were added to a suspension of platelets and platelet secretion or aggregation responses were measured; second, labelled collagen molecules or fragments were added to a platelet suspension and then the platelets were separated and the amount of radioactivity associated with them was measured; third, a suspension of platelets was passed through a column of Sepharose beads on the surface of which collagen fragments or molecules were immobilised, and the number of platelets adhering to the column was calculated. There are, however, two major problems associated with such experiments. Collagen molecules or even fragments of molecules readily self-associate either when free in solution or during attachment to a matrix such as sepharose beads; hence, variable amounts of 'pseudo' fibrils may be present. Also, most platelet–collagen binding experiments are open to the criticism that weak interactions probably do not survive the separation procedures. The conclusions from such studies were somewhat inconsistent; some workers claimed that platelets did not bind collagen monomers under physiological conditions (Gordon & Dingle, 1974; Kronick, 1975); others concluded that some small fragments of the collagen molecule could not only bind to platelets but also stimulate them, although at relatively high concentrations. For example, Kang, Beachey & Katzman (1974) found such activity in the denatured α 1 chain or the α 1 CB5 peptide fragment of chick collagen at about 1 mg ml^{-1}, compared with about 0.2 μg ml^{-1} for native type I fibrils; note that the difference in activity on a *molar* basis is even greater. There was, however, some inconsistency even here because the same fragment from rat collagen was ineffective. The two α1CB5 fragments are chemically very similar, but the chick fragment self-associates more readily, so again it appeared that the formation of an ordered quaternary structure might be important, even for the weak platelet-stimulating activity observed with collagen fragments. There was much experimental effort on this topic (for reviews see Jaffe, 1976; Michaeli & Orloff, 1977) and the current consensus is that monomers or fragments can bind to platelets, but rather weakly and with little stimulatory effect (Puett *et al.*, 1973; Brass & Bensusan, 1975).

These conclusions are consistent with the hypothesis advanced by Lüscher, Pfueller & Massini (1973) to explain platelet stimulation by polymers. This introduced the concept of platelets interacting weakly with monomers or submolecular fragments through binding sites of low specificity and affinity; however, when multiple simultaneous interactions were possible with several such sites in optimal alignment (that is, in an ordered polymer with regular repeating sequences) binding was much stronger and a cellular response could be induced.

Random polymerisation alone is not enough – for example, platelets are not stimulated by gelatin, which is polymerised but denatured collagen. It appears that the native fibril formed *in vivo* by collagen type I or type III exhibits an optimal configuration: platelet aggregation can be induced by 0.1–0.2 μg ml^{-1} of type I collagen in suspension. In contrast, suspensions of native type IV collagen (which has an amorphous structure *in vivo*) are inactive. However, if collagens type I and IV are dissolved and chemically reassembled as SLS ('segment long spacing') fibrils, they will both stimulate platelets with an activity threshold around 10 μg ml^{-1} (Barnes, M. J., Bailey, A. J., MacIntyre, D. E. & Gordon, J. L.; unpublished work). The native type I fibril has a repeating 67 nm period with asymmetric banding, and the SLS fibril has a repeating 268 nm period also with asymmetric banding. From such observations, it is clear that the steric arrangement of collagen molecules is more important than chemical differences between subtypes in regulating interaction with platelets. These results complement and support the observations (Muggli, 1978) that guinea pig skin collagen chemically reassembled into four different fibrillar structures showed different profiles of platelet-aggregating activity in each case. It should be emphasised, however, that although such differences in activity can be accurately compared on a weight basis, the degree of dispersion of the fibrils may vary, and this also affects activity (Muggli & Baumgartner, 1973).

Until recently, there was no suggestion of what the platelet 'receptor' for collagen might be, to supplant the glycosyl transferase hypothesis. A recent paper by Bensusan *et al.* (1978) has provided an alternative suggestion. These workers induced platelet–collagen adhesion by mixing a suspension of type I native collagen fibrils with well-washed platelets, then disrupted the adherent platelets by sonication and afterwards determined what platelet proteins remained adherent to the collagen fibrils. The main platelet membrane constituent remaining attached to the collagen was fibronectin, a glycoprotein of molecular weight of about 2×10^5 that is a major component of many cell membranes (Yamada & Olden, 1978). Also present were actin, myosin and tropomyosin, which suggested that fibronectin on the platelet membrane was involved in binding to collagen and was also connected to the platelet contractile proteins. The presence of fibronectin was confirmed immunologically and further evidence for its functional significance was obtained by demonstrating that collagen pretreated with fibronectin did not stimulate platelets, whereas anti-fibronectin antibodies induced platelet secretion.

Fibronectin therefore appears to be a possible candidate for the platelet 'receptor' for collagen, although the use of the term 'receptor' in this context is perhaps unfortunate; as usually employed (for example in the context of cholinergic, aminergic and other sites), the term implies a degree of specificity and affinity which is lacking in the case of fibronectin-collagen interaction. Other questions still remain: for example, plasma fibronectin can bind to triple helical or to denatured collagen (Dessau, Adelmann & Timpl, 1978), whereas the fibronectin-like protein isolated from platelets apparently binds to collagen fibres but not to gelatin – that is, it shows the same conformational specificity as for platelet–collagen adhesion. In addition, there remain to be determined the roles of other components, such as glycosaminoglycans, which modify collagen fibril formation (Silver, Yannas & Saltzmann, 1978), or the complement component Clq, which shows considerable structural homology with part of the collagen molecule and has been shown to affect platelet–collagen interaction (Cazenave *et al.*, 1976). The findings of Bensusan *et al.* (1978) do not prove that fibronectin mediates platelet–collagen adhesion, but they will undoubtedly stimulate further work, and if they are confirmed, the emphasis in future experiments on platelet–collagen interaction may turn to the mechanisms involved in collagen–fibronectin interaction, thus bringing this work in line with studies on cell adhesion in several different systems where fibronectin has been shown to be involved (Yamada & Olden, 1978).

Before leaving the subject of platelet–vascular adhesion, we should make a brief comparison between this phenomenon and that of granulocyte–vascular adhesion (an important event in the inflammatory process). The two processes can be compared using cells of the vascular wall cultured *in vitro*. Smooth muscle cells or fibroblasts cultured from pig aorta readily induce platelet adhesion, aggregation and secretion, whereas there is little or no interaction between platelets and pig aortic endothelial cells in culture. The non-reactivity of the endothelium is partly because these cells secrete prostacyclin (see section 6.2), which has an inhibitory effect on platelets, but also because endothelial cells secrete type IV collagen (which does not stimulate platelets in its native state) whereas smooth muscle cells apparently secrete only collagens type I and III.

This preferential adhesion to vascular smooth muscle rather than to endothelium is not observed with granulocytes, which readily adhere to endothelial cells *in vitro* (Lackie & de Bono, 1977) but have very little affinity for vascular smooth muscle cells (Beesley *et al.*, 1978; Pearson, Carleton, Beesley, Hutchings & Gordon, 1979). Having adhered to the cultured endothelium, granulocytes then migrate through the monolayer (Beesley *et al.*, 1978, 1979). The specificity of blood cell–vascular cell interaction observed in these experiments *in vitro* is consistent with observations made *in vivo*: as long as the endothelium remains intact platelets do not adhere in large numbers, but when endothelial damage occurs and the underlying connective tissue and smooth muscle cells beneath are exposed, then platelets adhere and form aggregates. In contrast, granulocyte adhesion to and migration through endothelium in small blood vessels is an important and regularly observed feature of the inflammatory response.

Although platelets adhere avidly to fibrillar collagen, platelet adhesion is not seen when several other biopolymers are tested – for example, proteoglycans, elastin, or preformed fibrin are not good substrates. Platelets can, however, adhere to many artificial surfaces and indeed the production of surfaces that are 'passive' to platelets is of considerable importance in the development of vascular prostheses and other appliances that must come in contact with blood. As with platelet–collagen interaction, the degree of platelet adhesion to such surfaces apparently depends on the arrangement of charged groups distributed on the surface of the artificial material.

5. PLATELET AGGREGATION

Platelet aggregation is defined simply as platelets adhering to other platelets, and it is induced by a wide range of biological stimuli, provided that an opportunity exists for platelets to come in contact with other platelets while they are exposed to the stimulus. Among the agents that can induce platelet aggregation are ADP, serotonin, some prostaglandins (and other products of arachidonate metabolism – see section 6.1.), catecholamines and thrombin. This diverse list of stimulatory agents is counterbalanced by a similar diversity of compounds that inhibit platelet aggregation. In general, these inhibitors can be divided into three classes: those agents that antagonise the platelet response to a specific stimulus by blocking the interaction of that stimulus with its receptor; those that specifically prevent the platelets from secreting constituents that promote further aggregation; and those agents that decrease platelet responsiveness generally, by interfering with the chain of intracellular events that leads to platelet activation – for example, agents such as prostacyclin and adenosine that stimulate adenylate cyclase and thus elevate the level of cyclic AMP in platelets. Inhibitors in this last group can inhibit platelet–vascular adhesion as well as aggregation, although higher concentrations are needed.

The sequence of events leading to platelet aggregation begins with the interaction of a stimulus with its specific receptor on the platelet plasma membrane. Most of the known receptors are glycoproteins, and some have been well characterised pharmacologically – for example, those for catecholamines and for serotonin (Drummond, 1976). The first detectable response of platelets to most stimulants is extrusion of thin pseudopodia – 'filopodia' – with a consequent loss of discoid shape. This change in shape, which is extremely rapid, is not apparently associated with any significant change in platelet volume and it facilitates cell–cell contact by virtue of the small diameter of the platelet filopodia. The central part of the cell rounds up, forming a spherical body. The membrane events that accompany the shape-change probably represent a triggering mechanism which leads in turn to the more profound cellular changes that regulate aggregation (and, subsequently, secretion). This concept is supported by the high velocity and high activation energy of the platelet shape-change (Born, 1970).

It is axiomatic that platelet aggregation requires platelet–platelet contact, and the aggregation is directly related to the number of collisions; therefore, any quantitative studies of aggregation, whether made *in vitro* or *in vivo*, must take account of the flow and shear rates in the experimental system used. As well as opportunities for platelet–platelet collisions, extracellular cofactors are usually required. Of these, calcium ions and fibrinogen are the most important: they are apparently involved in forming intercellular bridges between activated platelets (Mustard *et al.*, 1978). Much has been learned from observing the morphological effects of aggregating stimuli on platelets and determining the cofactor requirements for aggregation, but much more important from the mechanistic point of view is the sequence of intracellular events that follows the interaction of the stimulus with the platelets.

It appears that when platelets are activated (whether the response is shape-change, aggregation, secretion or all three) the activation depends on an increase in intracellular calcium ion concentration (see, for example, Feinman & Detwiler, 1974; Feinstein, Fiekers & Fraser, 1976). This can be brought about either by a downhill flow of calcium ions through a receptor-controlled gating system at the plasma membrane or through liberation of calcium ions from storage sites on the internal surface of the platelet plasma membrane (electron probe X-ray microanalysis has revealed such stores on the membrane, especially in the canalicular system; Daimon, Mizuhira & Uchida, 1978). In this respect, the activation of platelets resembles many other biological systems triggered by specific receptors, for example, muscarinic cholinergic, α-adrenergic, H1 histamine and serotonin. There is now a great deal of evidence to support the concept that calcium ions play a central role in the response of platelets (and of the other systems listed above) to stimuli. This evidence includes the inhibitory effects of calcium-chelating agents (both intracellular and extracellular) and observations that the effects of specific stimuli can be mimicked by calcium ionophores (e.g. compound A23187). These are agents which do not apparently act on any specific receptor but form pores in the cell membrane through which calcium ions can pass; also, these agents can liberate calcium from storage sites. Thus, they can bypass the receptor stage of the activation sequence. Further evidence is provided by calcium antagonists both inorganic (e.g. lanthanum) and organic (e.g. nifedipine) which have inhibitory actions on systems (such as platelet aggregation) that require extracellular calcium ions.

It is not yet clear what molecular events in the cell membrane follow receptor activation and lead to calcium liberation or influx, but some evidence strongly suggests that the removal of the phosphoinositol head group from phosphatidylinositol (a phospholipid present in small amounts in the cell membrane) may be involved. Not only does increased phosphatidylinositol breakdown accompany the activation of receptors whose responses are mediated by an elevation in cytoplasmic calcium ion concentration (Michell, 1975), but it appears that the phosphatidylinositol response actually precedes the calcium ion response. Three pieces of evidence support this concept. First, phosphatidylinositol turnover is stimulated by agonists even in cells that have been deprived of calcium ions and thus do not exhibit their normal physiological response to stimulation. Secondly, increasing cytoplasmic calcium ion concentrations by using ionophores (which bypass the receptor system) does not stimulate phosphatidylinositol turnover. Thirdly, stimulation of phosphatidylinositol turnover is not diminished by exposure of cells to calcium-antagonistic drugs which prevent the influx or liberation of calcium ions (Michell, Jones & Jafferji, 1977). Evidence at present available is compatible with the concept that when a stimulant interacts with its receptor, a coupled membrane-bound enzyme system hydrolyses phosphatidylinositol and this then results in the liberation of bound calcium, or the influx of extracellular calcium, or both.

There may be another step immediately following receptor occupation: activation of a membrane-bound protease has been implicated as an early event in platelet activation by some stimuli (Henson, Gould & Becker, 1976). This is probably linked to subsequent events, such as arachidonate metabolism, but it would be wrong to assume that all the early biochemical events that have been recorded during platelet activation are causally related. Platelets have activation pathways that can operate independently – for example, although arachidonate metabolism usually plays an integral and important role, activation can proceed even when this pathway is blocked. This plurality reflects the numerous potential consequences of elevating cytoplasmic calcium.

Having established that an increase in cytoplasmic calcium mediates a response to stimulation, it is then necessary to consider what may control the response: in other words, what inhibitory mechanisms exist. In the platelet, agents that elevate cyclic AMP, such as adenosine (Haslam & Rosson, 1975) and prostacyclin (Best, Martin, Russell & Preston, 1977) all inhibit activation, and it appears that this effect of cyclic AMP is mediated at least partly through the nucleotide promoting the sequestration of calcium back into storage sites on the plasma membrane. This is consistent with the concept that cyclic AMP has only an inhibitory role in platelets and is not apparently involved in mediating the initial response of the cell to stimulation (Haslam, Davidson & Desjardins, 1978), except possibly in specialised cases such as the response of platelets to the stimulatory prostaglandin endoperoxides (Salzman, MacIntyre, Steer & Gordon, 1978). Cyclic GMP at one stage rivalled cyclic AMP as the most popular candidate for an intracellular messenger in platelets, because it increases in concentration as platelets respond to at least some stimuli, but careful investigation has revealed that although it may play a minor role in platelet activation it is not causally related to it (Haslam, 1975).

In platelets, as in many other cells, some of the intracellular mechanisms following the elevation of calcium ions are not well understood. Responses such as aggregation and secretion are, however, accompanied by a change in the contractile proteins in the platelet and the polymerisation and interaction of actin and myosin are calcium-dependent processes. These contractile proteins are important in the intracellular regulation of platelet activation, and it has also been suggested that they are directly involved in intercellular events through actin exposed on the membrane of one platelet interacting with myosin exposed on the membrane of another (Booyse & Rafelson, 1972). This hypothesis is not generally accepted, mainly because the immunological evidence suggesting surface localisation of myosin can be explained by antigenic heterogeneity (Pollard, Fujiwara, Handin & Weiss, 1977).

The intracellular roles of the contractile proteins do, however, merit some comment. The molar ration of actin to myosin in platelets is > 100, compared with about 6 in skeletal muscle, which means that the maximum force developed by the platelet contractile system is much less than that in muscle, but that the apparatus is capable of extreme shortening. During the shape-change a significant proportion of the platelet mass is rapidly converted into long, thin filopodia containing bundles of unipolar actin filaments, anchored to the tip. Platelet filopod formation is analogous to that of sperm during the acrosomal reaction – specifically, it is probable that actin polymerisation at sites on the platelet periphery drives the growth of the filopod. The bulk of the platelet myosin apparently remains in the central spherical body.

When filopodia from different platelets have adhered to each other, internal contraction produces a compact platelet aggregate. The extent of this contraction is best illustrated by considering clot retraction, a process whereby platelets (constituting only about 0.2% of the blood volume) cause a clot to retract by about 50%. This is achieved by filopodia binding to polymerising fibrin (though platelets do not adhere to *preformed* fibrin) and subsequently retracting the fibrin strands by intracellular contraction. Each platelet, only about 7 μm^3 in volume, squeezes out about 2000 μm^3 of serum from the clot during this process, by generating filopodia with a total length of more than 100 μm, thus making the *effective* platelet volume more than 4000 μm^3. Because of the arrangement of the contractile elements in the activated platelets, the centrally located bipolar myosin filaments can cross-link antiparallel actin filaments from filopodia on opposite sides of the platelet, thus forming a transplatelet contractile unit with the same geometrical features as the sarcomere of striated muscle, whereby the actin-myosin arrangement pulls the Z-lines in towards the centre (Pollard *et al.*, 1977). The compaction of a platelet aggregate involves the same processes as in clot retraction, except that interfilopodial contacts, rather than filopodia–fibrin contacts, are made.

In any discussion on platelet aggregation and the mechanisms that regulate it, it is important to distinguish between 'primary aggregation', which is caused by the exogenous stimulus to which the platelets have been exposed, and 'secondary aggregation', which follows the primary response if the stimulus is powerful enough. This secondary platelet aggregation is caused by constituents secreted from the platelets initially activated by the exogenous stimulus – for example, ADP and thromboxane A_2. Thus, secondary aggregation is dependent on secretion by the platelets, and as such is more appropriately discussed in the next section.

6. SECRETION

6.1. *Secretion from platelets*

There are two types of secretory response observed in platelets: degranulation and stimulated synthesis. Although in other cell types a variety of compounds, including proteins, may be synthesised and secreted in response to a stimulus, in the platelet stimulated synthesis is restricted to products of arachidonate metabolism – the so-called prostaglandin synthase pathway. Products of this pathway and constituents released by degranulation both play important roles in modulating platelet activation and hence platelet adhesion and aggregation also.

Degranulation in platelets can include the α-granule population as well as the dense granules. Lysosomal enzymes contained in the α-granules are less readily secreted than other constituents: stimulants such as ADP or stimulatory prostaglandins can release dense granule contents without significant secretion of lysosomal glycosidases (Mills, Robb & Roberts, 1968; MacIntyre, McMillan & Gordon, 1977), whereas stronger stimuli such as thrombin and collagen induce secretion of lysosomal enzymes also (Holmsen & Day, 1970; Gordon, 1975). However, when platelet prostaglandin synthase is blocked by drugs such as aspirin, the secretion of lysosomal enzymes is inhibited, which implies that prostaglandin synthase is necessary (though not *per se* sufficient) for this secretory response (MacIntyre *et al.*, 1977). Some constituents of the heterogeneous α-granule population, such as β-thromboglobulin and the mitogenic factor, are released by ADP and are also more easily released by thrombin than are the dense granule contents (Witte *et al.*, 1978). In contrast, the dense granules are a homogenous population, and their contents are apparently all released together. The dense granule contents are important promoters of platelet activation: the role of calcium ions has been discussed above; ADP is one of the most powerful platelet stimulants known; and serotonin, although not a potent platelet stimulant by itself, has synergistic effects with other stimuli such as ADP or certain prostaglandins.

There is an intimate relationship between platelet aggregation and secretion. As indicated above, the aggregation response to an exogenous agent ('primary aggregation') does not necessarily lead to degranulation and prostaglandin synthesis, although these secretory responses usually follow if the initial stimulus is powerful enough. The close contact of platelet membranes during aggregation is itself an additional stimulus to secretion (i.e. the initial membrane perturbation induced by the agent is reinforced by the physical apposition of the cells). Most stimuli (e.g. ADP, adrenaline, prostaglandins, thromboxane A_2) cannot induce secretion if aggregation is prevented by using a system in which interplatelet contact is impossible, although thrombin is an exception. Thus, secretion usually follows aggregation (or adhesion, in the case of collagen), and further aggregation ('secondary aggregation') follows secretion. It is, therefore, not surprising that the cellular mechanisms regulating platelet degranulation are very similar to those that induce aggregation, and they can be summarised as follows. When a stimulant activates its receptor on the plasma membrane, a local perturbation follows, probably involving phosphatidylinositol hydrolysis and/or activation of a membrane-bound protease, which liberates calcium and/or induces calcium influx. This increase in cytoplasmic Ca^{2+} activates the phospholipase (Rittenhouse-Simmons, Russell & Deykin, 1977) that liberates arachidonic acid from membrane phospholipids, thus initiating the prostaglandin synthase pathway. Calcium-dependent reorganisation of actin and myosin leads to the extrusion of filopodia, then to internal contraction that promotes fusion of

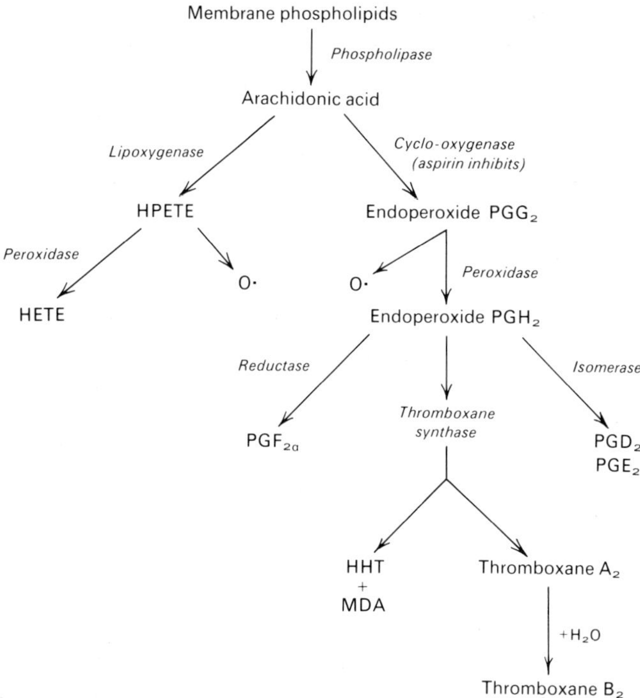

Fig. 2. Schematic outline of arachidonic acid metabolism in platelets.

granules with the plasma membrane, usually at invaginations of the canalicular system, where there are calcium stores (Daimon *et al.*, 1978). This granule fusion, which results in extrusion of the granule contents, is presumably facilitated by calcium ions neutralising the repulsive negative charges on the plasma and granule membrane and by the diacylglycerol formed when phosphatidylinositol is hydrolysed (Allan & Michell, 1975).

Some of the specific reactions that follow the increase in cytoplasmic Ca^{2+} have now been characterised: there is, for example, calcium-dependent phosphorylation of several discrete proteins, including one of 20000 molecular weight that may be the light chain of platelet myosin. Platelets contain phosphorylase kinase and several protein kinases, some activated by Ca^{2+} and some by cyclic AMP, and the specific, but different, patterns of protein phosphorylation that are induced by platelet activators and inhibitors suggest that these phosphorylation reactions may play important roles in the regulation of platelet function (Haslam, Lynham & Fox, 1979).

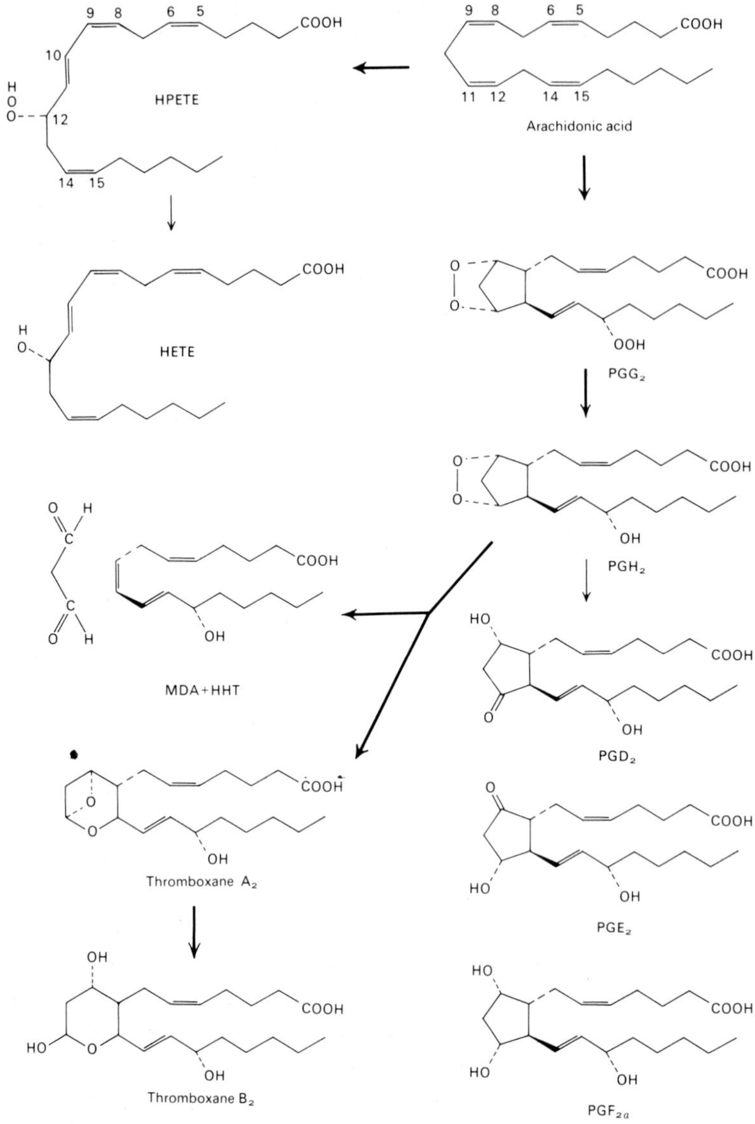

Fig. 3. Structures of arachidonic metabolites formed by platelets.

The other reactions modulating platelet secretion that have been well characterised are those of arachidonate metabolism. This pathway is initiated by Ca^{2+}-dependent activation of a phospholipase, possibly itself associated with a stimulus-activated membrane-bound protease (Feinstein *et al.*, 1977), and the main features of the pathway are outlined in Fig. 2, with structures of the major compounds in Fig. 3. Activation of a phospholipase in the platelet membrane liberates arachidonic acid, and this 20-carbon fatty acid is converted by membrane-bound enzymes either into hydroxy-acids (lipoxygenase pathway) or into labile intermediate compounds known as the endoperoxides PGG_2 and PGH_2 (cyclo-oxygenase pathway) (Hamberg, Svensson & Samuelsson, 1974). The endoperoxides, which can themselves induce platelet aggregation and secretion, are converted mainly to thromboxane $(Tx)A_2$, which is highly labile ($t_{\frac{1}{2}}$ approximately 30 s) but is an extremely potent platelet stimulant and vasoconstrictor (Hamberg, Svensson & Samuelsson, 1975). The mechanism by which these agents stimulate platelets has not been unequivocally established, but it appears that stimulation could result at least in part from a calcium ionophoric action (Gerrard, Peterson, Townsend & White, 1976; Gerrard, Butler & White, 1978). Thus, TxA_2 could participate in an autocatalytic sequence in which Ca^{2+} activates phospholipase, liberating arachidonic acid for metabolism, and one of the major metabolites itself liberates more Ca^{2+}. Two questions then arise: what are the effects of other arachidonic metabolites on platelets, and what mechanisms exist for inhibiting this autocatalytic cycle of platelet activation?

The biological effects of most of the arachidonate metabolites have now been clarified, although some (such as TxA_2 and the endoperoxides) are difficult to study because of their lability. However, stable analogues of PGH_2 have been synthesised which have allowed some definition of the structural characteristics required for prostaglandins to stimulate platelets (MacIntyre & Gordon, 1977; MacIntyre, Salzman & Gordon, 1978). The breakdown product of TxA_2, known as TxB_2, is stable but inactive on platelets, although the naturally formed compound (but perhaps not a synthetic isomer) induces leukocyte chemotaxis (Boot, Dawson & Kitchen, 1976; Goetzl & Gorman, 1978). Thromboxane synthase, the enzyme that forms TxA_2, also produces, apparently in a bimolecular reaction, the 17-carbon hydroxyacid known as HHT, together with the 3-carbon fragment malonaldehyde (MDA). Although HHT and malonaldehyde are apparently without effect on platelets, malonaldehyde can easily be measured either colorimetrically (Stuart, Murphy & Oski, 1975) or fluorimetrically (McMillan, MacIntyre & Gordon, 1977) after combination with thiobarbituric acid, and thus serves as a convenient indicator of thromboxane synthase activity.

The amounts of the so-called stable prostaglandin (PGD_2, PGE_2 and $PGF_{2\alpha}$) that are formed by platelets represent only a small proportion (less than 1 %) of the total cyclo-oxygenase products. All these prostaglandins are vasoactive, but $PGF_{2\alpha}$ has virtually no effect on platelets, and PGE_2 exerts only a modest potentiatory effect when combined with agents that aggregate human platelets, although it can aggregate pig platelets directly (MacIntyre & Gordon, 1975). PGD_2 is an inhibitor, because it powerfully stimulates adenylate cyclase, and it may be important in modulating platelet activation.

The biological roles of products of the lipoxygenase pathway have not yet been fully clarified, but it is clear that the 12-hydroxy fatty acid HETE stimulates leukocyte chemotaxis (Turner, Tainer & Lynn, 1975), and the proportion of released arachidonic acid that is converted by the lipoxygenase pathway is high. Therefore, the possible actions of lipoxygenase products such as HETE, its hydroperoxy precursor HPETE, and other hydroxy-acid products recently described (Jones *et al.*, 1978) ought to be considered when the effects of arachidonate metabolites on cell activation and adhesion are being investigated.

One approach that has been used to investigate the potential importance of arachidonate metabolites in the activation of platelets or of other cells is to block prostaglandin production by inhibiting cyclo-oxygenase with drugs such as aspirin or indomethacin. These inhibitors do not, however, affect the lipoxygenase pathway. Phenidone is a compound that has been shown to block both the cyclo-oxygenase and the lipoxygenase pathways (Blackwell & Flower, 1978) and compounds with this spectrum of activity should be considered in future experiments designed to explore the role of arachidonic metabolites in cell activation or adhesion.

The last arachidonate metabolites that merit discussion are the oxygen radicals produced in some of the reactions (e.g. $PGG_2 \rightarrow PGH_2$; HPETE → HETE). Kuehl *et al.* (1977) suggested that these radicals play a pivotal role in the inflammatory process, and they can also inhibit some of the enzymes that metabolise arachidonic acid. Prostacyclin synthase is particularly susceptible to oxidants (Moncada, Gryglewski, Bunting & Vane, 1976*a*), while cyclo-oxygenase is somewhat less susceptible and thromboxane synthase is little affected. Thus, platelet cyclo-oxygenase generates a 'suicide' reaction – that is, one that produces its own poisoner (Egan, Paxton & Kuehl, 1976). This negative feedback can be regarded as the inhibitory counterpart of the TxA_2/Ca^{2+} positive reinforcement cycle discussed above.

In summary, activation of the prostaglandin synthase pathway in platelets results in the formation of numerous metabolites, some of which have stimulatory actions on platelets and on some other cells, and some of which inactivate the enzymes that formed them. One potential inhibitor (PGD_2) is also produced; this acts by stimulating adenylate cyclase and thus elevating cyclic AMP in the platelets, but it is produced only in very small amounts and its role in regulating platelet–vascular adhesion or platelet aggregation is probably subordinate to that of PGI_2 (prostacyclin) which is an even more potent stimulant of platelet adenylate cyclase and is secreted in large amounts by the cells of the vascular wall.

6.2. *Secretion from vascular cells*

Since the discovery of a labile prostaglandin produced from vascular cells that inhibited platelet activation (Moncada, Gryglewski, Bunting & Vane, 1976*b*) and its subsequent identification as PGI_2 or prostacyclin (Johnson *et al.*, 1976) there has been great interest in this compound as a possible modulator of platelet adhesion, aggregation and secretion. Indeed, it has been suggested that the reason why the platelets did not stick to vascular endothelium was that the endothelial production of prostacyclin prevented the platelets being activated and thus adhering. This hypothesis ignores the fact that platelets do not stick to other cells, such as erythrocytes, with which they come in frequent contact; erythrocytes do not synthesise any prostaglandins. None the less, prostacyclin may be important in regulating platelet–vascular adhesion. At one stage, it was suggested that when platelets were stimulated and PGH_2 was produced, this diffused into the endothelial cells which then utilised this intermediate compound to synthesise PGI_2. This hypothesis was based partly on the observation that vascular fragments in platelet-rich plasma synthesised much more PGI_2 than when they were incubated in physiological buffer alone (Moncada, Higgs & Vane, 1977), but some doubt was cast on the hypothesis when MacIntyre, Pearson & Gordon (1978) showed that cell-free plasma was as powerful as platelet-rich plasma at stimulating prostacyclin synthesis by endothelial cells. Subsequently, Needleman, Syche & Raz (1979) demonstrated that platelets did not donate PGH_2 to vascular cells for conversion to PGI_2 unless the platelet thromboxane synthesis pathway was blocked. Also, some vascular tissues were unable to convert exogenous PGH_2 to PGI_2 (Needleman *et al.*, 1978) although aortic segments (Needleman *et al.*, 1979) and cultured aortic smooth muscle cells (Tansik, Namm & White, 1978) and endothelial cells (Marcus, Weksler & Jaffe, 1978) could effect this conversion. It has been postulated that PGI_2 is continuously produced and is a circulating hormone (Moncada, Korbut, Bunting & Vane, 1978) but this hypothesis has been directly contradicted by other work (Smith, Ogletree, Lefer & Nicolau, 1978) and thus this question remains to be resolved. There is also controversy about the relative roles of endothelial and vascular smooth muscle cells in secreting PGI_2 (MacIntyre, Pearson & Gordon, 1978; Tansik, Namm & White, 1978; Pure & Needleman, 1979).

However, regardless of the precise site and mechanism of synthesis of PGI_2, it is generally recognised as the most powerful known inhibitor of platelet secretion and aggregation; it will also inhibit platelet–vascular adhesion, although higher concentrations are needed (Higgs, Moncada & Vane, 1978). Prostacyclin has been observed to decrease the numbers of granulocytes marginating in microvessels *in vivo* (Higgs *et al.*, 1978) although studies on granulocyte–endothelium interaction *in vitro* showed no such inhibitory effect – in fact, the interactions were slightly increased (Pearson, Carleton, Beesley, Hutchings & Gordon, 1979). Further work is therefore needed to clarify the role of prostacyclin in modulating adhesion of various cell types.

The mechanism by which PGI_2 inhibits platelet function deserves some comment. As with PGD_2, PGE_1 and adenosine, its main action is to stimulate adenylate cyclase and thus to elevate cyclic AMP concentrations in platelets, but the precise effects of cyclic AMP as an inhibitor of platelet function are still a subject of investigation and debate. PGE_1 induces phosphorylation of a group of membrane-bound-platelet proteins distinct from those phosphorylated in the calcium-dependent reactions discussed above – in fact, PGE_1 can reduce some of these calcium-dependent phosphorylations (Haslam *et al.*, 1979). It is probable that all adenylate cyclase stimulants inhibit platelets partly through cyclic AMP-dependent phosphorylation of proteins that facilitate the active transport of Ca^{2+} from the cytoplasm to storage sites in the membrane (Käser-Glanzmann, Jakábová, George & Lüscher, 1977, 1978) – a mechanism closely analogous to that previously described in cardiac muscle (Tada, Kirchberger,

Repke & Katz, 1974; Tada, Kirchberger & Katz, 1975). In addition, cyclic AMP may affect the liberation of Ca^{2+} from these sites, and can inhibit at least two of the enzymes involved in arachidonate metabolism. Both phospholipase (Minkes *et al.*, 1977) and cyclo-oxygenase (Malmsten, Granström & Samuelsson, 1976) are inhibited by cyclic AMP, but it seems likely that these actions are less important than its effect on calcium mobilisation (Gerrard, Peller, Crick & White, 1977; Gerrard *et al.*, 1978). In this connection, it is noteworthy that cyclic AMP is an effective inhibitor of phospholipase when the enzyme is activated by thrombin or collagen, but virtually inactive against the calcium ionophore A23187 (Feinstein *et al.*, 1977).

The emphasis on PGI_2 has tended to obscure other biologically active constituents secreted by vascular cells. Endothelium in culture also secretes PGE_2 (usually about $\frac{1}{5}$ to $\frac{1}{2}$ of the PGI_2 produced) and vascular smooth muscle cells in culture secrete large amounts of PGE_2 (up to one hundred times as much as endothelium), with relatively little PGI_2 production (Pearson, Ager, Trethevick & Gordon, 1979). The PGE_2 thus produced could potentiate platelet responses to other stimuli, and in some species induce platelet aggregation directly (q.v.).

Recent work in our laboratory has shown that endothelium and vascular smooth muscle in culture can secrete adenine nucleotides when exposed to potentially damaging stimuli such as thrombin or other proteolytic enzymes (Carleton, Gordon, Hutchings & Pearson, 1979; Pearson & Gordon, 1979) without any evidence of cell damage as revealed by vital dye uptake or release of cytoplasmic enzymes. The major nucleotide released was ATP but this was rapidly converted extracellularly to form enough ADP to induce platelet aggregation. Previous work established that endothelial cells possessed an ADPase (Heyns, van den Berg, Potgieter & Retief, 1974; Lieberman, Lewis & Peters, 1977), but this is only one of a complex of ectonucleotidases and kinases on endothelium and smooth muscle, including a very efficient ATPase. Thus, vascular cells can secrete adenine nucleotides when stimulated and also convert ATP released from damaged cells or secreted by platelets, producing ADP in sufficient quantities to induce platelet adhesion and aggregation. Hence, vascular cells can affect platelet activation not only by exerting an inhibitory action through secretion of prostacyclin, but also by stimulating platelets through the secretion of PGE_2 and the production of ADP.

7. SUMMARY AND CONCLUSIONS

The general phenomenon of platelet adhesion is usually subdivided into two categories: platelet aggregation (that is, platelets sticking to other platelets) and platelet adhesion to any other surface – such as the collagen in the subendothelial area of the vascular wall, or to artificial materials such as those used in vascular prostheses. This subdivision is paralleled by differences in some aspects of the mechanisms involved in regulating the two categories of platelet adhesion. Platelet aggregation is initiated by a specific stimulatory agent binding to a receptor on the cell membrane, inducing a membrane perturbation which leads to calcium ion liberation or influx, and thence to more profound cellular changes including the extrusion of filopodia, which facilitates adhesion, and the secretion of constituents (either from storage granules or from the metabolism of arachidonic acid) that can specifically stimulate other platelets in the vicinity and thus reinforce the aggregation process. In contrast, platelet adhesion to collagen (and probably most if not all other surfaces) is apparently a much less specific event: recognition sites of high specificity and affinity on the platelet membrane are apparently not involved, but the surface in question must contain an appropriate organisation of charged groups in optimal alignment and in regular repeating sequence. The multiple interactions thus produced then apparently allow membrane perturbations in the platelet that can result in the same sequence of intracellular regulatory mechanisms as those described above for aggregation, resulting once more in the secretion of constituents that promote the activation of other neighbouring platelets.

Several mechanisms exist by which platelet adhesion can be inhibited. For example, adhesion to a surface can be reduced by altering the distribution of charges on it (a topic of considerable potential clinical importance) and, in this regard, the relative affinity of a surface for different proteins can be important – for example, a surface coated with albumin is much more passive to platelets than a surface coated with fibrinogen (Packham, Evans, Glynn & Mustard, 1969). Additionally, the responsiveness of the platelets themselves may be reduced by elevating their intracellular level of cyclic AMP or by blocking their prostaglandin synthase pathway. Finally, platelet aggregation induced by specific stimuli can be blocked by masking the receptor to a specific stimulus (e.g. with a specific antagonist) or

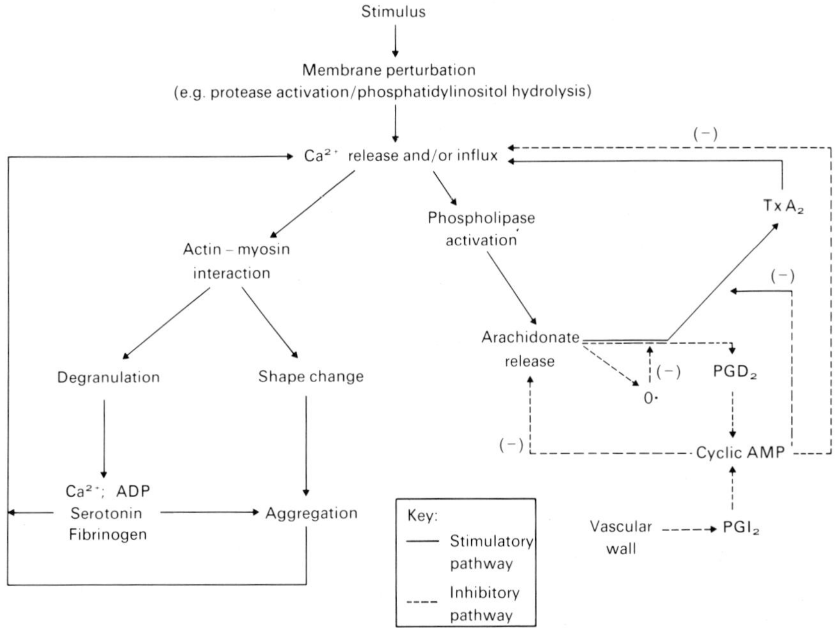

Fig. 4. Schematic representation of the major mechanisms regulating platelet activation. Stimulatory pathways are indicated by continuous lines and inhibitory pathways by interrupted lines. The effects of inhibiting *feedback loops* are shown as (−).

by removal of the stimulus (for example by uptake and/or metabolism). The regulatory mechanisms involved in platelet activation are outlined in Fig. 4.

The phenomenon of platelet adhesion is likely to remain a topic of considerable interest for some time for the following three reasons: first, it is important in physiology and pathology; second, it is a dramatic and rapid process, and relatively easily studied; and finally, it may be a valuable model of cell adhesion in several other systems. There is already a substantial amount of information on the mechanisms that regulate it, but there appears to be plenty of mileage left, and perhaps some workers who have studied cell adhesion in other systems may now be tempted to investigate platelets, and thus help not only to elucidate some of the processes important in haemostasis and vascular disease but also to further our understanding of intercellular adhesive processes in general.

REFERENCES

ALLAN, D. & MICHELL, R. H. (1975). Accumulation of 1,2-diacylglycerol in the plasma membrane may lead to echinocyte transformation of erythrocytes. *Nature, London*, **258**, 348–9.

BALLEISEN, L., GAY, S., MARX, R. & KUHN, K. (1975). Comparative investigation of the influence of human and bovine collagen types I, II and III on the aggregation of human platelets. *Klinische Wochenschrift*, **53**, 903–5.

BARNES, M. J., GORDON, J. L. & MACINTYRE, D. E. (1976). Platelet-aggregating activity of Type I and Type III collagens from human aorta and chicken skin. *Biochemical Journal*, **160**, 647–51.

BAUMGARTNER, H. R. & MUGGLI, R. (1976). Adhesion and aggregation: morphological demonstration and quantitation *in vivo* and *in vitro*. In *Platelets in Biology and Pathology*, ed. J. L. Gordon, pp. 23–60. Amsterdam: Elsevier/North Holland.

BAUMGARTNER, H. R., TRANZER, J. P. & STUDER, A. (1967. An electron microscopic study of platelet thrombus formation in the rabbit with particular regard to 5-hydroxytryptamine release. *Thrombosis et Diathesis Haemorrhagica*, **18**, 592–604.

BEESLEY, J. E., PEARSON, J. D., CARLETON, J. S., HUTCHINGS, A. & GORDON, J. L. (1978). Interaction of leukocytes with vascular cells in culture. *Journal of Cell Science*, **33**, 85–101.

BEESLEY, J. E., PEARSON, J. D., HUTCHINGS, A., CARLETON, J. S. & GORDON, J. L. (1979). Granulocyte migration through endothelium in culture. *Journal of Cell Science* **3,** 237–48.

BEHNKE, O. (1970). The morphology of blood platelet membrane systems. *Series Haematologica*, **3**(4), 3–16.

BENSUSAN, H. B., KOH, T. L., HENRY, K. G., MURRAY, B. A. & CULP, L. A. (1978). Evidence that fibronectin is the collagen receptor on platelet membranes. *Proceedings of the National Academy of Sciences, USA*, 75, 5864–8.

BEST, L. C., MARTIN, T. J., RUSSELL, R. G. G. & PRESTON, F. E. (1977). Prostacyclin increases cyclic AMP levels and adenylate cyclase activity in platelets. *Nature, London*, **267**, 850–2.

BLACKWELL, G. J. & FLOWER, R. J. (1978). 1-Phenyl-3-pyrazolidone: an inhibitor of cyclo-oxygenase and lipoxygenase pathways in lung and platelets. *Prostaglandins*, **16**, 417–25.

BOOT, J. R., DAWSON, W. & KITCHEN, E. A. (1976). The chemotactic activity of thromboxane B_2: a possible role in inflammation. *Journal of Physiology*, **257**, 47P.

BOOYSE, F. M. & RAFELSON, M. E. (1972). Mechanism and control of platelet–platelet interactions. III. A relaxation–contraction model for platelet aggregation. *Microvascular Research*, **4**, 207–13.

BORN, G. V. R. (1970). Observations on the change in shape of blood platelets brought about by adenosine diphosphate. *Journal of Physiology*, **209**, 487–511.

BRASS, L. & BENSUSAN, J. (1975). The platelet: collagen interaction. *Federation Proceedings*, **34**, 241.

BUCKMASTER, G. A. (1906). The blood platelets. *Science Progress*, **1**, 73–90.

CARLETON, J. S., GORDON, J. L., HUTCHINGS, A. & PEARSON, J. D. (1979). Secretion and extracellular metabolism of adenine nucleotides by endothelial cells in culture, *Journal of Physiology* **291**, 40P.

CAZENAVE, J. P., ASSIMEH, S. N., PAINTER, R. H., PACKHAM, M. A. & MUSTARD, J. F. (1976). Clq inhibition of the interaction of collagen with human platelets. *Journal of Immunology*, **116**, 162–3.

CHESNEY, C. M., HARPER, E. & COLMAN, R. W. (1972). Critical role of the carbohydrate side chains of collagen in platelet aggregation. *Journal of Clinical Investigation*, **51**, 2693–701.

CRAWFORD, N. (1976). Platelet microfilaments and microtubules. In *Platelets in Biology and Pathology*, ed. J. L. Gordon, pp. 121–58. Amsterdam: Elsevier/North Holland.

DAIMON, T., MIZUHIRA, V. & UCHIDA, K. (1978). Ultrastructural localization of calcium around the membrane of the surface-connected system in the human platelet. *Histochemistry*, **55**, 271–9.

DESSAU, W., ADELMANN, B. C. & TIMPL, R. (1978). Identification of the sites in collagen alpha-chains that bind serum anti-gelatin factor (cold insoluble globulin). *Biochemical Journal*, **169**, 55–9.

DRUMMOND, A. H. (1976). Interactions of blood platelets with biogenic amines: uptake, stimulation and receptor binding. In *Platelets in Biology and Pathology*, ed. J. L. Gordon, pp. 203–40. Amsterdam: Elsevier/North Holland.

EGAN, R. W., PAXTON, J. & KUEHL, F. A. (1976). Mechanism for irreversible self-deactivation of prostaglandin synthetase. *Journal of Biological Chemistry*, **251**, 7329–35.

FEINMAN, R. D. & DETWILER, T. C. (1974). Platelet secretion induced by divalent calcium ionophores. *Nature, London*, **249**, 172–3.

FEINSTEIN, M. B., BECKER, E. L. & FRASER, C. (1977). Thrombin, collagen and A23187 stimulated endogenous platelet arachidonate metabolism: differential inhibition by PGE_2, local anesthetics and a serine-protease inhibitor. *Prostaglandins*, **14**, 1075–93.

FEINSTEIN, M. B., FIEKERS, J. & FRASER, C. (1976). An analysis of the mechanism of local anesthetic inhibition of platelet aggregation and secretion. *Journal of Pharmacology and Experimental Therapeutics*, **197**, 215–28.

FRENCH, J. E., MACFARLANE, R. G. & SANDERS, A. G. (1964). The structure of haemostatic plugs and experimental thrombi in small arteries. *British Journal of Experimental Pathology*, **45**, 467–74.

GERRARD, J. M., BUTLER, A. M. & WHITE, J. G. (1978). Calcium release from a platelet calcium-sequestering membrane fraction by arachidonic acid and its prevention by aspirin. *Prostaglandins*, **15**, 703.

GERRARD, J. M., PELLER, J. D., KRICK, T. P. & WHITE, J. G. (1977). Cyclic AMP and platelet prostaglandin synthesis. *Prostaglandins*, **14**, 39–50.

GERRARD, J. M., PETERSON, D., TOWNSEND, D. & WHITE, J. G. (1976). Prostaglandins and platelet contraction. *Circulation*, **54**: Suppl. II, 196.

GERRARD, J. M., WHITE, J. G., RAO, G. H. R. & TOWNSEND, D.-W. (1976). Localization of platelet prostaglandin production in the platelet dense tubular system. *American Journal of Pathology*, **83**, 293–8.

Goetzl, E. J. & Gorman, R. R. (1978). Chemotactic and chemokinetic stimulation of human eosinophil and neutrophil polymorphonuclear leukocytes by 12-L-hydroxy-5,8,10-heptadecatrienoic acid (HHT). *Journal of Immunology*, **120**, 526–31.

Gordon, J. L. (1975). Blood platelet lysosomes and their contribution to the pathophysiological role of platelets. In *Lysosomes in Biology and Pathology*, ed. J. T. Dingle & R. T. Dean, pp. 3–32. Amsterdam: Elsevier/North-Holland.

Gordon, J. L. & Dingle, J. T. (1974). Binding of radio-labelled collagen to blood platelets. *Journal of Cell Science*, **16**, 157–66.

Hamberg, M., Svensson, J. & Samuelsson, B. (1974). Prostaglandin endoperoxides. A new concept concerning the mode of action and release of prostaglandins. *Proceedings of the National Academy of Sciences, USA*, **71**, 3824–8.

Hamberg, M., Svensson, J. & Samuelsson, B. (1975). Thromboxanes: a new group of biologically active compounds derived from prostaglandin endoperoxides. *Proceedings of the National Academy of Sciences, USA*, **72**, 2994–8.

Haslam, R. J. (1975). Roles of cyclic nucleotides in platelet function. In *Biochemistry and Pharmacology of Platelets*, pp. 121–51. Amsterdam: Elsevier-/North Holland.

Haslam, R. J., Davidson, M. M. L. & Desjardins, J. V. (1978). Inhibition of adenylate cyclase by adenosine analogues in preparations of broken and intact human platelets. Evidence for the unidirectional control of platelet function by cyclic AMP. *Biochemical Journal*, **176**, 83–95.

Haslam, R. J., Lynham, J. A. & Fox, J. E. B. (1979). Effects of collagen, ionophore A23187 and prostaglandin E_1 on the phosphorylation of specific proteins in blood platelets. *Biochemical Journal*, **178**, 397–406.

Haslam, R. J. & Rosson, G. M. (1975). Effects of adenosine on levels of adenosine 3′:5′-cyclic monophosphate in human blood platelets in relation to adenosine incorporation and platelet aggregation. *Molecular Pharmacology*, **11**, 528–44.

Henson, P. M., Gould, D. & Becker, E. L. (1976). Activation of stimulus-specific serine esterases (proteases) in the initiation of platelet secretion. *Journal of Experimental Medicine*, **144**, 1657–73.

Heyns, A. du P., van den Berg, D. J., Potgieter, G. M. & Retief, D. P. (1974). The inhibition of platelet aggregation by an aorta intima extract. *Thrombosis et Diathesis Haemorrhagica*, **32**, 417–31.

Higgs, E. A., Moncada, S. & Vane, J. R. (1978). Effect of prostacyclin (PGI_2) on platelet adhesion to rabbit arterial subendothelium. *Prostaglandins*, **16**, 17–22.

Holmsen, H. & Day, H. J. (1970). The selectivity of the thrombin-induced platelet release reaction: Subcellular localization of released and retained constituents. *Journal of Laboratory and Clinical Medicine*, **75**, 840–55.

Holmsen, H., Salganicoff, L. & Fukami, M. H. (1977). Platelet behaviour and biochemistry. In *Haemostasis: Biochemistry, Physiology and Pathology*, ed. D. Ogston & B. Bennett, pp. 239–319. New York, London: Wiley.

Hovig, T. (1963). Aggregation of rabbit blood platelets produced *in vitro* by saline 'extract' of tendons. *Thrombosis et Diathesis Haemorrhagica*, **9**, 248–63.

Hugues, J., Herion, F., Nusgens, B. & Lapiere, C. M. (1976). Type III collagen and probably not type I collagen aggregates platelets. *Thrombosis Research*, **9**, 223–31.

Jaffe, R. & Deykin, D. (1974). Evidence for a structural requirement for aggregation of platelets by collagen. *Journal of Clinical Investigation*, **53**, 875–83.

Jaffe, R. M. (1976). Interaction of platelets with connective tissue. In *Platelets in Biology and Pathology*, ed. J. L. Gordon. Amsterdam: Elsevier/North-Holland.

Jamieson, G. A., Smith, D. F. & Kosow, D. P. (1975). Possible role of collagen glucosyl transferase in platelet adhesion. Mechanistic considerations. *Thrombosis et Diathesis Haemorrhagica*, **33**, 668–71.

Jamieson, G. A., Urban, C. L. & Barber, A. J. (1971). Enzymatic basis for platelet: collagen adhesion as the primary step in haemostasis. *Nature, London*, **234**, 5–7.

Johnson, R. A., Morton, D. R., Kinner, J. H., Gurman, R. R., McGuire, J. C., Sun, F. F., Whittaker, N., Bunting, S., Salmon, J., Moncada, S. & Vane, J. R. (1976). The chemical structure of prostaglandin X (prostacyclin). *Prostaglandins*, **12**, 915–28.

Jones, R. L., Kerry, P. J., Poyser, N. L., Walker, I. C. & Wilson, N. H. (1978). Identification of trihydroxyeicosatrienoic acids as products from incubation of arachidonic acid with washed blood platelets. *Prostaglandins*, **16**, 583–9.

Kang, A. H., Beachey, E. H. & Katzman, R. L. (1974). Interaction of an active glycopeptide from chick skin collagen (alpha 1 CB 5) with human platelets. *Journal of Biological Chemistry*, **249**, 1054–9.

Käser-Glanzmann, R., Jakábová, M., George, J. N. & Lüscher, E. F. (1977). Stimulation of calcium uptake in platelet membrane vesicles by adenosine 3′,5′-cyclic monophosphate and protein kinase. *Biochimica et Biophysica Acta*, **466**, 429–40.

Käser-Glanzmann, R., Jakábová, M., George, J. N. & Lüscher, E. F. (1978). Further characterization of calcium-accumulating vesicles from human blood platelets. *Biochimica et Biophysica Acta*, **512**, 1–12.

Kronick, P. (1975). Binding of collagen to platelets. *Federation Proceedings*, **33**, 1536.

Kuehl, F. A., Humes, J. L., Egan, R. W., Ham, E. A., Beveridge, G. C. & van Arman, C. G. (1977). Role of prostaglandin endoperoxide PGG_2 in inflammatory processes. *Nature, London*, **265**, 170–3.

Lackie, J. M. & de Bono, D. P. (1977). Interactions of neutrophil granulocytes (PMNs) and endothelium *in vitro*. *Microvascular Research*, **13**, 107–12.

Lieberman, G. E., Lewis, G. P. & Peters, T. J. (1977). A membrane-bound enzyme in rabbit aorta capable of inhibiting adenosine-diphosphate-induced platelet aggregation. *Lancet*, ii, 330–2.

Lüscher, E. F., Pfueller, S. L. & Massini, P. (1973). Platelet aggregation by large molecules. *Series Haematologica*, **3**, 382–91.

MacIntyre, D. E., Allen, A. P., Thorne, K. J. I., Glauert, A. M. & Gordon, J. L. (1979). Endotoxin-induced platelet aggregation and secretion. I.

Morphological changes and pharmacological effects. *Journal of Cell Science*, **28**, 225–36.

MacIntyre, D. E. & Gordon, J. L. (1975). Calcium-dependent stimulation of platelet aggregation by PGE_2 *Nature, London*, **258**, 337–8.

MacIntyre, D. E. & Gordon, J. L. (1977). Discrimination between platelet prostaglandin receptors with a specific inhibitor of bisenoic prostaglandins. *Thrombosis Research*, **11**, 705–13.

MacIntyre, D. E., McMillan, R. M. & Gordon, J. L. (1977). Secretion of lysosomal enzymes by platelets. *Biochemical Society Transactions*, **5**, 1178–80.

MacIntyre, D. E., Pearson, J. D. & Gordon, J. L. (1978). Localization and stimulation of prostacyclin production by vascular cells. *Nature, London*, **271**, 549–51.

MacIntyre, D. E., Salzman, E. W. & Gordon, J. L. (1978). Prostaglandin receptors on human platelets. Structure–activity relationships of stimulatory prostaglandins. *Biochemical Journal*, **174**, 921–9.

McMillan, R. M., MacIntyre, D. E. & Gordon, J. L. (1977). Simple, sensitive fluorimetric assay for malondialdehyde production by blood platelets. *Thrombosis Research*, **11**, 425–8.

Majno, G. & Palade, G. E. (1961). Studies on inflammation. II. The site of action of histamine and serotonin on vascular permeability: an electron microscopic study. *Journal of Biophysical and Biochemical Cytology*, **11**, 571–605.

Malmsten, C., Granström, E. & Samuelsson, B. (1976). Cyclic AMP inhibits synthesis of prostaglandin endoperoxide (PGG_2) in human platelets. *Biochemical and Biophysical Research Communications*, **68**, 569–76.

Marcus, A. J., Weksler, B. B. & Jaffe, E. A. (1978). Enzymatic conversion of prostaglandin endoperoxide PGH_2 and arachidonic acid to prostacyclin by cultured human endothelial cells. *Journal of Biological Chemistry*, **253**, 7138–41.

Menashi, S., Harwood, R. & Grant, M. E. (1976). Native collagen is not a substrate for the collagen glucosyltransferase of platelets. *Nature, London*, **264**, 670–2.

Michaeli, D. & Orloff, K. G. (1977). Molecular considerations of platelet adhesion. In *Progress in Haemostasis and Thrombosis*, vol. 3, ed. T. H. Spaet, pp. 29–59. New York: Grune & Stratton.

Michell, R. H. (1975). Inositol phospholipids and cell surface receptor function. *Biochimica et Biophysica Acta*, **415**, 81–147.

Michell, R. H., Jones, L. M. & Jafferji, S. S. (1977). A possible role for phosphatidyl inositol breakdown in muscarinic cholinergic stimulus–response coupling. *Biochemical Society Transactions*, **5**, 77–81.

Mills, D. C. B., Robb, I. A. & Roberts, G. C. K. (1968). The release of nucleotides, 5-hydroxytryptamine and enzymes from human blood platelets during aggregation. *Journal of Physiology*, **195**, 715–29.

Minkes, M., Stanford, M., Chi, M., Roth, G. J., Raz, A., Needleman, P. & Majerus, P. W. (1977). Cyclic adenosine 3′,5′-monophosphate inhibits the availability of arachidonate to prostaglandin synthetase in human platelet suspensions. *Journal of Clinical Investigation*, **59**, 449–54.

Moncada, S., Gryglewski, R. J., Bunting, S. & Vane, J. R. (1976). A lipid peroxide inhibits the enzyme in blood vessel microsomes that generates from prostaglandin endoperoxides the substance (prostaglandin X) which prevents platelet aggregation. *Prostaglandins*, **12**, 715–37.

MONCADA, S., GRYGLEWSKI, R., BUNTING, S. & VANE, J. R. (1976). An enzyme isolated from arteries transforms prostaglandin endoperoxides to an unstable substance that inhibits platelet aggregation. *Nature, London*, **263**, 663–5.

MONCADA, S., HIGGS, E. A. & VANE, J. R. (1977). Human arterial and venous tissues generate prostacyclin (prostaglandin X), a potent inhibitor of platelet aggregation. *Lancet*, i, 18–21.

MONCADA, S., KORBUT, R., BUNTING, S. & VANE, J. R. (1978). Prostacyclin is a circulating hormone. *Nature, London*, **273**, 767–8.

MUGGLI, R. (1978). Collagen-induced platelet aggregation: native collagen quaternary structure is not an essential structural requirement. *Thrombosis Research*, **13**, 829–43.

MUGGLI, R. & BAUMGARTNER, H. R. (1973). Collagen-induced platelet aggregation: requirement for tropocollagen multimers. *Thrombosis Research*, **3**, 715–28.

MUSTARD, J. F., PACKHAM, M. A., KINLOUGH-RATHBONE, R. L., PERRY, D. W. & REGOECZI, E. (1978). Fibrinogen and ADP-induced platelet aggregation. *Blood*, **52**, 453–66.

NEEDLEMAN, P., BRONSON, S. D., WYCHE, A., SIVAKOFF, M. & NICOLAOU, K. C. (1978). Cardiac and renal prostaglandin I_2. Biosynthesis and biological effects in isolated perfused rabbit tissues. *Journal of Clinical Investigation*, **61**, 839–49.

NEEDLEMAN, P., SYCHE, A. & RAZ, A. (1979). Platelet and blood vessel arachidonate metabolism and interactions. *Journal of Clinical Investigation*, **63**, 345–9.

PACKHAM, M. A., EVANS, G., GLYNN, M. F. & MUSTARD, J. F. (1969). The effect of plasma proteins on the interaction of platelets with glass surfaces. *Journal of Laboratory and Clinical Medicine*, **73**, 686–697.

PEARSON, J. D., AGER, E. A., TREVETHICK, M. A. & GORDON, J. L. (1979). Prostaglandin production by cultured vascular cells. *Agents Actions* supplement **4**, 120–6.

PEARSON, J. D., CARLETON, J. S., BEESLEY, J. E., HUTCHINGS, A. & GORDON, J. L. (1979). Granulocyte adhesion to endothelium in culture. *Journal of Cell Science* **28**, 225–35.

POLLARD, T. D., FUJIWARA, K., HANDIN, R. & WEISS, G. (1977). Contractile proteins in platelet activation and contraction. *Annals of the New York Academy of Sciences*, **283**, 218–36.

PEARSON, J. D. & GORDON, J. L. (1979). Vascular endothelial and smooth muscle cells in culture selectively release adenine nucleotides. *Nature, London*, **281**, 384–6.

PUETT, D., WASSERMAN, B. K., FORD, J. D. & CUNNINGHAM, L. W. (1973). Collagen-mediated platelet aggregation: Effects of collagen modification involving the protein and carbohydrate moieties. *Journal of Clinical Investigation*, **52**, 2495–506.

PURÉ, E. & NEEDLEMAN, P. (1979). The effect of endothelial damage on prostaglandin synthesis by the isolated perfused rabbit mesenteric vasculature. *Journal of Cardiovascular Pharmacology* (in press).

RITTENHOUSE-SIMMONS, S., RUSSELL, F. A. & DEYKIN, D. (1977). Mobilization of arachidonic acid in human platelets. Kinetics and Ca^{2+}-dependency. *Biochimica et Biophysica Acta*, **488**, 370–80.

ROBB-SMITH, A. H. T. (1967). Why the platelets were discovered. *British Journal of Haematology*, **13**, 618–37.

ROSEMAN, S. (1970). The synthesis of complex carbohydrates by multiglycosyltransferase systems and their potential function in intercellular adhesion. *Chemistry and Physics of Lipids*, **5**, 270–97.

ROSS, R., GLOMSET, J., KARIYA, B. & HARKER, L. (1974). A platelet-dependent serum factor that stimulates the proliferation of arterial smooth muscle cells *in vitro*. *Proceedings of the National Academy of Sciences, USA*, **71**, 1207–10.

SALZMAN, E. W., MACINTYRE, D. E., STEER, M. L. & GORDON, J. L. (1978). Effect on platelet activity of inhibition of adenylate cyclase. *Thrombosis Research*, **13**, 1089–1101.

SANTORO, S. A. & CUNNINGHAM, L. W. (1977). Collagen-mediated platelet aggregation. Evidence for multivalent interactions of intermediate specificity between collagen and platelets. *Journal of Clinical Investigation*, **60**, 1054–60.

SILVER, F. H., YANNAS, I. V. & SALZMAN, E. W. (1978). Glycosaminoglycan inhibition of collagen-induced platelet aggregation. *Thrombosis Research*, **13**, 267–78.

SMITH, J. B., OGLETREE, M. L., LEFER, A. M. & NICOLAOU, K. C. (1978). Antibodies which antagonise the effects of prostacyclin. *Nature, London*, **274**, 64–5.

STUART, M., MURPHY, S. & OSKI, A. (1975). A simple, non-radioisotope technic for the determination of platelet lifespan. *New England Journal of Medicine*, **292**, 1310–13.

TADA, M., KIRCHBERGER, M. A., REPKE, D. I. & KATZ, A. M. (1974). The stimulation of calcium transport in cardiac sarcoplasmic reticulum by adenosine 3′ 5′-monophosphate-dependent protein kinase. *Journal of Biological Chemistry*, **249**, 6174–80.

TADA, M., KIRCHBERGER, M. A. & KATZ, A. M. (1975). Phosphorylation of a 22,000-dalton component of the cardiac sarcoplasmic reticulum by adenosine 3′ 5′-monophosphate-dependent protein kinase. *Journal of Biological Chemistry*, **250**, 2640–7.

TANSIK, R. L., NAMM, D. H. & WHITE, H. L. (1978). Synthesis of prostaglandin 6-keto $F_{1\alpha}$ by cultured aortic smooth muscle cells and stimulation of its formation by a coupled system with platelet lysates. *Prostaglandins*, **15**, 399–408.

TRANZER, J. P. & BAUMGARTNER, H. R. (1967). Filling gaps in the vascular endothelium with blood platelets. *Nature, London*, **216**, 1126–8.

TRELSTAD, R. (1974). Human aorta collagens: evidence for three distinct species. *Biochemical and Biophysical Research Communications*, **57**, 717–25.

TURNER, S. R., TAINER, J. A. & LYNN, W. S. (1975). Biogenesis of chemotactic molecules by the arachidonate lipoxygenase system of platelets. *Nature, London*, **257**, 680–1.

WHARTON-JONES, W. T. (1851). On the state of the blood and blood vessels in inflammation. *Guy's Hospital Reports*, Ser. 2, **7**, 1–25.

WHITE, J. G. (1971). Platelet morphology. In *The Circulating Platelet*, ed. S. A. Johnson, pp. 46–122. London, New York: Academic Press.

WILNER, G. D., NOSSELL, H. L. & PROCUPEZ, T. L. (1971). Aggregation of platelets by collagen: polar active sites of insoluble human collagen. *American Journal of Physiology*, **220**, 1074–9.

WITTE, L. D., KAPLAN, K. L., NOSSEL, H. L., LAGES, B. A., WEISS, H. J. & GOODMAN, De W. S. (1978). Studies on the release from human platelets of the growth factor for cultured human arterial smooth muscle cells. *Circulation Research*, 42, 402–9.

YAMADA, K. M. & OLDEN, K. (1978). Fibronectins – adhesive glycoproteins of cell surface and blood. *Nature, London*, **275**, 179–84.

ZUCKER, M. B. & BORRELLI, J. (1962). Platelet clumping produced by connective tissue suspensions and by collagen. *Proceedings of the Society for Experimental Biology and Medicine*, **109**, 779–87.

Interactions of leukocytes and endothelium

J. M. LACKIE AND R. P. C. SMITH

Department of Cell Biology, Glasgow University, Glasgow G12 8QQ, UK

Most studies of cell adhesion have been done on cells which form solid tissues and, after the early stages of morphogenesis, probably have stable adhesive properties. Even fibroblasts, which tend to be more isolated than epithelial cells and which undoubtedly retain the ability to move in adult animals, probably do not undergo marked changes in their adhesiveness, except transiently at mitosis. Leukocytes are found in the blood as non-adherent circulating cells, but their role, in many pathological processes, is to leave the bloodstream and migrate into damaged or infected tissues. Since blood vessels are lined with endothelial cells, the transition from a circulating phase to a tissue phase will inevitably involve an adhesive interaction with endothelial cells; not only must there be an alteration in the adhesive interaction between leukocytes and endothelium, but the adhesions formed must be suitable to support the locomotion of leukocytes over and between the endothelial cells in order that they may leave the lumen of the blood vessel. These adhesive and locomotory activities are important in the cellular defence mechanisms of all animals with a well-developed blood vascular system and provide an interesting example of controlled changes in adhesive properties on a short time-scale. In a sense the activities exhibited by leukocytes are the converse of those shown by metastasising neoplastic cells, local invasion followed by a blood-borne phase, and the study of leukocytes may provide some insight into these more sinister pathologies.

Various leukocyte–endothelial interactions have been studied in some detail: (1) the adhesion of leukocytes to endothelium in the cellular inflammatory response (2) the adhesion of lymphocytes to the high-endothelial venules of lymph nodes as part of the normal pathway of lymphocyte recirculation (3) the adhesion of leukocytes and platelets to endothelium or subendothelium in thrombosis. Other interactions undoubtedly exist, and an additional area of interest is the adhesion of neoplastic cells to endothelium in blood-borne metastasis. Only the acute inflammatory response will be considered in this report; those interested in the other interactions may find a recent review helpful (Wilkinson & Lackie, 1979).

THE CELLULAR INFLAMMATORY RESPONSE

There are two distinct aspects to the inflammatory response: the changes in permeability of capillaries and venules and various microvascular changes which lead to oedema, redness and a local temperature increase, and the cellular part of the response, which involves the sticking and emigration of leukocytes and, in extreme cases, to the extravasation of red blood cells. Although many of the non-cellular events are very important and have received much attention we will be concerned only with the question of leukocytic adhesion and emigration; some of the microvascular changes may be important to this phase and are discussed later. Most of the discussion will be about the acute inflammatory response, since it seems probable that in chronic inflammatory lesions a persistent stimulus maintains the cellular response and the difference is only one of time-scale.

All of the classes of leukocytes are involved in inflammatory responses, although lymphocytic involvement tends to occur at a later stage and be more conspicuous in longer-term inflammatory lesions which involve a complex antigen as the irritant. Thus tubercle bacilli will tend to give rise to lesions rich in lymphocytes, whereas simple irritants such as talc excite a reaction dominated by mononuclear phagocytes (Spector, 1974 for review). The early stages of the response are characterised by the adhesion and emigration of neutrophil granulocytes and blood monocytes; because neutrophils are present in blood in much higher numbers they are a more conspicuous feature of the response and have received a disproportionate amount of attention. Indeed it would be nice to be able to discuss the interactions of monocytes with endothelium but there is almost no information available, and so they will be largely ignored in this review. Eosinophil granulocytes are involved particularly in inflammatory lesions which are a response to multicellular parasites such as schistosomes (Goetzl & Austen, 1977) and basophils are the source of histamine which may mediate many of the non-cellular events in inflammation.

This review will concentrate upon the interaction between neutrophil leukocytes and endothelium, since this is the best-studied interaction. It should, however, be pointed out that in studies *in vivo* it is impossible to distinguish the leukocyte classes, and although most of the cells are probably neutrophils this is not certain. There is no particular reason to suppose that neutrophils and monocytes differ markedly in their interactions with endothelium, since both cell types respond in the same way and at a similar stage of the response, but for the other cell types there may be more specific interactions.

The nature of the response

Local damage, of whatever kind, leads to an inflammatory response, and the process of inflammation is an important part of the defence system of the body. At almost all times an inflammatory response is in progress in some part of the body, and the clearance of damaged tissue and the elimination of invasive organisms are clearly of major importance in homeostasis. The basic response has been known for many years and Cohnheim's (1882) description has a remarkably modern air about it. More detailed descriptions come from observations on the microvasculature in transparent tissues such as the tadpole's tail (Clark, Clark & Rex, 1936), mesentery (Janoff & Zweifach, 1964), the hamster cheek pouch (Atherton & Born, 1972) and in particular the regenerated vascular bed in rabbit ear chambers (Allison, Smith & Wood, 1955*a*). The ultrastructural studies of Florey & Grant (1961), among others, complete the simple descriptive picture. Excellent reviews of this descriptive work abound and Grant (1973) gives a good detailed review.

Damage leads to changes in blood flow in the adjacent microvasculature partly through vasodilation and partly through the opening of additional capillary vessels. An early increase in flow rate is followed by a reduction, and the changes are complex (Wells, 1973). After 15–20 min leukocytes, probably mostly neutrophils, begin to adhere to the endothelium of post-capillary venules. The adhesion of leukocytes (margination) may continue until the walls of the vessels are completely lined with leukocytes and is shortly followed by the emigration (diapedesis) of leukocytes from the vessel. Leukocytes move over the endothelium and push their way between

(a)

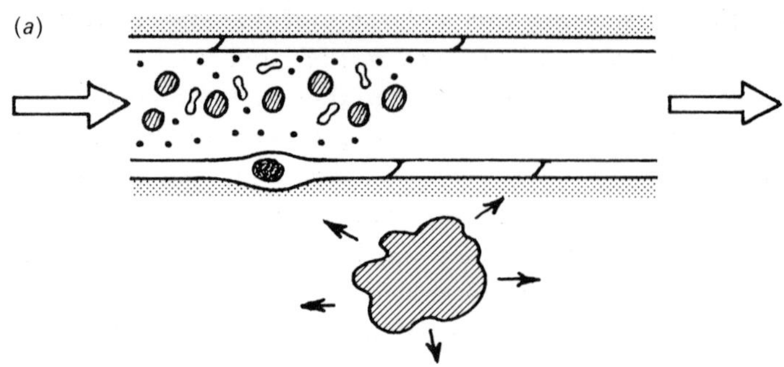

(b)

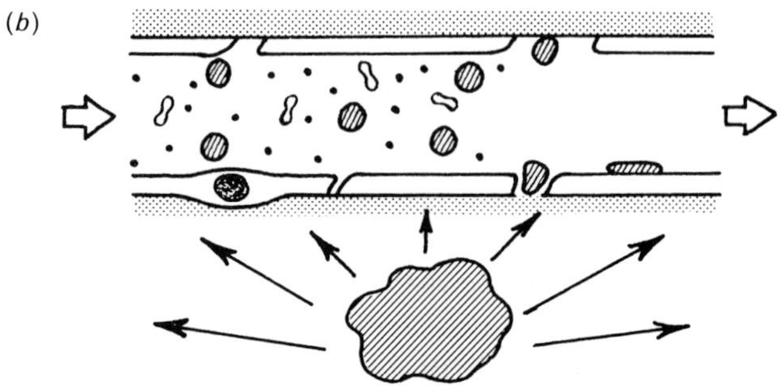

(c)

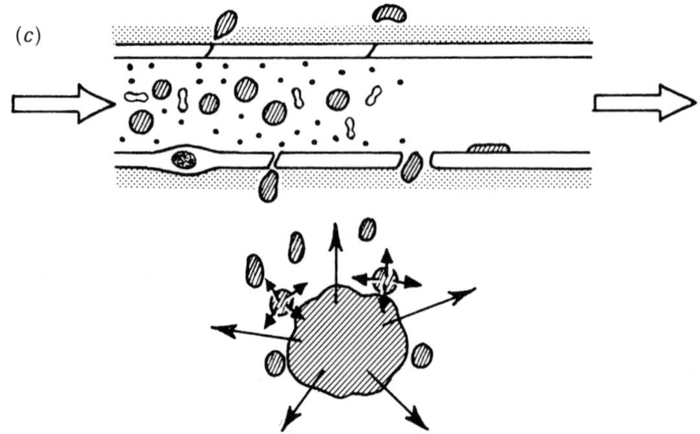

Key:

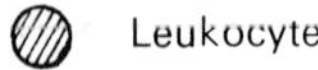

Leukocyte

Platelet

Red blood cell

endothelial cells, taking only 3–9 min to escape from the vessel (Clark, Clark & Rex, 1936). The emigration of leukocytes seems invariably to be between cells, even in the case of lymphocyte emigration (Schoefl, 1972), despite earlier suggestions to the contrary. The endothelial cells in the post-capillary venules, and to a lesser extent in the arterioles, do not have elaborate junctional complexes (Simionescu, Simionescu & Palade, 1975; 1976) and this undoubtedly makes emigration much easier.

The localised nature of the inflammatory response has frequently led to the assumption that the primary change must be in the endothelium of the blood vessel. This may not be the case: the overall reaction certainly is local, but there are several ways in which this localisation might come about and several possibilities may be distinguished.

The localisation of the response

Three main hypotheses might account for localisation.

(1) Local changes in endothelium–cell adhesiveness.

(2) Systemic changes in leukocyte adhesiveness but with local emigration.

(3) Local changes in blood flow, shifting the balance in favour of leukocyte adhesion with further localisation by restricted emigration.

Endothelial change. The first hypothesis might seem the most attractive, since the endothelium is the only static component and is much more likely to be influenced by the diffusible products of tissue damage or bacterial infection. Changes in the endothelium account for the local changes in capillary permeability and an observation frequently quoted in support of adhesive changes is that of Allison *et al.* (1955*a*). These authors found, in a rabbit ear chamber

Fig. 1. An inflammatory focus adjacent to a post-capillary venule. In (*a*) the blood flow (broad white arrow) is normal and the endothelium is undisturbed. Leukocytes, red blood cells and platelets are distributed in the stream according to size.

In (*b*) after 15–20 min, blood flow has diminished and margination has begun. Capillary permeability may still be high and a well-established gradient of chemotactic factors emanating from the lesion is suggested by the arrows. After 25–30 min (*c*), blood flow has returned to normal and emigration of leukocytes has occurred. Secondary gradients of chemotactic factors may arise from leukocytes near the focus.

preparation, that margination occurred unilaterally on the wall of the vessel adjacent to a very local experimentally-induced lesion. Their plates do indeed show unilateral margination but the interpretation is complicated by the complex geometry of the vessel, and leukocytes seem to be accumulating in 'slack-water'. Since margination requires only that the adhesive interaction between leukocytes and endothelium exceeds the shearing stresses tending to distract the cells, the local rheology must be taken into consideration.

Local adhesive changes may well be involved in reversed passive Arthus reactions (Ward & Cochrane, 1964) and in the local Schwartzman phenomenon (Stetson & Good, 1951), but in both these rather bizarre lesions there are likely to be many other complicating factors, and the margination phase has received less attention than the later neutrophil-mediated tissue damage (for a good review of these reactions see Cochrane & Janoff, 1974).

Systemic leukocytic change. The possibility that all leukocytes become adhesive, and that margination occurs throughout the body seems a drastic alternative model. To immobilise all the leucocytes or all the neutrophils in the circulation in response to every minute pin-prick seems an extravagant way of mounting a local response. Evidence for this model does, however, exist. Studies on the population kinetics of granulocytes (Athens *et al.*, 1961 *a* & *b*) show that the total blood granulocyte pool can be subdivided into two approximately equal compartments, the marginated granulocyte pool and the circulating pool. Cells can move freely from one compartment to another, and an increase in the circulating pool can be brought about at the expense of the marginated pool by adrenaline, which decreases the adhesiveness of neutrophils (see Table 1) and increases blood flow, and by exercise, which also increases blood flow. The marginated pool is greatly increased in transient neutropenias (depleting the circulating pool) which can be brought about by various agents which increase the adhesiveness of neutrophils. One of the most interesting ways of inducing such a neutropenia is by systemic complement activation, induced with cobra venom factor (McCall, De Chatelet, Brown & Lachmann, 1974) or by exposing blood to a foreign surface, as occurs in renal dialysis (Jensen *et al.*, 1973) and in filtration leukophoresis (Schiffer, Aisner & Wiernik, 1975; Hammerschmidt *et al.*, 1978). Experimental neutropenias can be induced by intravascular complement activation, can be partially inhibited by systemic complement depletion, and there is a good correlation with the effects of complement on neutrophil adhesiveness *in vitro* (Fehr & Jacob, 1977; O'Flaherty, Craddock & Jacob, 1978). Neutropenias are also induced by a variety of agents which affect neutrophil adhesiveness, notably endotoxin (Essex & Grana, 1949; Leusen & Essex, 1953; Athens *et al.*, 1961*b*) and agents which tend to decrease blood flow (Bassen, Etess & Rosenthal, 1952). The margination which brings about this neutropenia leads to a massive pulmonary sequestration of neutrophils (Toren, Goffinet & Kaplow, 1970, among others), scarcely surprising since the capillary bed of the lung is the largest in the body and the first encountered by cells which have been altered by agents injected into major veins, the normal inoculation site. Interestingly, neutropenias tend not to be accompanied by lymphocytic leukopenia (Bassen *et al.*, 1952) and this suggests that there may be cell-type-specific effects involved in lymphocyte recirculation. The neutropenias discussed above are transient, with a nadir at around 5 min, the circulating granulocyte count returning to normal within 60–120 min.

Such a systemic neutrophil margination will not localise the inflammatory response unless there is some local stimulus to emigration. Those neutrophils which have left the blood vessels do not re-enter the circulation, and those which have begun moving may have changed their morphology and be less liable to being swept back into circulation. The movement of emigrating cells towards the focus of infection may well involve chemotaxis, a phenomenon which leukocytes demonstrate unequivocally *in vitro* (Wilkinson, 1974*a*; Wilkinson & Lackie, 1979), and will be accelerated by chemokinesis. Many chemotactic factors have chemokinetic effects (for a discussion of the differences between these reactions see Keller *et al.*, 1977) and if chemotactic factors are released from a local lesion they will tend to cause massive local emigration on a time-scale which would be suitable to localise transiently marginated neutrophils.

The superimposition of a local stimulus for emigration on a generalised stimulus for adhesion would serve to generate a local response and it could be appropriate to activate the adhesion response following major inflammatory affronts. This model does not, however, seem sensible for responding to trivial injuries.

Local rheological change. The third possibility is a compromise between the very local model and the systemic. Local inflammation leads to changes in blood flow, hence the redness and heat associated with inflammation, and the pharmacological mediation of this phase of the response has received much attention (Willoughby, 1973; Zweifach, 1973, for reviews).

The microcirculation in tissues is complex and normally only a proportion of the capillaries are open (for review see Wells, 1973); opening all the capillaries will inevitably alter the flow rate in the vessels and the wall shear stress will vary as a function of flow rate. Detailed analysis is extremely difficult since the blood vessels are pulsatile (Zweifach, 1974) and the geometry of the microvasculature is complex, but in an undisturbed capillary bed there is an abrupt decline in flow rate, and therefore in wall shear stress, as the blood reaches the post-capillary venules (Mayrovitz, Tuma & Wiedeman, 1977; Mayrovitz, Wiedeman & Tuma, 1977). This probably accounts for the observation that margination and diapedesis normally occur in post-capillary venules rather than in capillaries or pre-capillary arterioles, and may also account for the differences in intercellular junctions in the endothelia of these different vessels. Only in the post-capillary venules will the wall shear stress drop sufficiently to permit the possibility of margination, and probably in an inflammatory response the local changes in blood flow, which will take some time to occur if the stimulus is a diffusible factor, will tip the balance even further in favour of margination. Immediately adjacent to the lesion, leukocytes may be stimulated to emigrate by diffusible chemokinetic or chemotactic factors and a further localisation of the response may occur. A contributory factor may be the effect of decreased flow rate on the distribution of cells in the stream. In an undisturbed laminar flow, particles will tend to distribute themselves so that the largest particles occupy the central region, the axial part of the flow, leading to a reduced cell number immediately adjacent to the vessel wall. Any tendency of red cells to clump will force leukocytes towards the marginal zone because they are now the smaller particles: many agents which induce neutropenias tend to affect red cell clumping (Fahraeus, 1929; Vejlens, 1938). The distribution of particles between axial and marginal parts of the stream will also depend upon flow rate, and decreased flow rate will increase the probability that leukocytes leave the axial stream and enter the marginal zone, a first step towards margination. If this third possibility, that local changes in blood flow lead to margination, is correct, we must suppose that small changes in the kinetics of leukocyte–endothelium collisions are sufficient to affect the probability of adhesion. This does not seem to be a difficult requirement: the balance between the circulating and marginated pools of granulocytes (see above) can be disturbed very easily and such sensitivity would be quite appropriate. Additional support comes from the careful in-vivo studies of Atherton & Born (1972; 1973), who attempted to quantify the interaction between leukocytes and endothelium in a hamster cheek-pouch system by measuring the speed at which leukocytes rolled along the walls of vessels. They found that the rate was sensitive to blood flow, not surprisingly perhaps, and commented that collisions seemed to be 'inelastic' in inflamed areas. Similarly, Giddon & Lindhe (1972) using the same in-vivo system, found that leucocyte–endothelium interactions were markedly affected by the topical application of local anaesthetics and that leukocytes which were already marginated would rejoin the circulating pool. Local anaesthetics do decrease leukocyte adhesiveness, but only to small extent (R. B. Allan, personal communication), also suggesting that small changes are sufficient to affect the interaction quite markedly. In a sense this is predictable: the adhesion formed by the neutrophil must not be so strong that movement is prevented – or the migration phase would be impossible.

It seems fairly clear from the observations *in vivo*, no matter which of the explanations for localisation is adopted, that we are dealing with a system which is exquisitely balanced: a small change in the strength of the adhesive interaction between leukocytes and endothelium will have marked consequences. Once the leukocyte has formed an adhesion then other factors come into play – and it is easy to see how chemokinetic and chemotactic factors might play a role in stimulating and directing the emigration of leukocytes – but until the leukocyte adheres, nothing will happen.

The inflammatory response poses an interesting problem for a cell biologist interested in cell adhesion, but only fairly recently have non-pathologists approached the system. Impressive though the in-vivo studies are, the limitations and uncertainties are formidable: we feel that new insights may well come from studies on isolated or recombined components in defined in-vitro systems.

In-vitro studies

These fall into two categories: studies on isolated components, particularly neutrophils, and studies of the interaction between leucokytes and endothelium. Although we will mostly be concerned with the latter, studies on the adhesiveness of neutrophils themselves are important if we are to determine which component changes its properties and attempt to distinguish between the models for localisation discussed above. Although both neutrophils and monocytes stick to endothelium in the early stages of the inflammatory response, attention has focused on neutrophils, which are present in much larger numbers in the bloodstream.

Neutrophil adhesion. A wide variety of adhesion assays have been used, and comparison of the results with different assay systems is often difficult. Some studies have used whole anti-coagulated blood or leukocyte-rich plasma and these, although more realistic models of the in-vivo situation, are more difficult to interpret since, for example, platelets may adhere first and modify the substratum. Glass bead columns (Garvin, 1968; Kvarstein, 1969*a*, *b*; Lorente, Fontan, Garcia Rodriguez & Ojeda, 1978) and nylon fibre columns (MacGregor, Spagnulo & Lentnek, 1964; Schiffer, Sanel, Young & Aisner, 1977) have been used, the retention of cells or the later ease of elution being taken as a measure of adhesiveness. Others have used glass capillary tubes, (Bryant & Sutcliffe, 1972) or coverslips (Lichtman & Weed, 1972) with distraction by centrifugation. Glass coverslips on the walls of modified Payling–Wright rotator flasks (Banks & Mitchell, 1973*a*, *b*, *c*) or static coverslips in wells or chambers (Gallin, Wright & Schiffmann, 1978; Gray, Tsan & Wagner, 1977) with distraction by passing through an air–medium interface (Lackie & de Bono, 1977) or by gravity (Smith, Hollers, Patrick & Hassett, 1979) have also been used. In addition, coverslip methods have been used for looking at adhesion to confluent monolayers of cells (see later section). Macrophage spreading on protein-coated glass (Rabinovitch & de Stefano, 1973*a*, *b*, 1974) is perhaps a measure of adhesiveness, although macrophages are not circulating cells and probably differ in many respects from monocytes.

Aggregation methods have recently begun to be used, although the methods of bringing about collisions between the cells vary (Talstad, 1972; Lackie, 1974; 1977; O'Flaherty, Kreutzer & Ward, 1977).

Table 1. *Adhesiveness of neutrophil leukocytes*

Agent/condition	Cell type/assay	Effect on adhesion	Reference
In vivo			
Heparin Warfarin	REC	None	Allison & Lancaster (1961)
Fibrin/fibrinogen	REC	None	Allison & Lancaster (1960)
Histamine	HCP	None	Atherton & Born (1972)
Divalent ion chelation	HCP	Decrease	Atherton & Born (1972)
	REC	Decrease	Thompson, Papadimitriou & Walters (1967)
Adrenaline Exercise Prednisone	Human leukocytosis	Decreased marginated pool	Athens *et al.* (1961*b*) Bishop *et al.* (1971)
Cortisone	REC	Decreased margination	Allison, Smith & Wood (1955*b*)
Local anaesthetic	HCP	Reverses pre-existent margination	Giddon & Lindhe (1972)
Casein[a] *E. coli* culture filtrate[a]	HCP	Increased margination	Atherton & Born (1972)
Gelatin and other erythrocyte agglutinins	Guinea pig mesentery	Increase	Fahraeus (1929)
Endotoxin[a]	Human neutropenia	Decreased circulating pool	Athens *et al.* (1961*b*)
Complement-activated plasma[a]	Rabbit and human neutropenia	Increase	O'Flaherty, Craddock & Jacob (1978)
Cobra venom factor	Rabbit	Increased lung sequestration	O'Flaherty, Kreutzer & Ward (1978*a*, *b*), McCall *et al.* (1974)
Rabbit neutrophil granule fraction	REC	Marked increase	Janoff & Zweifach (1964)
Formylated tripeptides[a] and other chemotactic factors	Rabbit neutropenia	Increase	O'Flaherty, Kreutzer & Ward (1977)
Damage (re-injected cells)	Human neutropenia	Irreversible margination	Athens *et al.* (1961*a*)

Table 1. (*cont.*)

Agent/Condition	Cell source	Effect on adhesion	Reference
In vitro. Glass bead columns and similar whole-blood studies			
Ca^{2+}	Rat peritoneal	None	Garvin, (1968)
	Human peripheral	None	Kvarstein, (1969*b*), Bryant & Sutcliffe (1972)
Cyanide and other inhibitors of oxidative phosphorylation	Human peripheral	None	Kvarstein (1969*a*), MacGregor, Spagnulo & Lentnek, (1974)
Heparin, Aspirin, Hydrocortisone	Human peripheral	None[b]	MacGregor *et al.* (1974)
Absence of divalent ions	Rat, human	Decrease	Garvin, (1968), Kvarstein, (1969*b*), Banks & Mitchell (1973*a*, *b*, *c*)
Aspirin, Indomethecin, Phenylbutazone, Prednisone, Colchicine	Human blood from recipients of drug	Decrease	MacGregor (1976)
Decreased complement C1 or C4 (Hereditary)	Human peripheral	Decrease	MacGregor, Negendank & Schreiber (1978)
Local anaesthetic	Human peripheral	Decrease	Schiffer *et al.* (1977)
Glycolytic inhibitors, Sulphydryl blocking agents	Human	Decrease	Kvarstein (1969*a*), Banks & Mitchell (1973*a*, *b*, *c*), Giordano & Lichtman (1973)
Increased cyclic AMP levels	Human leukocyte-rich plasma	Decrease	Bryant & Sutcliffe (1974)
Mg^{2+}, Mn^{2+} and other divalent ions	Various	Increase	Garvin (1968)
'Plasma factor'	Human peripheral blood	Increase	Lentnek, Schreiber & MacGregor (1976)
Platelet activators (ADP, 5HT etc.)	Human	Increase	Banks & Mitchell (1973*b*, & *c*)

Agent/condition	Cell source	Effect on adhesion	Reference
In vitro. Aggregation studies[c] (see also Tables 2 and 3)			
Anti-inflammatory agents, histamine, serotonin, Ca^{2+}	Rabbit peritoneal	No effect	Lackie (1974, 1977)
pH 6.5–7.8	Rabbit peritoneal	No effect	Lackie (unpublished)
Increased cyclic AMP levels	Rabbit peritoneal	Decrease	Lackie (1974)
Absence of divalent ions	Rabbit peritoneal	Decrease	
Colchicine	Rabbit peritoneal	Decrease	Lackie (1974)
Inhibitors of glycolysis	Rabbit peritoneal	Decrease	
BSA (slightly chemotactic)	Human peripheral	Decrease	Lackie (unpublished)
Local anaesthetics	Human peripheral	Decrease	R. B. Allan (unpublished)
Divalent ions other than Ca^{2+}	Sheep peritoneal	Increase	Wilkins, Ottewill & Bangham (1962*a*, *b*)
Cytochalasin B	Rabbit peritoneal	Increase	Lackie (1977)
$A23187 + Ca^{2+}$	Rabbit peritoneal	Marked increase	Lackie (1977)
Endotoxin	Rabbit peritoneal	Increase	Lackie (1977)
Particles to phagocytose	Rabbit peritoneal	Increase	Allison *et al.* (1963), Lackie (1977)
	Human peripheral	Increase	Talstad (1972)
Chemotactic factors	Rabbit and human	Complex	See later discussion
Temperature	Rabbit (adhesion to coverslip)	Sharp decrease below *c.* 25 °C	Lackie (unpublished)

REC = Rabbit ear chamber }
HCP = Hamster cheek pouch } Direct visual observation.

[a] Chemotactic.

[b] Unless taken by volunteer blood donor.

[c] Many of these results are similar to those on whole blood.

Direct visual scoring was used semi-quantitatively by Allison, Lancaster & Crosthwaite (1963), and Wilkins, Ottewill & Bangham (1962*a*, *b*) in an excellent study which is often overlooked.

Despite the diversity of the methods employed, the results obtained are generally in agreement and, although some assays appear to be more sensitive to small changes, particularly the aggregation assays, it is possible to generalise to a considerable extent. An extensive range of agents have been tested and in Table 1 we attempt to provide a summary. In many respects the adhesiveness of leukocytes is similar to that of other tissue cells, although they seem, surprisingly, rather more adhesive (Lackie & Armstrong, 1975).

Not only do neutrophils adhere to endothelium and migrate, but they are phagocytic and must therefore be able to adhere to bacteria as the first step in phagocytosis. The relative ease of phagocytosis of different classes of bacteria seems to depend critically upon the surface properties of the bacteria relative to those of the neutrophil (Van Oss & Gillman, 1972; Stendahl & Edebo, 1972; Edebo, this volume and Watt, this volume). This may be an important clue to the way in which adhesion is modulated in neutrophils, especially since many chemotactic factors seem to be amphiphilic molecules and the hydrophobicity of a denatured protein seems to correlate well with its chemotactic potency (Wilkinson, 1974*b*).

Another important and interesting aspect of neutrophil behaviour is their secretion of granule contents, which undoubtedly contributes to tissue damage in some inflammatory lesions (Buckley, 1963; Stewart, Ritchie & Lynch, 1974) and may serve to amplify or modulate other parts of the overall response (Hurley & Spector, 1961; Moses, Ebert, Graham & Brine, 1964; Janoff, Schaefer, Scherer & Bean, 1965). Phagocytosis and other 'membrane-activating events' stimulate the metabolic activity of neutrophils (Rossi, Patriarcha & Romeo, 1974), which depends mostly upon glycolysis and the hexose monophosphate pathway, and this may be linked with the stimulation of secretion. Two classes of granules are very conspicuous in neutrophils, the specific granules which contain mostly lysozyme and lactoferrin, and the azurophil granules which are typical primary lysosomes containing a wide range of degradative enzymes (Bretz & Baggiolini, 1974). The specific granules are probably released before the azurophils (Bainton, 1973; Leffel & Spitznagel, 1974) and contain a complement activator; the azurophils contain an inactivator of the complement component C5a which would be produced by this

Table 2. *The effects of various agents on the aggregation and release of secretory granule contents by neutrophil leukocytes*

Agent	Effect on aggregation[a]	Effect on release[a]
Catecholamines	–	–
Methylxanthines	– –[b]	–
Colchicine	–	–[c]
Indomethacin	–	–[d]
Phenylbutazone	– –[g]	– –[e]
Cholinergic agents	0	+[c]
Cytochalasin B	+	+
A23187 + Ca^{2+}	+ +	+ + +
Endotoxin	+ +	+ +
Phagocytic stimuli	+	+
High concentrations of f-Met-Leu-Phe	+[g]	+[f, g]
Low concentrations of f-Met-Leu-Phe	–[g]	Not known

Aggregation was measured as described by Lackie (1974) and the release of lysozyme and β-glucuronidase were taken as an indication of secretory granule release. Unless otherwise stated the data are from Lackie (1977).

[a] Inhibition (–), enhancement (+), no effect (0).

[b] Lackie (1974).

[c] Ignarro (1973) Human leukocytes stimulated with heat-aggregated gamma-globulin. Similar results with Cytochalasin B treated neutrophils (Zurier *et al.*, 1974).

[d] Northover (1977).

[e] Perper (1975).

[f] Gallin *et al.* (1978); Becker (1976).

[g] Unpublished.

activator (Wright & Gallin, 1977). This immediately suggests an autocatalytic effect, self-regulated and possibly analogous to the platelet cascade system. Complement components certainly affect adhesion of neutrophils, and lactoferrin, a major component of the otherwise rather uninteresting specific granules, stimulates neutrophil movement (Breckenridge, personal communication).

Many anti-inflammatory agents and modulators of the inflammatory response may act through their effect on secretion (Ignarro & George, 1974; Northover, 1977) and a good correlation exists between secretory activity and increased adhesiveness (Lackie, 1977; Gallin *et al.*, 1978; Table 2). Surprisingly, however, the increased

Table 3. *Effect of chemotactic factors on the aggregation of neutrophil granulocytes after 60 min at 37 °C*

Chemotactic factor	Concentration (M)	Percentage of control (= 100 %) Particle number ± SEM (n)
α_S-casein	4.2×10^{-5}	147.6 ± 3.4 (18) $p < 0.001$
β-casein	4.2×10^{-5}	143.3 ± 6.4 (24) $p < 0.001$
f-met-phe	10^{-5}	86.2 ± 1.2 (24) $p < 0.001$
f-met-leu-phe	10^{-9}	87.2 ± 5.2 (11) $p < 0.05$
f-met-leu-phe	10^{-7}	36.3 ± 7.0 (15) $p < 0.001$

n = total number of replicate assays. Using Student's t-test. P (difference from controls = 0) values as shown. Values greater than 100 % indicate an inhibition of adhesion.

adhesiveness seems to be a consquence of secretory activity and is not mediated by the contents of the granules. It should perhaps be pointed out that in these experiments (Lackie, 1977) no separation of granules was attempted, and it might be argued that the two granule types act in opposition and the net effect would be nil. Nevertheless, secretion is accompanied by the appearance of new membrane on the surface of the neutrophil, which may be associated with the altered binding of chemotactic factors (Gallin *et al.*, 1978), and which is associated with an increased amount of a large (molecular weight 150000) proteolytically-sensitive membrane component (Thorne, Oliver & Lackie, 1977). It is tempting to link the appearance of new membrane with increased adhesiveness and Stossel (1978) has suggested that granule membrane may have an important role in locomotion and chemotaxis.

A group of agents which are of particular interest in the context of inflammatory responses are chemotactic factors. Good circumstantial evidence exists that chemotaxis plays a part in directing the movement of leukocytes once they have left the blood vessel. This implies that a gradient of chemotactic factor exists from the vessel wall to the lesion, and this assumption is essential to the leukocyte-based model for localization discussed above. A simplifying hypothesis would be that chemotactic factors are also responsible for margination, and thus a single diffusible factor might control the whole cellular response. Since activated complement gives rise to a potent chemotactic factor, C5a, and the activation of complement will

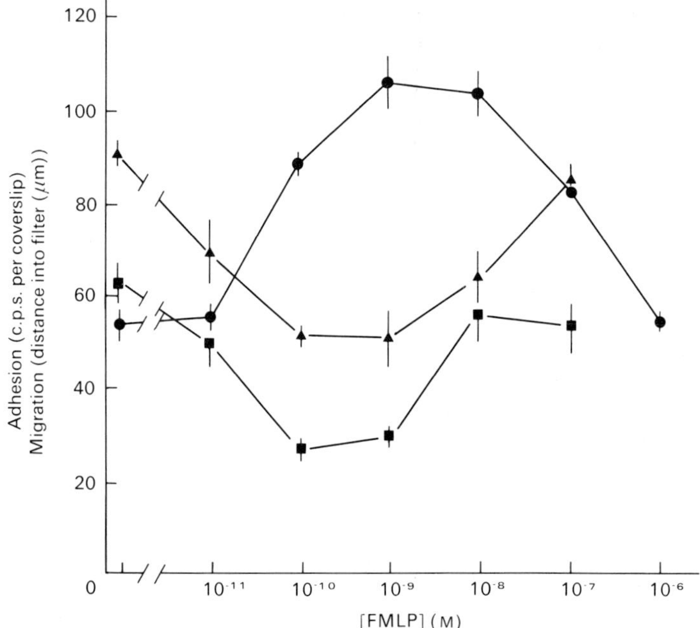

Fig. 2. The effect of the chemotactic factor f-met-leu-phe on adhesion of neutrophils at 30 min and their migration in filters. Key: ●——●, migration-±SEM; ▲——▲, adhesion to serum-coated glass±SEM; ■——■, adhesion to confluent endothelium, ±SEM. For adhesion, each point is the mean from four replicate coverslips. For migration, each point is the mean of five measurements on each of duplicate filters.

bring about a transient margination of neutrophils, the hypothesis does not seem too far-fetched. We have examined the effects of various chemotactic factors on the adhesiveness of neutrophils (Table 3) and on their adhesion to endothelial monolayers *in vitro* (see Table 6) and find a rather complex picture (Smith, Lackie & Wilkinson, 1979). In general chemotactic factors appear to decrease the adhesiveness of neutrophils, unless added at concentrations above the optimum for chemotaxis (Fig. 2). The effects appear to be even more complex when the time-dependence is considered. We have chosen to study the formylated tripeptides, well-defined and highly potent chemotactic agents, in detail, and some of the results are shown in Table 3 and Figs. 2 and 3. For formyl-methionyl-leucyl-phenylalanine (f-Met-Leu-Phe) the optimal chemotactic dose is approximately 10^{-9} M and at concentrations below this the peptide decreases adhesion, as assayed by aggregation or by collection onto

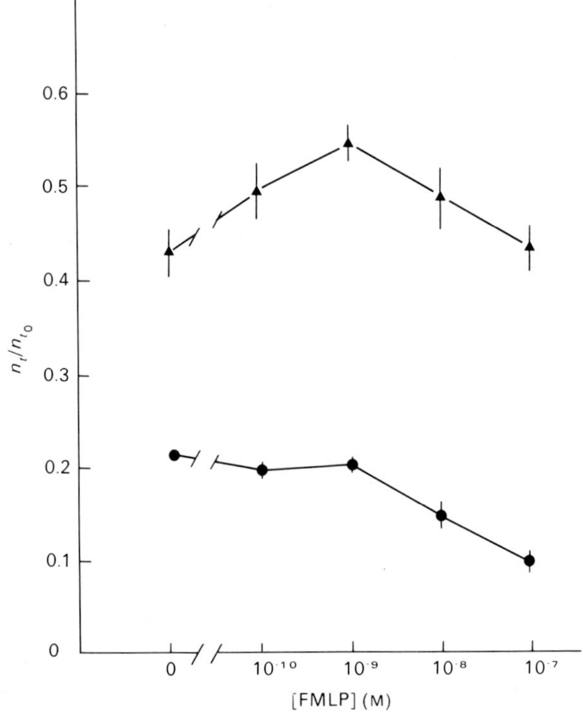

Fig. 3. The effect of the chemotactic factor, f-met-leu-phe, on the aggregation of neutrophils. Where omitted, standard error bars fall within the limits of the symbol. Each point is the mean from four replicate assays. Key: ▲——▲, $t = 30$ min; ●——●, $t = 60$ min. 10^6 cells at time n_{t^0}.

coverslips. After 30–60 min these low concentrations begin to stimulate adhesion, and high doses have an immediate stimulatory effect. At high ($\geqslant 10^{-7}$ M) doses, secretion is also stimulated (Becker, 1976; Smith, unpublished).

In terms of the time-course of the inflammatory response these observations on adhesion are more sensible than might first appear. Near the vessel wall chemotactic-factor concentrations will be low due to the proximity of the circulating blood, and reducing the adhesion may actually increase the rate of locomotion of neutrophils (see later discussion). Within 30 min the neutrophil will be moving up-gradient towards the focus of the lesion, and as the concentration of factor increases the cell will become trapped by its increased adhesion and will be stimulated to secrete granule contents which may attract other neutrophils (Zigmond & Hirsch, 1973) and also, slightly later, to secrete hydrolytic enzymes which will begin the clearance process.

Our results on the effects of chemotactic factors differ from those reported by O'Flaherty and co-workers (O'Flaherty, Kreutzer & Ward, 1977; 1978*a*, *b*). The discrepancies can, we feel, be explained partly on the basis of the differences in assay technique and partly by the complex dose- and time-dependence. They find that formylated peptides and activated complement induce a transient increase in the adhesiveness of rabbit and human neutrophils (O'Flaherty, Kreutzer & Ward, 1977, 1978*a*), the effect being lost within 8 min of exposure to the factor. Their assay involves rapid stirring of dense (5×10^6 cell ml^{-1}) suspensions and sample counting with a Coulter counter, and they have looked at the very early stages of the response. We have used longer times of incubation and much lower shearing-stress conditions (Smith, Lackie & Wilkinson, 1979). The response which O'Flaherty *et al.* observe is sensitive to temperature, inhibitors of glycolytic metabolism, cytochalasin B and divalent cations (O'Flaherty, Kreutzer, Showell & Ward, 1977; O'Flaherty, Showell & Ward, 1977) and in these respects resembles the spontaneous aggregation of neutrophils (see Table 1). The enhancement of aggregation by chemotactic peptides can be blocked by a competitive antagonist of binding (O'Flaherty, Showell, Kreutzer, Ward & Becker, 1978) and exposure to the peptides leads to desensitisation, although the cells will still respond to complement (O'Flaherty, Kreutzer, Showell, Becker & Ward, 1978). The effects of chemotactic factors can be mimicked by the calcium ionophore A23187 in the presence of external calcium, and the effect of the chemotactic factors is probably mediated through enhanced secretory activity and the appearance of new membrane (Gallin, Wright & Schiffmann, 1978; O'Flaherty, Kreutzer & Ward, 1978*a*). The optimum concentration of peptide to induce exocytosis is higher than the optimum for chemotaxis (Gallin *et al.*, 1978) and this may account for the complexity of the dose response.

The adhesion of neutrophils to non-cellular substrata also seems to increase when the cells are treated with chemotactic factors (Smith, Hollers, Patrick & Hassett, 1979). This seems somewhat paradoxical since increased adhesiveness seems to reduce the rate of movement of neutrophils, as these authors and others (Gallin *et al.*, 1978; see later) have observed, yet these chemotactic factors are potent chemokinetic agents. We have found that the addition of an optimum

Table 4. *The effect of an optimal chemotactic concentration of formyl-methionyl-leucyl-phenylalanine (10^{-9} M) on the locomotion of rabbit peritoneal neutrophils*

Conditions	No. anchored	No. moving	Σd per 100 frames	No. of frames
1 % BSA	12	28	244	400
1 % BSA + 10^{-9} M f-Met-Leu-Phe	0	20	318	750
0.5 % BSA	26	24	53	383
0.5 % BSA + 10^{-9} M f-Met-Leu-Phe	2	28	255	310

Time-lapse film was analysed and locomotion estimated by the summed net displacement (Σd) of all moving cells in arbitrary units per 100 frames. Anchored cells are those neutrophils which show vigorous locomotory activity but which fail to show a net displacement at any stage of the film. The medium was Hank's–Hepes basic salts solution with bovine serum albumin (BSA).

chemotactic concentration of f-met-leu-phe (10^{-9} M) decreases the proportion of immobilised cells on plane substrata used in time-lapse studies (Table 4) which seems inconsistent with increased adhesiveness. This apparent paradox may, however, be explained if the increased adhesiveness allowed greater traction and thus increased the probability of detaching the tail-adhesions which anchor the cells.

To some extent the present confusion in this area is perhaps a hopeful sign. A gradient of chemotactic factor(s) will be present between the lesion and the blood vessel and a complex dose-response may be entirely appropriate to modify the activities of neutrophils at different times and in different spatial positions relative to the focus. At the lower end of the gradient enhanced movement and little secretory activity are required, and at the upper end immobilisation, enhanced lysosomal enzyme release and possibly phagocytosis. If neutrophils are sensitive to minor alterations in environment, as one might expect a defensive system to be, then it is easy to see how differences in assay techniques might generate conflicting results.

These studies on neutrophil adhesion provide an essential background to a study of their adhesion to endothelial monolayers, the substratum to which they must adhere *in vivo*.

Neutrophil adhesion to endothelial monolayers. Within recent years it has become possible to grow primary endothelial cells *in vitro*, and thus endothelial monolayers have become available as a substratum for neutrophil adhesion studies. Various aspects of the biology of endothelial cells have received attention (Thorgeirsson & Robertson, 1978 for review). Unfortunately it is very difficult to get pure cultures of capillary endothelium, although a published method exists (Wagner & Matthews, 1975) and most workers have used aortic endothelium from pigs or calves, or human umbilical cord vein endothelium. We assume, although this may not be justifiable, that the cells from large vessels are basically similar to those in capillaries. One piece of evidence that supports this is the similarity in the results obtained by de Bono (1976), looking at lymphocyte adhesion to aortic endothelium *in vitro*, and Stamper & Woodruff (1976) investigating the adhesion of lymphocytes to fixed sections of lymph nodes. The latter system is particularly interesting in that it reveals that even after fixation the endothelium of the high-endothelial venules is a particularly adhesive substratum for lymphocytes. It is, of course, possible to argue that wall shear stress in large vessels will always exceed the adhesive interaction, and that margination does not occur in large vessels for this reason; the observation that endothelium grown under flow conditions is less adhesive for lymphocytes (de Bono, 1976) makes it possible that the endothelium in large vessels is indeed rather different.

Despite these problems it is possible to use endothelium as a substratum for neutrophil adhesion, and neutrophils adhere with high affinity to endothelial cell monolayers *in vitro*.

There is some variability in the data from different groups, some finding endothelium to be more adhesive than serum-coated glass or plastic (Lackie & de Bono, 1977; Hoover, Briggs & Karnovsky, 1978; MacGregor, Macarak & Kefalides, 1978; Beesley *et al.*, 1978), others that endothelium, although a good substratum is not quite as good as serum-coated glass (Smith & Lackie, unpublished). No simple explanation of these discrepancies has been found, despite the interchange of material between Beesley *et al.* and our group, and the apparently identical assay conditions used. This may arise from slight variations in culture conditions or subculturing technique, but should not be allowed to obscure the fact that endothelium is a very much better substratum for adhesion than other cell monolayers (Hoover *et al.*, 1978; MacGregor, Macarak & Kefalides, 1978) and is also a

Table 5. *The effects of a variety of agents on the adhesion of rabbit peritoneal exudate neutrophils to serum-coated glass and to confluent monolayers of porcine aortic endothelium in an in-vitro coverslip collection assay*

Agent	Concentration (M)	Effect ± standard deviation (number of replicate pairs) (control = 1.00)		Significance of difference between glass and endothelium
		Serum-coated glass	Endothelial monolaver	
Histamine	10^{-4}	0.80 ± 0.26 (23)**	0.81 ± 0.26 (15)*	n.s.[a]
Serotonin	10^{-5}	0.84 ± 0.10 (5)***	0.83 ± 0.27 (9)	n.s.
Bradykinin triacetate	10^{-6}	0.87 ± 0.14 (5)	0.78 ± 0.30 (9)	n.s.
Aspirin	10^{-3}	0.72 ± 0.15 (10)***	0.58 ± 0.21 (5)*	n.s.
ϵ-amino-n-caproic acid	10^{-2}	1.05 ± 0.14 (4)	1.07 ± 0.33 (6)	n.s
Hydrocortisone-21-Na succinate	10^{-5}	0.87 ± 0.18 (22)**	0.66 ± 0.12 (6)***	$p < 0.05$
Prednisolone-21-Na succinate	10^{-4}	0.93 ± 0.39 (16)	0.82 ± 0.24 (11)*	n.s
Colchicine	10^{-5}	0.82 ± 0.20 (7)	0.48 ± 0.12 (10)***	$p < 0.001$
Aminophylline	10^{-4}	0.79 ± 0.17 (13)***	0.68 ± 0.17 (5)*	n.s.
Formaldehyde fixation of endothelium			0.67 ± 0.20 (7)	

The effect shown is the ratio of experimental and control coverslips paired randomly within experiments. At least two separate experiments were done.

A paired-sample t-test was used to determine the statistical significance.

* p (difference from controls = 0) < 0.05; ** $p < 0.005$; *** $p < 0.001$.

[a] Not significantly different.

suitable substratum for locomotion (Lackie & de Bono, 1977). Migration of neutrophils through endothelial monolayers *in vitro* has also been observed by Beesley *et al.* (1979).

The ready adherence of neutrophils to endothelium *in vitro*, whether under static 'collecting lawn' conditions (Lackie & de Bono, 1977) or using a crude flow system (Beesley *et al.*, 1978) might suggest that cultured endothelial cells are 'inflamed' by the conditions and treatments which they suffer and that we are measuring stimulated adhesion. Differences between laboratories might then reflect very subtle differences in the condition of the endothelial monolayers. If the interaction between neutrophils and endothelium is finely balanced, as has been argued, then this would not be surprising. We have attempted to modify the interaction between neutrophils and endothelium *in vitro*, arguing that agents which are active in the inflammatory response should alter the stickness of the endothelium and that varying the culture conditions should also affect adhesion. Our results are generally in accord with observations made by other laboratories. Neither inflammatory mediators nor anti-inflammatory drugs affect the neutrophil–endothelium interaction, except in so far as they affect neutrophils directly.

These data are summarised in Table 5 and the effects of varying endothelial cell density are shown in Fig. 4. We have extended these studies to look at the effects of chemotactic factors, particularly f-met-leu-phe. Pretreatment of the endothelium is ineffective, whereas pretreatment of neutrophils diminishes adhesion to a similar extent as it affects neutrophil–neutrophil adhesion in parallel studies (Smith, Lackie & Wilkinson, 1979; Tables 6 & 7). Comparable studies by Hoover *et al.* (1978) indicated that pre-treatment of either component of the system affected adhesion, but their adhesion assays were run over a longer time-course and they may have been getting penetration of the monolayers by emigrating neutrophils: small changes in the locomotory activity of the neutrophils would be sufficient to affect the assay, and the high chemokinetic potency of many of these agents makes this a serious problem in such studies. Another indication of the relatively passive role of the endothelium is that fixation of the monolayer does not greatly affect the adhesiveness of lymphocytes (de Bono, 1976) or neutrophils (Table 5).

One difficulty in interpreting these data is the negative aspect of the evidence. We do not observe the changes in neutrophil–endothelium adhesion which we would expect if the localisation of

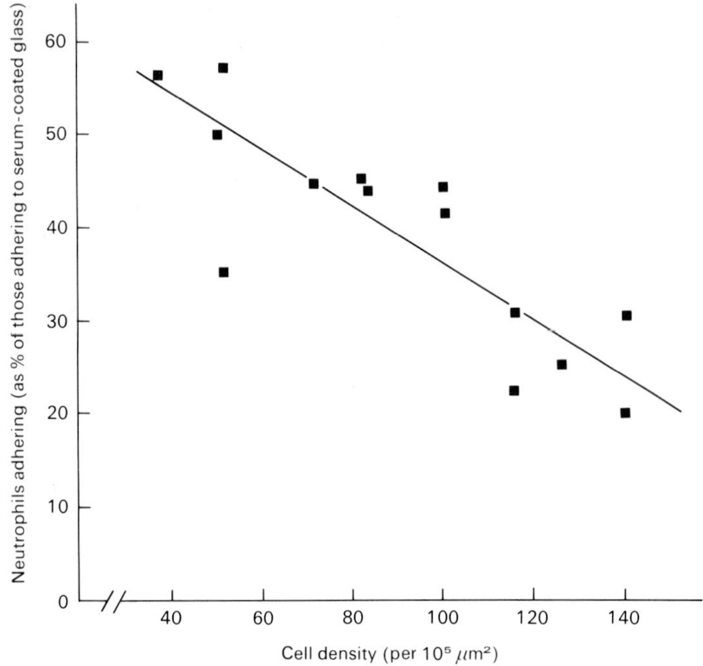

Fig. 4. The effect of varying endothelial cell density on the adhesion of neutrophils.

the inflammatory response derived primarily from changes in the endothelium. We do find changes which are consistent with the primary change being in neutrophils, but to argue from this that endothelial change does not occur is rather dangerous. It could be argued that because the endothelium we use is not derived from post-capillary venules and because it is in a wholly artificial environment, all we are observing is an abnormal interaction or a fully fledged inflammatory response which is not susceptible to modification. We cannot refute these criticisms but would suggest that the evidence from in-vivo studies does not point unequivocally to a primary endothelial change and that our findings are wholly consistent with the other models for localisation discussed earlier. Certainly, the observed effects of chemotactic factors can be rationlised on the emigration-based localisation models, and seem slightly paradoxical on the endothelial-change model. Cell-type specificity can be introduced into the generalised-margination models if cell-type-specific chemotactic factors exist, although the evidence for such factors is equivocal (see discussion in Wilkinson & Lackie, 1979).

Table 6. *Effect of chemotactic factors on the adhesion after 30 min at 37 °C of neutrophil granulocytes to confluent endothelium and serum-coated glass*

Chemotactic factor	Concentration (M)	Adhesion (compared to controls = 100 %) ± SEM (n) Endothelium	Serum-coated glass
α_s-casein	4.2×10^{-5}	43 ± 5 (6)	62 ± 3 (3)*
β-casein	4.2×10^{-5}	64 ± 4 (13)	49 ± 6 (6)
Alkali-denatured HSA	1.4×10^{-5}	57 ± 8 (12)	25 ± 6 (3)*
Native HSA[a]	1.4×10^{-5}	136 ± 7 (14)	143 ± 10 (8)
f-tri-tyr	10^{-9}	63 ± 7 (11)	60 ± 5 (16)
f-met-leu-phe	10^{-9}	46 ± 4 (22)	59 ± 5 (23)
f-met-phe	10^{-5}	43 ± 3 (30)	55 ± 4 (33)
met-phe[b]	10^{-5}	84 ± 6 (13)**	83 ± 5 (13)*

n = number of replicate coverslips.
Using Student's t-test, p (difference from controls = 0) < 0.001 for all values except * $p < 0.01$ ** $p < 0.025$.
[a] Not chemotactic.
[b] Weakly chemotactic.

Table 7. *Effect of pretreatment of neutrophil granulocytes, endothelium or serum-coated glass on neutrophil adhesion after 30 min at 37 °C*

Pretreatment with chemotactic factor of	Chemotactic factor (M)	Adhesion of neutrophils ± SEM (n) (compared to controls = 100 %) Endothelium	Serum-coated glass
Endothelium, 30 min at 37 °C	f-met-phe, 10^{-5}	94 ± 5 (15)	
	f-met-leu-phe, 10^{-9}	91 ± 10 (4)	
	β-casein, 4.2×10^{-5}	94 ± 17 (11)	
Serum-coated glass, 30 min at 37 °C	f-met-phe, 10^{-5}		94 ± 5 (16)
	f-met-leu-phe, 10^{-9}		105 ± 4 (4)
	β-casein, 4.2×10^{-5}		102 ± 2 (7)
Neutrophils, 15 min at 37 °C	f-met-phe, 10^{-5}	53 ± 3 (7)***	51 ± 2(10)***
	f-met-leu-phe 10^{-9}	74 ± 5 (4)*	71 ± 4 (4)**

n = number of replicate coverslips. Using Student's t-test p (difference from control = 0) *** < 0.001, ** < 0.01, * < 0.025.
All pretreatments were followed immediately by two washes in BSS.

Adhesive interactions and leukocyte emigration

Another aspect of the leukocyte–endothelium interaction is the ability of leukocytes to locomote over and squeeze between endothelial cells. Locomotion requires a specific type of adhesion to the substratum, and the adhesion formed must resist the contractile forces moving the cell body forward, but be transient so that movement will continue. That this is indeed the case is clear both from observations *in vivo* and *in vitro*. Neutrophils are invasive cells, and do not exhibit contact paralysis of locomotion when they collide with fibroblasts or endothelial cells. (Armstrong & Lackie, 1975; Lackie & de Bono, 1977). On fibroblasts, the observed overlapping of neutrophils is less than would be expected on the basis of chance; contact inhibition type II is exhibited. This is not, however, true for neutrophils and endothelial cells, probably because the adhesion of neutrophils to endothelial cells is of comparable strength to their adhesion to the serum-coated substratum. A simple explanation of this observation is that the leukocyte–substratum adhesion resists the contractile forces better than the leukocyte–fibroblast adhesion, and thus the cell body moves towards the substratum adhesion. If relative adhesion affects behaviour in this way, the adhesiveness should also affect the rate of locomotion quite markedly, and there should be a distinct optimum value for adhesiveness at which locomotion is most effective. In a well-adapted system, as we may suppose the leukocyte–endothelium interaction to be, emigration of leukocytes might be considerably enhanced if the adhesion is of optimal strength, and might be inhibited if the adhesion is too strong. Accordingly we have examined the interaction between adhesion and locomotion for neutrophils.

Most of our previous work has been on rabbit peritoneal neutrophils, cells which have undergone one margination and emigration phase but which are in most respects identical to blood neutrophils (Sabioncello, Dekaris, Veselic & Silobrcic, 1976, and our own comparative studies). Rabbit blood neutrophils can be prepared from centrifuged whole blood by discarding the buffy coat and platelet-rich plasma, resuspending the pellet and then sedimenting red cells with gelatin (Henson, 1971). This gives a moderately pure suspension of neutrophils which can be compared with peritoneal exudate neutrophils, although the handling procedures are very different and strict comparison is difficult. Blood neutrophils will adhere well to

Table 8. *Comparison of rabbit peritoneal neutrophils prepared as described in Lackie (1974) and rabbit blood neutrophils prepared by the method of Henson (1971)*

Blood	Peritoneal
Adhesion (number of cells per 100 μm^2 field; number of replicates in parentheses)	
38 ± 2 (5)	64 ± 3 (5)
30 ± 2 (5)	80 ± 5 (5)
101 ± 6 (6)	145 ± 6 (6)
81 ± 6 (6)	155 ± 7 (6)
82 ± 4 (10)	323 ± 7 (10)
Movement on plane substratum (mean square distance in unit time; number of cells tracked in parentheses)	
17.9 (40)	24.4 (31)
Migration into filters (μm; number of replicate counts in parentheses)	
42 ± 1.3 (20)	33 ± 1 (20)

Adhesion to coverslips was done in basic salts solution plus 1% BSA, distracting loose cells by passing through an air–medium interface. Filming and movement in filters were assayed as described in notes to Tables 4 and 9.

endothelial monolayers and may be more or less adhesive than peritoneal neutrophils. Since peritoneal neutrophils vary greatly in their adhesiveness, and since a variety of factors could affect the comparison – both physiological variation and differences in handling procedures – this is probably not of great significance. Using preparations of blood neutrophils, which were less adhesive than peritoneal neutrophils, we found (Table 8) that the locomotion of blood neutrophils over a plane substratum was slower than that of the more adhesive peritoneal neutrophils. Their locomotion through the three-dimensional matrix of micropore filters used in Boyden-type chemotaxis assays (Wilkinson, 1974*b*) was, however, faster. Thus, depending upon the gross topography of the substratum and the type of locomotory activity required, increasing adhesiveness may either increase or decrease the rate of locomotion.

Because of the difficulties of comparing blood and peritoneal neutrophils, which are prepared in different ways, we have repeated

Table 9. *Movement of rabbit peritoneal neutrophils on a plane substratum in medium with varying amounts of bovine serum albumin (BSA)*

BSA (%)	Mean square distance	No. of cells	% moving	% anchored	% dropping off
0	12.1	146	4	96	0
0.5	15.6	66	50	50	14
1.0	24.3	31	71	29	35
2.0	22.8	30	93	7	67
5.0	12.6	37	97	3	72

The BSA used was slightly chemotactic. Sample data from filming done on single batches of cells are shown, but data for three other batches are essentially the same, including one done using purely chemokinetic albumen. The filming was done on an inverted coverslip preparation and a proportion of cells drop off during the course of the film, giving a measure of their adhesiveness. Filming was done with a 10 s lapse interval in Hank's–Hepes basic salts solution and analysed using a stop-action projector, the mean square distance moved is calculated on the assumption of a two-dimensional random walk pattern of locomotion (Gail & Boone, 1970) in arbitrary units.

these studies using only peritoneal neutrophils, varying their adhesion by altering the composition of the medium. As judged by aggregation assays and by collection on substrata (Smith, C. W. *et al.*, 1979 and our own data) the addition of bovine serum albumin (BSA) will reduce the adhesiveness of neutrophils, and using this as a means to modify adhesiveness we find that there is a difference in the adhesion optimum *in vitro* for locomotion on a plane substratum (as is, *in vivo*, the blood vessel wall) and for locomotion through a three-dimensional matrix (as is, *in vivo*, moving between endothelial cells and through other tissues). On a plane substratum, stickier cells move faster, but in a micropore filter the optimum is lower and cells which move poorly on serum-coated glass move rather well through filters (Tables 9 and 10). This observation is supported by the data of Gallin, Wright & Schiffmann (1978), although they do not stress the point. At both ends of the adhesion range, immobilisation will take place: excessive adhesion, as on non-protein-coated glass, prevents movement and weak adhesion prevents locomotion, presumably because traction is lost (Table 9).

Again this can be taken as an argument for a fairly delicate balance

Table 10. *Migration of rabbit peritoneal neutrophils into micropore filters in a Boyden-type assay*

BSA (%)	distance moved into filter (μm) ± SEM
0	22 ± 1
0.5	50 ± 2
1.0	59 ± 2
2.0	49 ± 2
5.0	53 ± 3

No gradient was present, and distance moved was estimated using the leading-front method of Zigmond & Hirsch (1973).

in the neutrophil–endothelium interaction. Possibly the effect of low doses of chemotactic factors in decreasing adhesiveness operates when the neutrophils are leaving the blood vessels and beginning to move through a three-dimensional matrix, the overall effect being to enhance emigration. Notice that in Table 4 the effect of the chemotactic peptide is proportionately much greater at a low BSA concentration when the cells are more adhesive and it is reasonable to suppose that, depending upon the adhesiveness of the neutrophil relative to the optimum for the type of movement required, the effects of chemotactic factors may either be to reduce or to stimulate movement. It is also clear that chemotactic factors affect the locomotory machinery directly as well as the adhesiveness (Smith, C. W. *et al.*, 1979) and the interaction of enhanced locomotory activity and altered adhesiveness will be complex. An additional problem is that locomotion *in vivo* is on substrata which are likely to differ in, for example, their rigidity, from the solid substrata used in these studies. It is difficult, if not impossible, to predict how altering the traction will affect the ability of a cell to move over a substratum which has complex rheological properties (and which might be thixotropic) such as the surfaces of other cells and gels of connective tissue components. The problems facing an emigrating neutrophil are complex and immature neutrophils will marginate (Athens *et al.*, 1965) but not migrate (Boggs, 1960) possibly because they are insufficiently plastic to squeeze between endothelial cells (Lichtman & Weed, 1972).

IN CONCLUSION

The interaction between neutrophils and endothelium is well adapted and delicately balanced. It seems that neutrophils may normally remain in circulation because the wall shear stress in blood vessels just exceeds the adhesive forces, and that small changes in the adhesive interaction may have disproportionate effects in allowing neutrophils to marginate. For this reason we have considered the study of neutrophil adhesion to be important, the more so since the endothelium seems to be a relatively passive component, and we favour models for the localisation of the inflammatory response which do not depend upon a primary change in the endothelium. The localisation of the response probably occurs through the local stimulation of emigration, and chemotactic factors are likely mediators of this. The effects of chemotactic factors on the adhesion of neutrophils are beginning to be understood, although there is still much confusion, and changes in adhesiveness will affect locomotion of neutrophils.

A cell-biological approach to the pathology of acute inflammation seems to be on the verge of producing a better understanding of this common defence reaction and a study of the pathology seems to be generating some interesting cell biological problems.

REFERENCES

Allison, F. & Lancaster, M. G. (1960). Studies on the pathogenesis of acute inflammation. II. The relationship of fibrinogen and fibrin to the leucocytic sticking reaction in ear chambers of rabbits injured by heat. *Journal of Experimental Medicine*, **111**, 45–64.

Allison, F. & Lancaster, M. G. (1961). Studies on the pathogenesis of acute inflammation. III. The failure of anticoagulants to prevent the leucocytic sticking reaction and the formation of small thrombi in rabbit ear chambers damaged by heat. *Journal of Experimental Medicine*, **114**, 535–53.

Allison, F., Lancaster, M. G. & Crosthwaite, J. L. (1963). Studies on the pathogenesis of acute inflammation. V. An assessment of factors that influence *in vitro* the phagocytic and adhesive properties of leucocytes obtained from rabbit peritoneal exudate. *American Journal of Pathology*, **43**, 775–95.

Allison, F., Smith, M. R. & Wood, W. B. (1955*a*). Studies on the pathogenesis of acute inflammation. I. The inflammatory reaction to thermal injury as observed in the rabbit ear chamber. *Journal of Experimental Medicine*, **102**, 655–68.

Allison, F., Smith, M. R. & Wood, W. B. (1955*b*). Studies on the pathogenesis of acute inflammation. II. The action of cortisone on the inflammatory response to thermal injury. *Journal of Experimental Medicine*, **102**, 669–76.

Armstrong, P. B. & Lackie, J. M. (1975). Studies on intercellular invasion *in vitro* using rabbit peritoneal neutrophil granulocytes (PMNs) I. Role of contact inhibition of locomotion. *Journal of Cell Biology*, **65**, 439–62.

Athens, J. W., Raab, S. O., Haab, O. P., Mauer, A. M., Ashenbrucker, H., Cartwright, G. E. & Wintrobe, M. M. (1961*a*). Leukokinetic Studies. III. The distribution of granulocytes in the blood of normal subjects. *Journal of Clinical Investigation*, **40**, 159–64.

Athens, J. W., Haab, O. P., Raab, S. O., Mauer, A. M., Ashenbrucker, H., Cartwright, G. E. & Wintrobe, M. M. (1961*b*). Leukokinetic studies. IV. The total blood, circulating and marginal granulocyte pools and the granulocyte turnover rate in normal subjects. *Journal of Clinical Investigation*, **40**, 989–95.

Athens, J. W., Raab, S. O., Haab, O. P., Boggs, D. R., Ashenbrucker, H., Cartwright, G. E. & Wintrobe, M. M. (1965). Leukokinetic studies. X. Blood granulocyte kinetics in chronic myelocytic leukemia. *Journal of Clinical Investigation*, **44**, 765–77.

Atherton, A. & Born, G. V. R. (1972). Quantitative investigations of the adhesiveness of circulating polymorphonuclear leucocytes to blood vessel walls. *Journal of Physiology*, **222**, 447–74.

Atherton, A. & Born, G. V. R. (1973). Relationship between the velocity of rolling granulocytes and that of the blood flow in venules. *Journal of Physiology*, **233**, 157–65.

Bainton, D. F. (1973). Sequential degranulation of the two types of polymorphonuclear leukocyte granules during phagocytosis of micro-organisms. *Journal of Cell Biology*, **58**, 249–64.

Banks, D. C. & Mitchell, J. R. A. (1973*a*). Leucocytes and thrombosis. I. A simple test of leucocyte behaviour. *Thrombosis Diathesesis Haemorrhagia*, **30**, 36–46.

Banks, D. C. & Mitchell, J. R. A. (1973*b*). Leucocytes and thrombosis. II. Relationship between leucocyte behaviour and divalent cations, sulphydryl groups, red cells and adenosine diphosphate. *Thrombosis Diathesesis Haemorrhagia*, **30**, 47–61.

Banks, D. C. & Mitchell, J. R. A. (1973*c*). Leucocytes and thrombosis. III. Effect on white cell behaviour of substances which induce or inhibit platelet aggregation. *Thrombosis Diathesesis Haemorrhagia*, **30**, 62–71.

Bassen, F. A., Etess, A. D. & Rosenthal, R. L. (1952). Leukopenia following administration of nicotinic acid. *Blood*, **7**, 623–30.

Becker, E. L. (1976). Some interrelations of neuttrophil chemotaxis, lysosomal enzyme secretion, and phagocytosis as revealed by synthetic peptides. *American Journal of Pathology*, **85**, 385–94.

Beesley, J. E., Pearson, J. D., Carleton, J. S., Hutchings, A. & Gordon, J. L. (1978). Interactions of leucocytes with vascular cells in culture. *Journal of Cell Science* **38**, 237–48.

Beesley, J. E., Pearson, J. D., Hutchings, A., Carleton, J. S. & Gordon, J. L. (1979). Granulocyte migration through endothelium in culture. *Journal of Cell Science* (in press).

Bishop, C. R., Rothstein, G., Ashenbrucker, H. E. & Athens, J. W. (1971). Leukokinetic studies. XIV. Blood neutrophil kinetics in chronic, steady-state neutropenia. *Journal of Clinical Investigation*, **50**, 1678–89.

Boggs, D. R. (1960). The cellular composition of inflammatory exudates in human leukemias. *Blood*, **15**, 466–75.

Bretz, U. & Baggiolini (1974). Biochemical and morphological characterization of azurophil and specific granules of human neutrophilic polymorphonuclear leukocytes. *Journal of Cell Biology*, 63, 251–69.

Bryant, R. E. & Sutcliffe, M. C. (1972). A method for quantitation of human leucocyte adhesion to glass. *Proceedings of the Society for Experimental Biology and Medicine*, **141**, 196–202.

Bryant, R. E. & Sutcliffe, M. C. (1974). The effect of 3′, 5′-adenosine monophosphate on granulocyte adhesion. *Journal of Clinical Investigation*, **54**, 1241–4.

Buckley, I. K. (1963). Delayed secondary damage and leucocyte chemotaxis following focal aseptic heat injury *in vivo*. *Experimental and Molecular Pathology*, **2**, 402–17.

Clark, E. R., Clark, E. L. & Rex, R. O. (1936), Observations on polymorphonuclear leukocytes in the living animal. *American Journal of Anatomy*, **59**, 123–73.

Cochrane, C. G. & Janoff, A. (1974). The Arthus reaction: a model of neutrophil and complement–mediated injury. In *The Inflammatory Process*, ed. B. W. Zweifach, L. Grant & R. T. McCluskey, 2nd edn, vol. III, pp. 86–162. London, New York: Academic Press.

Cohnheim, J. (1882). *Lectures on General Pathology*, 2nd edn, pp. 242–382. Translated by A. B. McKee (1889). London: The New Sydenham Society.

de Bono, D. (1976). Endothelial–lymphocyte interactions *in vitro*. I. Adherence of nonallergised lymphocytes. *Cellular Immunology*, **26**, 78–88.

Essex, H. E. & Grana, A. (1949). Behaviour of the leukocytes of the rabbit during periods of transient leukopenia variously induced. *American Journal of Physiology*, **158**, 396–400.

Fahraeus, R. (1929). The suspension stability of the blood. *Physiological Reviews*, **9**, 241–74.

Fehr, J. & Jacob, H. S. (1977). *In vitro* granulocyte adherence and *in vivo* margination: two associated complement-dependent functions. Studies based on the acute neutropenia of filtration leukophoresis. *Journal of Experimental Medicine*, **146**, 641–52.

Florey, H. W. & Grant, L. H. (1961). Leucocyte migration from small blood vessels stimulated with ultraviolet light: an electron microscope study. *Journal of Pathology and Bacteriology*, **82**, 13–17.

Gail, M. H. & Boone, C. W. (1970). The locomotion of mouse fibroblasts in tissue culture. *Biophysical Journal*, **10**, 980–93.

Gallin, J. I., Wright, D. G. & Schiffmann, E. (1978). Role of secretory events in modulating human neutrophil chemotaxis. *Journal of Clinical Investigation*, **62**, 1364–74.

Garvin, J. E. (1968). Effects of divalent cations on adhesiveness of rat polymorphonuclear neutrophils *in vitro*. *Journal of Cellular Physiology*, **72**, 197–212.

Giddon, D. B. & Lindhe, J. (1972). *In vivo* quantitation of local anesthetic suppression of leukocyte adherence. *American Journal of Pathology*, **68**, 327–38.

Giordano, G. F. & Lichtman, M. A. (1973). The role of sulphydryl groups in human neutrophil adhesion, movement and particle ingestion. *Journal of Cellular Physiology*, **82**, 387–96.

Goetzl, E. J. & Austen, K. F. (1977). Cellular characteristics of the eosinophil compatible with a dual role in host defense in parasitic infections. *American Journal of Tropical Medicine and Hygiene*, **26**, 142–50.

Grant, L. (1973). The sticking and emigration of white blood cells in inflammation. In *The Inflammatory Process* ed. B. W. Zweifach, L. Grant & R. T. McCluskey, 2nd edn, vol. II, pp. 205–50. London, New York: Academic Press.

Gray, H. W., Tsan, M.-F. & Wagner, H. N. (1977). A quantitative study of leukocyte cohesion: effects of divalent cations and pH. *Journal of Nuclear Medicine*, **18**, 147–50.

Hammerschmidt, D. E., Craddock, P. R., McCullough, J., Kronenberg, R. S., Dalmasso, A. P. & Jacob, H. S. (1978). Complement activation and pulmonary leukostasis during nylon fiber filtration leukopheresis. *Blood*, **51**, 721–30.

Henson, P. M. (1971). The immunologic release of constituents from neutrophil leucocytes. *Journal of Immunology*, **107**, 1535–46.

Hoover, R. L., Briggs, R. T. & Karnovsky, M. J. (1978). The adhesive interaction between polymorphonuclear leukocytes and endothelial cells *in vitro*. *Cell*, **14**, 423–8.

Hurley, J. V. & Spector, W. G. (1961). Endogenous factors responsible for leucocytic emigration *in vivo*. *Journal of Pathology and Bacteriology*, **82**, 403–20.

Ignarro, L. J. (1973). Neutral protease release from human leukocytes regulated by neurohormones and cyclic nucleotides. *Nature, London*, **245,** 151–4.

Ignarro, L. J. & George, W. J. (1974). Hormonal control of lysosomal enzyme release from human neutrophils: elevation of cyclic nucleotide levels by autonomic neurohormones. *Proceedings of the National Academy of Sciences, USA*, **71**, 2027–31.

Janoff, A., Schaefer, S., Scherer, J. & Bean, M. A. (1965). Mediators of inflammation in leukocyte lysosomes. II. Mechanism of action of lysosomal cationic protein upon vascular permeability in the rat. *Journal of Experimental Medicine*, **122**, 841–51.

Janoff, A. & Zweifach, B. W. (1964). Production of inflammatory changes in the microcirculation by cationic proteins extracted from lysosomes. *Journal of Experimental Medicine*, **120**, 747–64.

Jensen, D. P., Brubaker, L. H., Nolph, K. D., Johnson, C. A. & Nothum, R. J. (1973). Hemodialysis coil-induced transient neutropenia and overshoot neutrophilia in normal man. *Blood*, **41**, 399–408.

Keller, H. U., Wilkinson, P. C., Abercrombie, M., Becker, E. L., Hirsch, J. G., Miller, M. E., Ramsey, W. S. & Zigmond, S. H. (1977). A proposal for the definition of terms related to locomotion of leucocytes and other cells. *Clinical and experimental Immunology*, **27**, 377–80.

Kvarstein, B. (1969*a*). Effect of some metabolic inhibitors on the adhesiveness of human leucocytes to glass beads. *Scandinavian Journal of Clinical and Laboratory Investigation*, **24**, 35–40.

Kvarstein, B. (1969*b*). Effects of proteins and inorganic ions on the adhesiveness of human leucocytes to glass beads. *Scandinavian Journal of Clinical and Laboratory Investigation*, **24**, 41–48.

Lackie, J. M. (1974). The aggregation of rabbit polymorphonuclear leucocytes: effects of anti-mitotic agents, cyclic nucleotides and methyl xanthines. *Journal of Cell Science*, **16**, 167–180.

Lackie, J. M. (1977). The aggregation of rabbit polymorphonuclear leucocytes (PMNs). Effects of agents which affect the acute inflammatory response and correlation with secretory activity. *Inflammation*, **2**, 1–15.

Lackie, J. M. & Armstrong, P. B. (1975). Studies on intercellular invasion *in vitro* using rabbit peritoneal neutrophil granulocytes. II. Adhesive interaction between cells. *Journal of Cell Science*, **19**, 645–52.

Lackie, J. M. & de Bono, D. (1977). Interactions of neutrophil granulocytes (PMNs) and endothelium *in vitro*. *Microvascular Research*, **13**, 107–12.

Leffel, M. S. & Spitznagel, J. K. (1974). Intracellular and extracellular degranulation of human polymorphonuclear azurophil and specific granules induced by immune complexes. *Infection and Immunity*, **10**, 1241–9.

Lentnek, A. L., Schreiber, A. D. & MacGregor, R. R. (1976). The induction of augmented granulocyte adherence by inflammation. Modulation by a plasma factor. *Journal of Clinical Investigation*, **57**, 1098–1103.

Leusen, I. R. & Essex, H. E. (1953). Leukopenia and changes in differential leucocyte counts produced in rabbits by dextran and acacia. *American Journal of Physiology*, **172**, 231–6.

Lichtman, M. A. & Weed, R. I. (1972). Alteration of the cell periphery during granulocyte maturation: relationship to cell function. *Blood*, **39**, 301–16.

Lorente, F., Fontan, G., Garcia Rodriguez, M. C. & Ojeda, J. A. (1978). A simple and reproducible method to evaluate granulocyte adherence. *Journal of Immunological Methods*, **19**, 47–51.

McCall, C. E., de Chatelet, L. R., Brown, E. & Lachmann, P. (1974). New biological activity following intravascular activation of the complement cascade. *Nature, London*, **249**, 841–2.

MacGregor, R. R. (1976). The effect of anti-inflammatory agents and inflammation on granulocyte adherence. Evidence for regulation by plasma factors. *American Journal of Medicine*, **61**, 597–607.

MacGregor, R. R., Macarak, E. J. & Kefalides, N. A. (1978). Comparative adherence of granulocytes to endothelial monolayers and nylon fiber. *Journal of Clinical Investigation*, **61**, 697–702.

MacGregor, R. R., Negendank, W. G. & Schreiber, A. D. (1978). Impaired granulocyte adherence in multiple myeloma: relationship to complement system, granulocyte delivery, and infection. *Blood*, **51**, 591–9.

MacGregor, R. R., Spagnulo, P. J., Lentnek, A. L. (1974). Inhibition of granulocyte adherence by ethanol, prednisone, and aspirin, measured with an assay system. *New England Journal of Medicine*, **291**, 642–6.

Mayrovitz, H. N., Tuma, R. F. & Wiedeman, M. P. (1977). Relationship between microvascular blood velocity and pressure distribution. *American Journal of Physiology*, **232**, H400–5.

Mayrovitz, H. N., Wiedeman, M. P. & Tuma, R. F. (1977). Factors influencing leukocyte adherence in microvessels. *Thrombosis and Haemostasis*, **38**, 823–30.

Moses, J. M., Ebert, R. H., Graham, R. C. & Brine, K. L. (1964). Pathogenesis of inflammation. I. The production of an inflammatory substance from rabbit granulocytes *in vitro* and its relationship to leucocyte pyrogen. *Journal of Experimental Medicine*, **120**, 57–82.

NORTHOVER, B. J. (1977). Effect of indomethacin and related drugs on the calcium ion-dependent secretion of lysosomal and other enzymes by neutrophil polymorphonuclear leucocytes *in vitro*. *British Journal of Pharmacology*, **59**, 253–9.

O'FLAHERTY, J. T., CRADDOCK, P. R. & JACOB, H. S. (1978). Effect of intravascular complement activation on granulocyte adhesiveness and distribution. *Blood*, **51**, 731–9.

O'FLAHERTY, J. T., KREUTZER, D. L., SHOWELL, H. S., BECKER, E. L. & WARD, P. A. (1978). Desensitization of the neutrophil aggregation response to chemotactic factors. *American Journal of Pathology*, **93**, 693–706.

O'FLAHERTY, J. T., KREUTZER, D. L., SHOWELL, H. J. & WARD, P. A. (1977). Influence of inhibitors of cellular function on chemotactic factor-induced neutrophil aggregation. *Journal of Immunology*, **119**, 1751–6.

O'FLAHERTY, J. T., KREUTZER, D. L. & WARD, P. A. (1977). Neutrophil aggregation and swelling induced by chemotactic agents. *Journal of Immunology*, **119**, 232–9.

O'FLAHERTY, J. T., KREUTZER, D. L. & WARD, P. A. (1978*a*). Chemotactic factor influences on the aggregation, swelling and foreign surface adhesiveness of human leukocytes. *American Journal of Pathology*, **90**, 537–50.

O'FLAHERTY, J. T., KREUTZER, D. L. & WARD, P. A. (1978*b*). The influence of chemotactic factors on neutrophil adhesiveness. *Inflammation*, **1**, 37–48.

O'FLAHERTY, J. T., SHOWELL, H. J., KREUTZER, D. L., WARD, P. A. & BECKER, E. L. (1978). Inhibition of *in vivo* and *in vitro* neutrophil responses to chemotactic factors by a competitive antagonist. *Journal of Immunology*, **120**, 1326–32.

O'FLAHERTY, J. T., SHOWELL, H. J. & WARD, P. A. (1977). Influence of extracellular Ca^{2+} and Mg^{2+} on chemotactic factor-induced neutrophil aggregation. *Inflammation*, **2**, 265–276.

PERPER, R. J. (1975). Mechanisms by which leukocytes emigrate and induce tissue destruction. In *Future Trends in Inflammation, II*, ed. J. P. Giroud, D. A Willoughby & G. P. Velo, pp. 232–42. Basle, Stuttgart: Birkhäuser Verlag.

RABINOVITCH, M. & DE STEFANO, M. J. (1973*a*). Macrophage spreading *in vitro*. I. Inducers of spreading. *Experimental Cell Research*, **77**, 323–34.

RABINOVITCH, M. & DE STEFANO, M. J. (1973*b*). Macrophage spreading *in vitro*. II. Manganese and other metals as inducers or as co-factors for induced spreading. *Experimental Cell Research*, **79**, 423–30.

RABINOVITCH, M. & DE STEGANO, M. J. (1974). Macrophage spreading *in vitro*. III. The effect of metabolic inhibitors, anesthetics and other drugs on spreading induced by subtilisin. *Experimental Cell Research*, **88**, 153–62.

ROSSI, F., PATRIARCHA, P. L. & ROMEO, D. (1974). Regulation of oxidative metabolism and function of phagocytes. In *Future Trends in Inflammation, I*, ed. G. P. Velo, D. A. Willoughby & J. P. Giroud, pp. 103–22. Padua, London: Piccin Medical.

SABIONCELLO, A., DEKARIS, D., VESELIC, B. & SILOBRCIC, C. (1976). A comparison of peritoneal exudate cells and peripheral blood leukocytes in direct and indirect migration inhibition tests as *in vitro* assays for tuberculin hypersensitivity in guinea pigs. *Cellular Immunology*, **22**, 375–83.

Schiffer, C. A., Aisner, J. & Wiernik, P. H. (1975). Transient neutropenia induced by transfusion of blood exposed to nylon fiber filters. *Blood*, **45**, 141–6.

Schiffer, C. A., Sanel, F. T., Young, V. B. & Aisner, J. (1977). Reversal of granulocyte adherence to nylon fibers using local anaesthetic: possible application to filtration leukopheresis. *Blood*, **50**, 213–25.

Schoefl, G. I. (1972). The migration of lymphocytes across the vascular endothelium in lymphoid tissue. A re-examination. *Journal of Experimental Medicine*, **136**, 568–88.

Simionescu, M., Simionescu, N. & Palade, G. E. (1975). Segmental differentiations of cell junctions in the vascular endothelium. The microvasculature. *Journal of Cell Biology*, **67**, 863–85.

Simionescu, M., Simionescu, N. & Palade, G. E. (1976). Segmental differentiations of cell junctions in the vascular endothelium. Arteries and veins. *Journal of Cell Biology*, **68**, 705–23.

Smith, C. W., Hollers, J. C., Patrick, R. A. & Hassett, C. (1979). Motility and adhesiveness in human neutrophils. Effects of chemotactic factors. *Journal of Clinical Investigation*, **63**, 221–9.

Smith, R. P. C., Lackie, J. M. & Wilkinson, P. C. (1979). The effects of chemotactic factors on the adhesiveness of rabbit neutrophil granulocytes. *Experimental Cell Research* **122**, 169–77.

Spector, W. G. (1974). Chronic inflammation. In *The Inflammatory Process*, ed. B. W. Zweifach, L. Grant & R. T. McCluskey, 2nd edn, vol. III, pp. 277–91. London, New York: Academic Press.

Stamper, H. B. & Woodruff, J. J. (1976). Lymphocyte homing into lymph nodes: *in vitro* demonstration of the selective affinity of recirculating lymphocytes for high-endothelial venules. *Journal of Experimental Medicine*, **144**, 828–33.

Stendahl, O. & Edebo, L. (1972). Phagocytosis of mutants of *Salmonella typhimurium* by rabbit polymorphonuclear cells. *Acta Pathologica et Microbiologica Scandinavica*, **B80**, 481–8.

Stetson, C. A. & Good, R. A. (1951). Studies on the mechanism of the Shwartzman phenomenon. Evidence for the participation of polymorphonuclear leucocytes in the phenomenon. *Journal of Experimental Medicine*, **93**, 49–64.

Stewart, G. J., Ritchie, W. G. M. & Lynch, P. R. (1974). Venous endothelial damage produced by massive sticking and emigration of leukocytes. *American Journal of Pathology*, **74**, 507–32.

Stossel, T. P. (1978). The mechanism of leukocyte locomotion. In *Leukocyte Chemotaxis: Methods, Physiology and Clinical Implications*, ed. J. I. Gallin & P. G. Quie, pp. 143–60. New York: Raven Press.

Talstad, I. (1972). The relationship between phagocytosis of polystyrene latex particles by polymorphonuclear leucocytes (PML) and aggregation of PML. *Scandinavian Journal of Haematology*, **9**, 516–23.

Thompson, P. L., Papadimitriou, J. M. & Walters, M. N.-I. (1967). Suppression of leucocytic sticking and emigration by chelation of calcium. *Journal of Pathology and Bacteriology*, **94**, 389–96.

Thorgeirsson, G. & Robertson, A. L. (1978). The vascular endothelium – pathobiological significance. A review. *American Journal of Pathology*, **93**, 802–48.

Thorne, K. J. I., Oliver, R. C. & Lackie, J. (1977). Changes in the surface properties of rabbit polymorphonuclear leucocytes, induced by bacteria and bacterial endotoxin. *Journal of Cell Science*, **27**, 213–25.

Toren, M., Goffinet, J. A. & Kaplow, L. S. (1970). Pulmonary bed sequestration of neutrophils during hemodialysis. *Blood*, **36**, 337–40.

Van Oss, C. J. & Gillman, C. F. (1972). Phagocytosis as a surface phenomenon. I. Contact angles and phagocytosis of non-opsonised bacteria. *Journal of the Reticuloendotheial Society*, **12**, 283–92.

Vejlens, G. (1938). The distribution of leucocytes in the vascular system. *Acta Pathologica et Microbiologica Scandinavica*, Supplement **33**.

Wagner, R. C. & Matthews, M. A. (1975). The isolation and culture of capillary endothelium from epididymal fat. *Microvascular Research*, **10**, 286–97.

Ward, P. A. & Cochrane, C. G. (1964). A function of bound complement in the development of Arthus reactions. *Federation Proceedings*, **23**, 509.

Wells, R. (1973). Rheologic factors in inflammation. In *The Inflammatory Process*, ed. B. W. Zweifach, L. Grant & R. T. McCluskey, 2nd edn, vol. II, pp. 149–60. London, New York: Academic Press.

Wilkins, D. J., Ottewill, R. H. & Bangham, A. D. (1962*a*). On the flocculation of sheep leucocytes. I. Electrophoretic studies. *Journal of Theoretical Biology*, **2**, 165–75.

Wilkins, D. J., Ottewill, R. H. & Bangham, A. D. (1962*b*). On the flocculation of sheep leucocytes. II. Stability studies. *Journal of Theoretical Biology*, **2**, 176–91.

Wilkinson, P. C. (1974*a*). *Chemotaxis and Inflammation*. Edinburgh: Churchill–Livingstone.

Wilkinson, P. C. (1974*b*). Surface and cell membrane activities of leukocyte chemotactic factors. *Nature, London*, **251**, 58–60.

Wilkinson, P. C. & Lackie, J. M. (1979). The adhesion, migration and chemotaxis of leucocytes in inflammation. In *Current Topics in Pathology*, ed. H. Z. Movat, pp. 47–88, Berlin, Heidelberg & New York: Springer-Verlag.

Willoughby, D. A. (1973). Mediation of increased vascular permeability in inflammation. In *The Inflammatory Process*, ed. B. W. Zweifach, L. Grant & R. T. McCluskey, 2nd edn, vol. II, pp. 303–34. London, New York: Academic Press.

Wright, D. G. & Gallin, J. E. (1977). A functional differentiation of human neutrophil granules: generation of C5a by a specific (secondary) granule product and inactivation of C5a by azurophil (primary) granule products. *Journal of Immunology*, **119**, 1068–76.

Zigmond, S. H. & Hirsch, J. G. (1973). Leukocyte locomotion and chemotaxis. New methods for evaluation and demonstration of cell-derived chemotactic factor. *Journal of Experimental Medicine*, **137**, 387–410.

Zurier, R. B., Weissmann, G., Hoffstein, S., Kammerman, S. & Tai, H. H. (1974). Mechanisms of lysosomal enzyme release from human leucocytes. II. Effects of cAMP and cGMP, autonomic agonists, and agents which affect microtubule function. *Journal of Clinical Investigation*, **53**, 297–309.

Zweifach, B. W. (1973). Microvascular aspects of tissue injury. In *The Inflammatory Process*, ed. B. W. Zweifach, L. Grant & R. T. McCluskey, 2nd edn vol. II, pp. 3–46. London, New York: Academic Press.

Zweifach, B. W. (1974). Quantitative studies of microcirculatory structure and function. I. Analysis of pressure distribution in the terminal vascular bed in cat mesentery. *Circulation Research*, **34**, 843–57.

Histocompatibility systems: cell recognition and cell adhesion

A. S. G. CURTIS

Department of Cell Biology, University of Glasgow, Glasgow G12 8QQ

The advantages of using mutant strains to investigate the biology of a process are well instanced for adhesion by the papers in this volume by Hughes and by Gerisch. Yet their respective papers do not tell us anything directly about the recognition properties involved in the interaction of two normal cell types in the morphogenesis of a multicellular organism, simply because any interaction between two cell types that may occur in a slime mould is still a matter of dispute (Garrod, 1978) and because the behaviour of drug-resistance mutants with normal cells has not been tested.

Nevertheless, we should bear in mind that much of the impetus to investigate cell adhesion lies in our hope that changes and differences in adhesion are responsible for the normal morphogenesis of animals. The four theories that have been proposed as explanations for the morphogenetic movements of cells utilise cell adhesion as an element to a greater or lesser extent. These theories are: (1) specific adhesion (see Moscona, 1962); (2) the differential adhesion hypothesis (Steinberg, 1962, 1970, 1978); (3) chemotaxis (see Holtfreter, 1939); (4) the interaction modulation hypothesis (Curtis, 1974, 1976, 1978). The involvement of adhesion in chemotaxis has been suggested by Dierich, Wilhelmi & Till (1977). The respective merits of each theory are discussed in the same publication by Burger, Burkart, Weinbaum & Jumblatt (1978), Steinberg (1978) and Curtis (1978). Despite this concentration on cell adhesion as a major factor in the control of cell positionings, I feel that we have to admit that this assumption is far from proven.

Thus it seems entirely reasonable that we should ask whether we can use the interaction of cells of different genotype to test whether certain processes in adhesion or positioning can be ascribed to certain

Table 1. *The* H-2 *complex; functions (restrictions) and display of antigens*

Locus	*K*	*IA*	*IB*	*IJ*	*IE*	*IC*	*S*	*G*	*D*	*L*
Function										
CML										
induction	++								++	++
effector reaction	++								++	++
				←Some effects→						
MLR	+	++	++			+		+	+	+
T–B cooperation		++	++							
T-cell-macrophage		←Effects→								
Suppressor T cell–receptor T cell interaction										
Antigens displayed by	T	T_C T_H	T_C	T_S	T_H	T_S			T	
		←Macrophages→								
		←Skin→								
		← B cells→			← B cells→					

+ = some effect of this locus; + + = marked effect of this locus.
CML = Cell-mediated lympholysis. MLR = Mixed lymphocyte reaction.
T_H = helper T cell, T_C = cytotoxic T cell, T_S = suppressor T cell.
Data from Katz (1977), Snell (1978) and Hayes & Bach (1978).

genetic loci. Obviously there may be advantages in using normal mutants from fully functioning animals so that we can investigate the normal process. Again it is clear that the differentiation that ensures that different cell types within one organism have different properties so that they position themselves in different sites, must depend on a number of genes in order that a sufficient number of cell types may be specified. Bodmer (1972) suggested that the major histocompatibility complex (MHC) might play this role in adult and late embryonic animals. Bennet (1975) (see also Jacob, 1978) has suggested that the *T/t* complex may play a similar role in early embryonic mice. Bodmer's suggestion seemed at the time to be one based to a great extent on the inadequacy of the alternative explanations for the real function of the MHCs rather than on direct evidence that they acted in the cell–cell recognition required for cell positioning in embryogenesis and in adult life. However it became clear in the next few years that the mouse MHC system, the *H-2* complex, played a major role in determining whether a range of cell interactions would take place. In all these experiments mouse cells of different genotype were combined and the matching at certain loci (or mismatching) required to obtain or abrogate the interaction was recorded. The use of congenic strains of mice provided essential experimental material for these experiments.

The major systems in which matching or mismatching at certain *H-2* loci is required to elicit the reaction are listed in Table 1. Zinkernagel & Doherty (1974) were the first to term this type of phenomenon '*H-2* restriction'. They observed that virally infected cells generated cytotoxic lymphocytes in a mouse, and that the cytotoxic reaction on virally infected tissue culture cells only occurred if these cells were matched at loci *K* or/and *D* with the cytotoxic lymphocytes. They advanced two alternative explanations of this phenomenon: (1) 'the altered-self hypothesis', which basically states that the antigen–receptor reaction involves the histocompatibility gene products of *K* or *D*, and (2) the 'intimacy' hypothesis, which simply suggests that if cells are mismatched for their *K* and *D* products they cannot come into close enough contact for cytolysis to ensue. It has already been shown (Martz, 1977) that adhesion is a prerequisite for cytolysis in this type of system. The other examples of *H-2* restriction of cell interaction do not appear to have been analysed appreciably in terms of cell contact behaviour, but it seems likely that cell contact may be involved.

Thus it is of interest to test whether the *H-2* system and other MHC systems play a role in controlling cell adhesion. This paper reviews some of the appropriate evidence and reports new relevant evidence.

H-2 RESTRICTION IN CELL ADHESION: RECENT DATA

Bartlett & Edidin (1978), Zeleny, Matousek & Lengerova (1978) and I (Curtis, 1978) all described examples of *H-2* restriction of adhesion in mouse cell cultures. The first two groups used the collecting-lawn assay for adhesion, introduced by Walther, Ohman & Roseman (1973). In this assay, cell suspensions are allowed to settle on to monolayers of the target cell for a short period (1 to 30 min). Then the unadherent cells are washed off and the proportion of cells bound from the suspension measured. This assay may suffer from the fact that it is usual to prepare the cell suspension by trypsinisation, so that adhesive mechanisms may be somewhat damaged. Roth (1968) suggested, having used a rather different assay, that the specificity of cell adhesion might be damaged for several hours after trypsinisation. Nevertheless, both groups found evidence for some degree of *H-2* restriction of adhesion for fibroblasts (Bartlett & Edidin, 1978) and for bone marrow and lymph node cells (Zeleny *et al.*, 1978). In the latter case the cell types involved were not precisely identified, and the effect on adhesion was a small diminution between strains B10.A(2R) and B10.A(5R) which have haplotypes *kkkkkdddb* and *bbbkkdddd*, respectively. Thus Zeleny *et al.* could not precisely identify which loci were involved, but could have concluded that loci *IJ*, *IE*, *IC*, *S* and *G* did not take part in the reaction. Bartlett & Edidin investigated several strain combinations and found that allogeneic combinations of cells sometimes showed slightly reduced adhesion when compared with syngeneic combinations. Their results suggested to them that the selectivity was controlled by the monolayer rather than by both the suspension and the monolayer cell type. In no case was the difference between allogeneic and syngeneic adhesion very marked, for collection rates varied from 1.15 in some combinations to 1.70 in others. Combinations mismatched extensively at *H-2* tended to have lower adhesion rates than other combinations. No systematic use of strains to identify the loci involved was carried out.

McClay & Gooding (1978) examined the collection of cells from suspensions by aggregates by using different strains (C3H/MeJ and C57B1/10SnJ, which are $H\text{-}2^k$ and $H\text{-}2^b$, respectively (but are also allogeneic for many other loci) of late embryonic liver and brain cells. No evidence was found for any allogeneic effect on adhesion. Both these authors and Bartlett & Edidin examined the effect of *H-2* antisera on adhesion, but results must be treated with reserve because this treatment may well modify adhesion mechanisms massively in unknown ways.

My own work (Curtis, 1978) used a different test system. Media conditioned by actively growing cells were extracted for fractions that contained activity in reducing the adhesiveness of syngeneic cells of different tissue type. In earlier work (Curtis & van de Vyver (1971) on strain types of the freshwater sponge, *Ephydatia fluviatilis*; Curtis

(1974) on chick embryonic cells (see also Curtis, 1978), and on mouse T and B lymphocytes, Curtis & de Sousa (1975)), evidence had been found for the production of low molecular weight factors that diminish the adhesion of unlike cell types. Thus B lymphocytes produce a low molecular weight protein or glycoprotein that diminishes the adhesion of T lymphocytes, leukocytes and macrophages, and T lymphocytes and thymocytes produce a glycoprotein that diminishes the adhesion of B lymphocytes, leukocytes and macrophages. These tests were carried out on syngeneic cells, though a few experiments were also carried out on xenogeneic cells. It occurred to me to ask whether effects would be greater or lesser on allogeneic cells. In the first experiments reported in full (Curtis, 1979) I found that the effects of these interaction modulation factors (IMFs) were enhanced if they acted on allogeneic cells rather than on syngeneic ones. Thus it became obvious to enquire as to whether this diminution of adhesion contained an element due to the genetic mismatch between IMF source and responding cell. The test system for this question is to examine whether these factors diminish the adhesion of allogeneic cells of the same tissue type. Thus do T cell IMFs from one strain diminish the adhesion of T cells from another strain? Preliminary results were published (Curtis, 1978) which showed clearly that IMF glycoprotein from one strain diminishes the adhesion of another congenic strain if the products of the *D* locus are not identical between the strains. Slight evidence for additional involvement of a non-*H-2* locus was found in the response of the A/WySn strain, but in combinations of other strains evidence was found for the involvement of *H-2 D* alone. this result was described more fully by Curtis (1979). Typical results are shown in Table 2.

Thus there is clear evidence that cells can produce substances that diminish the adhesion of allogeneic cells if a mismatch at *H-2 D* exists. The immediate questions are as to whether this effect exists between other cell types and whether the same locus is involved. Experimental results from two new systems are described in this paper before the situation is discussed.

Table 2. H-2 *Relationship of IMFs; allogeneic effects of IMFs on adhesion of thymus cells*

Target cell strain type	IMF origin	Adhesion	*H-2* mismatch	Congenic
A/WySn	A/WySn	9.4	None	–
	B10.AKM	0.03	*IC, S, G & D*	No
	B10.A (5R)	9.0	*K, IA & IB*	No
B10.S (7R)	B10.S (7R)	8.4	None	–
	B10.T (6R)	8.3	All except *D*	Yes
B10.G	B10.G	10.7	None	–
	B10.AKM	11.0	All except *D*	Yes
B10.A	B10.A	13.8	None	–
	B10.A (2R)	0.0	*D* only	Yes
	B10.A (4R)	2.1	All except *K & IA*	Yes
	A/WySn	4.8	None	No
B10.A (2R)	B10.A (2R)	27.4	None	–
	B10.A	0.0	*D* only	Yes
	B10.A (4R)	13.9	*IB, IJ, IE, IC, S & G*	Yes
B10.A (4R)	B10.A (4R)	11.8	None	–
	B10.A (2R)	14.9	*IB, IJ, IE, IC, S & G*	Yes
	B10.A	0.0	All except *K & IA*	Yes

IMFs were applied at 7×10^3 units ml^{-1} or as the yield of 1×10^7 cells ml^{-1} for those IMFs not precisely assayed.

Experimental methods

Two different experimental systems were used. In the first the experiments carried out on the thymocyte/T cell system were repeated on the B lymphocyte system. In the second an investigation was carried out to test whether contact inhibition of movement (Abercrombie & Heaysman, 1954) differed between allogeneic cell types and syngeneic types. Experiments were carried out on epithelia and fibroblasts.

Mouse strains B10, B10.A, B10.A(2R), B10.A(4R), B10.AKM, B10.BR, B10.G were used. These strains were maintained for us in the Department of Neurological Sciences, Southern General Hospital, Glasgow; their origins have been described in full by Curtis (1979).

B lymphocytes were prepared by three daily injections of rabbit anti-serum to mouse lymphocyte (0.5 ml, subcutaneously, Searle Laboratories, Batch 14). B cells were prepared from peripheral lymph nodes of these animals by the techniques described by Curtis & de Sousa (1975) and contained on all occasions more than 85 % Ig-positive cells.

The adhesion of the lymphocytes to each other in RPMI medium was measured by using the collision efficiency method (Curtis, 1969) at low shear rates over a 30 min period. Measurements were made with cells in RPMI alone and in the presence of B interaction factors prepared from syngeneic cells or from allogeneic cells.

Epithelia were grown in Ham's F10 medium plus 10 % foetal calf serum as outgrowths from explants. After three or four days, when outgrowths had met, the cultures were fixed, stained and the degree of contact inhibition assessed by measurement of the overlap index (Abercrombie & Heaysman, 1954). The epithelia were of adult kidney tubule origin, fibroblasts being removed by collagenase treatment of the explants (1 mg ml^{-1} collagenase in Ham's F10 medium for 1 h at 37 °C). The overlap index in confronted outgrowths was measured.

Results

(*a*) H-2 *restriction in the adhesion of B lymphocytes.* The adhesiveness of B cells from various congenic strains of mice was measured in the presence of syngeneic or allogeneic B IMF. The IMF fractions were prepared by the technique described by Curtis & de Sousa (1975) and applied at such a concentration that 1×10^6 cells ml^{-1} were in the presence of IMF derived from 1×10^6 cells. Results are shown in Table 3. Examination of these results shows that in some combinations B lymphocyte IMF diminishes the adhesion of allogeneic B cells. These combinations are those in which the strain producing the IMF and the responding strain are mismatched at the *K* locus or/and the *1A* subregion.

(*b*) H-2 *restriction in contact inhibition of movement.* These experiments were mainly carried out by Paul Rooney, see Curtis & Rooney (1979).

In those experiments which were originally intended to be preliminary ones in the investigation of whether there are genetic restrictions in contact inhibition of movement, a considerable number of explants from two allogeneic strains were placed in the same culture

Table 3. H-2 *relationship of B cell IMFs: allogeneic effect of IMFs on the adhesion of B lymphocytes*

Target cell strain type	IMF origin	Adhesion	*H-2* mismatch
B10.BR	B10.BR	11.6	None
	B10.AKM	15.3	*D* only
B10.AKM	B10.AKM	15.5	None
	B10.BR	14.8	*D* only
B10	B10	10.9	None
	B10.A(2R)	0.0	All except *D*
	B10.A(4R)	0.0	*K and IA*
B10.A(2R)	B10.A(2R)	12.9	None
	B10	0.0	All except *D*
	B10.A(4R)	13.7	*IB, IJ, IE, IC, S and G*
B10.A(4R)	B10.A(4R)	13.9	None
	B10	0.0	*K and IA*
	B10.A(2R)	8.3	*IB, IJ, IE, IC, S and G*

Adhesion measured as collision efficiency percentages. Measurements made with IMF yields of *c.* 1×10^6 cells on 1×10^6 cells.

dish. The expectation was that if there was any allogeneic effect on contact inhibition we would find at least two (possibly three) populations with different mean values of contact inhibition, and thus though the *H-2* type of any given explant would not be known, the results would demonstrate the existence of any effect that might occur. (Equal numbers of explants from each strain type were placed in the dish so it would be expected that 50 % of the confronted paired outgrowths between two explants would be between unlike cells.) The original plan was then to carry out an experiment with paired unlike explants in microwells. However the results of the preliminary experiment (see Table 4) are quite clear and show that the overlap index of all confronted cultures in each dish are reduced in those cultures of congenic pairs mismatched at *H-2 K* and *H-2 D*. Thus these results establish in one step that *H-2* restriction occurs in contact inhibition and that the effect is mediated by a soluble diffusible product of the cells. Results are shown in Table 4 (*a–d*). Incompatibilities at *K* and *D* produce the effect.

Table 4. *Contact inhibition of movement between histo-incompatible mouse epithelial cells*

(*a*)

Strain	*H-2* mismatch	Overlap index	
B10.G		0.27	0.12
B10.BR		0.31	0.24
B10.BR with B10.G	All loci	0.11	0.07

Significance of difference in means between mixed cultures and controls:
B10.G cf. Mixed t = 6.01, d.f. = 58, $p < 0.001$
B10.BR cf. Mixed t = 4.55, d.f. = 58, $p < 0.001$.

(*b*)

Strain	*H-2* mismatch	Overlap index	
B10.G		0.28	0.12
B10.AKM		0.23	0.13
B10.G with B10.AKM	All except *D*	0.08	0.06

Significance of difference in means between mixed culture and controls:
B10G cf. Mixed t = 7.47, d.f. = 58, $p < 0.001$
B10.AKM cf. Mixed t = 5.48, d.f. = 58, $p < 0.001$.

B10.BR		0.30	0.12
B10.AKM		0.23	0.13
B10BR with B10.AKM	*D* only	0.10	0.06

Significance of difference in means between mixed culture and controls:
B10.BR cf. Mixed t = 4.69, d.f. = 58, $p < 0.001$
B10.AKM cf. Mixed t = 4.75, d.f. = 58, $p < 0.001$.

(*c*)

Strain	*H-2* mismatch	Overlap index	
B10.A(2R)		0.31	0.06
B10.A(4R)		0.34	0.11
B10.A(2R) with B10.A(4R)	*IB, IJ, IE, IC, S & G*	0.34	0.19

Significance of difference in means between mixed cultures and controls:
B10.A(2R) cf. Mixed t = 0.72, d.f. = 58, $p < 0.1$
B10.A(4R) cf. Mixed t = 0.02, d.f. = 58, $p < 0.1$.

Table 4 (*cont.*)

B10 . HTT		0.29	0.17
ATL		0.36	0.12
B10 . HTT with ATL	*IA, IB*	0.30	0.12

Significance of difference in means between mixed cultures and controls:
B10 . HTT cf. Mixed t = 0.10, d.f. = 58, $p < 0.1$
ATL cf. Mixed t = 1.97, d.f. = 58, $p < 0.05$.
Note that these two strains are not congenic but subsidiary work has shown that non-*H-2* loci are not involved in this type of reaction

(*d*)

Strain	*H-2* mismatch	Overlap index	
B10		0.26	0.09
B10 . A(4R)		0.27	0.14
B10 with B10 . A(4R)	*K and IA*	0.10	0.08

Significance of difference in means between mixed cultures and controls:
B10. cf. Mixed t = 7.15, d.f. = 58, $p < 0.001$
B10 . A(4R) cf. Mixed t = 5.82, d.f. = 58, $p < 0.001$.

Discussion: cell intimacy

Zinkernagel & Doherty (1974) appreciated that there might be two substantially different explanations for the *H-2* restriction of cell-mediated lympholysis of those virally infected cells that matched the cytotoxic lymphocytes in the gene products of certain of the *H-2* loci. The first of these explanations has been termed the 'altered-self' theory, which simply states that the restriction is at the level of antigen stimulation and subsequent recognition by the effector lymphocytes. This idea, though most interesting in terms of theories of antigen recognition, has no implications about the control of cell adhesion. The intimacy theory simply suggests that cell must be alike at certain histocompatibility loci for close interaction to occur. This intimacy could be seen either in terms of some molecular complementarity between alike cells at the appropriate loci or alternatively in terms of some inhibition between unlike cells. The results described above strongly support the idea that *H-2* restriction might be due to the allogeneic inhibition of adhesion by the action of soluble factors of the IMF type.

Earlier work (Curtis, 1978, 1979) showed that:

(1) The T cell IMF glycoprotein is bound to a range of syngeneic and allogeneic target cells. Those of this series that have been tested have their adhesion reduced by the IMF.

(2) Cells recover their adhesion when the IMF factor is removed from the medium.

(3) The T cell IMF acts to control lymphocyte positioning *in vivo* (Davies & Curtis, 1979).

Thus this work suggests that a single glycoprotein acts in interactions between tissues of different type within one organism and also allogeneically between identical or different tissue types. This suggestion is crucial to the proposal that the normal function of the histocompatibility system is in the control of those interactions normally undertaken by the different cell types within an organism. If it were the case that the IMF preparations contain two or more glycoproteins it is then possible that one is concerned with syngeneic interactions and the other with allogeneic interactions, and thus that the two systems are unrelated. Purity of IMF preparations is discussed later.

The nature of the control of cell interactions in the immune system: allogeneic interactions. Katz (1978*a*, *b*) in reviewing the work of his own groups and others on T–B cell interaction concluded that the failure of T and B cells to cooperate across a major histocompatibility barrier in many situations was not due to:

(1) The possibility that T and B cell migration in reconstituted animals containing T and B cells of different *H-2* type might be abnormal so that the cell types never met. This possibility is disproven by the finding that *H-2* restriction can be demonstrated in *in vitro* cultures.

(2) A block to cell–cell interaction introduced by the presence of foreign histocompatibility specificities.

Yet the evidence from work on adhesion suggests that this block exists. It is appropriate at this point to consider the control of the T–B cell interaction in syngeneic systems before reconsidering Katz's evidence. De Sousa & Haston (1976), noting the fact that partial immunosuppression with ALS led to improved antibody response to antigenic stimulation, examined the proportion of B and T cell pairs in dissaggregates from spleens of normal and partily immunosuppressed animals. They found an increase in the number of B–T pairs, and pointed out that this was consistent with the idea that T cells

Table 5. *Immunosuppression by T and by B cell IMF glycoproteins*

System	PFU 10^{-6} spleen cells
Immunosuppression by T cell IMF	
(*a*) Syngeneic	
Control, no antigen	5590
Control, with antigen	50320
T IMF added; 4.6×10^4 units, 10^{-6} spleen cells	5958
(*b*) Allogeneic	
Control, no antigen	1297
Control, with antigen	24653
T IMF added 4.6×10^4 units, 10^{-6} spleen cells	2224
Immunosuppression by B cell IMF	PFU 10^{-6} spleen cells
(*a*) Syngeneic	
Control, no antigen	0
Control, with antigen	34056
B IMF added, 5×10^4 units, 10^6 cells	6950
(*b*) Allogeneic	
Control, no antigen	463
Control, with antigen	24554
B IMF added, 5×10^4 units, 10^6 spleen cells	0

Mishell–Dutton assays. T IMF Batch 128, activity 1.15×10^6 u mg^{-1}. Assay on fifth day, antigen SRBC 6×10^6 per 1×10^7 spleen cells. Syngeneic system, CBA cells. Allogeneic system, C57 B110/ScSn cells.

might produce substances that tended to suppress B cell adhesion. Curtis & de Sousa (1975) reported the presence of such substances. Bell & Shand (1975) also demonstrated that abnormally large T:B cell number ratios in reconstituted animals tended to lead to immunosuppression. Thus it is important to enquire whether the T or B cell IMF act as immunosuppressants. Simple *in-vitro* experiments (Table 5) show that this can be so. It may be relevant that Martineau & Johnson (1976) found that normal mouse serum contained an immunosuppressant that appeared to act by reducing cell adhesion. IMFs can be obtained from sera. Thus I should like to propose that all experiments comparing the extent of antibody synthesis in a given system with that in another need to take account of the levels of IMFs in the two systems and the extent of cell response to them. Thus it would be expected that at low T:B cell ratios interaction between histoincompatible strains might become more possible: as far as I know this seems not to have been tested yet.

Katz, Mamaoka & Benacerraf (1973) were able to obtain cooperation between T and B cells across a certain degree of mismatch by making reconstituted animals containing either hybrid (Balb/c × A/J) F_1 T cells and parental B cells or parental T cells and hybrid B cells. This result led to the suggestion that inhibitory systems could not act. I feel that this conclusion is over-sanguine in the light of the later evidence that B:T proportions and IMF levels control the extent of the cooperation.

Bechtol & McDevitt (1976) were able to obtain cooperation across *I*-region incompatibilities in chimaeric mice, with strain combinations that would not cooperate in *in vivo* or *in vitro* situations when adult cells were used. Thus there may be several situations such as presence in a chimaeric embryo, long-term residence in an animal in which histoincompatible reactions cannot occur (F1 hybrid) or residence in a homozygous graft in an irradiated T-reconstituted mouse (Fink & Bevan, 1978) (hybrid between graft and another strain) in which cooperation becomes possible between cells which are incompatible at some *H-2* locus in one of their chromosome sets, and which display both sets of *H-2* products. These observations refer of course not only to T–B interactions but also to the recognition of target cell by cytotoxic lymphocytes. Evidence such as this has been interpreted by Katz (1978) in terms of a theory of 'adaptive differentiation', namely a process that results in preferential interactions amongst cells that have undergone their differentiation in the same genotypic environment.

Once we begin to consider how a process of adaptive differentiation might develop we also may appreciate that some of the theories are entirely consistent with the idea that IMFs act in controlling cell interactions.

(1) If we regard the *H-2* gene products, in part at least, as cell interaction molecules and propose the simple model that they act in binding cell to cell by identical molecules being able to interact, then we would expect homozygous cells to be able to interact with heterozygous cells which shared alleles at the important cell interaction (*CI*) loci. Release of fragments of these molecules in a univalent form would give haptenic inhibition. However, this model has a number of disadvantages, chief amongst which is the fact that it does not explain the allogeneic inhibition of adhesion by IMF molecules.

(2) Alternatively we should perhaps pay attention to the fact that IMFs are bound both to syngeneic and to allogeneic target cells of

the appropriate type (Curtis, 1979) yet produce no effect on syngeneic cells. Therefore, we should concentrate our attention on the fact that there may be some feature of the syngeneic system which inhibits IMF action whether or not another gene product of that locus in a heterozygote may be also displayed in a hybrid cell at the same time.

Thus if a cell type expresses the product of an *H-2* locus which is homologous with that locus active in IMF production by a second cell-type there will be no effect of the IMF on the first cell type. It is perhaps worth raising the conjecture that IMFs do not attack the producer cell because they are secreted with an inhibitory sequence, in the same manner as some proteases (see Neurath, 1975), which has to be removed before activity develops and which can inhibit activity if present in sufficient concentration as a fragment. This model is consistent with both the intimacy theory and with the observations that have been used to argue against physiological inhibition of cooperation. Moreover, it may explain the observations that cooperation can develop in embryonic chimaerae where the proximity of two genotypes and the accumulation of large amounts of the inhibitor means that cells are protected against IMF effects even though they are of the incorrect haplotype.

(3) Effects of selection in the experiments with hybrids, varied B:T population ratios, and possible masking of reactive systems by excess IMFs should not be discounted.

Yet these experiments do not appear to tell us about the systems that control cell interactions within one normal organism, where though interactions between B and T cells, etc. are permitted, systems operate to separate B and T cells into different areas. However, the main question raised in this paper is whether the same system that controls allogeneic interactions is primarily responsible for interactions between cells of different tissue type within one organism.

Are histocompatibility gene interactions involved in cell interactions within one organism? The work just described demonstrates an association of the T cell and B cell IMFs with different MHC loci. The initial work carried out on T cell IMF demonstrated a remarkable decrease in adhesion induced by this glycoprotein (Curtis & de Sousa, 1975; Curtis, 1978) on syngeneic B cells. The glycoprotein is bound to the surface of both B and T lymphocytes, to leukocytes, macrophages and some other cell types. The T IMF acts in the experimental modification of cell positioning both when injected *in vivo* or when labelled cells are treated with the IMF externally and then injected into the animal, (Curtis, 1978; M. D. J. Davies and A. S. G. Curtis, in preparation). Thus if the T cell IMF preparations contain but a single glycoprotein it is suggestive that the same glycoprotein acts in both positioning in a syngeneic situation and in allogeneic interactions and that a histocompatibility-complex-related produce is directly involved in cell positioning in a single organism. The purity of the T IMF glycoprotein has been established by N-terminal amino acid analysis and the molecular weight measured by the PTH-method of Banks *et al.* (1976) as 10800. The N-terminal amino acid is phenylalanine and the penultimate one is valine. The detection of one amino acid at each stage of the sequencing establishes the purity.

One of the implications of this conclusion is that the basis of cell–cell recognition in the organism is the presence or absence of the expression of certain identical *H-2* products in two different cell types. In allogeneic situations, interactions with overt effects only appear if a particular gene product of one locus found in one cell type is entirely missing from the other. This system would form a good model for intertissue interactions within an organism. If each distinct tissue type displays, say, only seven to ten different MHC products out of a possible twenty, then there will be many different possible cell types. If several or all the loci are connected with the production of an IMF-like cell interaction molecule then one tissue will be able to affect the adhesion of another if , and only if, the second tissue lacks the expression of this gene. Even if it is supposed that each cell type produces only one IMF there will be a large number of different tissues with which that tissue will interact to produce a decrease in the adhesion of the target tissue.

However, is there evidence that there is this proposed difference in dispaly of the activity of MHC products by different tissues? Until recently methods and techniques were unrefined so that it was impossible to be sure whether this might be true. Work to date has been mainly confined to lymphoid cells but it is already suggested on limited evidence (Fathman *et al.* (1975) and David, Meo, McCormick & Shreffler (1976)) that certain *Ia* specificities are restricted to either T or B cells while others are common to both. For instance the *IJ* region specifies antigenic determinants found only on a subclass of suppressor T cells (Hayes & Bach, 1978). I would suggest that in the future it will be of considerable interest to look for such restrictions in MHC antigen display and to relate this as far as we are able to, to control of cell interactions both in cells of the immune system and in other cell types. Another method of examining this problem is to examine systems for production of *H-2*-associated molecules that can be shown to modify cell behaviour both *in vitro* and *in vivo*.

Recognition systems in cell adhesion. A number of theories of cell adhesion are specifically designed to account for the real or apparent specificity of selectivity seen in a variety of forms of cell behaviour. Chief amongst the types of behaviour are the sorting out of aggregates into regularly arranged aggregates (Steinberg, 1962), the sorting out of sponge cell types of different species into separate bodies, and various features of cell behaviour in embryos. Chief amongst the theories which propose a marked specificity in molecular interactions are those described in this book by Gerisch, and by Barondes, and elsewhere by Moscona (1962), Roseman (1970), Lilien (1969) and Burger *et al.* (1978). These hypotheses all suggest that the interacting cell surfaces have a number of fixed or bound components that effect the specific interaction. Yet, as Steinberg (1962) pointed out, this type of hypothesis fails to explain the observed fact that positioning takes place in the construction of an embryo, or the sorting out of an aggregate. Namely, positioning means that one cell type forms a tissue in a definite relationship to another. This is the prime observation that we have to explain. Steinberg devised the differential adhesion hypothesis to explain these positionings; it is not appropriate to discuss it at length here but those interested are referred to papers by Steinberg (1976, 1978) and by Harris (1976). The other type of hypothesis that has been advanced to explain the development of patternings is the suggestion that concentration gradients of diffusible substances control the behavioural reactions of cells, leading or herding them into particular positions. Two main subtheories of this have been suggested. The first is that chemotaxis (or chemokinesis) (see Holtfreter, 1941) acts. The second is the interaction modulation theory (Curtis & Van de Vyver, 1961; Curtis, 1974, 1976, 1978). This suggests that gradients of substances diminish the adhesion of suitably unlike cell types. Under certain circumstances the two theories merge.

Thus the finding that IMFs injected *in vivo* modify cell behaviour is support for the interaction modulation theory. Again it has been shown (Curtis, 1974) that concentration gradients of IMFs alter cell sorting-out patterns in aggregates. The IMF theory suggests that adhesion is controlled by a series of specific reactions probably involving MHC products: it does not necessarily yet tell us about the mechanism of that adhesion but suggests that further investigation in this area might be rewarding.

Cell behaviour and MHC involvement. Our finding that contact inhibition of movement between epithelia is increased in combinations of cells histocompatible at *K* or *D* is perhaps somewhat unexpected, though predicted by Medawar (1978). It might be expected that allogeneic combinations of cells might lead to increased invasion and thus to increased overlap of the cells, because malignant cells are presumably allogeneic to their host cells at least at one locus. However, it should be noted that if the effect is due to a reduction in adhesion when allogeneic cells meet, then less overlap might be expected since the cells would be insufficiently adhesive for each to move over the other. It is interesting to speculate on whether some of the features of failure of graft fusion in histoincompatible grafts are due to the operation of an increased contact inhibition of movement.
containing two strains of epithelia, whether between histocompatible types or between histoincompatible ones, showed the same low value for overlap index (i.e. high contact inhibition) indicates that the system operates by the secretion of diffusible factors by the cells. It is thus tempting to conclude that the close similarities between this system and that in lymphocytes implies that both operate with an IMF mechanism and that cell behaviour as well as cell adhesion is controlled in this manner.

Finally, it may be appropriate to spend a paragraph speculating about the systems that maintain normal tissues in their correct places. The IMF hypothesis would suggest that a continuing process operates to do this so that different tissue types have their adhesion reduced when they approach closely so that invasion fails. Cells in solid masses of tissue of one cell type show little movement (Gershmann & Drumm, 1975) while cells in aggregates composed of two or more cell types show considerable movement until sorting out is complete, presumably because when intermixed they lower each other's adhesion.

REFERENCES

ABERCROMBIE, M. & HEAYSMAN, J. E. M. (1954). Observations on the social behaviour of cells in tissue culture. II. 'Monolayering' of fibroblasts. *Experimental Cell Research*, **6**, 293–306.

BANKS, B. E. C., DOONAN, S., FLOGEL, M., PORTER, P. B., VERNON, A., WALKER, J. M., CROUCH, T. H., HALSEY, J. F., CHIANCONE, E. & FASELLA, P. (1976). An assessment of some of the methods available for the determination of molecular weights of proteins as applied to aspartate aminotransferase from pig heart. *European Journal of Biochemistry*, **71**, 469–73.

BARTLETT, P. F. & EDIDIN, M. (1978). Effect of the H-2 gene complex rates of fibroblast intercellular adhesion. *Journal of Cell Biology*, **77**, 377–88.

BECHTOL, K. B. & MCDEVITT, H. O. (1976). Antibody response of C3H⇔(C XB X(WB))F_1 tetraparental mice to poly-L (tyr, glu)-poly-DL-Ala-poly-*L*-Lys immunization. Co-operation across I region differences. *Journal of Experimental Medicine*, **144**, 123–44.

BENNETT, D. (1975). The T-locus of the mouse. *Cell*, **6**, 411–54.

BELL, E. B. & SHAND, F. L. (1976). Changes in lymphocyte recirculation and liberation of the adoptive memory response from cellular regulation in irradiated recipients. *European Journal of Immunology*, **5**, 1–7.

BODMER, W. F. (1972). Evolutionary significance of the HL-A system. *Nature, London*, **237**, 139–45 (plus references continued on p. 183).

BURGER, M. M., BURKART, W., WEINBAUM, G. & JUMBLATT, J. (1978). Cell–cell recognition: molecular aspects. Recognition and its relation to morphogenetic processes in general. In *Cell–Cell Recognition* (32nd Symposium of the Society for Experimental Biology) ed. A. S. G. Curtis, pp. 1–23. Cambridge University Press.

CURTIS, A. S. G. (1969). The measurement of cell adhesiveness by an absolute method. *Journal of Embryology and Experimental Morphology*, **22**, 305–25.

CURTIS, A. S. G. (1974). The specific control of cell positioning. *Archives de Biologie*, **85**, 105–21.

CURTIS, A. S. G. (1976). Le positionnement cellulaire et la morphogenèse. *Bulletin de la Société Zoologique de France*, **101**, 9–21.

CURTIS, A. S. G. (1978). Cell–cell recognition: positioning and patterning systems. In *Cell–Cell Recognition* (32nd Symposium of the Society for Experimental Biology), ed. A. S. G. Curtis, pp. 51–82. Cambridge University Press.

CURTIS, A. S. G. (1979). The H-2 histocompatibility system and lymphocyte adhesion. Interaction modulation factor involvement. *Journal of Immunogenetics*, **6**, 155–66.

CURTIS, A. S. G. & ROONEY, PAUL (1979). H-2 restriction of contact inhibition of epithelial cells. *Nature, London*, **281**, 222–3.

CURTIS, A. S. G. & VAN DE VYVER, G. (1971). The control of cell adhesion in a morphogenetic system. *Journal of Embryology and Experimental Morphology*, **26**, 295–312.

DAVID, C. S., MEO, T., MCCORMICK, J. & SHREFFLER, D. C. (1976). Expression of individual Ia specificities on T and B cells. I. Studies with mitogen-induced blast cells. *Journal of Experimental Medicine*, **143**, 218–24.

de Sousa, M. & Haston, W. (1976). Modulation of B-cell interactions by T cells. *Nature, London*, **260**, 429–30.

Dierich, M. P., Wilhelmi, D. & Till, G. (1977). Essential role of surface-bound chemoattractant in leukocyte migration. *Nature, London*, **270**, 351–2.

Fathman, C. G., Cone, J. L., Sharrow, S. O., Tyrer, H. & Sachs, D. (1975). Ia alloantigen(s) detected on thymocytes by use of a fluorescence-activated cell sorter. *Journal of Immunology*, **115**, 584–9.

Gershman, H. & Drumm, J. (1975). Mobility of normal and virus-transformed cells in cellular aggregates. *Journal of Cell Biology*, **67**, 276–86.

Harris, A. K. (1976). Is cell sorting caused by differences in the work of intercellular adhesion? A critique of the Steinberg hypothesis. *Journal of Theoretical Biology*, **61**, 267–85.

Hayes, C. E. & Bach, F. H. (1978). T-cell-specific murine Ia antigens: serology of I-J and I-E subregion specificities. *Journal of Experimental Medicine*, **148**, 692–703.

Holtfreter, J. (1939). Gewebeaffinität, ein Mittel der embryonalen Formbildung. *Archiv für Experimentale Zellforschung*, **23**, 169–209.

Jacob, F. (1978). Mouse teratocarcinoma and mouse embryo. *Proceedings of the Royal Society of London*, **B201**, 249–70.

Katz, D. H. (1977). *Lymphocyte Differentiation, Recognition, and Regulation*, ed. F. J. Dixon Jr. & H. G. Kunkel. London, New York: Academic Press.

Katz, D. H. (1978). Self-recognition as a means of cell communication in the immune system. In *Cell–Cell Recognition* (32nd Symposium of the Society for Experimental Biology), ed. A. S. G. Curtis, pp. 411–28. Cambridge University Press.

Katz, D. H., Hamaoka, T. & Benacerraf, B. (1973). Cell interactions between histincompatible T and B lymphocytes. II. Failure of physiologic cooperative interactions between T and B lymphocytes from allogeneic donor strains in humoral response to hapten–protein conjugates. *Journal of Experimental Medicine*, **137**, 1405–18.

Lilien, J. (1969). Towards a molecular explanation for specific cell adhesion. *Current Topics in Development Biology*, **4**, 169–93.

Martineau, R. S. & Johnson, J. S. (1978). Normal mouse serum immunosuppressive activity: action on adherent cells. *Journal of Immunology*, **120**, 1550–63.

Martz, E. (1977). Mechanism of specific tumour-cell lysis by alloimmune lymphocytes: resolution and characterization of discrete steps in the cellular interaction. *Contemporary Topics in Immunobiology*, **7**, 301–61.

McClay, D. R. & Gooding, L. R. (1978). Involvement of histocompatibility antigens in embryonic cell recognition events. *Nature, London*, **274**, 367–8.

Medawar, P. (1978). Philosophy of Ignorance (book review of *Encyclopedia of Ignorance*), *Nature, London*, **272**, 772–4.

Mascona, A. A. (1962). Analysis of cell recombinations in experimental synthesis of tissues *in vitro*. *Journal of Cellular and Comparative Physiology*, Suppl. I, **60**, 65–80.

Neurath, H. (1975). Limited proteolysis and zymogen activation. In *Proteases and Biological Control* (Cold Spring Harbor Conference on Cell Proliferation) vol. 2, ed. E. Reich, D. B. Rifkin & E. Shaw, pp. 51–64.

Roseman, S. (1970). The synthesis of complex carbohydrates by multiglycosyltransferase systems and their potential function in intracellular adhesion. *Chemistry and Physics of Lipids*, **5**, 270–97.

Roth, S. (1968). Studies on intercellular adhesive selectivity. *Development Biology*, **18**, 602–31.

Snell, G. D. (1978). T cells, T cell recognition structures, and the major histocompatibility complex. *Immunological Reviews*, **38**, 3–69.

Steinberg, M. S. (1962). On the mechanism of tissue reconstruction by dissociated cells. I. Population kinetics, differential adhesiveness, and the absence of directed migration. *Proceedings of the National Academy of Sciences, USA*, **48**, 1577–82.

Steinberg, M. S. (1970). Does differential adhesion govern self-assembly processes in histogenesis? Equilibrium configurations and the emergence of a hierarchy among populations of embryonic cells. *Journal of Experimental Zoology*, **173**, 395–434.

Steinberg, M. S. (1976). Adhesion-guided multicellular assembly: a commentary upon the postulates, real and imagined of the differential adhesion hypothesis, with special attention to computer simulations of cell sorting. *Journal of Theoretical Biology*, **55**, 431–4.

Steinberg, M. S. (1978). Specific cell ligands and the differential adhesion hypothesis: How do they fit together? In *Specificity of Embryological Interactions*, ed. D. Garrod, pp. 97–130. London: Chapman and Hall.

Walther, B. T., Ohman, R. & Roseman, S. (1973). A quantitative assay for intercellular adhesion. *Proceedings of the National Academy of Sciences, USA*, **70**(8), 1569–73.

Zeleny, V., Matousek, V. & Lengerova, A. (1978). Intracellular adhesiveness of H-2 identical and H-2 disparate cells. *Journal of Immunogenetics*, **5**, 41–7.

Zinkernagel, R. M. & Doherty, P. C. (1974). Immunological surveillance against altered self components by sensitised T-lymphocytes in lymphocytic choriomeningitis. *Nature, London*, **251**, 547–8.

Analysis of cell adhesion in *Dictyostelium* and *Polysphondylium* by the use of *Fab*

GÜNTHER GERISCH, HEIDE KRELLE, SALVATORE BOZZARO, EVELINE EITLE AND RICHARD GUGGENHEIM

Biozentrum der Universität Basel, Klingelbergstrasse 70, 4056 Basel, Switzerland and Laboratory of Scanning Electron Microscopy, Universität Basel, Bernoullistrasse 32, 4056 Basle, Switzerland.

Dedicated to Professor Kenneth B. Raper, the discoverer of *Dictyostelium discoideum*, who performed in 1941 the first sorting-out experiments with this organism.

TWO TYPES OF TARGET SITES OF ADHESION-BLOCKING *Fab*

Aggregating cells of *Dictyostelium discoideum* form streams in which they move towards aggregation centres (Figs. 1 & 2*a*). Within the streams, the cells are typically elongated. They adhere to each other most tightly at their ends, but are also associated by side-to-side adhesion. Early work has shown that these two types of cell assembly can be separated by EDTA: whereas the end-to-end adhesion is EDTA stable, the side-by-side adhesion is sensitive (Gerisch, 1961). Only cells which have reached the aggregation stage form EDTA-stable contacts. Nevertheless, growth-phase cells are already adhesive, as demonstrated by the ability of suspended cells to agglutinate in the absence of EDTA. Mutants blocked in an early stage of development exhibit only the EDTA-sensitive type of adhesion (Beug, Katz & Gerisch, 1973). These results have been interpreted to mean that adhesion of aggregating cells of *D. discoideum* is mediated by two types of membrane–membrane interactions which differ by their EDTA-sensitivity.

Confirmation of the double nature of the adhesion system in

Present addresses: G.G., Max-Planck-Institut für Biochemie, D-8033 Martinsried bei München, Germany; S.B., McCollum Pratt Institute and Department of Biology, Johns Hopkins University, Baltimore, Md 21218, USA.

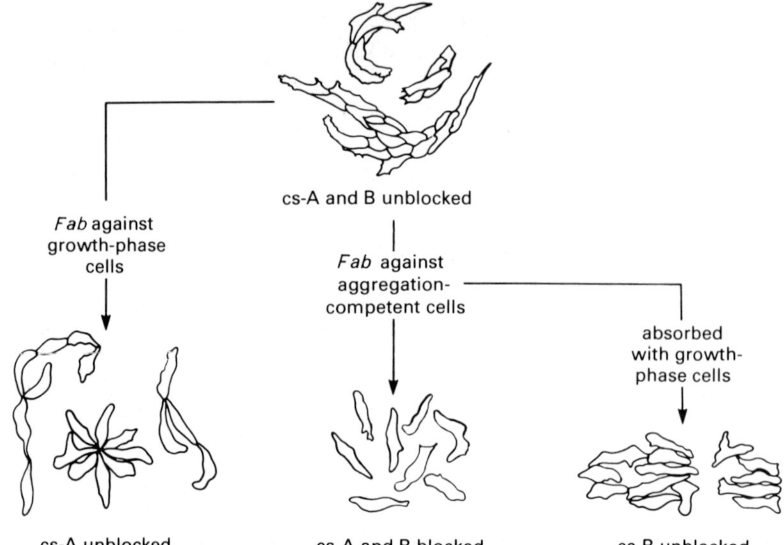

Fig. 1. Effects of adhesion-blocking *Fab* on cell assembly. During aggregation, streams of elongated cells are formed (top). Polyspecific *Fab* against membrane antigens of aggregation-competent cells completely and reversibly blocks intercellular adhesion (bottom, middle). The cells are still motile in the sense that they protrude pseudopods and change shape. Also, orientation of their long axis towards the source of chemotactic agents is still possible. There is, however, almost no translational movement, because the cell-to-substrate contact is also largely affected (Beug *et al.*, 1970).

More specific *Fab* preparations have characteristic effects on the pattern of cell assembly. Thus *Fab* against membrane antigens common to growth-phase and aggregation-competent cells blocks specifically side-by-side adhesion, allowing the cells to aggregate into chains or sometimes rosettes (left). On the other hand, *Fab* against membrane antigens which are characteristic of the aggregation stage, specifically suppresses the preference for end-to-end association (right). From Müller & Gerisch (1978) based on results of Beug, Katz & Gerisch (1973).

aggregating *D. discoideum* cells has been provided by immunochemical studies (Fig. 1) (Beug, Katz & Gerisch, 1973). Univalent antibody fragments (*Fab*) against membrane antigens of aggregation-competent cells completely block cell adhesion (Fig. 1*b*). *Fab* specific for membrane antigens of growth phase cells dissociates lateral connections without blocking end-to-end adhesion (Fig. 3*a–c*). In contrast, *Fab* against membrane antigens specific for aggregating cells selectively blocks the tight end-to-end adhesion. Cells treated with this *Fab* are still able to adhere to each other in an irregular array, often side-by-side (Fig. 3*d–f*). The *Fab* had been made specific for antigens of aggregating cells by absorption with growth cells.

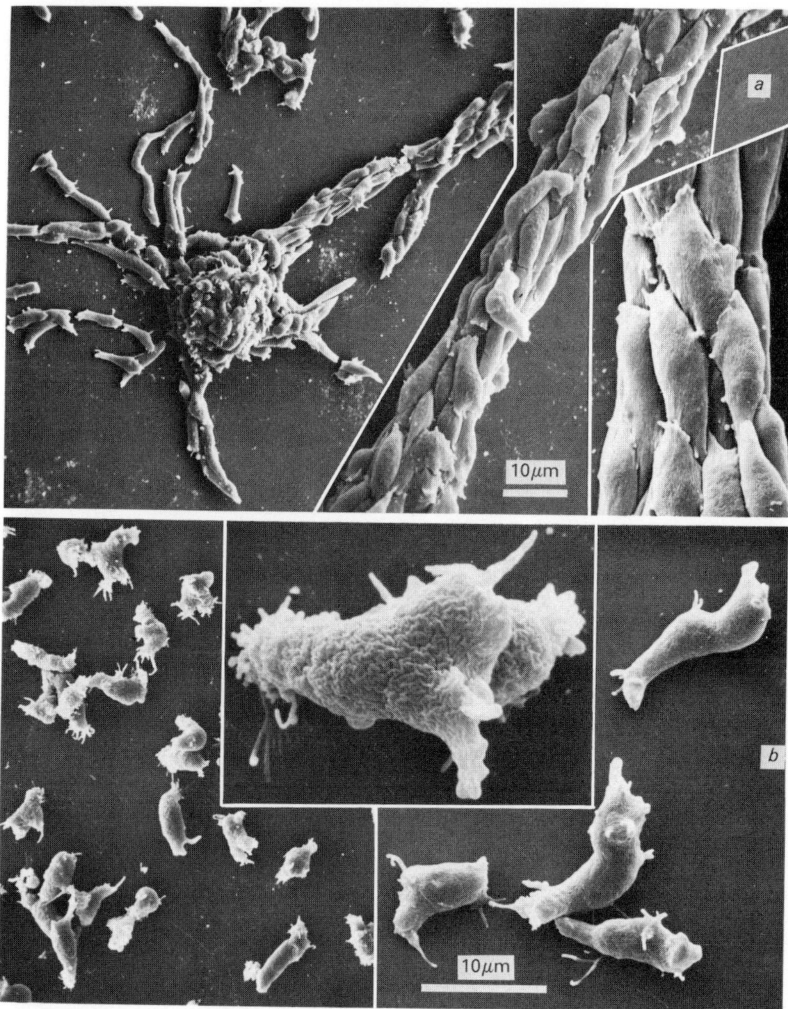

Fig. 2. Cohesion of aggregating cells and its blockage by *Fab*. (*a*) Aggregating cells and their assembly into streams by end-to-end and side-by-side contacts; (*b*) Complete dissociation of aggregation-competent cells by polyspecific anti-membrane *Fab*. The *Fab*-treated cells still retain the often elongated shape of aggregating cells and exhibit pseudopodial activity. They lose, however, not only their ability of forming intercellular adhesion, but also of attachment to glass. Exceptions are cells which have bundles of flat pseudopods closely associated with the substratum at their rear end, as shown at high-power magnification in the middle.

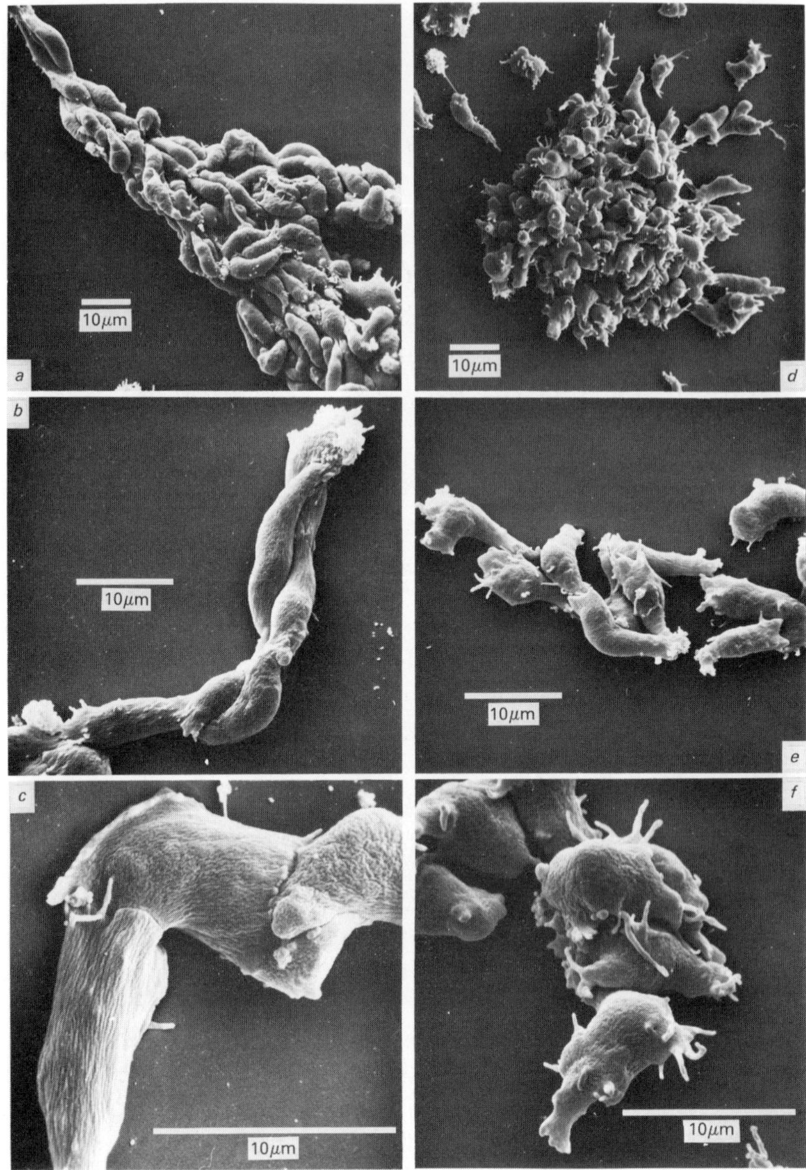

Fig. 3. Aggregation patterns of cells treated with *Fab* against membranes of growth-phase cells (*a–c*), and with *Fab* against membranes of aggregation-competent cells (*d–f*). The latter *Fab* had been absorbed with growth-phase cells in order to make it specific for developmentally regulated antigens. (*a*) Aggregate with mostly end-to-end connected cells; tight end-to-end cohesion is also shown in (*b*) and (*c*). (*d*) loose aggregate formation without a preference for end-to-end contacts. Similar loose aggregates are formed by mutants which lack contact sites A. Irregular, often side-by-side, assembly is also seen in (*e*) and (*f*). The latter shows side-by-side assembly mediated by microvilli-like extensions. The strain is v-12/M2.

The conclusion is that there are two types of target sites of adhesion-blocking *Fab*. Those already present on growth-phase cells have been called contact sites B. The other ones, expressed on aggregation-competent cells only, have been referred to as contact sites A. *Fab* blocks contact sites of either type independently of the other (Beug, Katz & Gerisch, 1973).

Contact sites A: a surface glycoprotein present in about 2×10^5 *copies per cell*

Blockage of cell adhesion is not a function of the number of *Fab* molecules bound to the cell surface. *Fab* against carbohydrate residues, mainly of glycosphingolipids (Wilhelms, Lüderitz, Westphal & Gerisch, 1974) has no effect on the activity of contact sites A, and only little on the activity of contact sites B (Beug, Katz, Stein & Gerisch, 1973). Binding of [^{3}H]-labelled *Fab* has revealed that nevertheless 2×10^6 *Fab* molecules are bound per cell. About the same number is bound in the case of polyspecific *Fab* against membrane antigens of aggregation-competent cells, completely blocking adhesion. After absorption of the *Fab* with growth-phase cells, its blocking activity against contact sites A is still preserved, although not more than 3×10^5 binding sites are exposed per cell. The *Fab* molecules bound under these conditions of complete blockage of contact sites A cover less than 2% of the total cell surface area.

Electron-microscopic studies using ferritin-labelled antibodies have confirmed that *Fab* absorbed with growth-phase cells binds to discrete, separate loci on the cell surface (Gerisch *et al.*, 1974). On the other hand, the non-blocking anti-carbohydrate *Fab* remains bound to the surfaces even at the areas of intimate adhesion at the ends of the cells, as shown by fluorescence labeling (Beug, Katz, Stein & Gerisch, 1973). This substantiates the idea that cell adhesion is brought about by the interaction of discrete sites dispersed over the cell surface. When bound to other sites on the cell surface, *Fab* molecules of 6 nm length are able to fit between contiguous cells without sterically inhibiting adhesion. Our working hypothesis is therefore that adhesion is mediated by the interaction of specific molecules which extend from adjacent cell surfaces far enough to bridge the intercellular gap. These molecules are present in relatively small number, not sufficient to form a dense surface coat.

Further work has been focused on the identification of contact sites

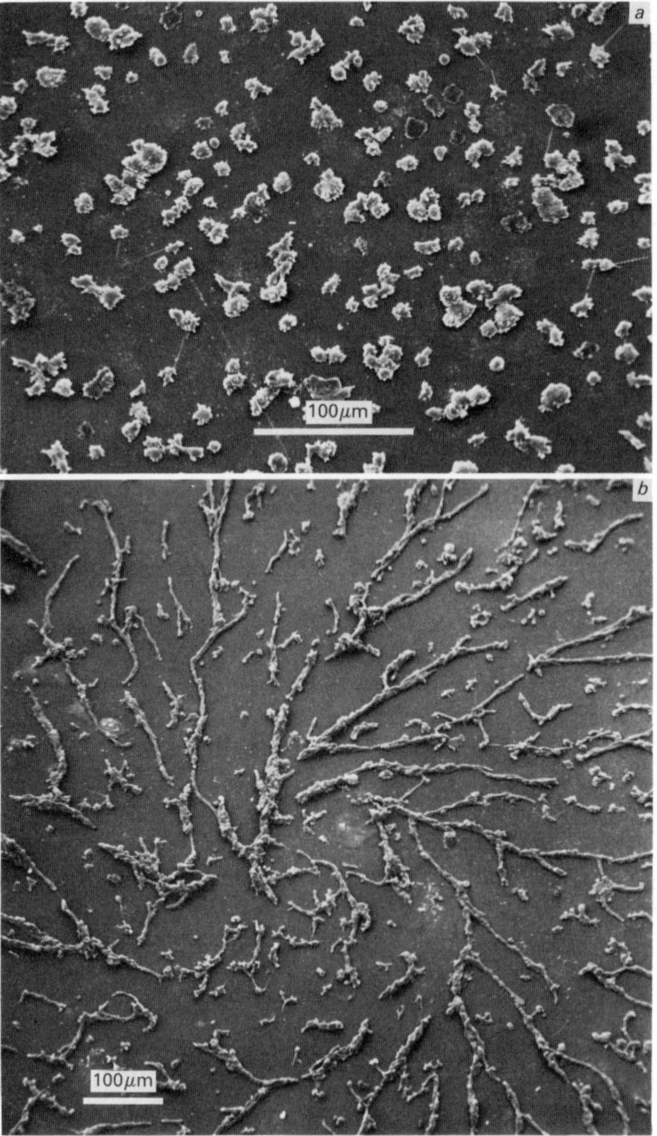

Fig. 4. Effect of cyclic-AMP pulses on the acquisition of aggregation competence. After the replacement of nutrient medium by phosphate buffer, cells of strain Ax-2 were cultivated for 4 h on a shaker either without stimulation (*a*) or under stimulation by cyclic-AMP pulses of 1×10^{-7} M amplitude, applied at intervals of 6–8 minutes (*b*). The cells were then washed and incubated on glass for 75 min. Whereas the control cells are still non-elongated and non-aggregating, the cyclic-AMP stimulated cells have become fully aggregation competent. They aggregate by chemotaxis and end-to-end adhesion into streams which, as in this case, are often arranged in a whirl-like fashion.

A, the developmentally regulated ones. Using *Fab* absorption as an assay, purification has ended up with one glycoprotein which completely absorbs the *Fab* that blocks contact sites A (Müller & Gerisch, 1978). This indicates that a single molecular species contains all the immunodeterminants of contact sites A. The purified glycoprotein is only one among about 35 different concanavalin A binding proteins in plasma membranes of aggregating *D. discoideum* cells (West & McMahon, 1977). Thus in accord with the labelling experiments discussed above, the purified material, which is immunologically identical with contact sites A, is not a major membrane component.

Contact sites A are regulated by periodic cyclic-AMP pulses

Cells of *D. discoideum* are able to release the chemotactic factor cyclic-AMP periodically in form of pulses. The periodicity of cyclic-AMP release is based on oscillatory changes of adenylate cyclase activity (Roos, Scheidegger & Gerisch, 1977). In certain strains of *D. discoideum*, NC-4 and its axenically growing derivative Ax-2, these pulses accelerate cell development from the growth phase to the aggregation-competent stage (Figs. 4 & 5). A continuous elevation of the extracellular cyclic-AMP concentration does not have the same effect as a pulsatile signal input, probably because the response system rapidly adapts to non-fluctuating concentrations of the stimulant. A continuous influx of extracellular cyclic-AMP inhibits, rather than stimulates, cell development, provided the cyclic-AMP concentrations do not exceed the nanomolar to micromolar range. A reasonable explanation is that a high cyclic-AMP background makes it difficult for the cells to detect their own cyclic-AMP pulses which stimulate development.

Acceleration of cell development by cyclic-AMP pulses is manifested in the precocious increase in activity and number of contact sites A. Activity has been tested by determining EDTA-stable cell adhesion, the number of contact sites A has been determined by *Fab* absorption and retitration of the adhesion blocking activity of the absorbed *Fab*. This can be achieved by using living cells in the presence of EDTA in order to eliminate an effect of contact sites B on the assay.

Regulation of contact site A expression by cyclic-AMP pulses is most pronounced in wild-type cells which have been removed from nutrient medium during the late stationary phase. These cells are

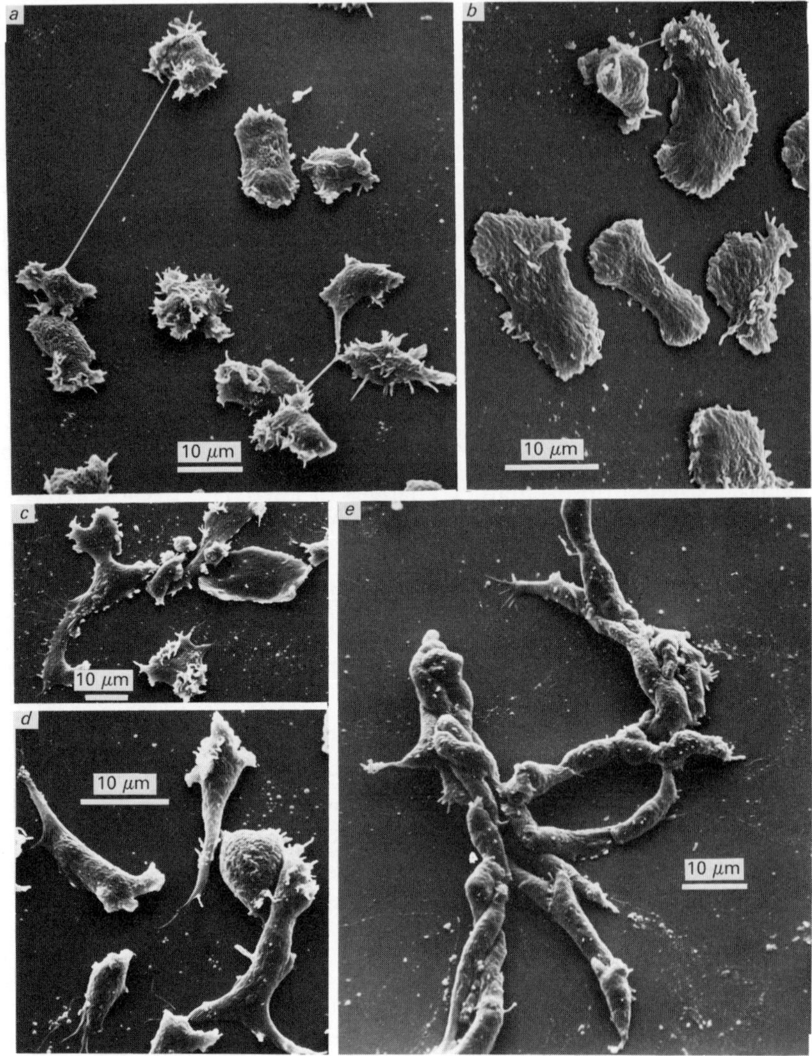

Fig. 5. Changes of cell shape during normal and cyclic-AMP stimulated development of strain Ax-2. (*a*, *b*) cells were washed free of nutrient medium, resuspended in 0.017 M phosphate buffer pH 6.0 and immediately plated on glass. *c*–*e*, the washed cells were shaken for 6 h in the buffer without (*c*, *d*) or with stimulation by cyclic-AMP pulses (*e*), then washed once again and plated on glass for 30 min. The amplitude of the cyclic-AMP pulses was 1×10^{-7} M, the period 6 min. Cells fixed shortly after the end of growth and subjected to scanning electron microscopy appear flattened and tightly adhering to the substratum (*a*, *b*). Often they are temporarily connected by long extensions (*a*). These are probably binucleate cells which divide into uninucleate ones. The beginning of shape changes associated with aggregation competence is shown in (*c*) and (*d*), where the cells are partly elongated, and contact to the substratum is diminished. At the same time, 6 h after the end of growth, cyclic-AMP stimulated cells have acquired full aggregation competence. As seen in (*e*), they aggregate into streams, leaving behind a coat on the substratum.

unable to acquire aggregation competence spontaneously. Stimulation by cyclic-AMP pulses makes them however fully aggregation competent (Gerisch, Fromm, Huesgen & Wick, 1975) (see Fig. 5). A similar effect has been obtained in certain non-aggregating mutants which are blocked in an early stage of development. These mutants also acquire aggregation competence under the influence of cyclic-AMP pulses (Darmon, Brachet & Pereira da Silva, 1975).

SIMILARITIES AND DIFFERENCES IN THE CONTACT-SITE SYSTEM OF *POLYSPHONDYLIUM PALLIDUM* AS COMPARED TO *D. DISCOIDEUM*

During aggregation, cells of *P. pallidum* completely sort out from cells of *D. discoideum*. This has prompted us to investigate the contact sites of *P. pallidum* in comparison to those of *D. discoideum*. One difference is that – at least in our hands – cell adhesion in growth-phase cells of *P. pallidum* is equally, or even more, resistant to EDTA as it is in aggregation-competent cells (Bozzaro & Gerisch, 1978). Therefore *Fab* absorption is the method of choice for the detection of stage-specific differences in the contact site system. Antibodies have been raised against aggregation-competent cells, and *Fab* has been prepared. The *Fab* completely inhibits cell adhesion in aggregation-competent cells as well as in growth-phase cells. After exhaustive absorption with growth-phase cells, only the adhesion of aggregation-competent cells is inhibited. This indicates the presence of developmentally regulated and non-regulated components of the contact site system in *P. pallidum*. Neither of these components of *P. pallidum* exerts immunological cross-reactivity with contact sites of *D. discoideum*; the adhesion-blocking activity of *Fab* is species-specific (Bozzaro & Gerisch, 1978). One target site of adhesion blocking *Fab* has been purified from *P. pallidum* membranes. Similarly to contact site A of *D. discoideum*, the material is a glycoprotein.

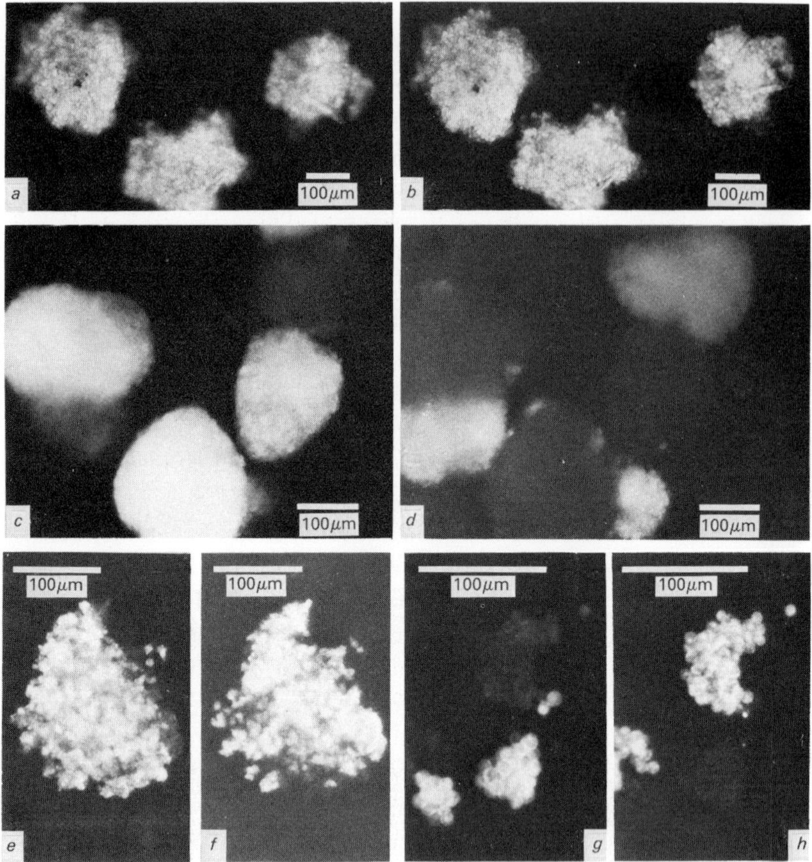

Fig. 6. Sorting out due to specific cell adhesion, and inhibition of sorting out by *Fab*. Suspended cells of *Dictyostelium discoideum* v-12/M2 and *Polysphondylium pallidum* WS 320 were mixed and gently shaken. Aggregation was stopped by the addition of Bouin's fixative (picric acid, acetic acid and formaldehyde). After careful washing, the aggregates were stained with a mixture of fluorescent-labelled antibodies against carbohydrate antigens of *D. discoideum* and *P. pallidum*. (*a*), (*c*), (*e*), (*g*) Fluorescence of TMRITC-anti-*D. discoideum* IgG; (*b*), (*d*), (*f*), (*h*) of FITC-anti-*P. pallidum* IgG.

(*a, b*) Cells of the two species were mixed immediately after the removal of nutrient bacteria, and allowed to aggregate for 1 h. No substantial sorting out is observed. (*c, d*) The same experiment performed with cells which had been mixed as single cells after 5 h of development in non-nutrient buffer. In this stage the cells also formed mixed aggregates, but they sorted out within 1 h into separate areas. (*e,f*) The same as before, but aggregated in the presence of anti-*P. pallidum* plus anti-*D. discoideum Fab* (2 plus 2 mg *Fab*). The *Fab* was specific for membrane antigens of aggregation-competent cells. Therefore it did not block the adhesion sites that are shared by growth-phase and aggregation-competent cells. This *Fab* blocked sorting out, indicating that specificity of cell recognition depended on developmentally regulated surface sites. (*g, h*) the same experiment as in (*c, d*) but with the addition of 3×10^{-5} M 2,4-dinitrophenol in order to round up the cells. The immobilised cells assembled into two distinct classes of aggregates, those containing predominantly *D. discoideum* cells, and others *P. pallidum* cells.

SPECIES-SPECIFIC SORTING OUT IS A CHARACTERISTIC OF AGGREGATION-COMPETENT CELLS AND STILL OCCURS AFTER THEIR IMMOBILISATION

When mixed cells of different species sort out during aggregation on a substratum, the mechanism that discriminates like from unlike cells can either reside in the chemotactic system or in membrane–membrane recognition associated with adhesion. Active, orientated cell movement as a mechanism of aggregate formation is eliminated in gently shaken suspensions. Under these conditions mixed cells of *D. discoideum* and *P. pallidum* form common aggregates (Fig. 6*a*, *b*), whereas they form separate ones when aggregating on agar plates. Similar results have been obtained in mixtures of other species (Raper & Thom, 1941; Shaffer, 1957; Bonner & Adams, 1958; Nicol & Garrod, 1978). This suggests that chemotaxis is one mechanism of sorting out.

Within the aggregates formed in mixed suspensions cells of *D. discoideum* and *P. pallidum* eventually sort out into separate areas (Fig. 6*c*, *d*) (Bozzaro & Gerisch, 1978). These areas are not arranged in concentric layers as has often been observed in tissue-specific sorting out of embryonal vertebrate cells, and interpreted to indicate differences in the strength rather than specificity of adhesion (Steinberg, 1975). When aggregation-competent cells of *D. discoideum* and *P. pallidum* are mixed, sorting out in suspended aggregates is observed within 1 h. At a final stage, the *D. discoideum* and *P. pallidum* areas within the aggregates often fall apart, so that eventually the suspension consists of virtually monospecific aggregates of the two species. If the same is done with growth-phase cells, sorting out is just starting after 4 h, i.e. at the time when cells of the *D. discoideum* strain v-12/M2 have acquired aggregation competence. The conclusion is that adhesion of growth-phase cells is non-selective.

Sorting out of aggregation-competent cells can be due to chemotactic movement of the cells within the aggregates, or to a developmentally regulated type of cell adhesion, or to both. Chemotaxis is much stronger in aggregation-competent cells than in growth-phase cells (Bonner *et al.*, 1972). In order to rule it out, cells have been immobilised by 2,4-dinitrophenol. At a critical concentration of $3–5 \times 10^{-5}$ M, 2,4-DNP completely and reversibly rounds up the cells without blocking cell adhesion (Gerisch, 1962). Under these conditions the surface of the cells is smooth; extensions like microvilli or filopodia are absent. When mixed in the presence of 2,4-DNP and gently shaken, aggregation-competent cells assemble into two kinds of groups, those containing a large majority of *D. discoideum* cells, and others containing almost exclusively *P. pallidum* cells (Fig. 6*g*, *h*). This result clearly demonstrates specificity of adhesion. It extends previous demonstrations of the unnecessity of chemotaxis for specific cell interactions in mixtures of different *Dictyostelium* species (Springer & Barondes, 1978). Under our conditions, any contribution of shape changes or spreading of cells on each other is eliminated. Growth-phase cells rounded up by 2,4-DNP still form loose aggregates but, as expected, no significant segregation of the two species is observed.

The above results suggest that the two types of cell adhesion – one active in growth-phase cells, the other only in aggregation-competent cells – strongly differ in their power of discrimination between like and unlike cells. Only adhesion attributed to contact sites A, and to their developmentally regulated counterparts on *P. pallidum* cells, appears to be associated with cell recognition. This conclusion is substantiated by the effects of *Fab* on cell sorting out. Aggregation competent cells have been preincubated with *Fab* of different specificities. For this purpose polyspecific *Fab* against membrane antigens of aggregation competent cells of *D. discoideum* or *P. pallidum* has been absorbed with homologous growth-phase cells. In the presence of the mixed *Fab* the cells still clump in suspension. This occurs because contact sites B and their equivalents on *P. pallidum* cells are unblocked. Sorting out within the cell groups is, however, substantially inhibited by the *Fab* which binds to membrane antigens typical of aggregation-competent cells (Fig. 6*e*, *f*). Controls have been performed by the use of *Fab* mixtures against carbohydrate residues of both species. These *Fab* molecules bind to the cell surfaces without blocking adhesion. They also do not detectably inhibit sorting out.

ARE CONTACT SITES RELATED TO CELL SURFACE LECTINS, AND HOW DO THEY FUNCTION?

On the surface of aggregating *D. discoideum* cells two lectins appear to be exposed, discoidin I and II. Similarly, aggregating *P. pallidum* cells possess two major lectins, pallidin I and II (Frazier, Rosen,

Reitherman & Barondes, 1976). The lectins of various species differ to some extent in their sugar specificity (Rosen, Reitherman & Barondes, 1975). It has been suggested that these lectins are implicated in adhesion and in species-specific cell recognition (Rosen & Barondes, 1978).

The lectins are not identical with the hitherto-purified target sites of adhesion-blocking *Fab*. The glycoprotein with contact site A specificity isolated from *D. discoideum* has been obtained in a state free of any discoidin (Müller & Gerisch, 1978). Similarly, a glycoprotein from *P. pallidum* which acts as a target site of adhesion-blocking *Fab* has been purified free of pallidins. Non-identity of contact sites A and discoidin is also indicated by the absence of a significant effect of cAMP pulses on the amount of discoidin, under conditions where cAMP pulses substantially increase contact sites A activity (Rossier, Eitle, van Driel & Gerisch, 1980).

The question is, then, whether contact sites interact with the lectins. In principle, the carbohydrate moieties of contact sites could act as lectin receptors. This has been tested for contact sites A, but hitherto evidence for their binding to one of the discoidins has not been obtained. Other attempts to relate the lectins to contact sites have been, in our hands, also unsuccessful. *Fab* against the pallidins only slightly inhibits adhesion of *P. pallidum* cells (Bozzaro & Gerisch, 1978). Similarly, in *D. discoideum*, no significant inhibition of cell adhesion by *Fab* against discoidin I and II has been observed. The same *Fab* preparations inhibit lectin-mediated erythrocyte agglutination, indicating that the bound *Fab* molecules interfere with the carbohydrate-binding activity of the lectins. Also, neither of the lectins detectably absorb adhesion blocking *Fab*. Finally, sorting out of *D. discoideum* and *P. pallidum* cells is still observed after preincubation of aggregation competent cells with a *Fab* mixture against the discoidins and pallidins.

Contact sites have been operationally defined as the target sites of adhesion blocking *Fab*, and supporting evidence has been accumulated for a participation of these sites in cell adhesion. Nevertheless, the way they act is unknown. The main question is whether the contact sites act as regulatory components of the adhesion system, or actually establish the links between contiguous cells (for a discussion see Gerisch, 1980). If the latter is true, it remains to be clarified whether cell linkage occurs by transmembrane dimer formation two identical contact site molecules, or by the interaction of contact sites with an as yet unidentified receptor. The fact that the two purified contact sites are glycoproteins raises another still unsolved question, that of a role of their carbohydrate moieties in cell interactions.

Acknowledgments. Our work has been supported by the Schweizerischer Nationalfonds and the Deutsche Forschungsgemeinschaft. We thank Mr J. Beltzer and Hp. Giuliani for valuable assistance in scanning electron microscopy.

REFERENCES

Beug, H., Gerisch, G., Kempff, S., Riedel, V. & Cremer, G. (1970). Specific inhibition of cell contact formation in *Dictyostelium* by univalent antibodies. *Experimental Cell Research*, **63**, 147–58.

Beug, H., Katz, F. E. & Gerisch, G. (1973). Dynamics of antigenic membrane sites relating to cell aggregation in *Dictyostelium discoideum*. *Journal of Cell Biology*, **56**, 647–58.

Beug, H., Katz, F. E., Stein, A. & Gerisch, G. (1973). Quantitation of membrane sites in aggregating *Dictyostelium* cells by use of tritiated univalent antibody. *Proceedings of the National Academy of Sciences, USA*, **70**, 3150–4.

Bonner, J. T. & Adams, M. S. (1958). Cell mixtures of different species and strains of cellular slime moulds. *Journal of Embryology and Experimental Morphology*, **6**, 346–56.

Bonner, J. T., Hall, E. M., Noller, S., Oleson, F. B., Jr. & Roberts, A. B. (1972). Synthesis of cyclic AMP and phosphodiesterase in various species of cellular slime molds and its bearing on chemotaxis and differentiation. *Developmental Biology*, **29**, 402–9.

Bozzaro, S. & Gerisch, G. (1978). Contact sites in aggregating cells of *Polysphondylium pallidum*. *Journal of Molecular Biology*, **120**, 265–79.

Darmon, M., Brachet, P. & Pereira da Silva, L. H. (1975). Chemotactic signals induce cell differentiation in *Dictyostelium discoideum*. *Proceedings of the National Academy of Sciences, USA*, **72**, 3163–6.

Frazier, W. A., Rosen, S. D., Reitherman, R. W. & Barondes, S. H. (1976). Multiple lectins in two species of cellular slime molds. In *Surface Membrane Receptors*, ed. R. A. Bradshaw, W. A. Frazier, R. C. Merrel, D. I. Gottlieb & R. A. Hogue-Angeletti, Nato Advanced Study Institutes, pp. 57–66. London, New York: Plenum Press.

Gerisch, G. (1961). Zellfunktionen und Zellfunktionswechsel in der Entwicklung von *Dictyostelium discoideum*. V. Stadienspezifische Zellkontaktbildung und ihre quantitative Erfassung. *Experimental Cell Research*, **25**, 535–54.

Gerisch, G. (1962). Zellfunktionen und Zellfunktionswechsel in der Entwicklung von *Dictyostelium discoideum*. VI. Inhibitoren der Aggregation, ihr Einfluss auf Zellkontaktbildung und morphogenetische Bewegung. *Experimental Cell Research*, **26**, 462–84.

Gerisch, G., Beug, H., Malchow, D., Schwarz, H. & v.Stein, A. (1974). Receptors for intercellular signals in aggregating cells of the slime mold,

Dictyostelium discoideum. In *Biology and Chemistry of Eucaryotic Cell Surfaces*, ed. E. Y. C. Lee & E. E. Smith, Miami Winter Symposium, vol. 7, pp. 49–66. London, New York: Academic Press.

GERISCH, G., FROMM, H., HUESGEN, A. & WICK, U. (1975). Control of cell-contact sites by cyclic AMP pulses in differentiating *Dictyostelium* cells. *Nature, London*, **255**, 547–9.

GERISCH, G. (1980). Univalent antibody fragments as tools for the analysis of cell interactions in *Dictyostelium*. In *Current Topics in Developmental Biology*, ed. M. Friedlander, A. Monroy & A. A. Moscona, in press. London, New York: Academic Press.

MÜLLER, K. & GERISCH, G. (1978). A specific glycoprotein as the target site of adhesion blocking Fab in aggregating *Dictyostelium* cells. *Nature, London*, **274**, 445–9.

NICOL, A. & GARROD, D. R. (1978). Mutual cohesion and cell sorting-out among four species of cellular slime moulds. *Journal of Cell Science*, **32**, 377–87.

RAPER, K. B. & THOM, C. (1941). Interspecific mixtures in the *Dictyosteliaceae*. *American Journal of Botany*, **28**, 69–78.

ROOS, W., SCHEIDEGGER, C. & GERISCH, G. (1977). Adenylate cyclase activity oscillations as signals for cell aggregation in *Dictyostelium discoideum*. *Nature, London*, **266**, 259–61.

ROSEN, S. D. & BARONDES, S. H. (1978). Cell adhesion in the cellular slime molds. In *Receptors and Recognition*, ed. D. R. Garrod, Series B, vol. 4, pp. 235–64. London: Chapman and Hall.

ROSEN, S. D., REITHERMAN, R. W. & BARONDES, S. H. (1975). Distinct lectin activities from six species of cellular slime molds. *Experimental Cell Research*, **95**, 159–66.

ROSSIER, C., EITLE, E., VAN DRIEL, R. & GERISCH, G. (1980). Biochemical regulation of cell development and aggregation. In *Dictyostelium discoideum. The Eukaryotic Microbial Cell*. 30th Symposium of the Society for General Microbiology, ed. G. W. Gooday, D. Lloyd & A. P. J. Trinci. Cambridge: Cambridge University Press.

SHAFFER, B. M. (1957). Aspects of aggregation in cellular slime moulds. I. Orientation and chemotaxis. *American Naturalist*, **91**, 19–35.

SPRINGER, W. R. & BARONDES, S. H. (1978). Direct measurement of species-specific cohesion in cellular slime molds. *Journal of Cell Biology*, **78**, 937–42.

STEINBERG, M. S. (1975). Adhesion-guided multicellular assembly: a commentary upon the postulates, real and imagined, of the differential adhesion hypothesis, with special attention to computer simulations of cell sorting. *Journal of Theoretical Biology*, **55**, 431–43.

WEST, C. M. & MCMAHON, D. (1977). Identification of concanavalin A receptors and galactose-binding proteins in purified plasma membranes of *Dictyostelium discoideum*. *Journal of Cell Biology*, **74**, 264–73.

WILHELMS, O.-H., LÜDERITZ, O., WESTPHAL, O. & GERISCH, G. (1974). Glycosphingolipids and glycoproteins in the wild-type and in a non-aggregating mutant of *Dictyostelium discoideum*. *European Journal of Biochemistry*, **48**, 89–101.

Developmentally regulated lectins in slime moulds and chick tissues–are they cell-adhesion molecules?

SAMUEL H. BARONDES M.D.

Department of Psychiatry, University of California, San Diego, La Jolla, California 92093, USA

Molecules that bind cells to other cells or to extracellular materials are often referred to as cell-adhesion molecules. Very little is known about them. Evidence I will present indicates that some cell adhesion molecules may be lectins. These polyvalent carbohydrate-binding proteins can be assayed as agglutinins of test erythrocytes that display complementary oligosaccharides on their surface. Their ability to agglutinate test cells suggests that they may bind appropriate cells together *in vivo*. The purpose of this presentation is two-fold: (1) to consider approaches to identification of cell-adhesion molecules; (2) to review studies of lectins in cellular slime moulds and chick tissues which implicate them as possible cell-adhesion molecules.

SOME GENERAL CONSIDERATIONS ABOUT CELL-ADHESION MOLECULES

Because of the versatility of proteins it is generally assumed that many cell-adhesion molecules are proteins, and that cell adhesion is mediated by interactions between a cell surface protein and some type of complementary cell surface receptor. It is also assumed that these interactions are of the same type as those mediating the binding of ligands to proteins in solution. Some common conceptions of this are illustrated in Fig. 1. In cases *a–c* (Fig. 1) the interacting cells are identical, but in case (*d*) they are not. Note also that the interacting molecules might be identical or different. Both interacting components could be proteins. If so they could be integral membrane proteins firmly rooted within the lipid bilayer (Fig. 1 *a*, *b*, *d*) or peripheral membrane proteins, like the complementary bivalent ligand shown in Fig. 1 *c*. It is easy to see how the latter, if isolated, could act as an

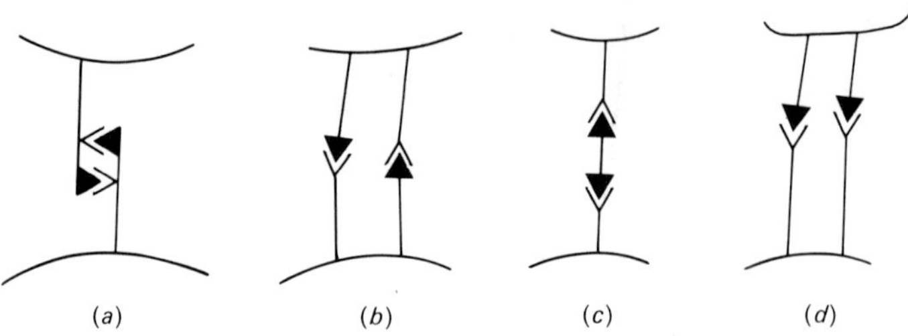

Fig. 1. Models of cell-adhesion molecules binding the surfaces of adjacent cells. (*a*) Single cell-adhesion molecule containing two different and complementary sites. (*b*) Two distinct complementary cell-adhesion molecules both bound in the lipid bilayer of the cell surface. (*c*) Two distinct complementary cell-adhesion molecules, one of which is outside the lipid bilayer. (*d*) As in (*b*), but each interacting cell contains only one type of cell-adhesion molecule.

agglutinin of any test cells which contained cell surface receptors resembling those to which it binds normally. In contrast, the ligands in cases *a*, *b* and *d* are depicted as univalent. If isolated in a univalent form they would not act as agglutinins (lectins). However, it remains possible that integral membrane proteins like those shown are bivalent and could agglutinate test cells after their solubilisation. It is also possible that univalent hydrophobic proteins might associate, after isolation, into bivalent aggregates.

The adhesive interactions shown in Fig. 1, although simple, could provide the basis for complex patterns of selective cellular association. For example, simply varying the number of cell-adhesion molecules on the cell surface or their time of appearance in development could provide some selectivity. Furthermore, the degree to which the specific molecules are ordered or clustered in the membrane could be of importance. Although it is possible that cell-adhesion molecules play no role in the specific cellular associations generally referred to as cellular recognition, it is easy to see how they could be employed for this purpose.

CRITERIA FOR IDENTIFICATION OF CELL-ADHESION MOLECULES

The first task in attempting to find cell-adhesion molecules is to establish criteria for their identification. Unfortunately, the phenomenon under consideration, binding of a cell to another cell or to an extracellular material, involves many more variables and is more complex than the association of two molecules in solution. For

Table 1. *Expected characteristics of (criteria for) a cell-adhesion molecule (CAM); and false positives and negatives*

Expected characteristics of CAM	Reasons for: False positive	False negative
Located on cell surface	Coincidental	Overlooked Inaccessible
Appearance of CAM correlates with development of adhesiveness	Coincidental	CAM present before adhesiveness but other factor is limiting
Complementary receptor on cell surface (if CAM is isolated as polyvalent molecule it may agglutinate test cells)	Binding has other function or is non-specific	CAM binds poorly when solubilised Receptors saturated
Binding univalent antibodies or haptens to CAM blocks adhesion	Non-specific effect	Affinity of univalent antibodies or haptens for CAM relatively low
Mutant with impaired cell adhesion has defective CAM	Indirect effect	Defect in CAM not detected by the measurements used

molecular analysis it may be convenient to consider that cells are billiard balls coated with varying amounts of adhesive material of varying degrees of mutual affinity. But cells are obviously more complex. Some characteristics that distinguish cells from such a model include: (1) irregular and changing shape; (2) motility, which may be influenced by positive and negative chemotactic systems; (3) continuous reorganisation of membrane structure in response to various cues. Failure to acknowledge these properties in evaluating actual experiments on cell-adhesion molecules may lead to serious misinterpretation.

Some commonly used criteria for identification of cell-adhesion molecules are listed in Table 1. Each of these criteria is compatible with all the models shown in Fig. 1 as well as more elaborate ones. A major point of this tabulation is to indicate that, in the experimental application of each one of these criteria, both false positive and false negative results can be obtained. For example, let us consider the most obvious criterion of a cell-adhesion molecule, the fact that it must be present on the cell surface. Clearly it is possible to have false positives. By that I mean that establishing that a particular molecule is present on the cell surface is not definitive evidence that it plays a role in cell adhesion. On the other hand the apparent absence of a molecule from the cell surface cannot be taken as incontrovertible evidence that it is not a cell-adhesion molecule. There are a number of ways in which this could come about. For example, the molecule might be relatively scarce compared with other cell surface molecules and could be overlooked by certain types of assays such as iodination of cell surface proteins with the lactoperoxidase technique. The molecule in question could also be hidden from such reagents, or others, such as antibodies raised against this molecule in a soluble form. It might be capable of interacting as a cell-adhesion molecule in this cryptic state – or it might become more exposed after a preliminary cellular interaction of another type. The point of this discussion is that failure to find a molecule on the cell surface could be a false negative in that it does not preclude its being a cell-adhesion molecule; and the presence of a molecule on the cell surface could be a false positive in that many cell surface molecules do not play a role in cell adhesion.

False positive and false negatives are also possible with each of the other criteria listed in Table 1. For example, a correlation between the time of synthesis of a putative cell-adhesion molecule during differentiation and the time of development of cellular adhesiveness could be coincidental. On the other hand, it is possible that the putative cell-adhesion molecule is present long before the cell converts from an non-adhesive to an adhesive form. One way this could come about is if the complementary substance with which it binds is the limiting factor – either its synthesis or its appearance on the cell surface. Another possibility is that some general property of the cell membrane is the limiting factor. Therefore, failure to demonstrate a correlation between the appearance of adhesiveness and the appearance of the putative cell-adhesion molecule could be a false negative

There could also be false positives and false negatives with another relatively straightforward criterion – the presence of complementary cell surface receptors for the putative cell-adhesion molecule. Demonstrating such receptors could be a false positive in that the molecule–receptor complex might be doing something other than binding cells together. For example, hormones bind to cells, triggering specific reactions – and yet hormones are not cell-adhesion molecules. Even if the putative cell-adhesion molecule is polyvalent and binding leads to agglutination of the cells, this could be a false positive. For example, basic proteins normally associated with nucleic acids can agglutinate cells by binding to acidic residues on the cell surface; but it is unlikely that these are cell-adhesion molecules. Failure to demonstrate binding of a putative cell-adhesion molecule to a cell or agglutination of cells by such molecules could also be misinterpreted. The molecule might bind very poorly once solubilised, since it might be in a somewhat denatured form. Another possibility is that receptors present on the cell surface of the cell are already saturated with cell-adhesion molecules, so that adding more does not lead to measurable binding.

The requirement that reagents that bind cell-adhesion molecules should block the adhesion reaction is also subject to misinterpretations. The reagents, be they univalent antibodies or specific ligands, could produce a false positive by non-specifically affecting cellular interactions. Or there might be a false negative, since the affinity of the antibodies or ligands might be so much lower than that of a normal complementary cell surface receptor that there is little competition and no measurable block in adhesion. Finally, even studies of mutants which show abnormal cell adhesion are subject to misinterpretation. A false positive could be produced if a mutation affected a critical step in differentiation which in turn blocked the appearance of both the cell-adhesion molecule and cell adhesion. Even a point mutation in the putative cell-adhesion molecule that is correlated with impaired adhesion of the mutant cells could be a false positive in that the effect could be indirect. By this I mean that the affected molecule may not directly bind cells together but might influence some other process which directly mediates the adhesion. A mutant cell could show impaired cell adhesion and could appear to have normal putative cell-adhesion molecules (false negative) since the molecular abnormality in these molecules might not be detectable by the measurement technique that is used. Only determination of the complete structural sequence of the potential cell-adhesion molecule might turn up the defect.

How then are we to conclusively identify cell-adhesion molecules? More specifically how will we be able to decide whether the candidates under consideration in this paper, lectins, are cell-adhesion molecules? Since satisfying any single criterion is not convincing, multiple experimental approaches are required. Even then interpretations must be made with caution. They should be heavily dependent on estimates of the validity of the cell-adhesion assays employed for many of these studies.

CRITERIA FOR AN ASSAY OF SPECIFIC CELL ADHESION

An important prerequisite for studying cell-adhesion molecules is to establish some quantitative measurement of cell adhesion which is valid and 'meaningful'. By this I mean that cell adhesion measured in the assay should be closely related to the cell adhesion which occurs *in vivo*. Establishing valid assays is probably the major stumbling block in studies of cell adhesion. In many cell-adhesion assays, a comparison is made either of the rate of binding of cells to each other or of their affinity for each other, calculated in various ways. In other assays, measurements are made of binding of cells to artificial surfaces or surfaces coated with biochemicals. Such assays have the merit of convenience and reproducibility. However, it is difficult to know the relationship between the parameters they measure and cell adhesion *in vivo*. How does binding in a gyrated suspension relate to binding *in vivo*? How can one decide if the affinity of a cell for a plastic dish is a measure of its adhesiveness *in vivo*? The fact that adhesion can be measured in these circumstances provides us with little confidence that it is related to the behaviour of the cells in the organism.

Some criteria which may be used in attempting to determine if a cell-adhesion assay is valid are listed in Table 2. Each criterion is the expectation of a correlation between adhesion as measured in the assay and some biological function of the cell. For example, if cells of two species of slime moulds segregate when mixed and form distinct colonies, then the demonstration of species-specific adhesion under the assay conditions is evidence that the assay is meaningful. By the same reasoning, if mixtures of liver cells and brain cells stick preferentially to like cells under the assay conditions, this is taken to mean that there is tissue specificity which validates the assay. Both species specificity and tissue specificity are frequently employed criteria. Yet their demonstration is not unequivocal evidence that the assay is valid. For example, in comparing two cell populations one might just happen to have properties, not normally related to cell adhesion, which make it relatively more adhesive in this assay. These

Table 2. *Criteria for a valid assay of cell adhesion*

Adhesion is
Species specific
Tissue specific
Specific to a certain stage of differentiation
Blocked by reagents that bind specific cell-adhesion molecules

properties might play no role in cell adhesion *in vivo*. One way of making this criterion more stringent is rank ordering of many cell types in which segregation in the assay is correlated with degree of segregation *in vivo*. Yet, interpretation of such studies may be difficult, since most tissues such as liver, brain, and muscle are not called upon to sort out in the same way *in vivo* as *in vitro*.

Developmental stage specificity is another criterion that has some value. For example, vegetative cellular slime mould cells observed in their natural environment show no association, whereas partially differentiated cells associate avidly. As Gerisch (1968) showed, experimental conditions can be devised such that vegetative slime mould cells associate poorly but partially differentiated cells associate well. This may be taken as evidence that the cell–cell adhesion observed between partially differentiated cells in such assays is biologically significant. It remains quite possible, however, that the adhesion measured in this assay is due to a property of the cell surface which appears with differentiation but which normally has nothing to do with cell–cell association; and the important cell-adhesion property might not be assayed under the experimental conditions. Thus, it remains possible that the criterion of developmental stage specificity, like the others already discussed, could lead to misinterpretations. The criterion of blockade by specific reagents could also be misinterpreted. For example consider the finding that antibodies (or their univalent derivatives) raised against a cell surface component blocks the cell–cell adhesion observed in a given assay. This does not necessarily validate the assay, since binding antibodies to cell surface molecules not normally involved in cell adhesion might be inhibitory. This objection might hold even if the cell surface antigen is tissue specific or specific for some stage in development.

Again we are left with the formidable problem posed by the complexity of cell adhesion. Although assays of cell adhesion are available, interpretations based on these assays must be made with caution.

STUDIES WITH CELLULAR SLIME MOULDS

Given the complexities of the problem of cell adhesion, it is particularly important to choose a favourable system for its analysis. The cellular slime moulds are becoming a popular choice for this work. To understand why it is first necessary to consider the life cycle of these simple eukaryotic cells (for details see Loomis, 1975). In the presence of bacteria which they feed upon, slime mould cells exist as unicellular amoebae which divide approximately every three hours. In this condition the cells are called vegetative cells; and this part of their life cycle may be called the non-social phase, since the cells show no tendency to associate. As long as there is ample food available and other conditions are favourable the amoebae remain in this state. However, when the food supply is exhausted the amoebae differentiate into aggregation-competent cells over the course of about 9–12 h. The cells then aggregate, in response to pulses of chemotactic agents, to form a multicellular structure containing up to 10^5 cells. In the ensuing 12 h this aggregate further differentiates into a fruiting body. About 20% of the cells in the aggregate become stalk cells and the rest become spore cells. The spore cells are in a dormant state, resistant to unfavourable environments. However, if the spores are exposed to a favourable environment they germinate and amoebae emerge and begin a new life cycle.

The cellular slime moulds therefore have a number of advantages for studying cell adhesion: (1) cells can be isolated in a non-adhesive form and at various stages of development of cell–cell adhesion; (2) a large number of cells at an identical stage of development can be raised in culture so that considerable material is available for biochemical studies; (3) the culture conditions are simple and resemble the natural environment of these organisms. Therefore slime mould cells show normal cell-association properties under defined conditions where cells from higher organisms might lose them; (4) there are a number of species of cellular slime moulds that have been shown to display species-specific cellular association raising the possibility of studying species-specific cell–cell adhesion (Raper & Thom, 1941; Bonner & Adams, 1958). In this section I will describe some properties of one assay that we use to study cell–cell adhesion in cellular slime moulds; and then consider the evidence that developmentally regulated lectins mediate cell adhesion in these organisms.

Measuring cell adhesion in slime moulds

Slime moulds show changes in cell association with differentiation and also show species-specific segregation. These characteristics provide two criteria for defining a meaningful assay of cell–cell adhesion. In one assay we have used, slime moulds are dissociated into single cells by vortexing and are permitted to adhere to each other by gyratory shaking under defined conditions (McDonough, Springer, & Barondes, 1979). The degree of adhesion can be measured in several ways. One method we use is to determine the number of single cells in the suspension at the outset of the experiment and at various times thereafter, with an electronic particle counter calibrated to record only particles with the dimension of one cell. When cells adhere they are no longer recorded as a particle by the counter, and the percentage of single cells that are no longer detected may be referred to as per cent agglutination. In Fig. 2 we compare per cent agglutination of vegetative cells and cells derived from loose aggregates, here referred to as 'cohesive cells'. The assay was performed at two gyration speeds and in the presence or absence of EDTA. High speeds of gyration shear cells apart and are presumed to be a measure of strength of adhesion. EDTA acts by an unknown mechanism, perhaps chelation of divalent cations important for adhesion. It has been used in many cell-adhesion assays and was shown by Gerisch (1968) to preferentially inhibit adhesion of vegetative slime mould cells. We have confirmed this, in that appropriate conditions can be selected under which the vegetative cells show no measurable agglutination but the cohesive cells show substantial agglutination (Fig. 2). Note that if this criterion were not used, conditions might be used which measured the type of cell–cell adhesion shown by vegetative cells which might have no biological significance.

Given the fact that this assay measures a cell–cell adhesion which is based on a developmentally regulated property of the cells, we then tested the possibility that the adhesiveness measured was also species-specific. To study this we chose two species of cellular slime moulds, *Dictyostelium discoideum* and *Dictyostelium purpureum*, which are

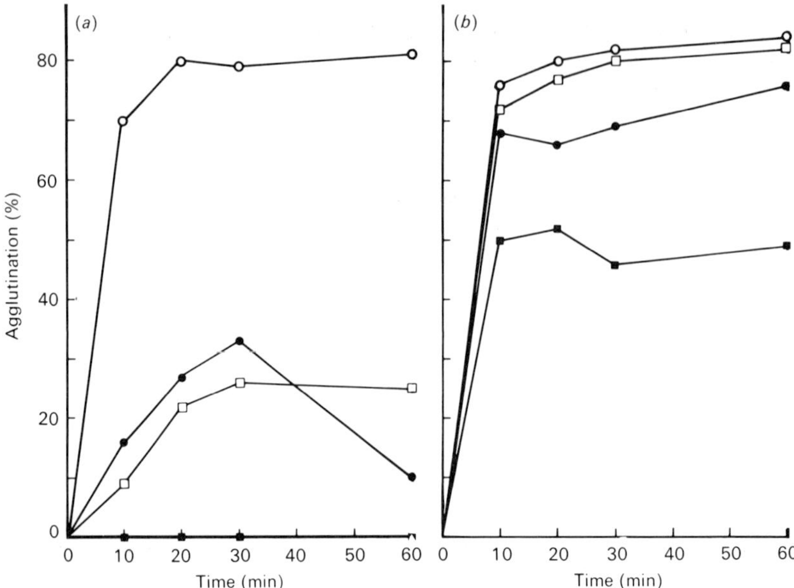

Fig. 2. Cell–cell adhesion, measured as per cent agglutination of (*a*) vegetative or (*b*) partially differentiated *D. discoideum* cells, observed under four assay conditions. Cells were harvested from bacterial growth plates after 40–48 h, washed with cold distilled water and either assayed immediately (vegetative cells) or differentiated on moist filter-pads until they were streaming into aggregates (cohesive cells) then harvested and assayed. Points are means from duplicates in two separate experiments. Conditions used were: gyration at 115 r.p.m. in dilute buffer (○——○); gyration at 115 r.p.m. in dilute buffer containing 10 mM EDTA (●——●); gyration at 200 r.p.m. in dilute buffer (□——□); gyration at 200 r.p.m. in dilute buffer containing 10 mM EDTA (■——■). (For details see McDonough *et al.*, 1979).

known to segregate completely when mixed together and allowed to differentiate (Raper & Thom, 1941; Bonner & Adams, 1978). We used cells of each species, which were partially differentiated and which showed cell–cell adhesion under the conditions shown in Fig. 2. We labelled one species green with fluorescein isothiocyanate and the other species red with tetramethylrhodamine isothiocyanate, mixed the cells and observed the agglutinates in the microscope with epifluorescent illumination. Under the conditions that we used we could readily determine the fluorescent colour of each individual cell in the agglutinates. We found that under the conditions employed the cell–cell adhesion was species-specific. Table 3 summarises the results of one such experiment. The results have been repeatedly confirmed with these and other species of cellular slime moulds.

Table 3. *Composition of mixed aggregates of labelled* D. discoideum *and* D. purpureum *compared with expectations if there were no species-specific cell adhesion*

Cells per aggregate	Number of aggregates of this size found	Mixed aggregates expected (%)	Number of mixed aggregates	
			Expected	Found
2	18	36	6	2
3	11	53	6	1
4	6	65	4	0
5	4	73	3	0
6	1	79	1	0
> 6	10	$\geqslant 84$	8	0
Total	50	56	28	3

Aggregates were formed by gyration of partially differentiated *D. discoideum* and *D. purpureum* for 10 min in buffer containing EDTA. The *D. discoideum* cells were labelled with fluorescein isothiocyanate and the *D. purpureum* cells were labelled with tetramethylrhodamine isothiocyanate. Fifty aggregates were observed, and the total number of cells per aggregate and the number of red and green cells was determined. A mixed aggregate contains at least one cell of each type. Per cent mixed aggregates expected was determined by using a binomial expansion corrected for the difference in adhesiveness of the two cell types as follows: After recording fifty aggregates, the fraction of cells of each type in all fifty aggregates was determined and used in the formula $f = 1-(a^n+b^n)$, where $f =$ the fraction of mixed cells expected if both cell types adhere equally well to each other, $a =$ the fraction of cell type a found in the total fifty aggregates, $b =$ the fraction of cell type b found in the total fifty aggregates, and $n =$ the aggregate size. For example, if equal numbers of cells were found in the fifty aggregates one would expect $1-(0.5^2+0.5^2) = 0.5$ or half of the aggregates containing two cells to be mixed; $1-(0.5^3+0.5^3) = 0.75$ or three quarters of the three-cell aggregates to be mixed and so on. From this fraction, and knowing the number of aggregates found in each class, we can calculate the number of mixed aggregates expected if no selective adhesion is involved. In the present case 77% of the cells in the fifty aggregates were *D. discoideum* and 23% were *D. purpureum*. The probability of the observed distribution of mixed aggregates being the same as the expected distribution was determined by calculating Chi-square. Therefore, the smaller p is the more likely it is that the cell adhesion is species-specific. In the present case $p < 0.005$. For details see McDonough, Springer & Barondes, (1979).

We conclude that the assay conditions that we employ measure developmentally regulated and species-specific cell cohesion, fairly strong evidence that the assay is measuring meaningful cell–cell adhesion.

Evidence that developmentally regulated lectins are cell-adhesion molecules in slime moulds

Extracts of cohesive forms of cellular slime moulds contain proteins that can be isolated in soluble form and assayed as agglutinins of erythrocytes (Rosen, Kafka, Simpson & Barondes, 1973). The agglutinins are carbohydrate-binding proteins, hereafter referred to as lectins, since agglutination can be blocked by specific sugars. Extracts of vegetative cellular slime moulds contain little or no detectable lectin activity. Lectins from a number of species of cellular slime moulds have been purified (Barondes & Haywood, 1979). In the case of *D. discoideum*, the species most carefully studied, two discrete lectins are synthesised with differentiation (Frazier, Rosen, Reitherman & Barondes, 1975). Each has been purified to homogeneity and is a homotetramer with subunit molecular weights of either 24,000 or 26,000. Other species also apparently contain more than one lectin (Barondes & Haywood, 1979). The lectins from all the species that have been studied are similar proteins but have discriminable physicochemical properties. The relative inhibitory potency of a number of simple saccharides on their hemagglutination activity is similar but not identical (Barondes & Haywood, 1979).

The developmentally regulated lectins in slime moulds are present on the cell surface. This has been shown in a number of ways. For example, antibodies raised against the purified proteins bind to the surface of partially differentiated cells, as shown by immunofluorescent and immunoferritin techniques (Chang, Reitherman, Rosen & Barondes, 1975; Chang, Rosen & Barondes, 1977). When the antibodies were applied to fixed permeable cells, most of the binding was intracellular (Chang *et al.*, 1977). Although precise quantitation of the amount of lectin on the cell surface is difficult, we found that about 2 % of the total cellular lection in *D. purpureum* (in the range of 2×10^5 molecules per cell) can be stripped from the surface of intact cells with appropriate sugars (W. R. Springer *et al.*, unpublished). After such stripping, about 8×10^5 molecules remain on the cell surface, as shown by binding specific antibodies, so that total cell surface lectin represents about 10 % of total cellular lectin. About 2×10^5 molecules of lectin per cell was found associated with a plasma membrane fraction from *D. discoideum* (Siu, Loomis & Lerner, 1978).

Evidence for cell surface receptors for slime mould lectins has been provided in a number of ways. First, fixed slime mould cells bind added exogenous slime mould lectin with affinities as high as 10^9 M^{-1} (Reitherman, Rosen, Frazier & Barondes, 1975). Such binding of slime mould lectins to slime mould cell surfaces can be blocked by appropriate sugars. Added lectin aggregated fixed slime mould cells which had lost their endogenous cell surface lectin activity during fixation (Reitherman *et al.*, 1975). Fixed partially differentiated slime mould cells were agglutinated at lower concentrations of slime mould lectin than fixed vegetative cells, but this was not the case with several plant lectins (Reitherman *et al.*, 1975). Lectins from one species of slime mould bound avidly to the surface of other species; but a small degree of selective lectin affinity has been observed (Reitherman *et al.*, 1975). Binding of ferritin-labelled lectin to the surface of fixed slime mould cells has also been shown (Chang *et al.*, 1977). Furthermore, living slime moulds which already contain considerable endogenous lectin on their cell surface, can bind a comparable amount of added slime mould lectin (Springer *et al.*, unpublished). This appears to be specific binding since it is blocked by specific hapten sugars.

The finding that slime moulds contain cell surface lectins and receptors at a stage when they display cell–cell adhesion raised the possibility that the lectins and receptors are cell-adhesion molecules. Were this the case then univalent antibody fragments which bind to the lectin, or haptens which bind to its active site, might be expected to inhibit cell–cell adhesion. These expectations have been met only in one specific set of experimental circumstances. In studies with *Polysphondylium pallidum* we found that developmentally regulated cohesiveness of the type shown above can be demonstrated under a number of conditions including the presence of hypertonic medium or metabolic inhibitors. Under these conditions cell–cell adhesion was blocked by two materials that bind the lectin – univalent antibodies or asialofetuin. The latter is a modified glycoprotein which attaches to the lectin's carbohydrate-binding site (Rosen, Chang & Barondes, 1977). However, these reagents had little or no effect when normal medium was used. The negative result with univalent antibody has been confirmed by Gerisch (this volume). In similar experiments with *D. discoiderum* and *D. purpureum* we have been unable to find any conditions where cell adhesion is measurable, and univalent immunoglobulin fragments have a specific inhibitory effect when compared with fragments from normal immunoglobulin (Springer *et al.*, unpublished). However, these results are difficult to interpret since most of the cell surface lectin does not bind the univalent antibodies (Springer *et al.*, unpublished). This lectin may be in a cryptic form, and yet might be capable of mediating cell adhesion.

In contrast to these ambiguous results, support for a possible role of cell surface slime mould lectins in cell adhesion has been given by studies with a mutant isolated by Ray, Shinnick & Lerner (1979) which shows a block of differentiation at the stage when the cells form aggregates. This mutant of *D. discoideum* has an abnormality in the major lectin from *D. discoideum* which comprises 90–95 % of the total cellular lectin and which is called discoidin I. The abnormal lectin is synthesised as the cells differentiate, and is detected by a radioimmunoassay. However its carbohydrate-binding site is abnormal since it fails to bind to Sepharose or to agglutinate appropriate test erythrocytes. An interesting property of this mutant is that, when mixed with normal slime mould cells, the mutant cells are incorporated into a normal fruiting body. Presumably the normal cells contain sufficient normal discoidin I on their surface to lead to aggregation with the mutant cells. Although evidence of this type might be a false positive, as considered above, it is a substantial addition to the case that the lectins in slime moulds are cell-adhesion molecules.

DEVELOPMENTALLY REGULATED LECTINS IN EMBRYONIC CHICK TISSUES

Because of these findings in slime moulds we sought developmentally regulated lectins in a higher organism, the embryonic chick. This organism was chosen for investigation since its development has been studied extensively and since it is readily available and inexpensive. We initially concentrated on developing muscle because it is a relatively simple and abundant tissue. We found that embryonic chick pectoral muscle contains two lectins, both of which show changes in activity with development. The first to be identified, which we designate lectin 1, is present at very low levels in pectoral muscle from 8-day-old chick embryos, rises to maximum in 16-day embryo muscle and then declines to very low levels in the adult (Nowak, Haywood & Barondes, 1976). At its maximum it constitutes about 0.1 % of the

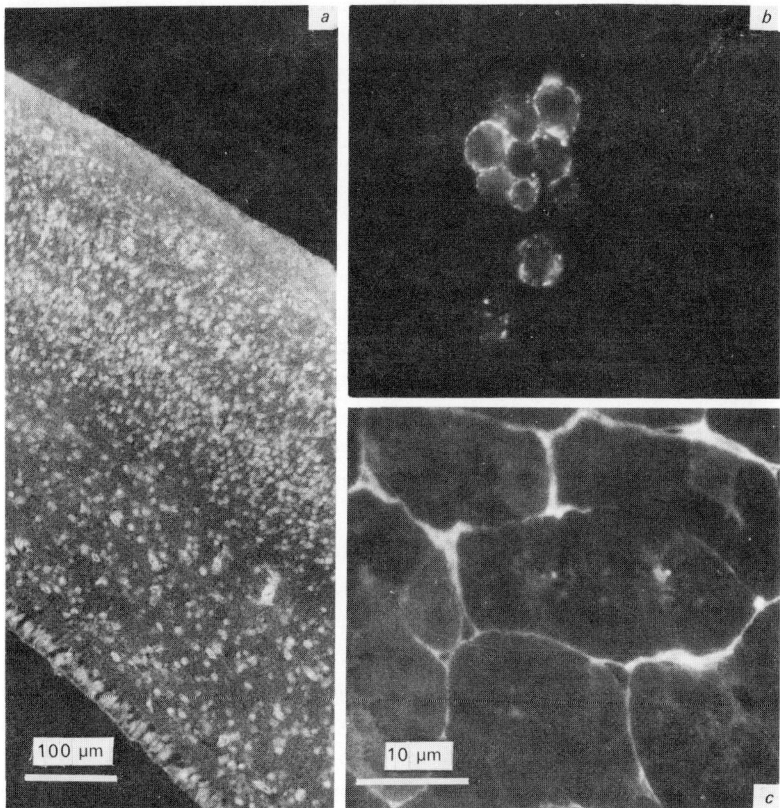

Fig. 3. Different localisation of endogenous chicken lectin 1 in different chicken tissues. Rabbit antibodies were raised against chicken lectin 1, purified from embryonic muscle by the method of Nowak *et al.* (1977). The antigen appeared homogenous and the antiserum gave a single line of identity with extracts of the three tissues studied here. After reacting the antiserum with cells or tissue sections, excess antibody was washed away, fluorescent goat anti-rabbit immunoglobulin was reacted, and observations were made with a fluorescence microscope. (*a*) Optic tectum section of 16-day chick embryo, showing intracellular labelling of layers of neurons and ependymal cells (for details see Gremo *et al.*, 1978). (*b*) Cultured myoblasts (unfixed), showing patches of antigen on the cell surface (for details see Nowak *et al.*, 1977). (*c*) Pancreas section from adult chicken, showing localisation of antigen to the extracellular space between lobules of secretory cells (for details see Beyer *et al.*, 1979). In all cases controls, using a large excess of normal rabbit serum, showed no significant staining.

protein in muscle extracts. It has been purified to homogeneity by affinity chromatography and is a dimer with subunit molecular weight of 15 000 (Den & Malinzak, 1977; Nowak, Kobiler, Roel & Barondes, 1977). Antibodies raised to it have been used to demonstrate that it is predominantly intracellular, although some is detectable on the surface of cultured myoblasts (Fig. 3). Existence of an unsaturated receptor on the cell surface is suggested by the finding (unpublished) that addition of exogenous lectin to cultured cells renders them more agglutinable by specific antibody raised against the lectin.

Thus far, there has been little direct study of the possible role of this lectin in cell adhesion. It has been reported that thiodigalactoside, which is a potent inhibitor of the hemagglutination activity of lectin 1, blocks fusion of myoblasts from an established rat cell line. (Gartner & Podleski, 1975). However, thiodigalactoside did not affect the fusion of chick myoblasts in primary cultures (Den, Malinzak & Rosenberg, 1976). Recently, evidence was presented that addition of purified lectin 1 to the medium of cultured myoblasts impairs fusion (R. G. MacBride and R. J. Przybylski, in preparation). However the specificity of this action is not yet clear.

Studies with other tissues suggest that lectin 1 may have been adapted for several functions. Lectin with indistinguishable physico-chemical, immunological and carbohydrate-binding properties has been purified from extracts of embryonic chick brain and liver (Kobiler, Beyer & Barondes, 1978) and is also found in some adult chicken tissues (Beyer, Tokuyasu & Barondes, 1979). Although lectin 1 may be identical in all these tissues, there are notable differences in its developmental regulation (Kobiler & Barondes, 1977) and site of localisation. Whereas the lectin in chick brain appears to show the same type of developmental regulation as in chick muscle, falling to low levels in the adult, levels of lectin 1 are very high in adult chicken liver, intestine and pancreas. Furthermore, there are differences in the distribution of the lectin in the various tissues. Lectin 1 is predominantly intracellular in embryonic neurons (Gremo, Kobiler & Barondes, 1978) (Fig. 3), myoblasts, and the goblet cells of the intestine (Beyer *et al.*, 1979). As indicated previously, it is also detectable on the cell surface of cultured myoblasts (Fig. 3) and has been detected on the surface of some neurons from the optic tectum. In contrast, in the pancreas it is localised to the extracellular space between the lobules (Beyer *et al.*, 1979) (Fig. 3).

Table 4. *Do lectins fit criteria for a cell adhesion molecule?*

Criterion	Lectins in Slime moulds	Lectins in Chick tissues
Cell surface location	Yes; but most is intracellular	Yes; but in several tissues most is intracellular
Appearance of lectin correlated with development of adhesiveness	Yes	Yes, in some tissues
Complementary cell surface receptor	Yes	Yes (not much data)
Binding antibodies or haptens to lectin blocks adhesion	Yes, under some conditions; generally no	Yes and no (controversy not yet resolved)
Mutant with impaired cell adhesion has defective lectin	Yes	Not studied

This localisation raises the question that lectin 1 may play some role in cell–matrix or matrix–matrix interactions in the pancreas.

Much less is known about the second lectin (lectin 2) from embryonic skeletal muscle. It too changes in activity with muscle development (Mir-Lechaire & Barondes, 1978), and is present in other tissues (Kobiler & Barondes, 1979). It differs strikingly from lectin 1, in that it agglutinates different test erythrocytes and is best inhibited by a specific group of glycosaminoglycans including dermatan sulphate, heparin and heparan sulphate (Kobiler & Barondes, 1979). It also differs in many ways from fibronectin (Ceri, Shadle, Kobiler & Barondes, 1979). It has not yet been purified to homogeneity, so that specific antibody has not yet been available as a reagent for its localisation. One clue to its function is that it is secreted into the medium of differentiating muscle cultures (Ceri *et al.*, 1979). This finding and the fact that it interacts with glycosaminoglycans, especially those found in the substrate attached material of cultured muscle (Ceri *et al.*, 1979), suggests an extracellular function, perhaps a role in adhesion of cells to extracellular matrix.

SUMMARY OF EVIDENCE THAT LECTINS ARE CELL-ADHESION MOLECULES

There is considerable evidence, summarised in Table 4, that the lectins in slime moulds play a role in cell–cell adhesion. Four out of our five criteria are met and failure to consistently meet the fifth criterion could be a false negative. Furthermore, the assay used for studying cell–cell adhesion in cellular slime moulds is probably valid, because it is both developmentally regulated and species-specific. Taken together the evidence that cell surface lectins in slime moulds play a role in cell adhesion is quite substantial.

Much less is known about the lectins in developing and adult chick tissue. However, some evidence is consistent with their role in cell–cell or cell–matrix interactions (Table 4). There is a distinct possibility that the same lectin functions differently in different tissues and at different stages of development.

Acknowledgements. This work was supported by grants from the USPHS (MH18282) and the McKnight Foundation. I thank Wayne Springer for reviewing the manuscript.

REFERENCES

Barondes, S. H. & Haywood, P. L. (1979). Comparison of developmentally regulated lectins from three species of cellular slime molds. *Biochimica et Biophysica Acta, Biomembranes*, **550**, 297–308.

Beyer, E. C., Tokuyasu, K. & Barondes, S. H. (1979). Immunohistological localization of endogenous lectin in adult chicken tissues. *Journal of Cell Biology* **82**, 565–71.

Bonner, J. T. & Adams, M. S. (1958). Cell mixture of different species and strains of cellular slime molds. *Journal of Embryology and Experimental Morphology*, **6**, 346–56.

Ceri, H., Shadle, P., Kobiler, D. & Barondes, S. H. (1979). Extracellular lectin and its glycosaminoglycan inhibitor in chick muscle cultures. *Journal of Supramolecular Structure* (in press).

Chang, C.-M., Reitherman, R. W., Rosen, S. D. & Barondes, S. H. (1975). Cell surface location of discoidin, a developmentally regulated carbohydrate-binding protein from *Dictyostelium discoideum*. *Experimental Cell Research*, **95**, 136–59.

Chang, C.-M., Rosen, S. D. & Barondes, S. H. (1977). Cell surface location of an endogenous lectin and its receptor in *Polysphondylium pallidum*. *Experimental Cell Research*, **104**, 101–9.

Den, H. & Malinzak, D. A. (1977). Isolation and properties of β-D-galactoside-specific lectin from chick embryo thigh muscle. *Journal of Biological Chemistry*, **252**, 5444–8.

Den, H., Malinzak, D. A. & Rosenberg, A. (1976). Lack of evidence for the involvement of a β-D-galactoside-specific lectin in the fusion of chick myoblasts. *Biochemical and Biophysical Research Communications*, **69**, 621–7.

Frazier, W. A., Rosen, S. D., Reitherman, R. W. & Barondes, S. H. (1975). Purification and comparison of two developmentally regulated lectins from *Dictyostelium discoideum*: discoidin I and II. *Journal of Biological Chemistry*, **250**, 7714–21.

Gartner, T. K. & Podleski, T. R. (1975). Evidence that a membrane bound lectin mediates fusion of L_6 myoblasts. *Biochemical and Biophysical Research Communications*, **67**, 972–8.

Gerisch, G. (1968). Cell aggregation and differentiation in *Dictyostelium*. In *Current Topics in Developmental Biology*, ed. A. A. Moscona & A. Monroy, vol. 3, pp. 159–97. New York, London: Academic Press.

Gremo, F., Kobiler, D. & Barondes, S. H. (1978). Distribution of endogenous lectin in the developing chick optic tectum. *Journal of Cell Biology*, **78**, 491–9.

Kobiler, D. & Barondes, S. H. (1977). Lectin activity from embryonic chick brain, heart and liver: changes with development. *Developmental Biology*, **60**, 326–30.

Kobiler, D. & Barondes, S. H. (1979). Lectin from embryonic chick muscle that interacts with glycosaminoglycans. *FEBS Letters*, **101**, 257–61.

Kobiler, D., Beyer, E. C. & Barondes, S. H. (1978). Developmentally regulated lectins from chick muscle, brain and liver have similar chemical and immunological properties. *Developmental Biology*, **64**, 265–72.

Loomis, W. F. (1975). *Dictyostelium discoideum: a developmental system*. New York, London: Academic Press.

McDonough, J. P., Springer, W. R. & Barondes, S. H. (1979). Species-specific cell cohesion in cellular slime molds: demonstration by several quantitative assays and with multiple species. *Experimental Cell Research* (in press).

Mir-Lechaire, F. & Barondes, S. H. (1978). Two distinct developmentally regulated lectins in chick embryo muscle. *Nature, London*, **272**, 256–8.

Nowak, T. P., Haywood, P. L. & Barondes, S. H. (1976). Developmentally regulated lectins in embryonic chick muscle and a myogenic cell line. *Biochemical and Biophysical Research Communications*, **68**, 650–7.

Nowak, T. P., Kobiler, D., Roel, L. E. & Barondes, S. H. (1977). Developmentally regulated lectin from embryonic chick pectoral muscle: purification by affinity chromatography. *Journal of Biological Chemistry*, **252**, 6026–30.

Raper, K. B. & Thom, C. (1941). Interspecific mixtures in the *Dictyosteliaceae*. *American Journal of Botany*, **28**, 69–78.

Ray, J., Shinnick, T. & Lerner, R. A. (1979). A mutation altering the function of a carbohydrate binding protein blocks cell–cell cohesion in developing *Dictyostelium discoideum*. *Nature, London*, **279**, 215–21.

Reitherman, R. W., Rosen, S. D., Frazier, W. A. & Barondes, S. H. (1975). Cell surface species-specific high affinity receptors for discoidin: developmental regulation in *Dictyostelium discoideum*. *Proceedings of the National Academy of Sciences, USA*, **72**, 3541–5.

Rosen, S. D., Kafka, J. A., Simpson, D. L. & Barondes, S. H. (1973). Developmentally regulated, carbohydrate binding protein in *Dictyostelium discoideum*. *Proceedings of the National Academy of Sciences, USA*, **70**, 2554–7.

Rosen, S. D., Chang, C.-M. & Barondes, S. H. (1977). Intercellular adhesion in the cellular slime mold *P. pallidum* inhibited by interaction of asialofetuin or specific univalent antibody with endogenous cell surface lectin. *Developmental Biology*, **61**, 203–13.

Siu, C. H., Loomis, W. F. & Lerner, R. A. (1978). Plasma membrane proteins in wild type and cascade-arrested mutant strains of *Dictyostelium discoideum*. In *The Molecular Basis of Cell–Cell Interactions*, ed. R. A. Lerner & D. Bergsma, pp. 439–58, New York: Alan R. Liss.

Cell surface glycoproteins in fibroblast adhesion

R. C. HUGHES, S. D. J. PENA* AND P. VISCHER

National Institute for Medical Research, Mill Hill, London NW7 1AA, UK

1. INTRODUCTION

Many events in cell differentiation are known to be accompanied by specific changes in cell surface membranes. In many cases differentiation appears to require or even consist of alterations in cellular adhesiveness. The adhesive properties of cells presumably play an important role in the regulation of cell division (Folkman & Moscona, 1978), in the outgrowth of cellular processes such as neurites from neuronal cells under the influence of nerve growth factor (Schubert, La Corbiere, Whitlock & Stallcup, 1978), in the control of cell motility (Abercrombie, 1961) and in the cell positioning taking place during embryogenesis. In most cases these complex processes probably involve modulation of cell–cell and cell–substratum interactions. For example, the folding of the neural plate into a neural tube may require the preferred adhesion of the neural tube cells with one another and a decrease in adhesion to surrounding mesodermal cells, while the migration of neural crest cells from the neural tube to specific areas of the developing embryo appears to involve adhesiveness to extracellular collagenous matrices (Weston, 1970) as in the laying down of the corneal stroma for example (Bard & Higginson, 1977). What mechanisms regulate the adhesiveness of one cell with another? Does cell–substratum adhesiveness regulate the specific pathways of movement and, if so, what structural surface components of the cells and their environment direct such movements? How are cells initiated into movement and what terminates this movement? Such questions leading to a thorough analysis of the surface-associated events accompanying adhesion and deadhesion of cells to one

* Present address: Montreal Neurological Institute, 3801 University Street, Montreal, Quebec H3A 2B4, Canada.

another, to connective tissue elements and other substrata could have important implications for various aspects of cell biology and pathology, for example, cancer biology.

A variety of biochemical approaches have been used to try to pinpoint surface membrane changes accompanying complex biological phenomena as a necessary first step in understanding their basis. The use of probes for cell surface carbohydrates has yielded considerable information of particular relevance to the present paper. The complex carbohydrates conjugated in glycoproteins, glycosphingolipids and proteoglycans are major constituents of cell surface membranes (Hughes, 1976*a*), whose carbohydrate moieties are exposed to the extracellular environment. It has long been speculated that the carbohydrate moieties of cell surfaces play some role in the control of cellular recognition, interaction and proliferation. Changes have been detected in surface carbohydrate composition on cell to cell contact, during the mitotic cell cycle and accompanying ontogeny and oncogenesis. Some of these complex carbohydrates serve as specific sites for ion transport and metabolite uptake and as receptors for hormones, growth factors, plant and bacterial toxins as well as infective organisms. In several cases the site of biological activity has been placed unequivocally in the carbohydrate moieties of the membrane components, encouraging the belief that carbohydrate structure makes an important contribution to determining the specificity of the cell surface.

In this paper we review briefly some of the experimental evidence obtained by ourselves and others implicating cell surface glycoproteins in various aspects of cell adhesion. Our own work has been concerned in particular with the adhesion of hamster fibroblasts in cell culture and, where appropriate, reference is made to other work using different systems. A hypothesis will be presented which emerges from changes observed in the adhesiveness of lectin-resistant mutants of hamster fibroblasts. Our aim is to characterise the surface molecules involved in cell–cell and cell–substratum interactions in this model system, which we believe has relevance to the behaviour of cells in physiologically more complex systems.

2. MODELS OF CELLULAR ADHESION

The differential affinity of two surfaces may stem from a variety of different mechanisms. Dan (1936) was among the first to show a requirement of divalent cations, particularly Ca^{2+} for cell–cell adhesion in chick embryonic cells, although later evidence (Urushihara, Takeichi, Hakura & Okada, 1976) showed this not to be a completely general phenomenon for all cells. It is interesting, therefore, that a number of calcium-binding cell membrane glycoproteins have been identified (see Hughes, 1976*a*), although to our knowledge none of these are present in surface membranes and in a position to mediate cell–cell adhesion via calcium bridges. A more widely applicable mechanism of cellular adhesion involving electrostatic interactions has received considerable attention (see Curtis, 1973, for discussion). It is well known that the cell surface has a net negative electric charge, due in considerable part to the surface carbohydrates, particularly *N*-acetylneuraminic acid residues of glycoproteins and gangliosides. On the other hand, it is certainly possible that patches of net positive charge of lipid or protein origin may be distributed over the cell surface, perhaps in a manner dictated by the defined distribution or clustering of acidic carbohydrate-containing membrane components. Adhesion may then be mediated at sites exhibiting a minimal ratio between two opposing electrical interactions. Hydrogen bonding between opposing carbohydrate-rich surfaces may also be sufficient to promote adhesive interactions provided the opposing surfaces approach sufficiently closely to allow molecular interactions to occur. Indeed, the carbohydrate-rich periphery of cells often extends considerable distances, and in view of the tight interchain hydrogen bonding exhibited by some polysaccharides (Morris, Rees, Thom & Welsh, 1977) this proposal remains an attractive hypothesis.

In 1970 Roseman formulated a radically different hypothesis of cellular adhesion involving cell-binding molecules, advanced many years earlier by Tyler (1946) and Weiss (1947). An important advantage of this mechanism lies in the predictions which are verifiable by relatively simple experiments and will occupy the remaining part of this paper. The idea of complementary surface-located molecules mediating cellular adhesiveness was well established by experiments with simple animal systems. When free-living salt-water sponges are dissociated by incubation in divalent-cation-free sea water and then re-incubated in undepleted salt water, the separated cells quickly re-aggregate. Mixing experiments showed that the dissociated cells of different sponges associate specifically with like cells. Although the specificity exhibited in such experiments is not surprising, in view of the widely disparate nature of the sponges used (often from different genera (see Curtis, 1973)), nevertheless these early experiments did suggest the presence on sponge cells of specific complementary molecules mediating their aggregation. More recent research has tended to confirm these predictions (Turner & Burger, 1973). In the more recent version of the original hypothesis of Tyler (1946) and Weiss (1947), Roseman (1970) proposed that the specific adhesion between cells may be mediated by the interaction of a carbohydrate sequence of a surface glycoprotein or glycosphingolipid of one cell and a carbohydrate-binding protein or glycoprotein of an adjacent cell. This model is capable of many interpretations. For example, the carbohydrate-binding component may be a glycosyl transferase located in the cell surface membrane as suggested originally (Roseman, 1970; Roth, McGuire & Roseman, 1971 *a*) or some other reagent, for example a lectin-like molecule (but see discussion by Gordon, this volume). Evidence is accumulating for the presence of lectins in animal cells (Ashwell & Morell, 1977) which may well be involved in the interactions envisaged by earlier workers (see Barondes, this volume).

The evidence obtained by Roseman and his colleagues which leads to this proposal was of two types. It was first shown by Roth *et al.* (1971 *b*) and independently by Vicker & Edwards (1972) that treatment of dissociated embryonic neural retina cells or hamster fibroblasts (BHK cells) with neuraminidases had little effect on, or even increased, their adhesive properties. However, subsequent treatment with β-galactosidase of the neural retina cells drastically reduced their adhesiveness to one another. The results obtained with neuraminidases could, of course, be explained by the effect of such enzymes in removing acidic sialic acid residues from the cell surface and a consequent lowering of net surface negative charge. However, the β-galactosidase experiment is difficult to explain in any terms other than the apparent important role of terminal exoglycosidase-sensitive galactose residues of surface glycoproteins, and possibly glycolipids, in cell–cell aggregation.

Corroboratory evidence for such a conclusion comes from the interesting experiments of Chipowsky, Lee & Roseman (1973). It was reported that mouse fibroblasts adhere tightly to carrier beads covalently coupled to a β-galactosyl ligand, while similar beads carrying β-*N*-acetylglucosaminyl ligands were inactive.

The complementarity between two cell surfaces proposed by Roseman (1970) may be mediated directly by a lock–key mechanism; a lock on one cell surface binds to a key on an opposing surface.

Alternatively, bridging glycoprotein molecules may form adhesive interactions by binding to identical or different molecules at the two cell surfaces. Moscona and his colleagues (Moscona, 1962; Hausman & Moscona, 1975) have described soluble factors released from cell cultures of chick embryo cells that apparently mediate intercellular adhesion in this fashion. In a sense the bridging molecules are part of an extracellular matrix. It is interesting, therefore, that glycosaminoglycans, which are major components of connective tissue and extracellular matrices, are released by a variety of cells in long-term culture and in some cases mediate cell–cell aggregation. Pessac and his colleagues, for example, showed that hyaluronic acid binds to protease-sensitive cell surface receptors (Pessac & Defendi, 1972*a*, *b*) to form aggregates from single-cell suspensions. Binding to cells of other mucopolysaccharides, e.g. heparin and heparan sulphate (Comper & Laurent, 1978; Culp, Rollins, Buniel & Hitri, 1978) as well as collagen (Grinnell & Minter, 1978) by protease-sensitive surface components has also been reported.

3. PURIFIED ADHESIVE FACTORS

Another matrix component is fibronectin (see Olden, this volume, and Rees, this volume). Fibronectin is not, apparently, an integral membrane glycoprotein (Stenman, Wartiovaara & Vaheri, 1977; Hedman, Vaheri & Wartiovaara, 1978; Hughes & Nairn, 1978; Yamada & Olden, 1978) and soluble, biologically active forms exist in plasma and other body fluids. Fibronectin can be readily released from cells by treatment with urea, reducing agents and, according to some reports, by cytochalasin B, suggesting that an intact microfilament system may be required for retention at the cell surface. Fibronectin is a major constituent of serum (0.3 g l^{-1}) probably by virtue of active secretion by endothelial cells, and when present in cell-bound form it may account for up to 3% of the total cellular proteins, in primary chick or hamster fibroblasts for example. The relationship of the soluble forms of fibronectin present in serum and other body fluids and in the extracellular medium of cultured cells to the cell-bound form is at present unclear. We have prepared fibronectins from hamster plasma, primary hamster embryo fibroblasts and the baby hamster kidney cell line (BHK) as well as from the spent culture fluids of BHK cells (S. D. J. Pena and R. C. Hughes, in preparation). These materials appear to have similar through not identical biological activity (Pena & Hughes, 1978*a*) and structure, e.g. molecular size. However, the cellular forms in particular contain smaller molecular fragments which presumably result from partial proteolysis occurring during culturing and extraction but nevertheless are active in biological tests to be discussed later. The amino acid and carbohydrate compositions of the hamster fibronectins are similar, and so are the patterns of radioactive peptides obtained by proteolysis of derivatives radioiodinated in tyrosine residues (our unpublished observations; see also Hughes & Nairn, 1978). Similarities of serum and cell fibronectins have been reported also by others (Vuento, Wrann & Ruoslahti, 1977; Yamada & Kennedy, 1979).

The fibronectin molecule is arranged as a static, fibrillar array at the cell surface (Mautner & Hynes, 1977; Schlessinger *et al.*, 1977; Hedman *et al.*, 1978). These sites are also enriched in procollagen and glycosaminoglycans in fibroblasts and smooth muscle cells (Bornstein & Ash, 1977; Comper & Laurent, 1978; Vaheri *et al.*, 1978), suggesting a close interaction between the cell surface, fibronectin and other matrix constituents. Affinity of fibronectin with collagens (Engvall & Ruoslahti, 1977; Dessau, Jilek, Adelmann & Hormann, 1978) and glycosaminoglycans or proteoglycans (Stathakis & Mosesson, 1977) have also been demonstrated *in vitro*. Since soluble or solubilised fibronectins agglutinate erythrocytes (Yamada, Yamada & Pastan, 1975; Yamada & Olden, 1978) and fibroblasts and increase the attachment and spreading of many cell types to a plastic substrate (Grinnell, 1978; Grinnell & Hays, 1978; Pena & Hughes, 1978*a*) or to collagen (Pearlstein, 1976; Grinnell & Minter, 1978; Pena & Hughes, 1978*b*), it seems clear that this molecule is an active adhesive factor interacting directly with cell surfaces as well as with other matrix components. The nature of such interactions and the role of carbohydrate are considered in later sections.

4. LECTIN-RESISTANT HAMSTER FIBROBLASTS

We chose to work with baby hamster kidney fibroblasts (BHK), an established cell line that has been used extensively for study of glycoprotein structure and biosynthesis (Warren, Buck & Tuszynski, 1978) and various aspects of cell adhesion (Vicker & Edwards, 1972; Edwards, 1973). These cells bind a variety of purified plant lectins, including ricin and abrin, PHA and concanavalin A and to a lesser

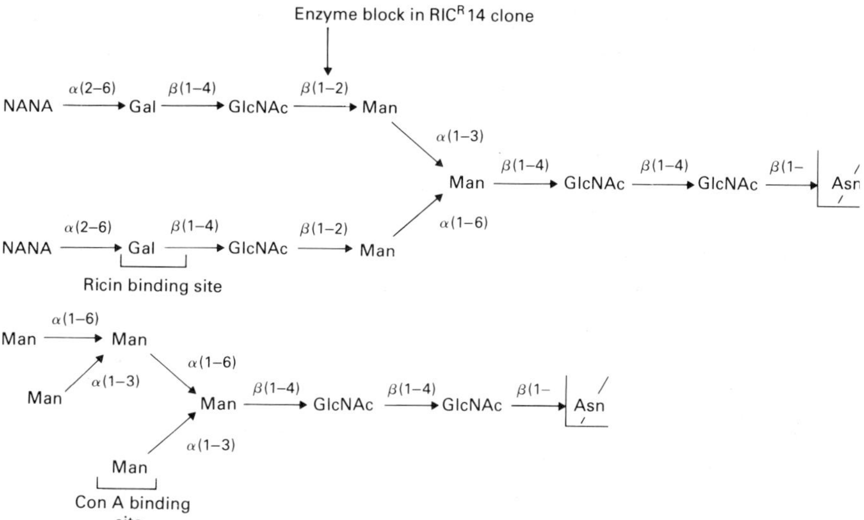

Fig. 1. A general structure of N-glycosidic chains of glycoproteins showing the enzyme deficiency in a ricin-resistant mutant of baby hamster kidney (BHK) fibroblasts, and the shortened chains formed as a result of the deficiency. Note the absence of ricin-binding sequences in the latter while concanavalin A binding is not reduced.

Abbreviations: NANA, *N*-acetylneuraminic acid; Gal, galactose; GlcNAc, *N*-acetylglucosamine; Man, mannose; Asn, asparagine residues of polypeptide moieties.

extent wheat germ agglutinin (Meager *et al.*, 1976). All of these lectins are cytotoxic and the initial binding to cell surface carbohydrates is in each case an essential step in cytotoxicity (Hughes, 1979). When cells are treated with sufficient lectin to inhibit bulk cell growth, cloned cell lines are obtained readily, although at low frequency (Meager, Ungkitchanukit, Nairn & Hughes, 1975; Meager, Ungkitchanukit & Hughes, 1976). These mutant cells are genetically stable in resistance to the particular lectin used. As expected, in many cases these cells show alterations in lectin-binding capacity, suggesting more or less profound changes in cell surface carbohydrate groups. The ricin-resistant BHK cell lines (Ric^{R} cells) also have been shown by experiments *in vitro* to lack the glycosyl transferases presumably responsible for assembly of the cell surface carbohydrate chains of glycoproteins to which ricin binds in the parental cells (Meager *et al.*, 1975; Hughes, 1976*a*, *b*; P. Vischer & R. C. Hughes, in preparation). Ricin binds to peripheral galactose residues in chains of the N-glycosidic type (Fig. 1). Two basic mechanisms could affect the assembly of carbohydrate chains to block ricin binding. First, the superabundance of glycosyl transferases attaching sugar residues onto peripheral galactose residues would prevent binding since it is known that ricin binding is strongest to terminal galactose units of carbohydrate chains. Therefore, addition of one or more sialic acid or fucose residues, which commonly are found attached to galactose in glycoproteins, to the termini of these carbohydrate chains would effectively block ricin binding. It is interesting, therefore, that increased (150–200%) activities of sialyl and fucosyl transferases are indeed detected in certain of the ricin-resistant cell lines (P. Vischer and R. C. Hughes, unpublished observations). Secondly, the lack of a glycosyl transferase operating in a step prior to the addition of the peripheral galactose residues would prevent assembly of the ricin-binding sequence. A profound loss (greater than 90%) of a specific *N*-acetylglucosaminyl transferase was detected in clone Ric^R14, and a more modest loss of this enzyme as well as galactosyl transferases were found in some of the other resistant clones (P. Vischer & R. C. Hughes, unpublished observations). Such defects would terminate all nascent carbohydrate chains of this type prematurely and before assembly of the galactose-containing sequence. Such a change in surface carbohydrate structure can be detected by lectin affinity chromatography of isolated surface membrane glycoproteins, by direct carbohydrate analysis or by (Pena, Mills & Hughes, 1979) isoelectric focusing techniques in conjunction with SDS–polyacrylamide gel electrophoresis. In general, the surface glycoproteins of Ric^R14 cells, for example, are less negatively charged and of lower molecular weight than the same glycoproteins present on parental cells, since the terminal sialic acid rich sequences are not attached and the carbohydrate chains are of smaller size.

These cells therefore provide a useful test of the hypothesis that surface carbohydrate residues, in particular galactose, play a role in cellular adhesion. First, however, one important requirement of this assumption was checked experimentally. It is known that in some (Hickman, Kulczycki, Lynch & Kornfeld, 1977) but not all (Hughes, Meager & Nairn, 1977; Olden, Pratt & Yamada, 1978; Damsky, Levy-Benshimol, Buck & Warren, 1979) cases, the migration of nascent membrane glycoproteins to the cell surface from intracellular sites of biosynthesis requires complete glycosylation. It could be argued, therefore, that any effects on cellular properties of a loss in glycosylating enzymes may be secondary to the normal insertion of active glycoprotein molecules into the surface membrane. We reasoned that surface molecules involved in cell–cell adhesion or in substrate attachment would represent a significant part of the total surface glycoprotein composition. The surface glycoproteins of parental and resistant cells were therefore labelled by lactoperoxidase-catalysed iodination of intact cells and subjected to proteolysis. Fingerprints of the radioiodinated tyrosine-containing peptides were indistinguishable for parental compared with carbohydrate-deficient cell lines (our unpublished observations) suggesting strongly that the surface glycoproteins of the latter cells are inserted normally into the surface membrane and the most striking differences are indeed in the carbohydrate moieties of these protein molecules expressed at the cell surface.

5. CELL-CELL AGGREGATION OF GLYCOSYLATION-DEFICIENT FIBROBLASTS

In 1976 we reported (Edwards, Dysart & Hughes, 1976) on the basis of experiments carried out in Glasgow by J. G. Edwards and J. McK. Dysart, that many of the ricin-resistant cell lines exhibited a reduced rate of aggregation (see Fig. 2). The assay used suffers, as do most other fibroblast-adhesion assays, from the necessity to dissociate cells from monolayer cultures by brief trypsinisation. The single cells, dispersed in medium containing serum, are rotated in culture flasks for several hours, during which time the size of the aggregates formed is monitored and the rate of disappearance of single cells is measured by electronic particle counting. It is assumed that these measurements are related to intercellular adhesiveness. However, other events almost certainly are taking place during the incubation period including surface repair and replacement of protease-sensitive molecules, which presumably contribute to the kinetics of cell reassociation. However, we have as yet been unable to detect any significant differences between the various mutant lines compared with parental BHK cells in the rate of membrane repair. Replacement of surface sialic acid residues after treatment of intact cells with neuraminidase followed by incubation in fresh medium proceeds at similar rates for mutant or parental cells. This method gives an idea of the bulk rate of turnover of surface glycoprotein molecules (Hughes, 1976*a*). Furthermore, an important example of

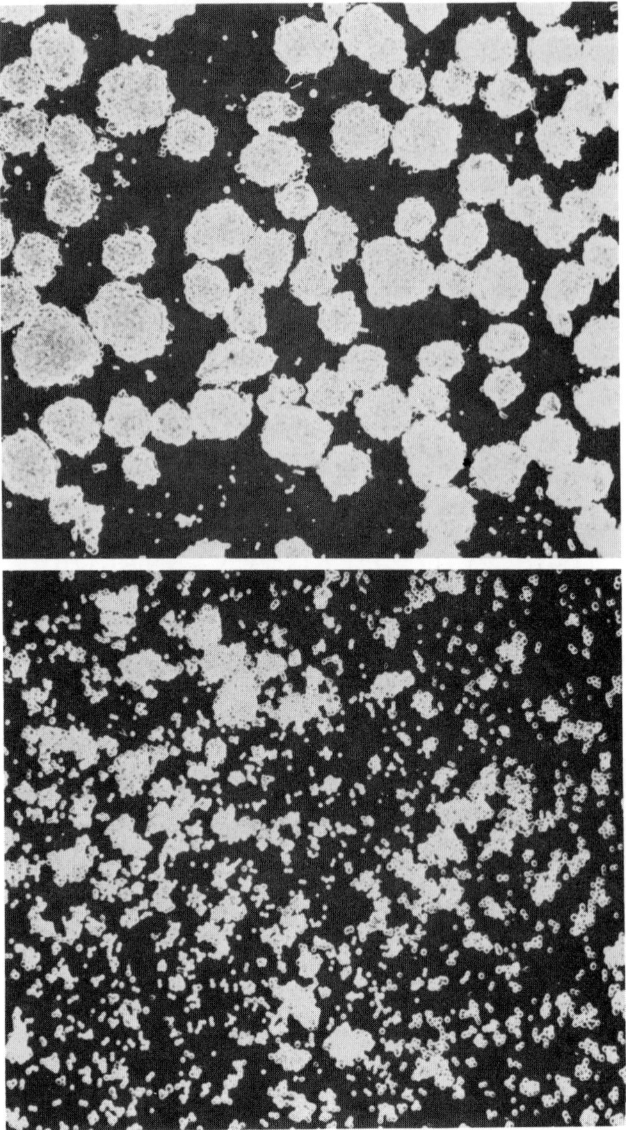

Fig. 2. Aggregation of trypsinised BHK cells (*top panel*), and of ricin-resistant mutant cells deficient in surface galactose terminal residues (*bottom panel*). Photographs (kindly supplied by Dr John Edwards) were taken after 24 h gyratory shaking at 37 °C.

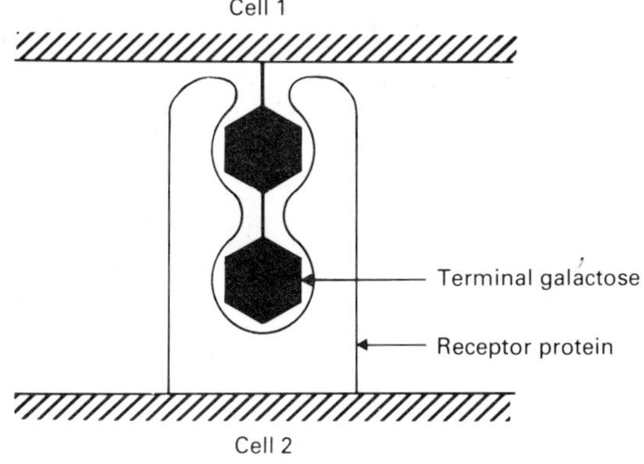

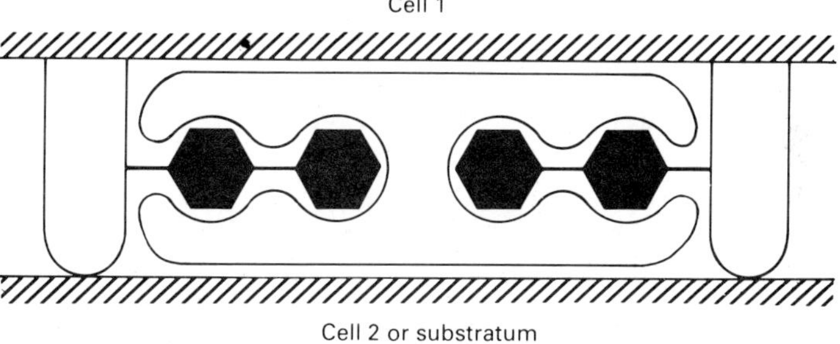

Fig. 3. Adhesion of hamster fibroblasts mediated through interaction of galactose-containing sequences of membrane glycoproteins and membrane lectins. Ricin-resistant mutants lacking terminal galactose residues interact poorly with lectin molecules and exhibit reduced adhesiveness. The lectin may be membrane-bound or secreted as a multivalent ligand to mediate cell–cell and cell–substratum adhesion. (From Edwards *et al.* (1976).

such a protease-sensitive adhesive surface component is fibronectin, and, as we shall see, this component may in some circumstances be replaced by similar factors present in the serum included in the adhesion assays, which obviates the requirement for cell replacement, even in short-term assays. Most importantly, from our point of view, the aggregation of trypsinised cells in gyratory cultures is well established as being sensitive to glycosidase treatments of the cells (Vicker & Edwards, 1972) indicating that at least in part the aggregation is controlled by the carbohydrate content of protease-resistant surface glycoproteins. Therefore, it seems that the reduced adhesiveness of the ricin-resistant cell lines is most likely due to changes in cell surface carbohydrates. This view was strengthened by the finding that ricin-resistant clones, Ric^{R}19 and 22, which exhibit no detectable alteration in surface carbohydrate structure and have a near-normal complement of glycosyl transferases and presumably are resistant to the toxin at a site subsequent to surface binding, appear to aggregate under the same conditions at a rate indistinguishable from the parental BHK fibroblasts. The Ric^R22 cells differ from normal BHK cells only in forming smaller aggregates, the significance of which is unclear. However, similar findings were reported by Ede & Flint (1975) with limb bud cells of the chick *talpid* mutant which appear to be more, not less, adhesive to each other than the normal chick cells, yet also form smaller clumps of aggregated cells.

We conclude that a deficiency in surface galactose receptors leads to a loss of intercellular adhesiveness of the hamster fibroblasts. The most reasonable interpretation is that glycoproteins of the cell surface interact with suitable receptors on apposing cell surfaces, or with adhesive factors, some of which may be provided by serum, to mediate cellular aggregation (Fig. 3).

6. INTERACTION OF FIBROBLASTS WITH FIBRONECTIN

The notion that the adhesion between cells, or between a cell and its growth surface or other substratum, may be mediated by basically similar mechanisms comes from recent progress in the identification of the adhesive molecule, fibronectin. Fibronectin exists in significant amounts between cells, at sites of contact as well as in the adhesive plaques formed by cells attached to a substratum. In cultured fibroblasts, fibronectin appears to be present in an extracellular compartment and not integrated into the plasma membrane proper. Thus, a lipid-poor fraction staining well with ruthenium red appears to contain the bulk of the fibronectin of cultured hamster fibroblasts (Graham, Hynes, Davidson & Bainton, 1975). It is interesting, therefore, that the most intense area staining with ruthenium red in cultured fibroblast monolayers is found at cell–cell contacts and underneath the cells where a fibrillar array of material passing from the cell surface to the growth surface can be discerned (Stomatoglou, 1977).

It was first shown unequivocally by Culp and Grinnell and their colleagues (Culp & Buniel, 1976; Grinnell, 1978) that serum factors stabilise the attachment of fibroblasts to a (plastic) substratum. This requirement is readily monitored by observation of the morphological

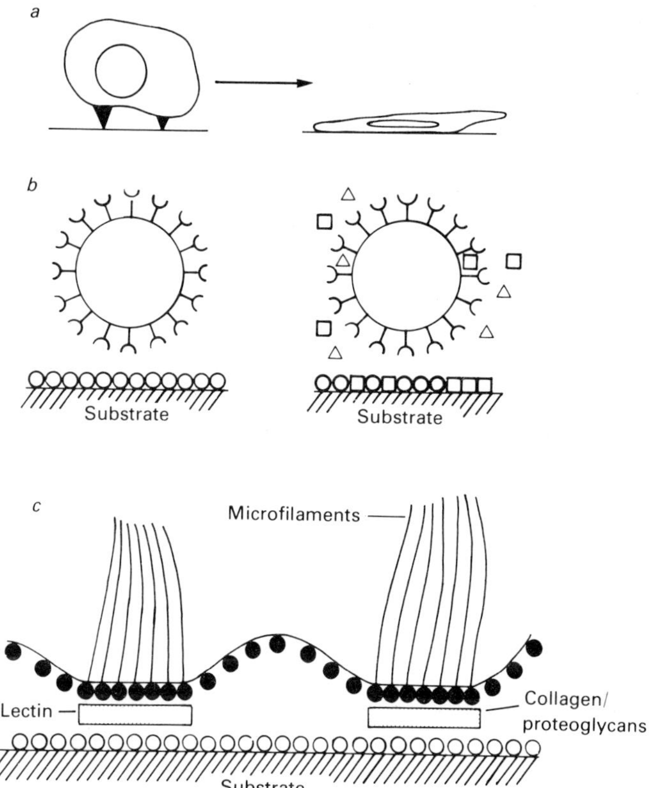

Fig. 4. Hypothetical mechanisms for cell attachment and flattening (*a*) on various substrata. In (*b*) absorbed fibronectin molecules (○) interact directly with cell surface receptors (carbohydrate chains of glycoproteins). Subsequently, flattening is induced (as in *a*) by the formation of many such contacts. The interaction is relatively specific since adsorbed albumin molecules (□) are inactive and prevent cell attachment. Neither albumin nor fibronectin (△) in the solution phase inhibit interaction of the cells with adsorbed fibronectin molecules. Either fibronectin undergoes, or is stabilised in, some conformational change to a biochemically active form after adsorption or the interaction of a cell with fibronectin-coated substrata requires cooperative binding of multiple contacts, each of individually weak affinity. In *c* interaction between substratum-adsorbed fibronectin molecules (○) and cell surface glycoproteins (●) is mediated by a multivalent lectin (*left*) or other substances, e.g. collagen and/or proteoglycans (*right*) with affinity both for cell surface glycoprotein receptors and fibronectin. The accumulation of contacts between cell surface receptors and adsorbed fibronectin molecules form localised adhesive plaques. Cytoplasmic microfilaments may be involved in formation of these stabilised plaques.

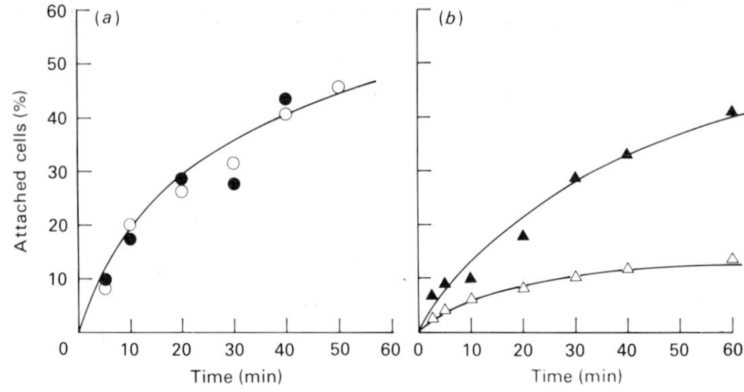

Fig. 5. Adhesion of BHK cells to plastic tissue culture dishes (*a*) or to denatured collagen (Type 1) (*b*) in the presence (●, ▲) or absence (○, △) of hamster plasma fibronectin (10 μg ml^{-1}). After dispersion by brief trypsin treatment, cells from monolayer cultures labelled by 16 h growth at 37 °C in medium containing [^{3}H]leucine (10 μCi ml^{-1}) were suspended in Eagle's medium plus foetal calf serum depleted of fibronectin by passage through a gelatin–Sepharose affinity column, washed once with serum-free Eagle's medium and suspended finally in this medium (Pena & Hughes, 1978*b*). Cells (2×10^5) were added in 1 ml serum-free Eagle's medium, fibronectin-supplemented as indicated, and the non-adherent cells removed after various times. Attached cells were dispersed for radioactive counting. Care must be taken to remove trace serum contaminants from the trypsinised cells before performing the adhesion assays.

changes occurring rapidly over one or two hours. The cells rapidly attach as round cells even in the absence of serum (Fig. 4*a*) but flatten down ('spread') only when a layer of adsorbed serum proteins is present (Fig. 4*b*), the most important component having characteristics very similar to purified serum fibronectin. Pearlstein (1976) also demonstrated a requirement for fibronectin in both the attachment and spreading of fibroblasts on a collagen film, a result confirmed by ourselves (Fig. 5*b*) and others (Grinnell & Minter, 1978). Subsequent studies have amply confirmed these conclusions for established fibroblastic cell lines using purified serum or cellular (Pena & Hughes, 1978*a*) fibronectins, although some uncertainty still exists concerning the exact requirements for fibronectin in the adhesion of cells to collagen fibrils. As Fig. 5*b* shows, without fibronectin BHK cells do attach (and flatten out) significantly on collagen polymerised by exposure to NH_4OH vapour, although added fibronectin clearly enhances adhesion. Following Pearlstein (1976) we find that this level of adsorption of cells is reduced substantially after treatment of the collagen gels with urea, perhaps due to removal of endogenous fibronectin in the collagen or the further denaturation and alteration of the fibrillar structure of the collagen substratum. Indeed, Grinnell & Minter (1978) have described substrata of native collagen fibrils to which BHK cells attach and spread apparently without *any* requirement for exogenous fibronectin, a result which Kleinman, McGoodwin, Rennard & Martin (1979) attribute to non-specific adhesion mediated by high levels of phosphate used in the former work. It is suggested that cells may bind to deposits of calcium phosphate rather than to the collagen. There is no doubt, however, that cells including BHK cells do show a significant requirement for exogenous fibronectin for adhesion and spreading onto collagen under defined conditions as well as for spreading on inert plastic surfaces.

We have shown recently that a threshold amount of fibronectin must be adsorbed to the substratum in order to promote the flattening of BHK cells (Hughes, Pena, Clark & Dourmashkin, 1979). The threshold appears to be of the order of 15×10^{-9} g cm^{-2} of surface (approximately 8×10^{11} molecules cm^{-2}) corresponding to a matrix of fibronectin molecules separated by approximately 75 nm, assuming an equal molecular distribution of the adsorbed protein on the surface. This distribution is very different to the closely packed array of fibronectin molecules achieved at saturation as determined by the Langmuir adsorption isotherm, which corresponds to an intermolecular centre-to-centre centre spacing of about 11.3 nm. Our conclusions are that a minimum number of contacts, about 45000 for BHK cells, must be formed between a cell and the substratum, as mediated by adsorbed fibronectin molecules to allow cell flattening. It has been shown repeatedly that the initial contacts between cells and a substratum are made by numerous filopodia, followed presumably by activation of the cellular contractile system (Heath & Dunn, 1978) to flatten down the cell onto the substratum (see Fig. 6*a*, *b*). The overall process is rather like pitching a bell-tent, i.e. establishment of anchor points followed by tightening of the guy-ropes. Little is known at present, however, of the additional factors controlling this process. One puzzling feature, for example, is the role of cell-derived fibronectin. Almost all of our work has been done with trypsinised cells, which therefore lack surface fibronectin. A requirement for exogenous serum-derived fibronectin in the early interactions with a substratum is therefore not unreasonable. However, we also find that BHK cells detached from monolayer cultures by EDTA and retaining the modest levels of surface fibronectin exhibited by these cells (Hughes & Nairn,

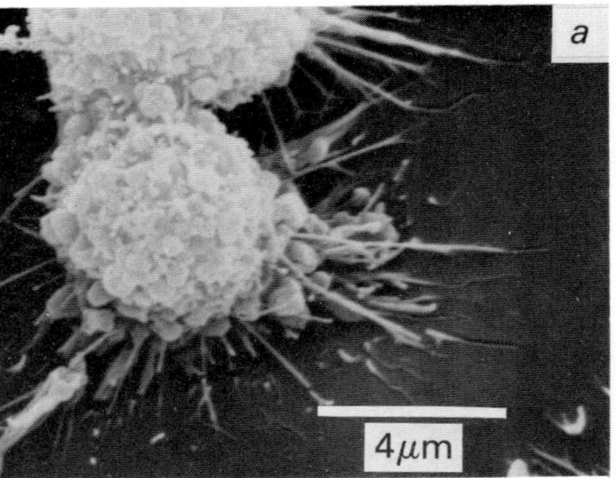

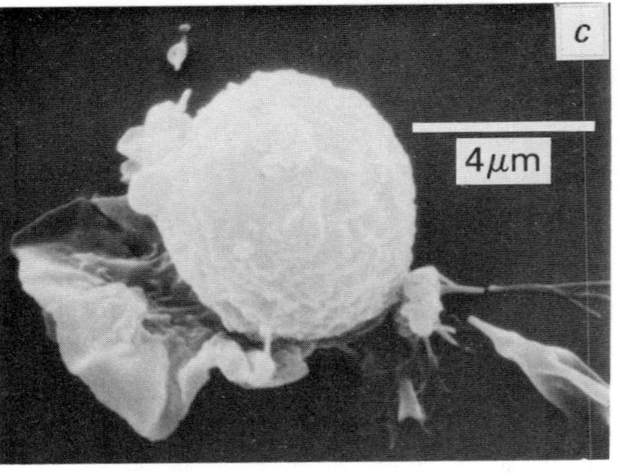

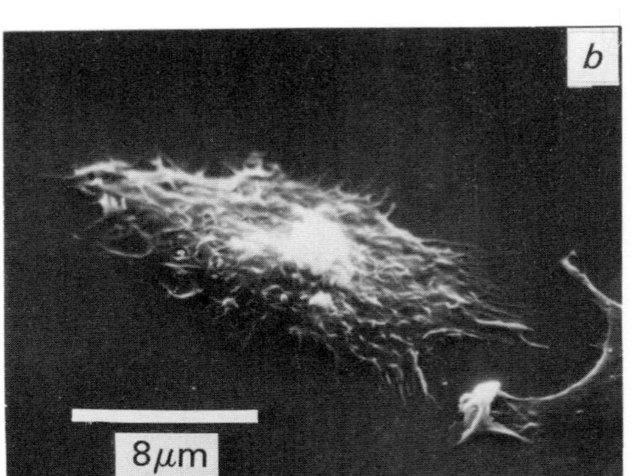

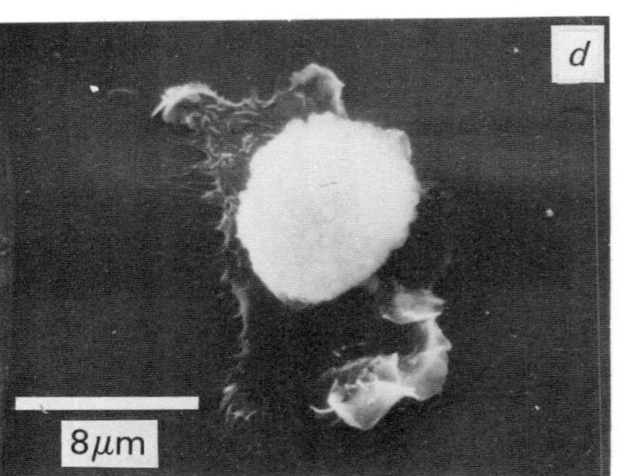

Fig. 6. Scanning electron micrographs of BHK cells (*a*, *b*) or ricin-resistant mutant cells (*c*, *d*) deficient in surface galactose terminal residues adhering to plastic coverslips coated with hamster plasma fibronectin after 5 min (*a*) or 90 min (*b*, *d*) incubation at 37 °C.

1977; Pena *et al.*, 1979) also require the addition of exogenous fibronectin to attach and spread onto a substratum. It is not that the fibronectin synthesised by BHK cells is inactive, since products extracted from these cells or from the spent culture fluids of these cells are as active as serum fibronectin in adhesion assays (Pena & Hughes, 1978*a*). Other studies with primary fibroblastic cells, for example human skin fibroblasts (Schwartz, Hellerqvist & Cunningham, 1978) or bovine lung fibroblasts (R. C. Hughes and Y. Courtois, unpublished observations) have shown that these cells attach and spread well onto various substrata without additional fibronectin even in the short term (1–2 h). Either these cells carry sufficient amounts of surface fibronectin, or provide by rapid synthesis *de novo* sufficiently high levels of fibronectin in the extracellular medium to promote cell adhesion, unlike established cell lines.

Since the flattening of trypsinised BHK cells onto a fibronectin-coated substratum proceeds in the presence of inhibitors of nucleic acid and protein synthesis (Pena & Hughes, 1978*b*), clearly replacement or insertion *de novo* of surface molecules is not necessary during the short (60–90 min) time required for flattening. The process does require a functional microfilament system, however, and is powerfully inhibited by cytochalasin B. Microtubules may not be as important, at least in these early events, and colchicine or vinblastine has little effect.

The attachment and spreading of BHK cells on a fibronectin-coated surface is, therefore, a complex series of events, including the presumptive interaction of substrate-adsorbed fibronectin molecules with the cell surface, the multiplication of such contacts and finally the triggering of a cytoplasmic response to reorder cytoskeletal components leading to cell flattening. Presumably, since our experiments can be done without macromolecular synthesis, subsequent events occurring on adhesion of cells to a substratum such as the induction of specific cellular protein synthesis (Folkman & Moscona, 1978) are not of direct concern, although their relevance to subsequent differentiation of adherent cells is of obvious importance.

Using the simplified system, the role of carbohydrate in the interaction of cells with fibronectin-coated substrates can be examined relatively simply. The following sugars alone or in combination have no effect on the attachment and spreading of BHK cells either to fibronectin-coated plastic or collagen, even at high concentrations; D-galactose, *N*-acetyl-D-galactosamine, lactose, D-fucose, thiodiga-

Table 1. *Effects of various substances on fibronectin-mediated adhesion*

Substance	Concentration	Appearance of cells on fibronectin-coated plastic or collagen	Attachment to collagen (%)
None	–	Spread	58
Ricin	130 μg ml^{-1}	Round	52
	5 μg ml^{-1}	Spread	61
Ricin B chain	150 μg ml^{-1}	Round	50
	3 μg ml^{-1}	Spread	59
Lysine	71 mM	Round	27
	3.5 mM	Spread	50
D-galactosamine[a]	71 mM	Round	15
	3.5 mM	Spread	56
N-acetyl-D-galactosamine	71 mM	Spread	55
	3.5 mM	Spread	56
Lactose[b]	71 mM	Spread	55
	3.5 mM	Spread	54

See legend to Fig. 5 for experimental details. Ricin-resistant mutant Ric^{R} 19 was used and incubations were at 37 °C for 90 min.

[a] D-glucosamine and D-mannosamine gave similar results.

[b] D-galactose, D-fucose, thio-di-galactoside at similar concentration were also inactive.

lactoside (see Table 1). We used these compounds as potent inhibitors of ricin binding to BHK cells. We reasoned, therefore, that if ricin-binding carbohydrate sequences are involved in binding to fibronectin, suitable hapten inhibitors may influence the response of the cells. D-fucose is a potent inhibitor of a membrane lectin of BHK cells (Dysart & Edwards, 1977) which may also be implicated in adhesive interactions involving carbohydrate moieties. These negative findings, although surprising, are none the less reconcilable with the hypothesis that any cell surface interactions involving fibronectins may be more specific than simple binding to monosaccharides or even disaccharide sequences. The inhibitory effect of D-galactosamine on BHK cell adhesion (Table 1) was expected from the observations of Yamada *et al.* (1975) who reported inhibition of haemagglutination by the fibronectin of chick fibroblasts. Similarly, as reported by Yamada *et al.* (1975), other primary amines, for example lysine and arginine, interact strongly with fibronectins, and we have found that these prevent attachment and spreading of BHK to fibronectin-coated substrata (Table 1).

In an alternative approach, we examined the effects of ricin itself or the sugar-binding subunit B of ricin on BHK cell–fibronectin interactions. Ricin or the B subunit at high concentration do inhibit the spreading of BHK cells on fibronectin-coated plastic substrata (Table 1). We conducted these experiments with the ricin-resistant cell line Ric^R19 (Meager *et al.*, 1976) which is unaffected over the short periods of our assays (1–2 h) by high ricin concentrations, as measured by incorporation of radioactive amino acids into total cellular protein. We had shown earlier (Edwards *et al.*, 1976) that Ric^R19 cells, which have a normal complement of surface carbohydrates binding ricin (Meager *et al.*, 1976), exhibit near-normal adhesive properties and therefore provide a valid alternative to using the parental, ricin-sensitive BHK cells. The parental BHK cells would, of course, be rapidly affected by such high ricin concentrations, which induce cell death. By contrast, the effects of ricin on the resistant mutant cells attached to a fibronectin-coated surface are reversible. Addition of lactose (10 mM), a hapten inhibitor of ricin, to cells bound to fibronectin-coated plastic in the presence of ricin causes a remarkable change in morphology of the cells from a rounded shape, with many surface projections and filopodia, to a well-flattened form, similar to control cells spread onto a coated surface in the absence of the lectin. The implications of these experiments are as follows: ricin binds to cell surface glycoproteins and alters somehow the interaction of at least some of these glycoproteins with the fibronectin-coated plastic surface, and thereby inhibits cell spreading. When ricin binding to cell surface receptors is reversed by lactose, stable contacts rapidly form and the cells spread normally. The possibility that ricin binding inhibits cell spreading by, in some way, freezing the surface glycoproteins in a cross-linked aggregate preventing the ordered rearrangement of those glycoproteins occurring during spreading was considered unlikely, since ricin is a poor agglutinin of cells, including BHK cells, even at high concentration (our unpublished observations). Agglutination of cells is usually considered to require the lateral rearrangement of surface molecules induced through cross-linking by multivalent ligands. Neither ricin nor the B subunit appears to exist in solution in multimeric form, and since each monomer contains only one sugar-binding site it would appear unlikely that these reagents function as multivalent ligans.

In order to clarify the effect of ricin we studied the adhesion of cells to collagen. As mentioned previously, fibronectin is required for attachment to as well as spreading on collagen films. As Table 1 shows, ricin or the B subunit has little effect on the attachment to collagen mediated by fibronectin, but again inhibits the spreading of attached cells. We suggest that the spreading phenomenon is more sensitive to ricin, since many contacts with the fibronectin-coated substratum are required, whereas less extensive interactions may be sufficient to facilitate simple attachment. Indeed, when the ricin B subunit is added to cells spread onto a fibronectin-coated collagen substratum we observe that the cells gradually, over the first 1–2 h, round up, and only after prolonged incubation (6–8 h) do the cells become detachable by shaking. Again, after addition of lactose at this point the cells re-attach and spread rapidly.

Further insight on the role of carbohydrates carrying the ricin-binding residues in interactions of cells and fibronectin-coated substrata was obtained using the mutant lines deficient in surface ricin-binding sites. Earlier, it was shown (Edwards *et al.*, 1976) that freshly dispersed cells of Ric^R19 and 22 adhere to a film of adsorbed calf serum or to collagen in the presence of trace amounts of serum to the same extent as parental BHK cells, while receptor-deficient clones, e.g., Ric^R14 and 17, adhere less well. We have now repeated and extended these observations using fibronectin-coated substrata. Time-lapse films (Pena & Hughes, 1978*b*) and scanning electron microscopy (Fig. 6) show that the mutant cells attaching to both types of substrates begin to flatten rather like BHK cells, but the flattened morphology appears to be unstable. The highly extended forms of BHK cells seen 1–2 h after seeding onto fibronectin-coated plastic, for example, do not appear in the mutant cells. Extensive areas of the cell surface remain highly convoluted and cells, even when spread, revert to a rounded morphology with high frequency (Pena & Hughes, 1978*b*). Thus, the general appearance of the mutant cells even after several hours (Fig. 6*c*, *d*) resembles more closely that of BHK cells freshly seeded onto the coated surface (Fig. 6*a*) than BHK cells kept in contact with the surface for comparable periods (Fig. 6*b*). Furthermore, the extent of attachment of these mutant cells to collagen, which can be readily quantitated, appears to be significantly reduced compared to parental BHK cells or the Ric^R19 ricin-resistant mutant cells (Pena & Hughes, 1978*b*).

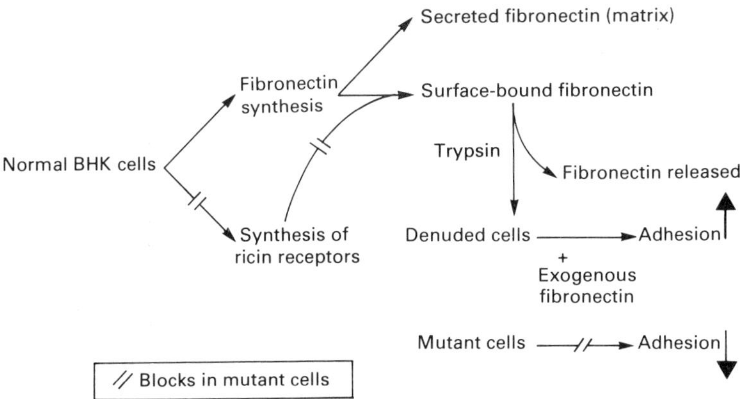

Fig. 7. Proposed relationship of the synthesis of cell surface glycoproteins (ricin receptors) to cell–fibronectin interactions involved in adhesion. See text for discussion of evidence for this hypothetical scheme.

Our findings are summarised in Fig. 7. We propose that the interaction of BHK cells with substrata mediated by fibronectin involves carbohydrate moieties, probably of glycoproteins. When these carbohydrates are blocked or their synthesis is defective, as in certain lectin-resistant mutants, cellular adhesiveness as measured by a variety of methods is reduced. The relevant carbohydrate moieties are part of glycoproteins of the cell plasma membrane, as shown by the relatively poor response of the mutant cells to exogenous serum fibronectin in the adhesion assays. Furthermore, fibronectin is synthesised and secreted in normal amounts by glycosylation-deficient mutant cells, and in this form is fully active biologically (Pena & Hughes, 1978*b*), showing that changes in the carbohydrate moiety of this molecule can be tolerated. A defect in cell surface glycoproteins also appears to reduce the surface expression of fibronectin in long-term monolayer cultures of mutant cells. Thus, glycosylation-deficient mutants either lack surface fibronectin completely as Ric^{R}17 cells or partially as Ric^{R}14 compared to normal BHK cells or certain mutants, e.g. Ric^{R}19 and 22 as detected by surface labelling (Pena *et al.*, 1979) or by immunofluorescence using specific antibody (our unpublished observations). In other words, these carbohydrate-defective mutant cells produce and secrete fibronectin but are unable to fix it at the surface, in agreement with their reduced adhesiveness on the fibronectin-coated substrata. It should be emphasised, however, that there is as yet no evidence to suggest that identical or even similar mechanisms are involved in these two interactions. Little is known concerning the interrelationships of surface-bound and secreted fibronectins, and it is entirely possible that these represent non-exchangeable pools, perhaps with different functions. Our experiments do suggest, however, that a lectin-like activity requiring a normal complement of surface galactose-containing sequences is involved in the adhesion of cells to substrata mediated by fibronectin, either directly in the interactions occurring between adsorbed fibronectin molecules and cell surface carbohydrates (Fig. 4*b*) or indirectly to form adhesive bridges (Fig. 4*c*). Alternatively, the interaction of other protease insensitive extracellular components such as heparan sulphate or collagen with affinity for both fibronectin and cell surface receptors may be involved, though there is as yet no evidence to suggest a requirement for surface carbohydrate groups in the latter.

7. PHYSIOLOGICAL SIGNIFICANCE OF FIBROBLAST ADHESION

Fibroblasts can no longer be considered as undifferentiated cells with no precise biological origin or function. The well-established migratory properties of fibroblasts in culture (Abercrombie, 1961) appear to be related to the laying down *in vivo* of the extracellular macromolecules, of connective tissue, collagen and proteoglycans, during development or tissue repair. For example, the migration of fibroblasts into sites of damage during the inflammatory response precedes formation of new connective tissue which constitutes the scar. The movement of fibroblasts in culture or during wound healing is likely to be analogous to the behaviour of most cells participating in morphogenetic movements. The mechanisms regulating such movements and leading to specific macromolecular synthesis must involve cell–cell and cell–substratum contacts mediated by cell surface molecules and perhaps by diffusible molecules emitted by cells at or remote from the appropriate site. Impressive evidence for the important role of the substratum in orienting cellular migration was obtained many years ago and termed contact guidance. A simple demonstration of the alignment of fibroblasts along collagen deposits mediated by one soluble, potentially diffusible substance, fibronectin, is shown in Fig. 8. Could such a phenomenon apply to the intact organism, for example in the migration of pigment cells or nerve axons along blood vessels, the movement of cardiac cells on an oriented endodermal substratum or of neural crest cells along the

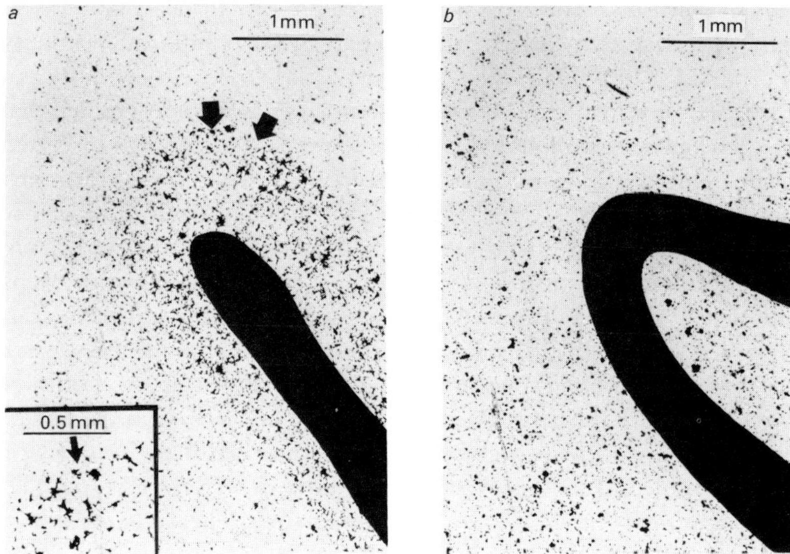

Fig. 8. Adhesion of BHK cells to collagen mediated by fibronectin. (*a*) A plastic tissue culture dish was coated by incubation at 37 °C for 60 min with bovine serum albumin (10 mg ml^{-1} in saline) to render the surface poorly adhesive for trypsinised cells. A collagen solution (Type I, 0.25 % in 0.2 % acetic acid) was streaked across the well-washed plate as indicated by the solid dark lines, and dried. The plate was washed again with serum-free Eagle's medium supplemented with fibronectin (10 μg ml^{-1}), then cells in Eagle's medium were added and the arrangement of attached cells was photographed after 90 min incubation at 37 °C. Note the sharp boundary (arrows) of well-spread cells defining the limits of the collagen deposit. (*b*) An identical experiment, but carried out in the absence of fibronectin. Note the relatively low and uniform background of attached cells over areas with or without collagen deposits.

neural tube (Weston, 1970; Bard & Higginson, 1977)? Could similar mechanisms regulate the expansion of the free edge of an epithelial cell sheet such as occurs very commonly during embryogenesis? In vertebrates, the majority of the epithelial and endothelial cellular mass is adjacent to collagenous layers contributed by cells of mesodermal origin. The appearance of intercellular adhesive substances such as fibronectin, as well as proteoglycans with affinity for collagen, clearly relates to the formation and maintenance of such an organisation. Little is known as yet, however, of the role of basement membrane components including fibronectin in the organisation of epithelial cell layers. These substances can be regarded either as extracellular or as superficial components of the surface membrane, held in place by a specific chemistry of the cell surface. We should like to propose that carbohydrate is part of this chemical specificity as outlined in the preceeding sections.

In migration, cells must make contacts with the substratum that are sufficiently strong and numerous to provide anchorage to flatten the cell and to pull it in the given direction. Our findings that many thousands of contacts must be provided for hamster fibroblasts to flatten out onto a fibronectin matrix is in accord with this notion. On the other hand, the adhesive interactions must be sufficiently weak and therefore reversible not to immobilise the cell completely. If excessively weak, the cells would remain in contact with the growth surface as immobile, rounded cells. The strength of the adhesive contacts require the intimate interactions of cell surface components with the substratum as well as the reorganisation of the underlying cytoplasmic contractile and shape-supporting systems of microfilaments and microtubules which presumably is triggered in some way by the initial contacts.

Our experiments are beginning to suggest mechanisms by which such changes in cellular adhesiveness may be manipulated. Whether similar manipulations occur *in vivo* remains speculative. It is interesting, however, that changes in cell surface carbohydrate structure have been established as occurring during the cell cycle. Mitotic cells differ from interphase cells, for example, in their complement of surface sialic acid (Warren *et al.*, 1978), are less adhesive and more readily detached from the substratum. Indeed, commonly used techniques to isolate mitotic cells utilise their reduced adhesiveness in this way. Similarly, anaplastic cells resemble mitotic cells in a rounded, highly convoluted surface morphology, an enhanced surface sialic acid content and are less adhesive to various substrata than flattened interphase normal fibroblasts.

Although analysis of fibroblast adhesion is as yet in an early stage, it has already revealed features that may apply rather generally. It is for this reason that the mechanisms of fibroblast adhesion under controlled culture conditions have clear relevance to differentiation and histogenesis.

Acknowledgements. We thank Drs R. Dourmashkin and J. Clark for helpful discussions and the electron microscopy, and G. Mills for help with many of the experiments reported. S. D. J. Pena was a recipient of an MRC (Canada) Fellowship award and P. Vischer is an EMBO Fellow. We also thank Ms Lydia Pearson for excellent handling of the manuscript.

REFERENCES

ABERCROMBIE, M. (1961). The bases of the locomotory behaviour of fibroblasts. *Experimental Cell Research*, Suppl. **8**, 188–98.

ASHWELL, G. & MORELL, A. G. (1977). Membrane glycoproteins and recognition phenomena. *Trends in Biochemical Science*, **2**, 76–9.

BARD, J. B. L. & HIGGINSON, K. (1977). Fibroblast–collagen interactions in the formation of the secondary stroma of the chick cornea. *Journal of Cell Biology*, **74**, 816–27.

BORNSTEIN, P. & ASH, J. F. (1977). Cell surface-associated structural proteins in connective tissue cells. *Proceedings of the National Academy of Sciences, USA*, **74**, 2480–4.

CHIPOWSKY, S., LEE, Y.-C. & ROSEMAN, S. (1973). Adhesion of fibroblasts to insoluble analogues of cell surface carbohydrates. *Proceedings of the National Academy of Sciences, USA*, **70**, 2309–12.

COMPER, W. D. & LAURENT, T. C. (1978). Physiological function of connective tissue polysaccharides. *Physiological Reviews*, **58**, 255–315.

CULP, L. A. & BUNIEL, J. F. (1976). Substrate attached serum and cell proteins in adhesion of mouse fibroblasts. *Journal of Cellular Physiology*, **88**, 89–106.

CULP, L. A., ROLLINS, B. J., BUNIEL, J. & HITRI, S. (1978). Two functionally distinct pools of glycosaminoglycans in the substrate adhesion site of murine cells. *Journal of Cell Biology*, **79**, 788–801.

CURTIS, A. S. G. (1973). Cell adhesion. In *Progress in Biophysics and Molecular Biology*, vol. 27, ed. J. A. V. Butler & D. Noble, pp. 315–86. Oxford: Pergamon Press.

DAMSKY, C. H., LEVY-BENSHIMOL, A., BUCK, C. A. & WARREN, L. (1979). Effect of tunacamycin on the synthesis, intracellular transport and shedding of membrane glycoproteins in BHK cells. *Experimental Cell Research*, **119**, 1–13.

DAN, K. (1936). Electrokinetic studies of marine ova. III. *Physiological Zoology*, **9**, 43–57.

DESSAU, W., JILEK, F., ADELMANN, B. C. & HORMANN, H. (1978). Similarity of antigelatin factor and cold insoluble globulin. *Biochimica et Biophysica Acta*, **533**, 227–37.

DYSART, J. McK. & EDWARDS, J. G. (1977). A membrane-bound haemagglutinin from cultured hamster fibroblasts. *FEBS Letters*, **75**, 96–104.

EDE, D. A. & FLINT, O. P. (1975). Intercellular adhesion and formation of aggregation in normal and *talpid*^{3} mutant chick limb mesenchyme. *Journal of Cell Science*, **18**, 97–111.

EDWARDS, J. G. (1973). Intracellular adhesion. In *New Techniques in Biophysical and Cell Biology*, vol. 1, ed. R. H. Pain & B. J. Smith, pp. 1–27. New York, London: Wiley.

EDWARDS, J. G., DYSART, J. McK. & HUGHES, R. C. (1976). Cellular adhesiveness reduced in ricin-resistant hamster fibroblasts. *Nature, London*, **264**, 66–8.

ENGVALL, E. & RUOSLAHTI, E. (1977). Binding of soluble form of fibroblast surface protein, fibronectin, to collagen. *International Journal of Cancer*, **20**, 1–5.

FOLKMAN, J. & MOSCONA, A. (1978). Role of cell shape in growth control. *Nature, London*, **273**, 345–9.

GRAHAM, J. M., HYNES, R. O., DAVIDSON, E. A. & BAINTON, D. F. (1975). The location of proteins labeled by the ^{125}I-lactoperoxidase system in the NIL8 hamster fibroblast. *Cell*, **4**, 353–65.

GRINNELL, F. (1978). Cellular adhesiveness and extracellular substrata. *International Review of Cytology*, **58**, 65–144.

GRINNELL, F. & HAYS, D. G. (1978). Cell adhesion and spreading factor. Similarity to cold insoluble globulin in human serum. *Experimental Cell Research*, **115**, 221–9.

GRINNELL, F. & MINTER, D. (1978). Attachment and spreading of baby hamster kidney cells to collagen substrata: effects of cold insoluble globulin. *Proceedings of the National Academy of Sciences, USA*, **75**, 4408–12.

HAUSMAN, R. E. & MOSCONA, A. (1975). Purification and characterization of the retina-specific cell-aggregating factor. *Proceedings of the National Academy of Sciences, USA*, **72**, 916–20.

HEATH, J. P. & DUNN, G. A. (1978). Cell to substratum contacts of chick fibroblasts and their relation to the microfilament system. A correlated interference–reflexion and high voltage electron microscopic study. *Journal of Cell Science*, **29**, 197–212.

HEDMAN, K., VAHERI, A. & WARTIOVAARA, J. (1978). External fibronectin of cultured human fibroblasts is predominantly a matrix protein. *Journal of Cell Biology*, **76**, 748–60.

HICKMAN, S., KULCZYCKI, A., LYNCH, R. G. & KORNFELD, S. (1977). Studies of the mechanism of tunicamycin inhibition of IgA and IgE secretion by plasma cells. *Journal of Biological Chemistry*, **252**, 4402–8.

HUGHES, R. C. (1976*a*). Membrane glycoproteins: a review of structure and function. Boston, London: Butterworth.

HUGHES, R. C. (1976*b*). Cell surface membranes of animal cells as the sites of recognition of infectious agents and other substances. In *The Specificity of Plant Diseases*, ed. R. K. S. Wood & A. Graniti, pp. 77–99. New York, London: Plenum Press.

HUGHES, R. C. (1979). Cell surface carbohydrates in relation to receptor activity. In *Glycoconjugate Research* (proceedings of the Fourth International symposium on Glycoconjugates), ed. J. D. Gregory & R. W. Jeanloz, 985–1005. New York, London: Academic Press.

HUGHES, R. C., MEAGER, A. & NAIRN, R. (1977). Effect of 2-deoxyglucose on the cell surface glycoproteins of hamster cells. *European Journal of Biochemistry*, **72**, 265–273.

HUGHES, R. C. & NAIRN, R. (1978). Labelling, solubilization and peptide mapping of fibroblast surface glycoprotein. *Proceedings of the National Academy of Sciences, USA*, **312**, 192–206.

HUGHES, R. C., PENA, S. D. J., CLARK, J. & DOURMASHKIN, R. (1979). Molecular requirements for the adhesion and spreading of hamster fibroblasts. *Experimental Cell Research* **121**, 307–14.

Kleinman, H., McGoodwin, E. B., Rennard, S. I. & Martin, G. R. (1979). Preparation of collagen substrates for cell attachment: effect of collagen concentration and phosphate effect. *Analytical Biochemistry*, **94**, 308–13.

Mautner, V. & Hynes, R. O. (1977). Surface distribution of LETS protein in relation to the cytoskeleton of normal and transformed cells. *Journal of Cell Biology*, **75**, 743–68.

Meager, A., Ungkitchanukit, A. & Hughes, R. C. (1976). Variants of hamster fibroblasts resistant to *Ricinus communis* toxin (ricin). *Biochemical Journal*, **154**, 113–24.

Meager, A., Ungkitchanukit, A., Nairn, R. & Hughes, R. C. (1975). Ricin resistance in baby hamster kidney cells. *Nature, London*, **257**, 137–9.

Morris, E. R., Rees, D. A., Thom, D. & Welsh, E. J. (1977). Conformation and intermolecular interactions of carbohydrate chains. *Journal of Supramolecular Structure*, **6**, 259–74.

Moscona, A. (1962). Analysis of cell recombinations in experimental synthesis of tissues *in vitro*. *Journal of Cellular and Comparative Physiology*, Suppl. 1, **60**, 65–80.

Olden, K., Pratt, R. M. & Yamada, K. M. (1978). Role of carbohydrates in protein secretion and turnover; effects of tunacamycin on the major cell surface glycoprotein of chick embryo fibroblasts. *Cell*, **13**, 461–73.

Pearlstein, E. (1976). Plasma membrane glycoprotein which mediates adhesion of fibroblasts to collagen. *Nature, London*, **262**, 497–500.

Pena, S. D. J. & Hughes, R. C. (1978*a*). Fibroblast to substratum contacts mediated by the different forms of fibronectin. *Cell Biology, International Reports*, **2**, 339–44.

Pena, S. D. J. & Hughes, R. C. (1978*b*). Fibronectin–plasma membrane interactions in the adhesion and spreading of hamster fibroblasts. *Nature, London*, **276**, 80–3.

Pena, S. D. J., Mills, G. & Hughes, R. C. (1979). Two-dimensional electrophoresis of surface glycoproteins of normal BHK cells and ricin resistant mutants. *Biochimica et Biophysica Acta*, **550**, 100–109.

Pessac, B. & Defendi, V. (1972*a*). Cell aggregation: role of acid mucopolysaccharides. *Science*, **175**, 898–900.

Pessac, B. & Defendi, V. (1972*b*). Evidence for distinct aggregation factors and receptors in cells. *Nature, London*, **238**, 13–15.

Roseman, S. (1970). The synthesis of complex carbohydrates by multiglycosyltransferase systems and their potential function in intercellular adhesion. *Chemistry and Physics of Lipids*, **5**, 270–97.

Roth, S., McGuire, E. J. & Roseman, S. (1971*a*). Evidence for cell surface glycosyltransferases. Their potential role in cellular recognition. *Journal of Cell Biology*, **51**, 536–47.

Roth, S., McGuire, E. J. & Roseman, S. (1971*b*). An assay for intercellular adhesive specificity. *Journal of Cell Biology*, **51**, 525–35.

Schlessinger, J., Barak, L. S., Hammes, G. G., Yamada, K. M., Pastan, I., Webb, W. W. & Elson, E. L. (1977). Mobility and distribution of a cell surface glycoprotein and its interaction with other membrane components. *Proceedings of the National Academy of Sciences, USA*, **74**, 2909–13.

Schubert, D., La Corbiere, M., Whitlock, C. & Stallcup, W. (1978). Alterations in the surface properties of cells responsive to nerve growth factor. *Nature, London*, **273**, 718–23.

Schwartz, C. F., Hellerqvist, C. G. & Cunningham, L. W. (1978). A collagenous component of the microexudate carpet secreted by attaching human fibroblasts. *Annals of the New York Academy of Sciences*, **312**, 450–2.

Stathakis, N. E. & Mosesson, M. W. (1977). Interactions among heparin cold insoluble globulin and fibrinogen in formation of the heparin-precipitate fraction of plasma. *Journal of Clinical Investigation*, **60**, 855–62.

Stenman, S., Wartiovaara, J. & Vaheri, A. (1977). Changes in the distribution of major fibroblast protein, fibronectin during mitosis and interphase. *Journal of Cell Biology*, **74**, 453–67.

Stomatoglou, S. C. (1977). Ultrastructural relationship between cell and substrate coats in serum-free and serum-supplemented cultures. *Journal of Ultrastructural Research*, **60**, 203–11.

Turner, R. S. & Burger, M. M. (1973). The cell surface in cell interactions. *Ergebnisse der Physiologie*, **68**, 122–55.

Tyler, A. (1946). An auto-antibody concept of cell structure, growth and differentiation. *Growth*, **10** (Symposium 6) 7–19.

Urushihara, H., Takeichi, M., Hakura, A. & Okada, T. S. (1976). Different cation requirements for aggregation of BHK cells and their transformed derivatives. *Journal of Cell Science*, **22**, 685–95.

Vaheri, A., Kurkinen, M., Letho, V.-P., Linder, E. & Timpl, R. (1978). Codistribution of pericellular matrix proteins in cultured fibroblasts and loss in transformation: fibronectin and procollagen. *Proceedings of the National Academy of Sciences, USA*, **75**, 4944–8.

Vicker, M. G. & Edwards, J. G. (1972). The effect of neuraminidase on the aggregation of BHK 21 cells and BHK 21 cells transformed by polyoma virus. *Journal of Cell Science*, **10**, 759–68.

Vuento, M., Wrann, M. & Ruoslahti, E. (1977). Similarity of fibronectins isolated from human plasma and spent culture medium. *FEBS Letters*, **82**, 227–31.

Warren, L., Buck, C. A. & Tuszynski, G. P. (1978). Glycopeptide changes and malignant transformation. A possible role for carbohydrate in malignant behaviour. *Biochimica et Biophysica Acta*, **516**, 97–127.

Weiss, P. (1947). The problem of specificity in growth and development. *Yale Journal of Biology and Medicine*, **19**, 235–78.

Weston, J. (1970). The migration and differentiation of neural crest cells. In *Advances in Morphogenesis*, vol. 8, ed. M. Abercrombie, J. Brachet & T. King, pp. 41–114. New York, London: Academic Press.

Yamada, K. M. & Kennedy, D. W. (1979). Fibroblast cellular and plasma fibronectins are similar but not identical. *Journal of Cell Biology*, (in press).

Yamada, K. M. & Olden, K. (1978). Fibronectins: adhesive glycoproteins of cell surface and blood. *Nature, London*, **275**, 179–85.

Yamada, K. M., Yamada, S. S. & Pastan, I. (1975). The major cell surface glycoprotein of chick embryo fibroblasts is an agglutinin. *Proceedings of the National Academy of Sciences*, USA, **72**, 3158–62.

Fibronectin: properties and role in cellular morphology and adhesion

KENNETH OLDEN, LIANG-HSIEN E. HAHN AND KENNETH M. YAMADA

Laboratory of Molecular Biology, National Cancer Institute, Bethesda, Maryland 20205, USA

INTRODUCTION

The participation of cell surface glycoproteins in cellular adhesion, malignancy, haemostasis, host defence by the reticuloendothetial system, organisation of connective tissue, control of cell surface architecture, morphology, hormone binding and endocytosis has been extensively documented in recent years (reviewed in Hynes, 1976; Yamada & Pastan, 1976; Vaheri & Mosher, 1978; Yamada & Olden, 1978; Yamada, Olden & Pastan, 1978). During the past five years our laboratory has expended considerable energy to elucidate the role of cellular fibronectin in determining the surface architecture, morphology, and adhesive properties of fibroblastic cells in culture. Fibronectin (from the Latin *fibra*, fibre, and *nectere*, to bind, tie) is the major glycoprotein found on the surface of many cultured cells; it was discovered by labelling cell surface proteins or carbohydrates with radioisotopes or with a specific antibody coupled to fluorescein or rhodamine. This class of glycoproteins is also known as cold-insoluble globulin (Morrison *et al.*, 1948), LETS protein (Hynes, 1973) and CSP (Yamada & Weston, 1974), and has an apparent subunit molecular weight of between 200000 and 250000 by SDS–polyacrylamide gel electrophoresis. Fibronectin was first discovered in human plasma in 1948 by Morrison and associates, and was called 'cold insoluble globulin' because it precipitated in the cold when complexed with fibrin or fibrinogen (Morrison *et al.*, 1948). Other closely related proteins or factors, named according to origin or biological properties, have also been described.

This chapter reviews and correlates the information that has

* Present address: Department of Oncology, Howard University Hospital, Washington, DC. 20060, U.S.A.

accumulated concerning fibronectin, with emphasis on data reported from our laboratory. We discuss the mechanism of regulation of this glycoprotein, its molecular properties, function, role of carbohydrates in its secretion and biological activity, and its interaction with collagen.

ISOLATION AND PURIFICATION

Biologically active fibronectin has been isolated from the surface of intact cells by extraction with low concentrations of urea (0.2–1.0 M) (Yamada & Weston, 1974, Yamada *et al.*, 1975). This procedure is so gentle that it detaches fibronectin from the outer surface of the plasma membrane without disrupting membrane integrity. Various chelating agents, salts, non-ionic and ionic detergents, guanidine, and dithiothreitol are much less effective in selectively extracting fibronectin or also result in more substantial cell lysis.

Although the fibronectin extracted from chick embryo fibroblasts by 1.0 M urea is relatively pure by the criterion of sodium dodecyl sulphate gel electrophoresis (Yamada & Weston, 1974; Yamada *et al.*, 1978), a more homogeneous preparation can be obtained by precipitation with 70% ammonium sulphate followed by resuspension in CAPS (cyclohexylaminopropane sulphonic acid) buffer (pH 11.0) and chromatography on Sepharose CL/4B. Fibronectin can also be further purified by affinity absorption to formalin-fixed erhthrocytes (Yamada *et al.*, 1975, 1978) or immobilised gelatin (Engvall & Ruoslahti, 1977). According to this isolation protocol, fibronectin has been purified to constant specific activity and is pure by the additional criteria that the reduced protein fraction migrates as a single band on (1) SDS or 8 M urea gels, (2) molecular sizing columns, and (3) in ultracentrifugation sedimentation analyses.

The efficiency of urea extraction is significantly reduced with human diploid fibroblasts, but homogeneous preparations can also be obtained according to the protocol described for chick cells. Cellular fibronectin has also been isolated by other protocols from hamster fibroblasts (Ali *et al.*, 1977, Carter & Hakomori, 1977; Pena & Hughes, 1978).

LOCATION AND DISTRIBUTION

Fibronectin is a major protein found on the cell surface of fibroblastic cells explanted into tissue culture. It constitutes 1–3% of total cellular protein (Yamada *et al.*, 1977*a*) and 40–50 % of total membrane protein of such cells *in vitro*.

The quantities of this glycoprotein are often decreased or absent from the cell surface after malignant transformation (Hynes, 1973; Hogg, 1974; Yamada & Olden, 1978). Similarly, it has been reported that the amounts of this protein are decreased upon the establishment of many permanent cell lines (Yamada *et al.*, 1977*b*).

Fibronectin has also been detected on the surface of astroglial and epithelial cells, in connective tissue, in primitive mesenchyme, basement membranes, in plasma, and in amniotic and cerebrospinal fluids and on platelets (Vaheri *et al.*, 1976; Chen *et al.*, 1977*a*; Birdwell *et al.*, 1978; Crouch *et al.*, 1978; Jaffe & Mosher, 1978). It is present in human plasma at a concentration of approximately 0.3 mg ml^{-1} (Mosesson & Umfleet, 1970), and is known as cold insoluble globulin because it binds to fibrinogen to form a cryoprecipitate (Stathakis & Mosesson, 1977) and its concentration is therefore decreased in serum. The amounts of fibronectin in blood are also decreased after extensive trauma, in the disseminated intravascular coagulation syndrome, and after reticuloendothelial system blockade following injection of colloids into animals (Aronsen *et al.*, 1972; Kaplan & Saba, 1975; Kaplan *et al.*, 1976; Blumenstock *et al.*, 1977*b*; Mosesson, 1978). In contrast, its levels are often elevated in hepatic diseases which affect the liver's protein-degradative capacity, such as cirrhosis.

Cell surface fibronectin has been shown to be most abundantly located in an aggregated fibrillar arrangement in cultured cells by direct or indirect immunofluorescence with affinity-purified anti-fibronectin antibodies (Wartiovaara *et al.*, 1974; Vaheri *et al.*, 1976; Yamada & Pastan, 1976; Bornstein & Ash, 1977; Chen *et al.*, 1977*b*; Mautner & Hynes, 1977; Schlesinger *et al.*, 1977; Stenman *et al.*, 1977; Bornstein *et al.*, 1978; Chen *et al.*, 1978; Crouch *et al.*, 1978; Hedman *et al.*, 1978; Vaheri & Mosher, 1978; Yamada, 1978). Small amounts of fibronectin are also diffusely distributed over the surface of the plasma membrane (Stenman *et al.*, 1977; Chen *et al.*, 1978; Crouch *et al.*, 1978; Furcht *et al.*, 1978). In addition, large amounts of fibronectin are also continuously secreted or sloughed from cells into the culture media (Yamada & Weston, 1974; Critchley *et al.*, 1976; Muir *et al.*, 1976; Sear *et al.*, 1976; Vaheri *et al.*, 1976; Ali *et al.*, 1977; Baum *et al.*, 1977; Mosher *et al.*, 1977; Olden & Yamada, 1977; Crouch *et al.*, 1978). Fibronectin on cells in culture is primarily localised between cells, and between cells and the tissue culture substratum (Chen *et al.*, 1976; Vaheri *et al.*, 1976; Yamada & Pastan, 1976; Bornstein & Ash, 1977; Mautner & Hynes, 1977; Yamada & Olden, 1978). This localisation is consistent with the proposed role in cell–cell and cell–substratum adhesiveness. When cells are rendered permeable to fluorescently labelled antibodies by acetone treatment, granular perinuclear intracytoplasmic staining is evident. This intracellular location of fibronectin is apparently due to pools of newly synthesised protein destined for secretion, because the intracellular straining can be abolished in pulse-chase experiments with inhibitors of protein synthesis (Yamada *et al.*, 1978; Yamada & Olden, 1978).

The synthesis and distribution of cellular fibronectin is closely related to cell differentiation. In immunofluorescence studies of the 3-day-old chick embryo, fibronectin is found in close proximity to the primitive mesenchymal cells and in several basement membranes (Linder *et al.*, 1975; Stenman & Vaheri, 1978). When nephrogenic mesenchymal cells differentiate into epithelial cells of the kidney tubule, fibronectin is lost from the cells and becomes detectable in basement membrane formed around the tubules (Wartiovaara *et al.*, 1976; Stenman & Vaheri, 1978; Zetter & Martin, 1978). Cultures of L8 and L6 myogenic cell lines contain fibronectin. (Teichberg *et al.*, 1974; Gartner & Podleski, 1975; Hynes, 1976; Nowak *et al.*, 1976; Chen, 1977; Yamada & Olden, 1978; Podleski *et al.*, 1979), organised in a fibrillar pericellular arrangement. When the myoblasts differentiate and fuse to form myotubes, fibronectin levels are decreased (Chen, 1977; Podleski *et al.*, 1979), the pericellular arrangement is lost and the remaining fibronectin is distributed diffusely over the cell surface (Chen, 1977; Podleski *et al.*, 1979). More recent studies have shown that a high level of fibronectin blocks fusion, and its reduction at the appropriate time stimulates fusion of L6 myoblasts (Podleski *et al.*, 1979). Similarly, addition of fibronectin to chondrocytes results in reversion to the mesenchymal phenotype.

A direct or indirect transmembrane association between surface fibronectin and the cytoskeletal elements is suggested by several observations. First, treatment of cells with low concentrations of proteases or anti-fibronectin results in cell rounding and loss of surface fibronectin and microfilament bundle organisation (Pollack & Rifkin, 1975; Yamada, 1978). Second, treatment of transformed cells, deficient in both fibronectin and microfilament bundles, with fibronectin results in their restoration (Ali *et al.*, 1977; Willingham *et al.*, 1977). Third, the overall organisation of cell surface fibronectin into fibrillar patterns parallel to the long axis of the cell roughly coincides with that of intracellular microfilament bundles (Mautner & Hynes, 1977; Willingham *et al.*, 1977). Fourth, treatment of cultured cells with cytochalasin B results in release of fibronectin from the surface, suggesting that the integrity of the microfilament bundles is required to maintain fibronectin on the cell surface (Ali & Hynes, 1977; Kurkinen *et al.*, 1978). Finally, the anti-fibronectin-induced redistribution of fibronectin to form surface 'caps' is inhibited by a combination of colchicine and cytochalasin (Yamada, 1978), suggesting that microfilaments are required for the redistribution to occur.

Other cell surface components are more mobile than the fibrillar network of cell surface fibronectin. For example, the concanavalin A receptor migrates with a diffusion constant of 4×10^{-11} cm^2 s^{-1}, while fibronectin's diffusion constant is 5×10^{-12} (Schlessinger *et al.*, 1977). Since cellular fibronectin fibrillar network is relatively immobile, it might impair the mobility of other surface components. However, mobility studies with lipid probes and various surface antigens suggests that their mobilities are not affected by the presence of fibronectin fibrils (Schlessinger *et al.*, 1977).

STRUCTURE, COMPOSITION AND CHEMICAL PROPERTIES

Cell surface fibronectin exists as disulphide-bonded dimers and multimers with small amounts of 220 000 molecular weight monomeric form (Hynes & Destree, 1977; Keski-Oja *et al.*, 1977; Yamada *et al.*, 1977*a*). On molecular sizing columns, unreduced fibronectin elutes as three distinct protein peaks (Yamada *et al.*, 1977*a*). After disulphide reduction of each molecular form, only the monomer is recovered. This indicates that isolated fibronectin is not disulphide bonded to other proteins of the cell surface or extra-cellular matrix and that the 220 000 molecular weight monomer is the smallest subunit.

Plasma fibronectin is a disulphide-bonded dimer of two polypeptides of molecular weight 220 000. The subunits of the plasma fibronectin may not be identical, since the reduced molecules do not comigrate when analysed by sodium dodecyl sulphate polyacrylamide gel electrophoresis (Mosesson *et al.*, 1975; Iwanaga *et al.*, 1978). The

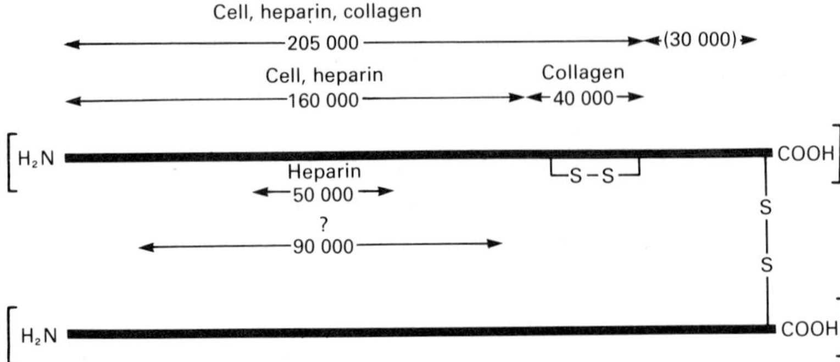

Fig. 1. Current model of fibronectin structure.

interchain disulphide bond appears to be restricted to the carboxyl terminus of the molecule (Chen *et al.*, 1977*c*; Hynes *et al.*, 1978*a*, *b*; Iwanaga *et al.*, 1978; Jilek & Hormann, 1978; Yamada *et al.*, 1978). The presence of the disulphide bonds is required for the cell-to-substratum adhesive properties of cellular fibronectin (Ali & Hynes, 1978). Both plasma and cellular fibronectin are either unfolded or elongated rather than globular, based on the finding of a Stoke's radius of 11 nm for the 220000 molecular weight monomer (Yamada *et al.*, 1977*a*). Both the monomeric and multimeric forms of fibronectin are highly asymmetric. The current model for the structure of the fibronectin is shown in Fig. 1.

The fibronectins apparently possesses very little regular secondary or tertiary structure, since circular dichroism yields very low optical activity (Hynes & Wyke, 1975; Alexander *et al.*, 1979). Nevertheless, tryptophan fluorescence spectra indicate that some of the tryptophan residues are buried in hydrophobic regions of the molecule, suggesting that some folded regions are present within the structure (Alexander *et al.*, 1979). These tryptophan residues can be exposed by heat denaturation; however, the agglutinating activity of cellular fibronectin is lost (Yamada *et al.*, 1975; Alexander *et al.*, 1979). It therefore appears that fibronectins are floppy, asymmetric molecules with localised regions of folding that are important for the expression of its adhesive activity.

From studies in which they have been directly compared cellular and plasma fibronectins appear to be very similar (Yamada & Olden, 1978). For example, their mobility in SDS–polyacrylamide gels, and some biological properties, are similar. Both cellular and plasma fibronectins also serve as substrate for the transglutaminases (Mosher, 1975). These enzymes catalyse the formation of lysine cross-linkages between proteins (Iwanaga *et al.*, 1978). The monomers of plasma fibronectin are progressively cross-linked to form multimers when the isolated protein is incubated with the activated blood coagulation factor XIII (plasma transglutaminase) even in the presence of disulphide-reducing reagents. Plasma fibronectin can also be cross-linked to fibrin; but in contrast, cellular fibronectin cannot be cross-linked to form larger multimers (Yamada & Olden, 1978).

However, several reports have appeared which show slight differences between plasma and cellular fibronectin with respect to solubility and electrophoretic mobilities in various gel systems (Yamada & Olden, 1978; Yamada & Kennedy, 1979), and marked differences with respect to haemagglutinating and adhesive activity (Yamada & Kennedy, 1979). For example, cellular fibronectin migrates as a single broad band, whereas plasma fibronectin migrates as a doublet in some gel systems; plasma fibronectin is soluble at pH 7.0, whereas the cellular form can only be solubilised by increasing the pH to 11.0 or by treatment with strong denaturing agents (Yamada & Olden, 1978); and cellular fibronectin is 50-fold more active in restoring a more normal morphology to transformed cells and 150-fold more active in promoting haemagglutination of fixed erythrocytes than plasma fibronectin (Yamada & Kennedy, 1979).

The amino acid composition of the isolated dimer of cellular fibronectin is fairly typical of cellular protein (Yamada *et al.*, 1977*a*; Yamada & Olden, 1978). It does not contain an unusual number of hydrophobic residues, and while it does contain slightly more proline and glycine than do many proteins, its overall composition is quite different from that of collagen, another protein of extracellular matrices.

The amino acid compositions of cellular and plasma fibronectins are very similar, although perhaps not identical (Mosesson *et al.*, 1975; Mosher, 1975; Muir *et al.*, 1976). The amino terminus of cellular fibronectin may be acetylated since it does not react with specific end-group reagents and is not removed by pyroglutamate amino peptidase. In contrast, the end-group of plasma fibronectin appears to be pyrolidone carboxylic acid (Mosesson *et al.*, 1975; Yamada *et al.*, 1977*a*).

It has been suggested that these differences in solubility and electrophoretic mobilities may be due to the presence of an extra peptide sequence on the amino-terminal end of cellular fibronectin which mediates its attachment to a receptor in the plasma membrane (Yamada & Olden, 1978; Yamada & Kennedy, 1979). The current speculation for these similarities and differences is that both firms of fibronectin are the products of the same gene, but that the cellular fibronectin from some strains of human skin fibroblasts is not larger than plasma fibronectin, and hence is probably not processed. Alternatively, it is also plausible that the two forms of fibronection are separate gene products which originated from gene duplication.

DECREASE AFTER MALIGANT TRANSFORMATION

Fibronectin levels are considerably reduced after transformation of many fibroblastic cells by oncogenic viruses or carcinogens (Hynes, 1973; Yamada & Olden, 1978), and its loss generally correlates with tumorigenicity following injection into immunosuppressed or immunocompatible animals. The decreases in amounts of fibronectin after malignant transformation are 5–7-fold in chick embryo fibroblasts (Olden & Yamada, 1977; Adams *et al.*, 1977) and more than 10-fold in human fibroblasts and glial cells (Hunt *et al.*, 1975; Vaheri & Ruoslahti, 1975). These findings stimulated enormous interest in fibronectin because of its possible direct involvement in malignancy. However, recent studies of closely related transformed cells grown in semi-solid matrix have cast doubt on this relationship (Marciani *et al.*, 1976; Yamada & Olden, 1978). Instead, decrease in fibronectin appears to correlate more with tendency of cells to metastasise (Chen *et al.*, 1978). In addition, studies in our laboratory have shown that normal growth properties are not restored when fibronectin is reconstituted on the surface of transformed fibronectin-deficient cells.

Previous studies have suggested that the decrease in fibronectin after transformation may be due to increased degradation (Yamada & Pastan, 1976), to decreased synthesis or loss of association with plasma membranes, or to faulty retention by cells, resulting in loss in the culture medium (Vaheri & Ruoslahti, 1975). We investigated the regulation of fibronectin by determining the sizes of its intracellular and extracellular pools, its rates of biosynthesis, transit times to the cell surface, and its rates of degration before and after transformation (Stone *et al.*, 1974; Adams *et al.*, 1977). We showed that the depletion of fibronectin on the cell surface after malignant transformation cannot be accounted for by increased cell surface protease activity (Stone *et al.*, 1974). In chick fibroblasts transformed by several strains of tumour viruses, the major factor responsible for the decrease was found to be decreased biosynthesis (Stone *et al.*, 1974; Adams *et al.*, 1977; Olden & Yamada, 1977). The decreased biosynthesis is due to decreased amounts of translatable messenger RNA (Adams *et al.*, 1977). However, other factors also contribute to the decrease in fibronectin, including an increase rate of proteolytic degradation (Stone *et al.*, 1974; Robbins *et al.*, 1974; Hynes & Wyke, 1975) and a decreased capacity of the glycoprotein to bind to the cell surface. In hamster cells, the loss of fibronectin after transformation is due to reduced biosynthesis and increased degradation (Hynes *et al.*, 1978*a*, *b*). However, the decrease in human fibroblasts is reported to be primarily due to decreased capacity of the transformed cells to retain this protein on the cell surface (Vaheri & Ruoslahti, 1975; Vaheri *et al.*, 1976).

Other factors also regulate quantities of cell surface fibronectin such as the phase of the cell cycle, growth conditions, hormones and state of embryonic differentiation (see review by Yamada & Olden, 1978). The amounts of cell surface fibronectin vary as cells move through the cell cycle. For example, the levels of fibronectin are lowest on cells during division. Non-proliferating cells in the G1 phase of the cell cycle generally have larger amounts of fibronectin than rapidly growing cells; however, this is not always the case. For example, cells prevented from dividing by serum deprivation can have either elevated or substantially reduced amounts of cell surface fibronectin (see review by Yamada & Olden, 1978). These findings are of interest since mitotic cells mimic transformed cells with respect to morphology, agglutinability by lectins, glycopeptide size, and loss of contact inhibition by movement. Mouse 3T3 fibroblasts grown in medium containing reduced concentrations of serum have little fibronectin, but normal cell surface levels of this glycoprotein are restored following treatment with epidermal growth factor. Similarly, the production of fibronectin is stimulated by treatment with thrombin. Levels of fibronectin on the surface of transformed human fibroblasts can be increased substantially by treatment with corticosteroids (see review by Yamada & Olden, 1978).

FUNCTION OF FIBRONECTIN

Since fibronectin is a major cell surface glycoprotein it is reasonable to assume that it might participate in cellular adhesion. This assumption is consistent with the finding that both the amount of fibronectin and the adhesiveness of cells to substrata are usually decreased in malignant cells.

We have examined the biological activity of fibronectin by means of four approaches. First, we directly measured the effects of fibronectin on cell–cell adhesiveness in a model assay system which measures haemagglutinating activity. Second, we performed reconstitution experiments in which fibronectin was added to transformed cells. Third, we evaluated fibronectin in assays for cell attachment to collagen and cell spreading. Finally, we used fibronectin antibody inhibition experiments to determine the effects of blocking fibronectin action on the cell surface on cellular behavior.

We found that fibronectin purified from cell surfaces of chick embryo fibroblasts can serve as an agglutinin that will aggregate formalinised sheep erythrocytes at concentrations as low as 1 μg ml^{-1} protein (Yamada *et al.*, 1975), and fibronectin can be recovered intact from the surface of the fixed erythrocytes. Treatment with 1 μg ml^{-1} trypsin or chymotrypsin or with 10–15 μM EDTA or EGTA prevents the expression of the haemagglutination activity. These chelating agents are known to disrupt cellular adhesion. The haemagglutination activity of fibronectin is also inhibited by urea, amino sugars, amines, and hypertonic salt solutions; but not by treatment with DNase or hyaluronidase.

Fibronectin also promotes aggregation of living, dissociated human, baby hamster kidney, Chinese hamster ovary, or chick embryo cells in suspension. For example, addition of isolated fibronectin to dissociated cells mixed continuously on a gyratory shaker according to a modification of the cell aggregation assay of Moscona (1973) promotes cell aggregation within 30–60 min (Fig. 2). Furthermore, fibronectin enhances the attachment of ^{14}C-labelled BHK cells to monolayers of BHK cells by approximately 2-fold in 30 min in the cell–cell adhesion assay procedure of Roseman *et al.* (1974.

Similarly, treatment of several transformed cell lines with cellular fibronectin also increases the extent of attachment, or conversely the

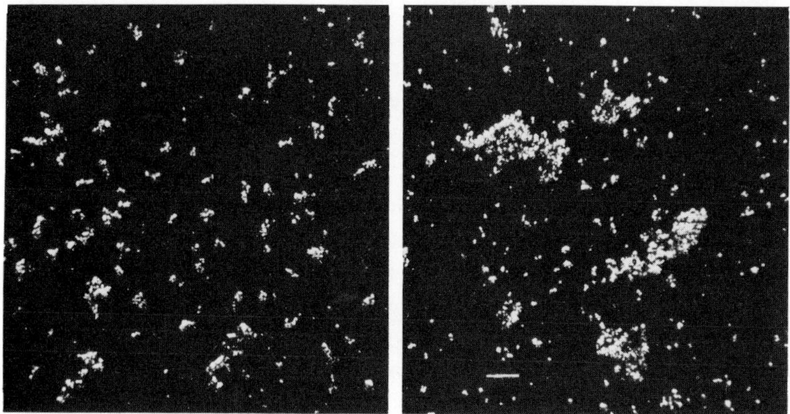

Fig. 2. Cellular aggregation by fibronectin. Check embryos (48 h old) were dissociated mechanically in calcium- and magnesium-free PBS by vigorous pipetting, then shaken on a gyratory shaker in serum-free Dulbecco–Vogt medium at 37 °C for 60 min. *Left*, control; *right*, plus 50 μg ml^{-1} of fibronectin. Bar = 50 μm.

resistance to detachment, to plastic tissue culture substrata (Yamada *et al.*, 1977*a*; Ali *et al.*, 1977). It has been reported that fibronectin purified from BHK or chick embryo fibroblasts increases the attachment of dissociated transformed BHK or SV40-3TS cells to collagen (Pearlstein, 1976; Yamada *et al.*, 1978). The fibronectin-mediated attachment of cells to collagen or spreading on to plastic substrate requires divalent cations, such as calcium, and the expenditure of cellular metabolic energy (Klebe, 1975; Pearlstein, 1976; Grinnell, 1976*a*, *b*; Grinnell *et al.*, 1977).

The serum factors required for cells to attach and spread on tissue culture dishes or for the attachment of certain cells to collagen-coated dishes are probably fibronectin. In addition, such serum factors can bind directly to collagen or tissue culture dishes in the absence of cells (Klebe, 1975; Pearlstein, 1976; Grinnell, 1976*a*, *b*). Furthermore, serum spreading factors are not required for the spreading of certain cells with high levels of cell surface fibronectin, and such cells can utilise the surface fibronectin in lieu of serum spreading factors to mediate their spreading and attachment to collagen (Yamada & Kennedy, 1979).

Both plasma and cellular fibronectins have been purified by affinity chromatography on columns to which denatured collagen has been covalently coupled (Engvall & Ruoslahti, 1977; Hynes *et al.*, 1978*a*;

Yamada *et al.*, 1978). The native collagens are not as effective as the denatured collagens of types I, II, and III in binding plasma fibronectin (Dessau *et al.*, 1978; Engvall *et al.*, 1978). Type III is the most effective native collagen in binding plasma fibronectin (Dessau *et al.*, 1978; Engvall *et al.*, 1978). Amino acid residues No. 643–819 of the collagen a-chain are responsible for the strong binding of plasma fibronectin (Kleinman & Goodwin, 1976; Engvall *et al.*, 1978) and several weaker binding sites also exist.

Recently the collagen-binding fragments of cellular and plasma fibronectins have been identified and isolated from chymotrypsin digests. Both were found to be single polypeptide chains with apparent subunit molecular weight of 40000 (Balian *et al.*, 1979; Hahn & Yamada, 1979). Plasma fibronectin may be involved in the removal of collagenous and other debris from blood following injury mediated by the fixed macrophages of the reticuloendothetial system. For example, several studies have identified an oposonic protein in plasma that participates in the reticuloendothelial system clearance of gelatin-coated colloidal particles form blood (Allen *et al.*, 1973; Kaplan & Saba, 1975; Kaplan *et al.*, 1976). In addition, an anti-gelatin factor has been identified in serum which binds to gelatin and is required for macrophage ingestion of denatured collagen (Maurer, 1954; Wolff *et al.*, 1967; Hopper *et al.*, 1970). Both the opsonic protein and the anti-gelatin factor have been isolated recently, and both appear to be identical to plasma fibronectin (cold-insoluble globulin) (Blumenstock *et al.*, 1977*a*, *b*; Dessau *et al.*, 1978; Saba *et al.*, 1978).

While both plasma and cellular fibronectin are equally effective in promoting cell spreading (Yamada & Kennedy, 1979) and attachment to collagen (Yamada & Kennedy, 1979), their biological activities may not be identical. For example, cellular fibronectin is more effective in haemagglutination and in restoring adhesiveness and normal morphology to transformed cells (Yamada & Kennedy, 1979).

Another approach to evaluate the biological function of fibronectin is to reconstitute it on the surfaces of transformed cells on which it is absent or greatly diminished (Hynes, 1973; Yamada & Weston, 1974; Adams *et al.*, 1977; Olden & Yamada, 1977). Such experiments were used to determine which normal fibroblastic properties were lost as a consequence of the loss of fibronectin. For these studies, fibronectin was added to the culture media of a wide variety of transformed fibroblastic cells. Approximately 15 % of the exogenous fibronectin spontaneously attaches to the cell layer; the absolute amount is approximately equivalent to that found on freshly explanted fibroblasts in culture (Yamada *et al.*, 1978).

The attachment of fibronectin to the cell layer is actually to the cell surface, as demonstrated by immunofluorescence localisation and by the accessibility of fibronectin to protease hydrolysis. The distribution on the surfaces of the transformed cells also appears to be similar to that of the non-transformed cells, as determined by immunofluorescence and by a normal pattern of redistribution following addition of antibodies. Fibronectin-treated cells exhibit increased adhesiveness to the substratum, as shown by their increased resistance to detachment from the substratum (Yamada *et al.*, 1976).

The structural and morphological alternations induced by the addition of fibronectin include the reappearance of prominent intracellular bundles of microfilaments containing actin and myosin, cell flattening, elongation of processes, parallel alignment of cells, and the surface becoming more smooth, with reduction in microvilli, blebs and ruffles. In general, cells revert to a phenotype characteristic of normal fibroblastic cells.

The presence of microfilament bundles appears to correlate best with increased cellular adhesion to the substratum. The restoration of the normal parallel alignment of fibronectin-reconstituted transformed cell cultures may be due to increased cell adhesion or to changes in specific cell-recognition events such as contact inhibition of movement (Willingham *et al.*, 1977). For example, normal fibroblastic cells tend to align in parallel array, whereas transformed cells are more antisocial in that they tend to grow in relatively random patterns with cell bodies criss-crossing one another. Fibronectin-treated cells also migrate more rapidly, both as single cells or as masses of cells migrating out from cell aggregates; this may be due to increased cell-to-substratum traction (Pouysségur *et al.*, 1977). Reconstitution experiments with a Balb/c3T3 mutant (AD6) cell line defective in cell–substratum adhesion further implicates fibronectin in cellular motility (Pouysségur *et al.*, 1977). These mutants have a substantially decreased rate of locomotion, as determined by time-lapse photomicroscopy. The addition of fibronectin to the AD6 mutant cell culture increases the rate and directionality of AD6 locomotion to normal (Pouysségur *et al.*, 1977).

Similar results have been obtained with the tumour cell line SV1, which also migrates slowly. Treatment of these cells' aggregates with

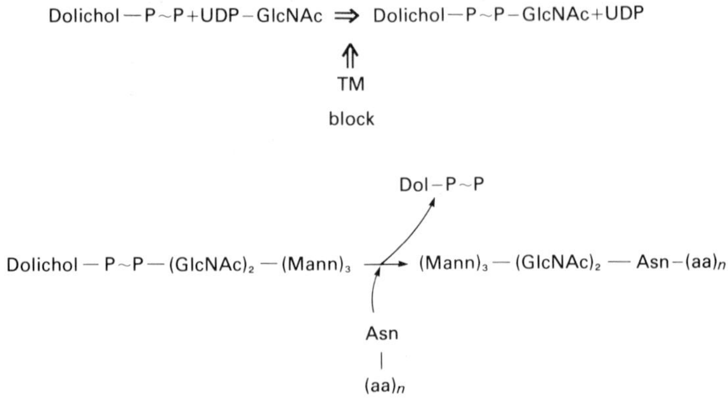

$$\text{Dolichol—P}\sim\text{P+UDP–GlcNAc} \Rightarrow \text{Dolichol—P}\sim\text{P–GlcNAc+UDP}$$

⇑
TM
block

$$\text{Dolichol — P}\sim\text{P — (GlcNAc)}_2\text{ — (Mann)}_3 \longrightarrow \text{(Mann)}_3\text{ — (GlcNAc)}_2\text{ — Asn–(aa)}_n$$

(Dol–P~P released; Asn–(aa)$_n$ enters)

Fig. 3. Site of tunicamycin inhibition.

exogenous fibronectin added to the culture media results in marked increase in the rate and extent of cellular migration from the aggregates.

The addition of cellular fibronectin did not restore normal growth rates, nutrient transport, or cyclic AMP levels to transformed cells. Other experiments also demonstrate that fibronectin does not play a direct role in growth regulation. For example, treatment with papain or thermolysin completely removes all detectable fibronectin from the cell surface, but does not stimulate growth of embryo fibroblasts (Zetter *et al.*, 1976). Conversely, treatment of cells with thrombin or insulin stimulates cell division without decreasing cellular fibronectin levels (Teng & Chen, 1975; Blumberg & Robbins, 1975).

ROLE OF CARBOHYDRATES IN TURNOVER, PROCESSING, AND BIOLOGICAL ACTIVITY

Several laboratories are currently investigating the role of oligosaccharides in intracellular processing and secretion of glycoproteins. According to a current hypothesis, the covalent attachment of carbohydrates to proteins is obligatory for the transfer of proteins across the hydrophobic, lipid membrane bilayer of the endoplasmic reticulum or for secretion in general (Eylar, 1965; Melchers, 1973). This hypothesis appears to be partially substantiated by more recent studies using mutant cells defective in glycosylation (Pouysségur & Pastan, 1976, 1977; Trowbridge *et al.*, 1978) or 2-deoxy-D-glucose, an inhibitor of glycosylation (Hughes *et al.*, 1977), in which it was

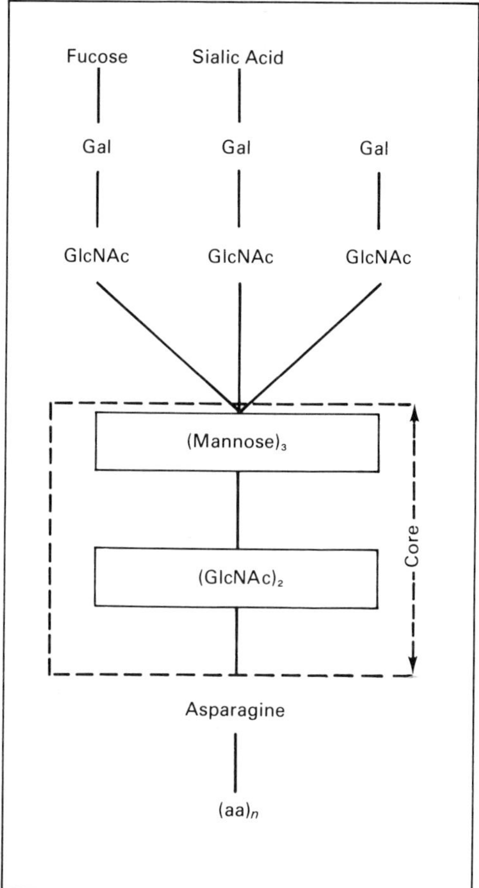

Fig. 4. Hypothetical structure of the carbohydrate moiety of fibronectin.

suggested that the extent of glycosylation affected the exposure of certain plasma-membrane-associated glycoproteins on the cell surface.

We have investigated the role of glycosylation in the synthesis, processing, secretion, degradation, distribution, and biological activity of cellular fibronectin, using secondary cultures of chick embryo fibroblasts (CEF). For our studies we have utilised the glucosamine-containing antibiotic called tunicamycin, an inhibitor of the synthesis of *N*-acetylglucosaminyl pyrophosphoryl polyisoprenol (Fig. 3) (Kuo & Lampen, 1974; Olden *et al.*, 1978). Since this reaction is required for the synthesis of the core sequence for N-glycosidically linked

Table 1. *Incorporation of leucine and sugars into CSP and trichloroacetic acid-insoluble fraction of cells*

	Control	(i) Tunicamycin (0.05 μg ml^{-1})	(ii) Tunicamycin (0.5 μg ml^{-2})	(iii) Tunicamycin (2 μg ml^{-1})	Inhibition (%) (i)	(ii)	(iii)
	(*a*) Incorporation into total protein (c.p.m. per μg protein)						
L-[U-^{14}C]leucine	1257 ± 103	1029 ± 118	842 ± 77	754 ± 121	18 ± 3	33 ± 1	40 ± 5
D-[U-^{14}C]glucosamine	1065 ± 88	351 ± 71	338 ± 54	288 ± 63	67 ± 4	68 ± 3	73 ± 4
D[2-^{3}H]mannose	438 ± 47	43 ± 15	29 ± 7	22 ± 11	91 ± 4	93 ± 1	95 ± 2
	(*b*) Incorporation into CAP (c.p.m. per μg protein)						
L[U-^{14}C]leucine	34 ± 3	16 ± 6	14 ± 3	12 ± 3	54 ± 12	59 ± 6	65 ± 6
D[U-^{14}C]glucosamine	29 ± 3	4 ± 1	4 ± 2	3 ± 2	86 ± 2	86 ± 5	90 ± 6
D[2-^{3}H]mannose	11	0.88	0.33	0.22	92	97	98

Cells were incubated with the radioisotopes for 24 h at 37 °C. Leucine and glucosamine were present at 2 μCi ml^{-1}, and mannose at 5 μCi ml^{-1}. Incorporation into total protein was determined by precipitation with 10% trichloroacetic acid, and into fibronectin by cutting the band from the polyacrylamide gel, dissolving in 1.5 ml 30% hydrogen peroxide, diluting to 20 ml with Aquasol and counting by liquid scintillation spectrophotometry.

oligosaccharides, tunicamycin treatment results in the synthesis of glycoproteins deficient in asparagine-linked oligosaccharides (Struck & Lennarz, 1977; Olden *et al.*, 1978). We chose this drug because the composition and chemical properties of the carbohydrate associated with fibronectin are characteristic of the 'complex' type of oligosaccharide linked by N-glycosidic bonds to asparagine (Fig. 4) (Yamada *et al.*, 1977*a*; Olden & Olden, 1979). Studies suggest that there are probably an average of 4–6 such oligosaccharide groups per molecule containing terminal galactose, sialic acid, and fucose residues (Yamada *et al.*, 1977*a*; Olden *et al.*, 1978; Olden & Olden, 1979). The carbohydrate composition of the various fibronectins is very similar (Mosesson *et al.*, 1975; Yamada *et al.*, 1977*a*; Carter *et al.*, 1978; Olden *et al.*, 1978), and the structure of the oligosaccharides on cellular fibronectin is being determined.

Treatment of cells with the antibiotic tunicamycin specifically blocked the glycosylation of glycoproteins of the N-glycosidic type (Table 1). Chick fibroblasts treated with tunicamycin produced only non-glycosylated fibronectin, and total amounts of this protein were substantially decreased (Olden & Olden, 1979; Vuento *et al.*, 1977; Olden *et al.*, 1979*b*; Pratt *et al.*, 1979). The decrease was not due to altered synthesis or secretion, but instead to increased proteolytic degradation (Olden *et al.*, 1978; Olden & Olden, 1979; Pratt *et al.*, 1979). Normal levels of fibronectin were partially restored by growing the cells in medium containing the protease inhibitor leupeptin (K. Olden, unpublished observation). These results suggest that for cellular fibronectin, carbohydrates are not necessary for synthesis, processing, or secretion, but instead may function to protect the protein from proteolytic attack by cellular proteases.

These results left the possibility that the biological activities, e.g. the cell surface adhesive interactions affected by fibronectin, might require the carbohydrate moiety. To examine the biological activity of cellular fibronectin, we compared the activities of the glycosylated and non-glycosylated species of this glycoprotein, utilising four in-vitro assay procedures. First, we directly measured the capacity of purified non-glycosylated fibronectin to agglutinate formalised sheep erythrocytes. Haemagglutination has been used as a model system to estimate cell–cell adhesiveness. Secondly, we performed reconstitution experiments in which fibronectin was added to transformed cells (SV40–3T3) deficient in the glycoprotein, and measured its capacity to restore the fibroblastic morphology characteristic of

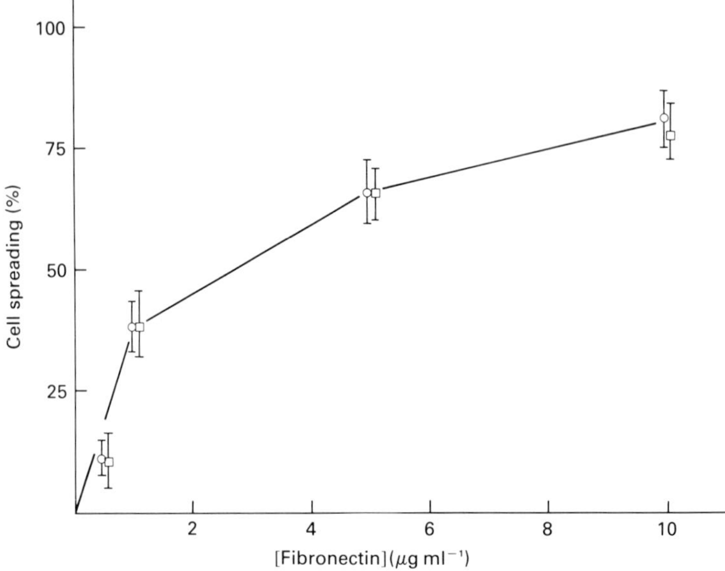

Fig. 5. Effect of isolated fibronectin on the spreading of non-transformed cells *in vitro*. BHK cells grown in monolayer cultures were removed from the dish by treatment with trypsin, then serum was added to inhibit further proteolysis. The cells were collected by centrifugation, washed twice in adhesion medium (150 mM NaCl, 3 mM KCl, 1 mM $CaCl_2$, 0.5 mM $MgCl_2$, 6 mM Na_2HPO_4, and 1 mM KH_2PO_4), and 0.5×10^6 cells in 1.0 ml adhesion medium were transferred to 35 mm culture dishes and incubated for 45 min in medium containing glycosylated (□) or non-glycosylated (○) fibronectin. Cell spreading at each concentration was scored by counting 250 cells and calculating the percentage of cells that were fully spread (no longer refractile by phase-contrast microscopy and surrounded by a continuous zone of lamellar cytoplasm).

non-transformed cells. Finally, glycosylated and non-glycosylated fibronectin were compared with respect to their capacity to promote (1) the spreading of cells (baby hamster kidney – BHK) on the surface of plastic tissue culture dishes, and (2) the binding of cells (Chinese hamster ovary – CHO) to collagen (type I) coated dishes.

We found that non-glycosylated fibronectin was as effective as the glycosylated protein in (1) promoting cell spreading (half-maximal at a protein concentration of 1 μg ml^{-1}) (Fig. 5), (2) mediation of attachment of cells to collagen (Fig. 6), (3) in restoration of the normal fibroblastic phenotype. Therefore, we conclude that the carbohydrate moiety of fibronectin is also not required for a variety of biological activities mediated by this glycoprotein.

Previous studies on the effects of tunicamycin on fibronectin

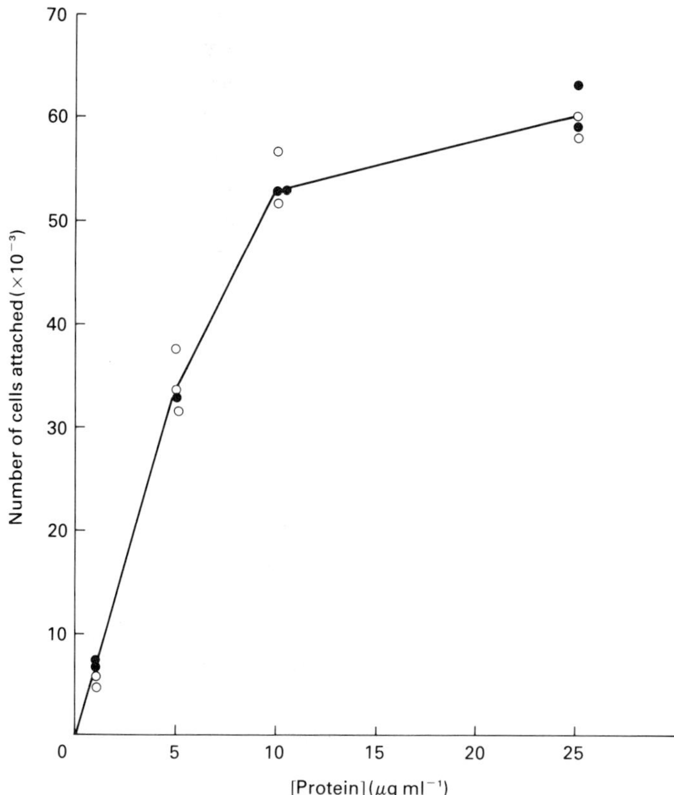

Fig. 6. Effect of isolated fibronectin on attachment of cells to collagen. Monolayer cultures of CHO cells were trypsinised, washed, and plated at a concentration of 10^5 cells ml^{-1} of alpha-modified Eagle's minimal essential medium in 35 mm plastic bacteriological dishes coated with 10 μg type I collagen.

suggested that the carbohydrate moiety was required to stabilise this protein against proteolytic degradation (Loh & Gainer, 1978; Olden *et al.*, 1978). To determine whether non-glycosylated fibronectin is inherently more sensitive to proteases than the glycosylated protein, we incubated both species of the isolated protein with protease (Pronase). The rate of release of TCA-soluble [^{14}C]leucine from each preparation is shown in Fig. 7. It is apparent that the non-glycosylated protein fraction is degraded to soluble peptides and amino acids approximately twice as rapidly as the glycosylated protein. The apparent increase in degradation of the non-glycosylated species of fibronectin may be due to an altered tertiary structure because of the absence of carbohydrates, or to decreased protection from proteases

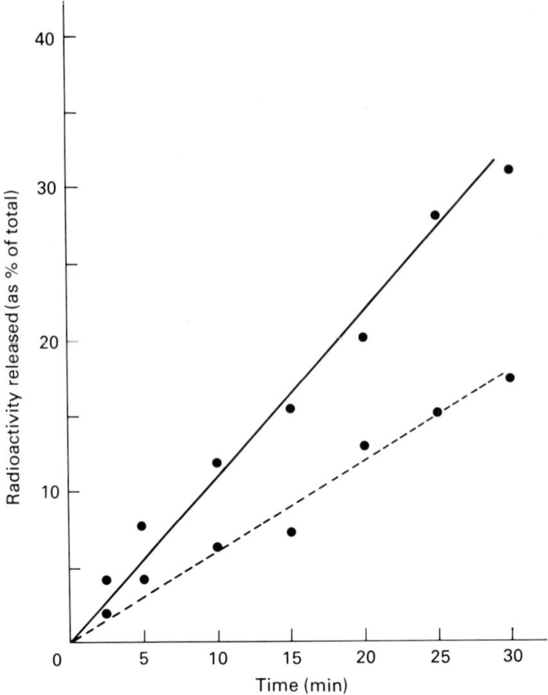

Fig. 7. Fibronectin (50 μg) isolated by antibody affinity chromatography as described in Methods, was incubated with 0.05 μg ml^{-1} of the broad spectrum protease Pronase in 0.1 M sodium phosphate buffer (pH 7.4) at 37 °C. Solid line shows non-glycosylated fibronectin, broken line glycosylated. Aliquots of 100 μl were removed at the indicated intervals and the radioactivity in both the soluble and precipitable fractions was determined after precipitation with trichloroacetic acid.

at the bends (β-turns) in the polypeptide backbone believed to occur at glycosylation sites in glycoproteins. Several laboratories have reported that defective proteins are degraded more rapidly than their normal counterparts (Goldberg *et al.*, 1975; Olden & Goldberg, 1978) and that the solubility properties of glycoproteins are markedly altered by inhibition of glycosylation (Leawitt *et al.*, 1977). Furthermore, elegant studies by Ashwell and associates have shown that the removal of a specific terminal sugar can result in a marked degradation of circulating plasma glycoproteins due to enhanced rate of uptake of such glycoproteins into the liver via highly specific hepatic lectins (Lunney & Ashwell, 1976). While the function of carbohydrates in protein secretions is not entirely resolved, elegant studies in other laboratories also suggest that glycosylation of protein is not a prerequisite for their secretion (Loh & Gainer, 1978, 1979; Struck *et al.*, 1978),

Other evidence for the role of carbohydrates in glycoprotein proteolysis has been reported for the viral precursor glycoproteins (haemagglutination and envelope protein E-2) (Leavitt *et al.*, 1977), the glycosylated ACTH–endorphin precursor of *Xenopus laevis* (Loh & Gainer, 1978, 1979), and the T25 glycoprotein of lymphoma cells (Trowbridge *et al.*, 1978). These investigators also found that the absence of the carbohydrate moiety results in more rapid intracellular degradation.

FIBRONECTIN VERSUS OTHER ADHESIVE MOLECULES

We compared the adhesive properties of fibronectin with those of collagen and polylysine by determining their effects on the morphology and cellular alignment of the mouse tumour cell line SV1.

For these studies the tissue culture dishes were precoated with either rat skin type I collagen, adhesive polycation, polylysine, or with fibronectin. Treatment with either collagen or polylysine induced cell spreading and flattening; however, in neither case was it as extensive as shown by treatment with fibronectin. It is important to note that precoating of dishes with either collagen or polylysine does not result in cellular elongation or alignment at confluence as demonstrated with fibronectin. If fibronectin is added to the culture medium of collagen- or polylysine-coated dishes, cellular alignment at confluency is now observed. In fact the alignment is more extensive than with fibronectin alone. However, the converse is not true; that is, the addition of collagen or polylysine to the culture medium of fibronectin-treated cultures does not potentiate cell flattening, elongation, or alignment (Yamada *et al.*, 1978).

We conclude that fibronectin has a dual function. First it directly enhances the adhesiveness of cells to the substratum; secondly, it controls cellular elongation or alignment at confluence. These functions of fibronectin are potentiated on collagen substrata.

SUMMARY

Several surface proteins of normal cells are altered by transformation. The most prominent of these alternations is the reduction in the amount of cellular fibronectin. Fibronectin is the major cell surface glycoprotein of chick embryo fibroblasts. This protein has been isolated, purified and reconstituted on the surfaces of 14 transformed cell lines from several species. It reverts the transformed phenotype to the normal fibroblastic morphology, adhesiveness, cell surface architecture, microfilament bundle organisation, motility, and alignment at confluence. However, it does not restore normal growth rates or uptake of nutrients. Several of the features characteristic of transformed cells can be induced by treatment of normal cells with antibody raised against fibronectin.

The mechanisms responsible for the decrease in fibronectin in transformed cells have been investigated in chick embryo fibroblasts transformed by various strains of Rous Sarcoma viruses. It was shown that the decreased quantities of fibronectin is due primarily to a five- to seven-fold reduction in fibronectin biosynthesis, although increased proteolytic degradation and shedding from the cell surface also contribute. The decrease in biosynthesis is apparently due to a five-fold reducting in translatable mRNA for fibronectin.

Fibronectin exists as a relatively immobile, fibrillar network on the cell surface. The molecules are highly asymmetric, disulphide cross-linked dimers and multimers. It contains exclusively asparagine-linked oligosaccharides. Cellular and plasma fibronectin are similar in composition and function but differ chemically at the amino termini and in their solubility properties.

The carbohydrate moiety of fibronectin is not required for its secretion or export from the intracellular site of synthesis to the cell surface, and also is not required for the biological activity. However, it appears to be required for protection against proteolytic degradation.

In general the studies described here have been conducted with fibroblasts in culture as in-vitro models of malignant transformation. We conclude that the decrease in cellular fibronectin accounts for several of the phenotypic characteristics of transformed cells in culture. However, we still do not know how this protein functions *in vivo*, its role in tumorigenicity, metastasis or host defense. It would also be of interest to elucidate the precise structural differences and regulation between plasma and cellular fibronectin.

Acknowledgement. We thank Dorothy Kennedy for invaluable technical assistance, Elizabeth Lovelace and Annie Harris for aid in cell culture, Raymond Steinberg for photographic services, Wilma Davis for typing the manuscript, and Susan Yamada, Ira Pastan and Robert Pratt for valuable collaboration. Kenneth Olden acknowledges the assistance of Valerie Hunter in the preparation of this manuscript.

REFERENCES

Adams, S. L., Sobel, M. E., Howard, B. H., Olden, K., Yamada, K. M., DeCrombrugghe, B. & Pastan, I. (1977). Levels of translatable mRNA's for cell surface protein, collagen precursors, and two membrane proteins are altered in *Rous sarcoma* virus-transformed chick embryo fibroblasts. *Proceedings of the National Academy of Sciences, USA*, **74**, 3399–403.

Alexander, S. S. Jr., Colonna, G. & Edelhoch, H. (1979). The structure and stability of human plasma cold-insoluble globulin. *Journal of Biological Chemistry*, **254**, 1501–5.

Alexander, S. S. Jr., Colonna, G., Yamada, K. M., Pastan, I. & Edelhoch, H. (1978). *Journal of Biological Chemistry*, **253**, 7787–90.

Ali, I. U. & Hynes, R. O. (1977). Effects of cytochalasin B and colchicine on attachment of a major surface protein fibroblasts. *Biochimica et Biophysica Acta*, **471**, 16–24.

Ali, I. U. & Hynes, R. O. (1978). Role of disulfide bonds in the attachment and function of LETS glycoprotein at the cell surface. *Biochimica et Biophysica Acta*, **510**, 140–50.

Ali, I. U., Mautner, V., Lanza, R. & Hynes, R. O. (1977). Restoration of normal morphology, adhesion, and cytoskeleton in transformed cells by addition of a transformation sensitive surface protein. *Cell*, **11**, 115–26.

Allen, C., Saba, T. M. & Molnar, J. (1973). Isolation, purification, and characterization of opsonic protein. *Journal of the Reticuloendothelial Society*, **13**, 410–23.

Aronsen, K. F., Ekelund, G., Kindmark, C. O. & Laurell, C. B. (1972). Sequential changes of plasma proteins after surgical trauma. *Scandinavian Journal of Clinical and Laboratory Investigation*, Suppl. **124**, 127–36.

Balian, G., Click, E. M., Crouch, E., Davidson, J. M. & Bornstein, P. (1979). Isolation of a collagen binding fragment from fibronectin and cold-insoluble globulin. *Journal of Biological Chemistry*, **254**, 1429–32.

Baum, B. J., McDonald, J. A. & Crystal, R. G. (1977). *Biochemical and Biophysical Research Communications*, **79**, 8–15.

Birdwell, C. R., Gospodarowicz, D. & Nicolson, G. L. (1978). Identification localization, and role of fibronectin on cultured bovine and endothelial cells. *Proceedings of the National Academy of Sciences, USA*, **75**, 3273–7

BLUMBERG, P. M. & ROBBINS, P. W. (1975). Effect of proteases an activation of resting chick embryo fibroblasts and on cell surface proteins. *Cell*, **6**, 137–47.

BLUMENSTOCK, F., WEBER, P. & SABA, T. M. (1977*a*). Isolation and biochemical characterization of α-2-opsonic glycoprotein from rat serum. *Journal of Biological Chemistry*, **252**, 7156–62.

BLUMENSTOCK, F., WEBER, P., SABA, T. M. & LAFFIN, R. (1977*b*). Electroimmunoassay of alpha-2-opsonic protein during reticuloendothelial blockade. *American Journal of Physiology*, **232**, 80–7.

BORNSTEIN, P. & ASH, J. F. (1977). Cell surface-associated structural proteins in connective tissue cells. *Proceedings of the National Academy of Sciences, USA*, **74**, 2480–4.

BORNSTEIN, P., DUKSIN, D., BALIAN, G., DAVIDSON, J. M. & CROUCH, E. (1978). Organization of extracellular proteins on the connective tissue cell surface: Relevance of cell-matrix in vitro and in vivo. *Annals of the New York Academy of Sciences*, **312**, 93–105.

CARTER, W. G., FUKUDA, M., LINGWOOD, C. & HAKOMORI (1978). Chemical composition, gross structure, and organization of transformation sensitive glycoproteins. *Annals of the New York Academy of Sciences*, **312**, 160–77.

CARTER, W. G. & HAKOMORI, S. (1977). Isolation and partial characterization of 'galactoprotein b' from hamster embryo fibroblasts. *Biochemical and Biophysical Research Communications*, **76**, 299–308.

CHEN, L. B. (1977). Alteration in cell surface LETS protein during myogenesis. *Cell*, **10**, 393–400.

CHEN, A. B., AMRAMI, D. L. & MOSESSON, M. W. (1977*c*). Heterogenicity of the cold-insoluble globulin of human plasma (CIg), a circulating cell surface protein. *Biochimica et Biophysica Acta*, **493**, 310–22.

CHEN, L. B., BURRIDGE, K., MURRAY, A., WALSH, M. L., COPPLE, C. D., BUSHNELL, A., MCDOUGALL, J. K. & GALLIMORE, P. H. (1978). Modulation of cell surface glycocalyx: studies on large, external transformation sensitive protein. *Annals of the New York Academy of Sciences*, **312**, 366–381.

CHEN, L. B., MAITLAND, N., GALLIMORE, P. H. & MCDOUGALL, J. K. (1977*a*). Detection of the large external transformation-sensitive protein on some epithelial cells. *Experimental Cell Research*, **106**, 39–46.

CHEN, L. B., MOSER, F. G., CHEN, A. B. & MOSESSON, M. W. (1977*b*). Distribution of cell surface LETS protein in co-cultures of normal and transformed cells. *Experimental Cell Research*, **108**, 375–83.

CHEN, A. B., MOSESSON, M. W. & SOLISH, G. I. (1976). Identification of the cold insoluble globulin of plasma in amniotic fluid. *American Journal of Obstetrics and Gynecology*, **7**, 958–61.

CRITCHLEY, D. R., WYKE, J. A. & HYNES, R. O, (1976). Cell surface and metabolic labeling of the proteins of normal and transformed chicken cells. *Biochimica et Biophysica Acta*, **436**, 335–52.

CROUCH, E., BALIAN, G., HOLBROOK, K., DUKSIN, D. & BORNSTEIN, P. (1978). Amniotic fluid fibronectin: characterization and synthesis by cells in culture. *Journal of Cell Biology*, **78**, 701–15.

DESSAU, W., ADELMAN, B. C., TIMPL, R. & MARTIN, G. R. (1978). Identification of the sites in collagen alpha chains that bind serum antigelatin factor (cold insoluble globulin). *Biochemical Journal*, **169**, 55–9.

Dessau, W., Jilek, F., Adelman, B. C. & Hormann, H. (1978). Similarity of antigelatin factor and cold insoluble globulin. *Biochimica et Biophysica Acta*, **532**, 227–37.

Engvall, E. & Ruoslahti, E. (1977). Binding of soluble form of fibroblast surface protein, fibronectin, to collagen, *International Journal of Cancer*, **20**, 1–5.

Engvall, E., Ruoslahti, E. & Miller, E. J. (1978). Affinity of fibronectin to collagens of different genetic types and to fibrinogen. *Journal of Experimental Medicine*, **147**, 1584–95.

Eylar, E. H. (1965). On the biological role of glycoproteins. *Journal of Theoretical Biology*, **10**, 89–113.

Furcht, L. T., Mosher, D. F. & Wendelschafer-Crabb, G. C. (1978). Differences in the ultrastructural localization of fibronectin on normal and transformed cells. *Annals of the New York Academy of Sciences*, **312**, 426–31.

Furcht, L. T., Mosher, D. F. & Wendelschafer-Crabb, G. (1978). Immunocytochemical localization of fibronectin (LETS protein) on the surface of L6 myoblasts: light and electron microscopic studies. *Cell*, **13**, 263–71.

Gartner, T. K. & Podleski, T. R. (1975). *Biochemical and Biophysical Research Communications*, **67**, 972–8.

Grinnell, F. (1976*a*). The serum dependence of baby hamster kidney cell attachment to a substratum. *Experimental Cell Research*, **97**, 265–374.

Grinnell, F. (1976*b*). Cell spreading factor. *Experimental Cell Research*, **102**, 51–62.

Grinnell, F. (1978). Cellular adhesiveness and extracellular substrata. *International Review of Cytology*, **58**, 65–144.

Grinnell, F., Hays, D. G. & Minter, D. (1977). Cell adhesion and spreading factor: partial purification and properties. *Experimental Cell Research*, **110**, 175–90.

Goldberg, A. L., Olden, K. & Prouty, W. F. (1975). Studies on the mechanisms and selectivity of protein degradation in *E. coli*. In *Intracellular Protein Turnover*, 1st edn, ed. R. T. Schimke & N. Katunuma, 17–55. New York, London: Academic Press.

Hahn, L-H. E. & Yamada, K. M. (1979). Identification and isolation of a collagen binding fragment of the adhesive glycoprotein fibronectin. *Proceedings of the National Academy of Sciences, USA*, **76**, 1160–3.

Hedman, K., Vaheri, A. & Wartiovaara, J. (1978). External fibronectin of cultured human fibroblasts is predominantly a matrix protein. *Journal of Cell Biology*, **76**, 748–60.

Hogg, N. M. (1974). A comparison of membrane proteins of normal and transformed cells by lactoperoxidase labeling. *Proceedings of the National Academy of Sciences, USA*, **71**, 489–92.

Hopper, K. E., Adelmann, B. C., Gentner, G. & Gay, S. (1976). Recognition by guinea pig peritoneal exude cells of conformationally different states of the collagen molecule. *Immunology*, **30**, 249–59.

Hughes, R. C., Meager, A. & Nairn, R. (1977). Effect of 2–deoxy-D-glucose on the cell surface glycoproteins of hamster fibroblasts. *European Journal of Biochemistry*, **72**, 265–73.

Hunt, R. C., Gold, E. & Brown, J. C. (1975). Cell cycle dependent exposure of a high molecular weight protein on the surface of mouse L cells. *Biochimica et Biophysica Acta*, **413**, 453–8.

HYNES, R. O. (1973). Alteration of cell-surface proteins by viral transformation and by proteolysis. *Proceedings of the National Academy of Sciences, USA*, **70**, 3170–4.

HYNES, R. O. (1976). Cell surface protein and malignant transformation. *Biochimica et Biophysica Acta*, **458**, 73–107.

HYNES, R. O., ALI, I. U., DESTREE, A. T., MAUTNER, D. V., PERKINS, M. E., SENGER, D. R., WAGNER, D. D. & SMITH, K. K. (1978*a*). A large glycoprotein lost from the surfaces of transformed cells. *Annals of the New York Academy of Sciences*, **312**, 317–42.

HYNES, R. O., ALI, I. U., MAUTNER, V. & DESTREE, A. (1978*b*). LETS Glycoprotein: Arrangement and function at the cell surface. *Birth Defects*, **14**, 139–53.

HYNES, R. O. & DESTREE, A. (1977). Extensive disulfide bonding of at the mammalian cell surface. *Proceedings of the National Academy of Sciences, USA*, **74**, 2855–9.

HYNES, R. O. & WYKE, J. A. (1975). Alterations in surface proteins in chicken cells transformed by temperature sensitive mutants of *Rous sarcoma* virus. *Virology*, **64**, 496–504.

IWANAGA, S., SUZUKI, K. & HOSHIMOTO, S. (1978). Bovine plasma cold insoluble globulin: its gross structure and function. *Annals of the New York Academy of Sciences*, **312**, 56–73.

JAFFE, E. A. & MOSHER, D. F. (1978). Synthesis of fibronectin by cultured human endothelial cells. *Journal of Experimental Medicine*, **147**, 1779–91.

JILEK, F. & HORMANN, H. (1978). Cold-insoluble globulin (Fibronectin), IV [1–3]. Affinity to soluble collagen of various types. *Zeitschrift für Physiologische Chemie (Hoppe-Seyler)*, **359**, 247–50.

KAPLAN, J. E. & SABA, T. M. (1975). Humoral deficiency and reticuloendothelial depression after traumatic shock. *American Journal of Physiology*, **230**, 7–14.

KAPLAN, J. E., SABA, T. M. & MOLNAR, J. (1976). Comparative disappearance and localization of isotopically labeled opsonic protein and soluble albumin following surgical trauma. *Journal of the Reticuloendothelial Society*, **20**, 375–84.

KESKI-OJA, J., MOSHER, D. F. & VAHERI, A. (1977). Dimeric character of fibronectin, a major-cell surface associated glycoprotein. *Biochimica et Biophysica Acta*, **74**, 699–706.

KLEBE, R. J. (1975). Cell attachment to collagen: The requirement for energy. *Journal of Cell Physiology*, **86**, 231–6.

KLEINMAN, H. K. & MCGOODWIN, E. B. (1976). Localization of the cell attachment region in types I and II collagens. *Biochemical and Biophysical Research Communications*, **72**, 426–32.

KUO, S. C. & LAMPEN, J. O. (1974). Tunicamycin an inhibitor of yeast glycoprotein synthesis. *Biochemical and Biophysical Research Communications*, **58**, 287–95.

KURKINEN, M., WARTIOVAARA, J. & VAHERI, A. (1978). Cytochalasin B releases a major surface associated glycoprotein, fibronectin, from cultured fibroblasts. *Experimental Cell Research*, **111**, 127–37.

LEAWITT, R., SCHLESINGER, S. & KORNFELD, S. (1977). Impaired intracellular migration and altered solubility of nonglycosylated glycoproteins of vesicular stomatitis virus and Sindbis Sindbis virus. *Journal of Biological Chemistry*, **252**, 9018–23.

LINDER, E., VAHERI, A., ROUSLAHTI, E. & WARTIOVAARA, J. (1975). Distribution of fibroblast surface antigen in the developing chick embryo. *Journal of Experimental Medicine*, **142**, 41–9.

LOH, PENG, Y. & GAINER, HAROLD. (1978). The role of glycosylation on the biosynthesis, degradation, and secretion of the ACTH-B-Lipotropin common precursor and its peptide products. *FEBS Letters*, **96**, 269–72.

LOH, P. Y. & GAINER, H. (1979). The role of the carbohydrate in the stabilization, processing, and packaging of the glycosylated ACTH-endorphin common precursor in toad pituitaries. *Endocrinology*, (in press).

LUNNEY, J. & ASHWELL, G. (1976). A hepatic receptor of avian origin capable of binding specifically modified glycoproteins. *Proceedings of the National Academy of Sciences, USA*, **73**, 341–3.

MARCIANI, D. J., LYONS, L. B. & THOMPSON, E. B. (1976). Characteristics of cell membranes from somatic cell hybrids between rat hepatoma and mouse L-cells. *Cancer Research*, **36**, 2937–44.

MAURER, P. H. (1954). Antigenicity of oxypolygelatin and gelatin in man. *Journal of Experimental Medicine*, **100**, 497–513.

MAUTNER, V. & HYNES, R. O. (1977). Surface distribution of LETS protein in protein in relation to the cytoskeleton of normal and transformed cells. *Journal of Cell Biology*, **75**, 743–66.

MELCHERS, F. (1973). Biosynthesis, intracellular transport, and secretion of immunoglobulins. Effect of 2-deoxy-D-glucose in tumor plasma cells producing and secreting immunoglobulins G1. *Biochemistry*, **12**, 1471–6.

MORRISON, P., EDSALL, R. & MILLER, S. G. (1948). Preparation and properties of serum and plasma proteins. XVIII. *Journal of the American Chemical Society*, **70**, 3103–8.

MOSCONA, A. A. (1973). Cell aggregation. In *Cell Biology and Medicine*, 1st edn, ed. E. Biltar, pp. 571–91. New York: Wiley Interscience.

MOSESSON, M. W. (1977). Cold-insoluble globulin (CIg), a circulating cell surface protein. *Thrombosis and Haemostasis*, **38**, 742–50.

MOSESSON, M. W. (1978). Structure of human plasma cold-insoluble globulin and the mechanism of its precipitation in the cold with heparin or fibrin-fibrinogen complexes. *Annals of the New York Academy of Sciences*, **312**, 1–10.

MOSESSON, M. W., CHEN, A. B. & HUSEBY, R. M. (1975). The cold insoluble globulin of human plasma: studies of its essential structural features. *Biochimica et Biophysica Acta*, **386**, 509–24.

MOSESSON, M. W. & UMFLEET, R. A. (1970). The cold-insoluble globulin of human plasma. *Journal of Biological Chemistry*, **245**, 5728–36.

MOSHER, D. F. (1975). Cross-linking of cold insoluble globulin by fibrin-stabilizing factor. *Journal of Biological Chemistry*, **250**, 6614–21.

MOSHER, D. F., SAKSELA, O., KESKI-OJA, J. & VAHERI, A. (1977). Distribution of a major surface-associated glycoprotein, fibronectin, in cultures of adherent cells. *Journal of Supramolecular Structure*, **6**, 551–63.

MUIR, L. W., BORNSTEIN, P. & ROSS, R. (1976). A presumptive subunit of elastic fiber microfibrils secreted by arterial smooth muscle cells in culture. *European Journal of Biochemistry*, **64**, 105–14.

Nowak, T. P., Haywood, P. L. & Barondes, S. H. (1976). Developmentally regulated lectin in embryonic chick muscle and a myogenic cell line. *Biochemical and Biophysical Research Communications*, **68**, 650–7.

Olden, K. & Goldberg, A. L. (1978). Energy coupling for intracellular proteolysis in *Escherichia coli*, *Biochimica et Biophysica Acta*, **542**, 385–398.

Olden, K. & Olden, A. T. (1979). Role of carbohydrates in protein secretion. In *Recent Advances in Cancer and Molecular Biology*, 1st edn, ed. W. A. Guillory, pp. 19–41. Utah University Press.

Olden, K. Pratt, R. M., Jaworski, C. & Yamada, K. M. (1979*a*). Evidence for role of glycoprotein carbohydrates in membrane transport: Specific inhibition by tunicamycin. *Proceedings of the National Academy of Sciences, USA*, **76**, 791–5.

Olden, K., Pratt, R. M. & Yamada, K. M. (1978). Role of carbohydrates in protein secretion, and turnover: Effects of tunicamycin on the major cell surface glycoprotein of chick embryo fibroblasts. *Cell*, **13**, 461–73.

Olden, K., Pratt, R. M. & Yamada, K. M. (1979*b*). The role of carbohydrate in biological function of the adhesive glycoprotein fibronectin. *Proceedings of the National Academy of Sciences, USA*, (in press).

Olden, K. & Yamada, K. M. (1977). Mechanism of the decrease in the major cell surface protein of chick embryo fibroblasts after transformation. *Cell*, **11**, 957–69.

Pearlstein, E. (1976). Plasma membrane glycoprotein which mediates adhesion of fibroblasts to collagen. *Nature, London*, **262**, 497–500.

Pena, S. D. J. & Hughes, R. C. (1978). Fibroblast to substratum contacts mediated by the different forms of fibronectin. *Cell Biology, International Reports*, **2**, 1339–44.

Podleski, T. R., Nichols, S., Randin, P. & Salpeter, M. M. (1979). Myoblast fusion. *Developmental Biology*, (in press).

Pollack, R. & Rifkin, D. (1975). Actin-containing cables within anchorage dependent rat embryo cells are dissociated by plasmin and trypsin. *Cell*, **6**, 495–506.

Pouysségur, J. M. & Pastan, I. (1976). Mutants of Balb/c 3T3 fibroblasts defective in adhesiveness to substratum: evidence for alteration in cell surface proteins. *Proceedings of the National Academy of Sciences, USA*, **73**, 544–8.

Pouysségur, J. M. & Pastan, I. (1977). Mutants of mouse fibroblasts altered in the synthesis of cell surface glycoproteins. *Journal of Biological Chemistry*, **252**, 1639–46.

Pouysségur, J., Willingham, M. C. & Pastan, I. (1977). Role of cell surface carbohydrates and proteins in cell behavior: Studies on the biochemical reversion of an *N*-acetylglucosamine deficient fibroblast mutant. *Proceedings of the National Academy of Sciences, USA*, **74**, 243–7.

Pratt, Robert, M., Yamada, K., Olden, K., Ohanian, S. H. & Hascall, V. C. (1979). Tunicamycin-induced alterations in the synthesis of sulfated proteoglycans and cell surface morphology in the chick embryo fibroblasts. *Journal of Experimental Cell Research*, **118**, 245–352.

ROBBINS, P. W., WICKUS, G. G., BRANTON, P. E., GAFFNEY, B. J., HIRSCHBERG, C. B., FUCHS, P. & BLUMBERG, P. M. (1974). The chick fibroblast cell surface after transformation by *Rous sarcoma* virus. *Cold Spring Harbor Symposia on Quantitative Biology*, **39**, 1173–80.

ROSEMAN, S., ROTTMANN, W., WALTHER, B., OHMAN, R. & UMBREIT, J. (1974). Measurement of cell–cell interactions. *Methods in Enzymology*, **32 B**, 597–611.

SABA, T. M., BLUMENSTOCK, F. A., WEBER, P. & KAPLAN, J. E. (1978). Physiologic role of cold-insoluble globulin in systemic host defense: Implications of its characterization as the opsonic alpha-2-surface binding glycoprotein. *Annals of the New York Academy of Sciences*, **312**, 43–55.

SCHLESSINGER, J., BORAK, L. S., HAMMES, G. G., YAMADA, K. M., PASTAN, I., WEBB, W. W. & ELSON, E. L. (1977). Mobility and distribution of a cell surface glycoprotein and its interaction with other membrane components. *Proceedings of the National Academy of Sciences, USA*, **74**, 2909–13.

SEAR, C. H. J., GRANT, M. E. & JACKSON, D. S. (1976). Identification of a major extracellular non-collagenous glycoprotein synthesized by human skin fibroblasts in culture. *Biochemical and Biophysical Research Communications*, **71**, 379–84.

STATHAKIS, N. E. & MOSESSON, M. W. (1977). Interactions among heparin, cold-insoluble globulin, and fibrinogen in formation of heparin precipitable fraction of plasma. *Journal of Clinical Investigation*, **60**, 855–65.

STENMAN, S. & VAHERI, A. (1978). Distribution of a major connective tissue protein, fibronectin, in normal human tissues. *Journal of Experimental Medicine*, **147**, 1054–64.

STENMAN, S., WARTIOVAARA, J. & VAHERI, A. (1977). Changes in the distribution of a major fibroblast protein, fibronectin, during mitosis and interphase. *Journal of Cell Biology*, **74**, 453–67.

STONE, K. R., SMITH, R. E. & JOKLIK, W. K. (1974). Changes in membrane polypeptides that occur when chick embryo fibroblasts and NRK cells are transformed with avian sarcoma viruses. *Virology*, **58**, 86–100.

STRUCK, D. K. & LENNARZ, W. J. (1977). Evidence for the participation of saccharide lipids in the synthesis of the oligosaccharide chain of ovalbumin. *Journal of Biological Chemistry*, **252**, 1007–13.

STRUCK, D. K., SUITA, P. B., LANE, M. D. & LENNARZ, W. J. (1978). Effect of tunicamycin on the secretion of serum proteins by primary cultures of rat and chick hapatocytes. *Journal of Biological Chemistry*, **253**, 5332–7.

TEICHBERG, V. I., SILMAN, I., BEUTSCH, D. D. & RESHEFF, G. (1974). *Proceedings of the National Academy of Sciences, USA*, **72**, 1383–7.

TENG, N. N. H. & CHEN, L. B. (1975). The role of surface proteins in cell proliferation as studied with thrombin and other proteases. *Proceedings of the National Academy of Sciences, USA*, **72**, 413–17.

TROWBRIDGE, I. S., HYMAN, R. & MAZAUSKAS, C. (1978). The synthesis and properties of T25 glycoprotein in Thy-1-negative mutant lymphoma cells. *Cell*, **14**, 21–32.

VAHERI, A. & MOSHER, D. F. (1978). High molecular weight, cell surface-associated glycoprotein (fibronectin) lost in malignant transformation. *Biochimica et Biophysica Acta*, **516**, 1–25.

Vaheri, A. & Ruoslahti, E. (1975). Fibroblast surface antigen produced but not retained by virus transformed human cells. *Journal of Experimental Medicine*, **142**, 530–5.

Vaheri, A., Ruoslahti, E., Westermark, B. & Ponten, J. (1976). A common cell-type specific surface antigen in cultured human glial cells and fibroblasts: Loss in malignant cells. *Journal of Experimental Medicine*, **143**, 64–72.

Vuento, M., Wrann, M. & Ruoslahti, E. (1977). Similarity of fibronectins isolated from human plasma and spent fibroblast culture medium. *FEBS Letters*, **82**, 227–31.

Wartiovaara, J., Keski-Oja, J., Kurkinen, M. & Stenman, S. (1976). Interactions of fibronectin, a cell type specific surface associated glycoprotein. In *Cell Interactions in Differentiation*, ed. M. Karkinen-Jaaskselainen, L. Saxen & L. Weiss. New York, London: Academic Press.

Wartiovaara, J., Linder, E., Ruoslahti, E. & Vaheri, A. (1974). Distribution of fibroblast surface antigen. Association with fibrillar structures of normal cells and loss upon viral transformation. *Journal of Experimental Medicine*, **140**, 1522–33.

Wartiovaara, J., Stenman, S. & Vaheri, A. (1976). Changes in expression in fibroblast surface antigen (SFA) during cytodifferentiation and heterokaryon formation. *Differentiation*, **5**, 85–9.

Willingham, M. C., Yamada, K. M., Yamada, S. S., Pouyssegur, J. & Pastan, I. (1977). Microfilament bundles and cell shape are related to adhesiveness to substratum and are dissociable from growth control in cultured fibroblasts. *Cell*, **10**, 375–80.

Wolff, I., Timpl, R., Pecker, I. & Steffen, C. (1967). A two-component system of human serum agglutination gelatine-coated erythrocytes. *Vox Sanguinis*, **12**, 443–56.

Yamada, K. M. (1978). Immunological characterization of a major transformation sensitive fibroblast cell surface glycoprotein: localization, redistribution, and role in cell shape. *Journal of Cell Biology*, **78**, 520–41.

Yamada, K. M. & Kennedy, D. W. (1979). Fibroblast cellular and plasma fibronectins are similar but not identical. *Journal of Cell Biology*, **80**, 492–8.

Yamada, K. M. & Olden, K. (1978). Fibronectins-adhesive glycoproteins of cell surface and blood. *Nature, London*, **275**, 179–84.

Yamada, K. M., Olden, K. & Pastan, I. (1978). Transformation sensitive cell surface protein: isolation, characterization, and role in cellular morphology and adhesion. *Annals of the New York Academy of Sciences*, **312**, 256–77.

Yamada, K. M. & Pastan, I. (1976). Cell surface protein and neoplastic transformation. *Trends in Biochemical Science*, **1**, 222–4.

Yamada, K. M., Schlesinger, D. H., Kennedy, D. W. & Pastan, I. (1977*a*). Characterization of a major fibroblast cell surface protein. *Biochemistry*, **16**, 5552–9.

Yamada, K. M. & Weston, J. A. (1974). Isolation of a major cell surface glycoprotein from fibroblasts. *Proceedings of the National Academy of Sciences, USA*, **71**, 3492–6.

Yamada, K. M., Yamada, S. S. & Pastan, I. (1975). The major cell surface glycoprotein of chick embryo fibroblasts is an agglutinin. *Proceedings of the National Academy of Sciences, USA*, **72**, 3158–62.

Yamada, K. M., Yamada, S. S. & Pastan, I. (1977*b*). Quantitation of a transformation sensitive, adhesive cell surface glycoprotein: decrease on several untransformed permanent cell lines. *Journal of Cell Biology*, **74**, 649–54.

Zetter, B. R., Chen, L. B. & Buchanan, J. M. (1976). Effects of protease treatment on growth, morphology, adhesion, and cell surface proteins of secondary chick embryo fibroblasts. *Cell*, **7**, 407–12.

Zetter, B. R. & Martin, G. R. (1978). Expression of a high molecular weight cell surface glycoprotein (LETS) by preimplantation mouse embryo and teratocarcinoma stem cells. *Proceedings of the National Academy of Sciences, USA*, **75**, 2324–8.

Relationships between actomyosin stress fibres and some cell surface receptors in fibroblast adhesion

D. A. REES, R. A. BADLEY AND A. WOODS

Unilever Research, Colworth Laboratory, Sharnbrook, Bedford, MK44 1LQ, UK

INTRODUCTION

Non-muscle cells contain a variety of structures based on F-actin, often with other muscle-associated proteins such as myosin, tropomyosin, α-actinin and filamin. The extent to which the individual structures interconvert and the mechanisms by which they do so are not yet understood, nor are the distinctions between their different functions. Some redirection of actomyosin activity appears to be invoked by contacts at the outer cell surface leading to such events as contact retraction, cell spreading, phagocytosis or patching or capping with concomitant shape change. These phenomena seem to point to the existence of transmembrane relationships by which the contacts outside the cell can be expressed in the activity of internal actomyosin.

One very striking system in which cell contacts influence cellular actomyosin is in the coupled formation in fibroblasts of focal adhesions and the so-called stress fibres or actomyosin bundles, as discussed in another contribution to this volume (see the chapter by Dunn). We will describe some further observations on external features which seem to be related to the state of organisation of stress fibres, and which we hope will have a bearing on mechanisms of actomyosin response. We have worked with fibroblasts from several different sources and illustrations will be chosen from cell types (listed in the captions to the figures)' which give particularly clear demonstrations of the general effects.

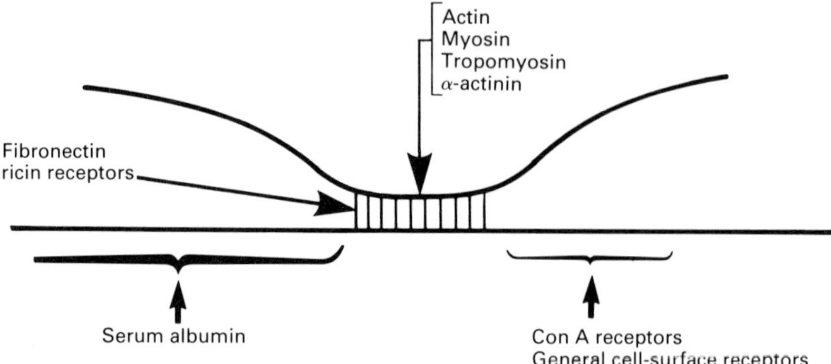

Fig. 1. Diagrammatic representation of the distribution of components at and around focal adhesions, as revealed by immunochemical analysis.

TRANSMEMBRANE RELATIONSHIPS AT AND AROUND FOCAL ADHESIONS

Focal adhesions are the specialised contact areas between membrane and substratum which develop when anchorage-dependent fibroblasts spread for growth on glass or plastic. We have described a method for the isolation of these structures from the cell bodies and an investigation of their composition by immunochemical methods (Rees *et al.*, 1978; Badley *et al.*, 1978). On the inner membrane face were located actin, myosin, tropomyosin and α-actinin as major components whereas, external to the cell, fibronectin and receptors for ricin were sandwiched between the plasma membrane and the glass (Fig. 1). The localisation of α-actinin in the cytoplasmic part of the adhesion plaque has been confirmed by immunoelectron microscopy using colloidal gold-labelled antibodies (Fig. 2; R. A. Badley, *et al.*, 1979). The ultrastructure of the adhesion shows side-to-side periodicity both inside and outside the membrane (Fig. 2; see also Rees, *et al.*, 1978). Because the value of the spacing of this periodicity, the protein composition and the other ultrastructural details are all strongly reminiscent of the Z-line structure of skeletal muscle, we have used this analogy to formulate an outline model for the focal adhesion (Fig. 3). This is based on the known dimensions of the proteins present and is an adaptation of a model proposed on the basis of better ultrastructural evidence for the Z-line structure in fish skeletal muscle (Franzini-Armstrong, 1973). The actin filaments are proposed (Fig. 3) to connect to the vertices of a regular pyrimidal

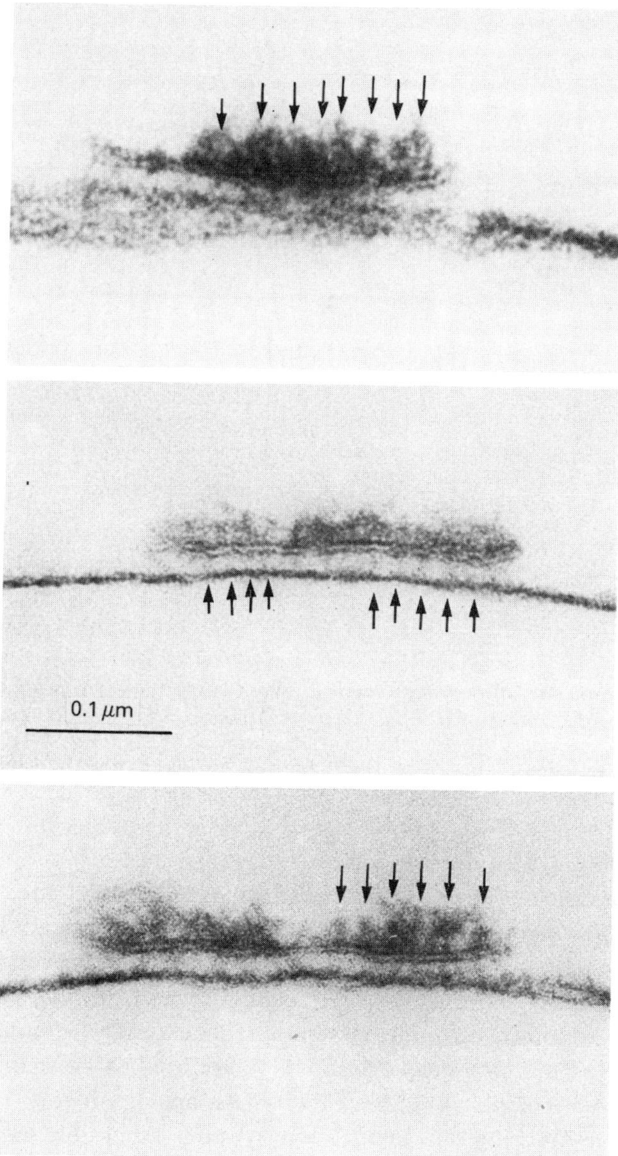

Fig. 2. Electron micrographs of vertical sections of focal adhesions isolated from 16C cells to show side-to-side periodicity (arrowed) on both membrane faces. Isolation and processing of adhesions were as previously described (Badley *et al.*, 1978, 1979).

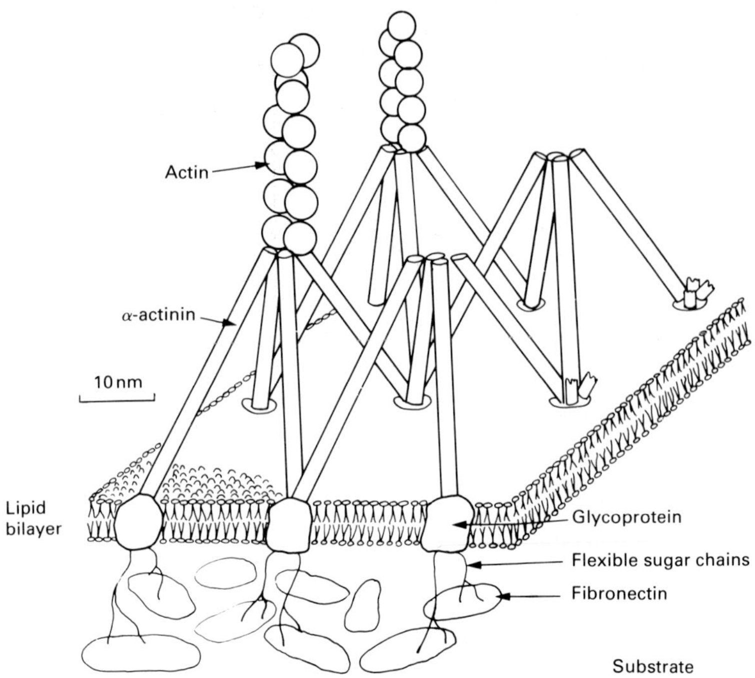

Fig. 3. Minimum outline model for features of supramolecular organisation at focal adhesions, based on the features seen in our electron micrographs, the components shown by immunochemical analysis to be present, a model previously proposed (Franzini-Armstrong, 1973) for the muscle Z-line, and the known molecular dimensions of α-actinin.

assembly of α-actinin rods which is itself stabilised by anchorage through transmembrane glycoprotein complexes to substratum-bound fibronectin. We suggest that the external chains are flexible, partly based on ultrastructural evidence and partly to explain how the submembranous assembly with its definite and regular geometry might be stabilised through attachment to a variety of amorphous substrates of nonbiological origin which are unlikely to share with it many features of matching periodicity. An analogy for this stabilisation is the mooring of a boat to a jetty: this is possible, regardless of the arrangement of the mooring points, provided that the ropes are long and flexible. The evidence from our own work is insufficient to show whether the chains which carry receptors for ricin bind directly to adsorbed fibronectin, and also whether these are identical to the flexible strands in our model. Another contribution to this

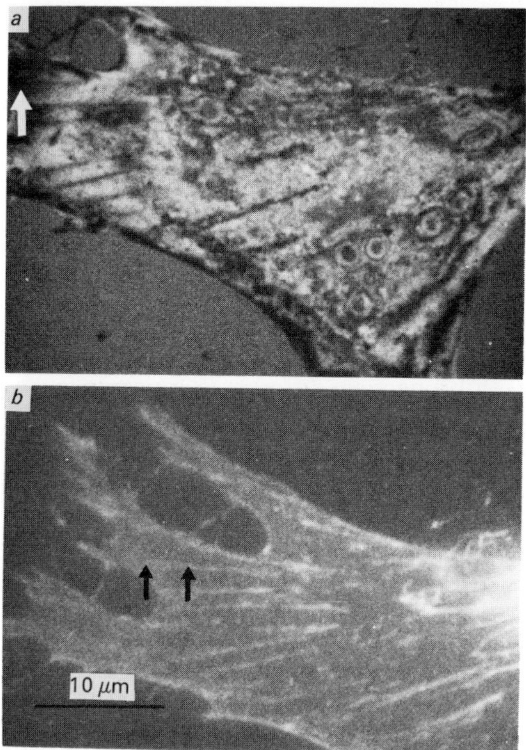

Fig. 4. Comparison of (*a*) interference reflection and (*b*) FITC-ricin staining of fixed (3.5% formaldehyde in PBS, 20 min at 37 °C) secondary chick embryo fibroblasts. Arrows show ricin streaks extending towards the cell nucleus from focal adhesions. Methods were as given by Badley *et al.* (1978).

volume (see the chapter by Hughes), does indicate, however, that the binding sites for ricin and fibronectin occur on the same molecule.

Further investigation of the distribution of the ricin-binding sites by fluorescence microscopy on whole cells, gives more information about their association with stress fibres. For example, Fig. 4 shows focal adhesions and the associated staining with FITC-ricin on the same chick cells. The adhesions themselves are larger in area and more elongated in shape than have been frequently noted in a variety of other cells. We have shown (see also p. 397) that such apparently enlarged adhesions can at least sometimes represent clusters of smaller focal adhesions, with each of the small adhesions terminating a sub-group of filaments of the larger bundle. The fluorescent streaks of bound FITC-ricin can be seen (Fig. 4) to extrapolate from the dark streaks in the interference reflection image, suggesting that they have a transmembrane relationship with stress fibres not only at their ends at focal adhesions but also at neighbouring parts of the cell surface that are separated from substratum. Indeed, bound FITC-ricin is seen in a fluorescence over most of the membrane, suggesting the possibility that similar actin–receptor relationships might exist in lower concentrations over the whole cell surface.

The ricin-visualised streaks are quite distinct from the fibrillar fibronectin deposits that are shown up elsewhere on the cells with the same reagent, and have been shown by Hynes & Destree (1978) and Heggeness, Ash & Singer (1978) to correlate with stress fibres in spreading cells. Fibronectin is different in appearance – being thicker, more kinked and more branched – and is readily distinguished also by double immunofluorescence and by differential susceptibility to EGTA, trypsin and cytochalasin B (see below). Also, fibronectin is confined to the fibrillar deposits, whereas ricin receptors occur more generally over the cell membrane.

COUPLED CHANGES IN STRESS FIBRES AND SURFACE FEATURES

We will describe first the dispersal of stress fibres by EGTA. This well-known and widely used detachment agent for fibroblasts acts by first rounding up the cells in a process that involves inner rather than outer structures because the rate of detachment can be altered by drugs which are believed to act on the cytoskeleton (Shields & Pollock, 1976; compare also the results for EDTA, Johnson & Pastan, 1972), and focal adhesions can be shown to remain intact during this phase (Rees, Lloyd & Thom, 1977; Rosen & Culp, 1977). Only later are the adhesions peeled away or pinched off. We therefore studied the changes in the different components of the cytoskeleton in the EGTA-rounding of BHK fibroblasts. Indirect immunofluorescence showed that the distributions of microtubules and 10 nm filaments seemed to fold into the cell body in parallel with cell rounding, with

---

Fig. 5. Actin (*a*, *b*); tubulin (*c*, *d*) and 10 nm filament (*e*, *f*) distributions in BHK cells and FITC-ricin staining (*g*, *h*) of chick embryo fibroblasts. Control cells (*a*, *c*, *e*, *g*) and those treated with EGTA 0.05 % in Ca^{2+}, Mg^{2+}-free phosphate-buffered saline (PBS) for 10 min at 37 °C (*b*, *d*, *f*, *h*) were fixed and processed for immunofluorescence using affinity purified antibodies and procedures as described (Badley *et al.*, 1978; Weber, Rathke & Osborn, 1978).

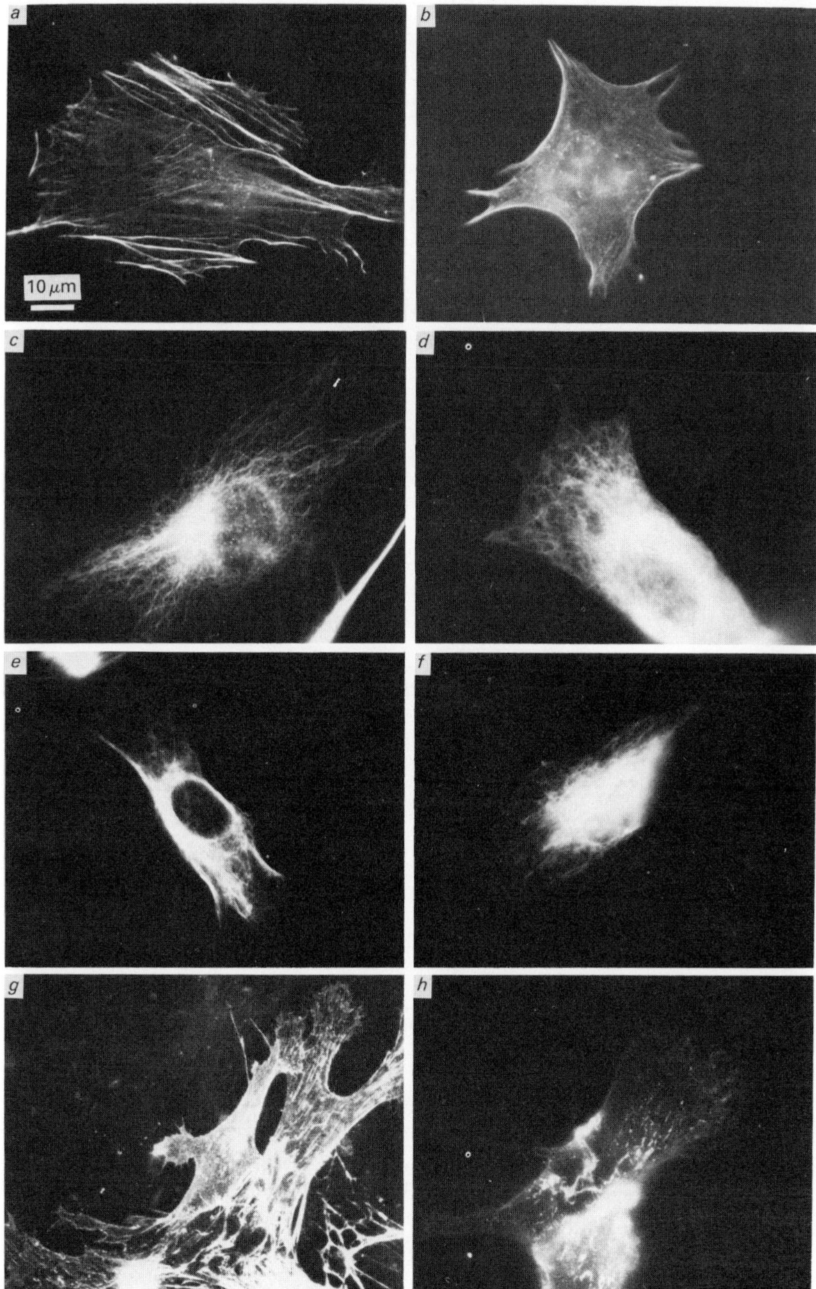

For legend see opposite.

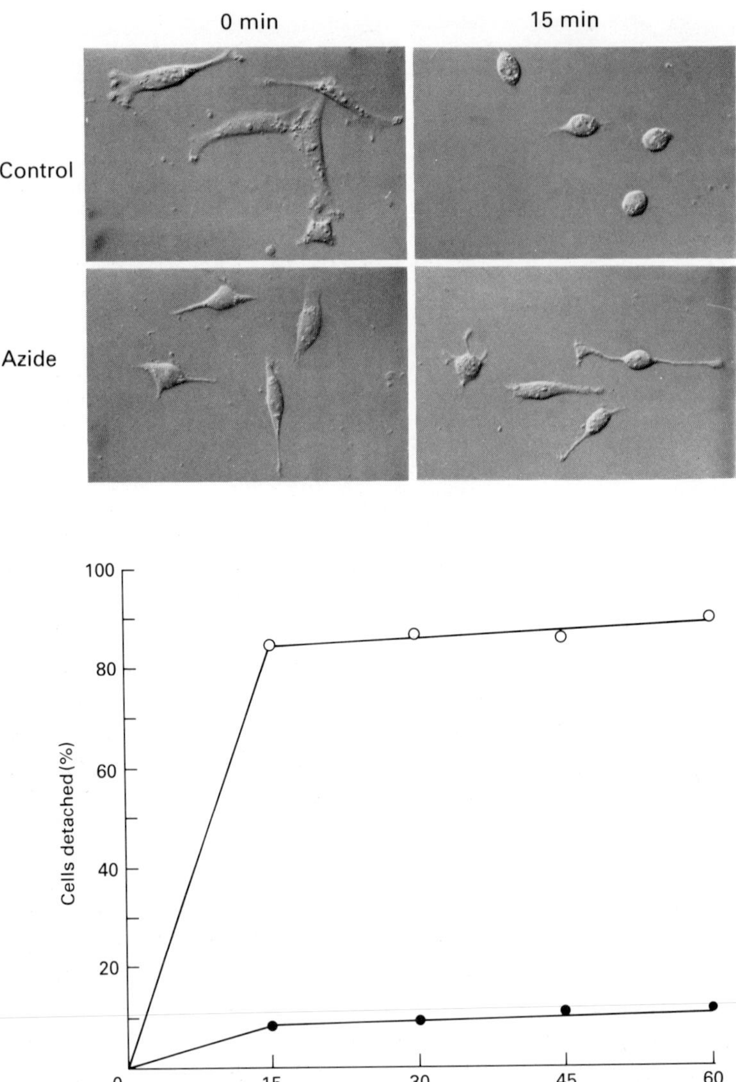

Fig. 6. (Below) Rate of detachment of 16C cells by EGTA (0.05 % in Ca^{2+} and Mg^{2+}-free PBS), in the absence (○) and presence (●) of sodium azide (10 mM). Cells were pre-incubated with PBS or 10 mM sodium azide in PBS, respectively (20 min at 37 °C). Methods were as described previously (Rees, Lloyd & Thom, 1977). (*Above*) morphology of cells under interference contrast optics, at the stages indicated.

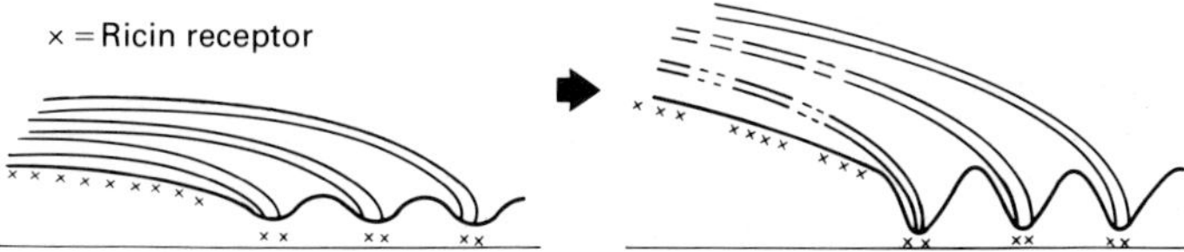

Fig. 7. Diagrammatic interpretation of the changes seen in the early stages of EGTA action on spread fibroblasts. Initially (*left*) large focal adhesions are observed which are actually clusters of small adhesions at each of which terminates a fraction of the total bundle of actin filaments; between the membrane and substratum at the adhesions, and also on parts of the membrane adjacent to the actin bundles remote from adhesions, receptors for FITC-ricin (denoted by 'X') are seen as linear streaks. On exposure to EGTA (*right*), the actin bundles begin to break up at a series of independent points along their length in parallel with a corresponding break-up in the ricin streaks; the associated energy-dependent contraction draws the membrane back around the adhesions to reveal the subunit structure.

little if any indication of preliminary depolymerisation (Fig. 5; R. A. Badley, A. Woods, L. Carruthers & D. A. Rees, in preparation). The simplest interpretation would be that these two structures merely draw in to accommodate shape changes that were initiated by an action on another component. Such a primary target could well be the more densely organised part of the actomyosin system, especially the stress fibres, because these broke up in advance of the shape change (Fig. 5). At intermediate stages of dispersal, their breakdown seemed to have started independently at a series of points along the length (Fig. 5).

In close parallel with these changes in the stress fibres, the distribution pattern of the class of receptors visualised with FITC-ricin was found to convert to a series of dots which maintain their alignment but have lost the continuity (Fig. 5). In the interference reflection image also, the sub-unit structure (see above, p. 393) became visible in the large adhesion areas. The overall process was energy-dependent because detachment (Fig. 6) was substantially arrested when azide was added to 10 mM (D. A. Rees & R. J. Safford, unpublished). The cell shape showed an initial retraction with azide, presumably as protrusive activity was inhibited, but EGTA then caused relatively little further withdrawal (Fig. 6). All these early events in the action of EGTA on spread fibroblasts may be summarised and related to each other, through the schematic representation shown in Fig. 7. This shows that the stress fibres begin to disperse at separate points along their length in a process that is coupled to the clustering of their associated receptors and to an energy-dependent withdrawal of membrane from substratum at areas around the adhesion zones. Presumably this active withdrawal is driven by the actin meshwork but we do not know whether it occurs merely because of the loss of the opposing, flattening influence of the stress fibres or whether actin is released from the stress fibres into a form having an activity directed towards cell rounding.

When stress fibres are broken up by treatment of the cells with trypsin or cytochalasin B, a similar conversion is seen in the pattern of the associated surface receptors from a continuous to a punctate distribution. With trypsin we have found, in confirmation of earlier work (Pollack & Rifkin, 1975), that the dispersal of stress fibres again runs ahead of the cell shape change. Further similarities to the action of EGTA were that stress fibre dispersal started at separate points along the length, and that the changes in microtubule distribution and 10 nm filament distribution could be interpreted merely as a folding back into the cell body as the shape changed (R. A. Badley, A. Woods, L. Carruthers & D. A. Rees, in preparation). Trypsin differed from EGTA in that it altered the structure at focal adhesions (Rees, Lloyd & Thom, 1977; Rees *et al.*, 1978) but a detailed discussion of these differences would be beyond the scope of this paper.

In an earlier discussion of cell adhesion (Rees, Lloyd & Thom, 1977) we distinguished between the contributions from physical forces at the outer surface ('stick') and the contributions from the cytoskeleton ('grip'). The grip component was proposed to act by bracing the cell to the substratum to diminish the stored elastic energy and make the cell less peelable as well as presenting a lower profile to shearing liquid, and maintaining surface macromolecules in a favourable configuration for adhesion. From the experiments described above we can say that the condensed actomyosin component of the cytoskeleton is chiefly responsible for grip since (*a*) it is the primary target for trypsin and EGTA leading to cell rounding in an energy-dependent contraction (*b*) the transmembrane interactions at the terminus of the stress fibre are likely to contribute to the stability of the total assembly at the adhesive contact.

ACTOMYOSIN CONSTRAINTS ON RECEPTORS

Having seen that the perturbation of internal stress fibres can influence the distribution of a class of receptors on the lower cell surface, we now describe some experiments of a quite different type

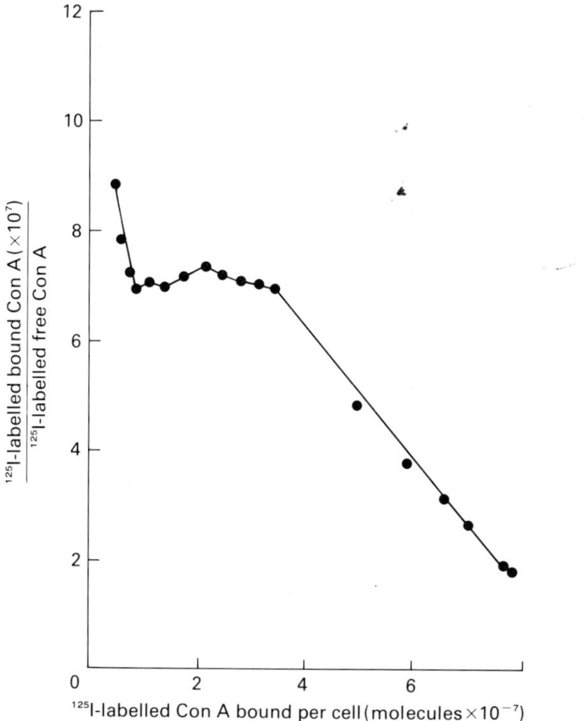

Fig. 8. Binding of con A to rat dermal fibroblasts (16C), plotted according to Scatchard (1949). Further details are given by Thom *et al.* (1979).

which give evidence for transmembrane influences at the upper surface.

We observed (Thom, Cox, Safford & Rees, 1979) that the binding of ^{125}I-labelled con A to spread fibroblasts (16C line) is a complex process which shows several distinct phases in a Scatchard plot (Fig. 8). In one region of the curve – between approximately 10^7 and 3×10^7 molecules of con A per cell on Fig. 8 – the slope is positive. This is an indication of cooperative binding – that is, a phase in which the binding of each molecule of con A seems to facilitate the binding of the next. Such behaviour has been observed before for a variety of cell types in suspension (Cuatracasas, 1973; Bornens, Karsenti & Avrameas, 1976; Gachelin, Buc-Caron, Lis & Sharon, 1976; Karsenti, Bornens & Avrameas, 1977; Reisner, Lis & Sharon, 1976; Prujansky, Ravid & Sharon, 1978) but always with the important difference that the phase with positive slope was seen at the very start of binding rather than, as with our spread fibroblasts, after an initial phase with a negative slope (0–10^7 molecules of bound con A per cell

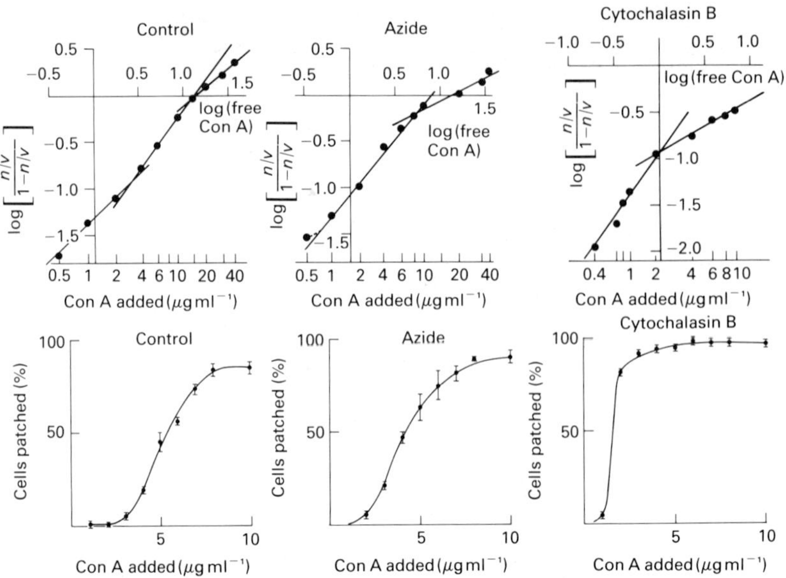

Fig. 9. Comparison of Hill (1913) plots of ^{125}I-labelled con A binding (*above*) to 16C fibroblast monolayers and the proportion of the cell population showing patching of bound con A (*below*) 20 min after lectin addition. Pretreatment of cells and binding (for detailed methods see Thom *et al.*, 1979) and patching of con A were performed at 37 °C in the presence of the indicated drug. Bound con A was visualised with goat anti-con A followed by rabbit anti-goat-FITC. Controls were incubated in HBSS; sodium azide treatment was 10 mM in PBS for 20 min; cytochalasin B treatment was 10 μg ml^{-1} in HBSS for 1 h. In the Hill plots n represents bound con A (molecules/cell) and v represents the number of free con A receptors per cell.

on Fig. 8). For reasons argued in detail elsewhere (Thom *et al.*, 1979) we interpret this as showing that for the spread fibroblasts it is necessary to achieve a certain level of binding – which we designate the nucleation threshold – before the cooperative phase can start. The existence of a threshold is not surprising because analogous effects are common in cooperative events of all types including, for example, crystallisation phenomena and conformation changes in biopolymers. It might be important that this threshold has so far been observed in spread fibroblasts and not in suspended cells, and especially that it is shifted to lower con A concentrations when the cells are pretreated with azide or cytochalasin B (Fig. 9). We shall return to discuss the significance of these observations below.

The reason for the cooperativity was deduced from studies of the distribution of the bound con A using indirect immunofluorescence.

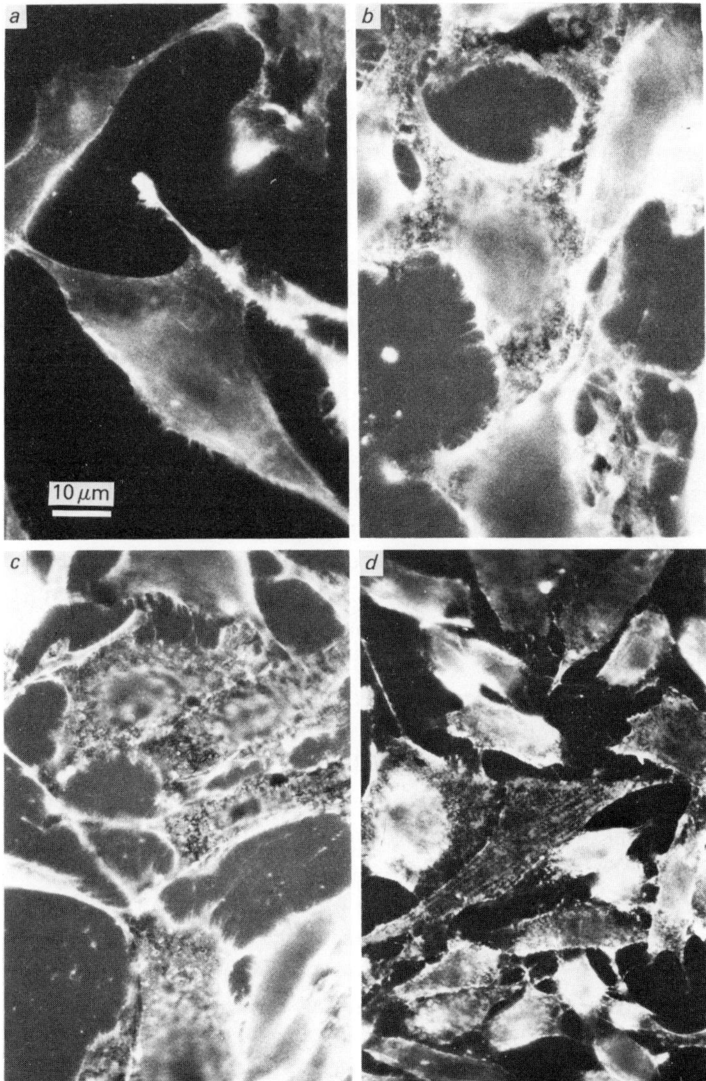

Fig. 10. Distribution of con A bound to 16C cells after incubation with (*a*) con A, 2 μg ml^{-1} in HBSS; (*b, c*) con A, 20 μg ml^{-1}; (*d*) succinyl con A, 100 μg ml^{-1} followed by goat anti-con A (Miles, diluted 1:40 with PBS). Each incubation was for 20 min at 37 °C. After washing with PBS (3 × 1 min) and fixation as for Fig. 4, (*a*)–(*c*) were incubated with goat anti-con A, washed with PBS (3 × 5 min) and (*a*)–(*d*) were stained with rabbit anti-goat-FITC (diluted 1/40 with PBS). Note patching visible in (*b*)–(*d*). Cells fixed prior to con A or after succinyl con A but prior to goat anti-con A (not shown), show no evidence of patching (compare (*a*) shown).

At low concentrations of con A, the distribution on the cell surface was uniform; however, at progressively higher concentrations we saw a rising proportion of cells in which the fluorescence had begun to appear mottled showing the onset of patching (Fig. 10). The progress of this patching was found to parallel very closely the cooperative binding, even in cells treated with cytochalasin B or azide each of which altered both the patching and the binding (Fig. 9). There can therefore be little doubt that the cooperativity and the patching are manifestations of the same underlying events. This would be understandable if the starting distribution of cell surface receptors were to offer binding possibilities for con A that could be improved to satisfy more valencies of each bound molecule by displacement or distortion of receptors. It would then be likely that con A would be more and more tightly bound as the receptors clustered, which is precisely the meaning of our observation that cooperativity parallels the visible patching. The origin of the nucleation threshold is therefore in an energy barrier to receptor clustering, which can only be surmounted when a sufficient number (apparently about 10^7) of con A molecules have bound to build up a sufficient driving force.

Patching is often described as a metabolically passive process because it occurs in the presence of metabolic inhibitors, as distinct from capping which requires energy and actomyosin activity. However, our results lead to the paradoxical conclusion that patching is actually *facilitated* by agents which impair actomyosin function, in the sense that the nucleation threshold which precedes the cooperative phase of binding (which is equivalent to the minimum con A concentration required for visible patching) is lowered by pretreatment of the cells with azide or especially cytochalasin B. This seems to imply that actomyosin components somehow 'hold the receptors in place' to create the energy barrier to receptor clustering. Two observations suggest that the functional state of actomyosin here might be a form that is characteristic of spread rather than rounded cells, such as the bundles. Firstly, the distinct nucleation threshold has been observed only in spread fibroblasts, which characteristically differ from suspended cells in having a visible proportion of cellular actomyosin recruited into bundles. Secondly cytochalasin B, an agent which is known to impair actomyosin bundles as well as other functions, seems to abolish the nucleation threshold; in contrast the effect of azide is less pronounced even though its superior ability to inhibit EGTA induced rounding and detachment (Fig. 6; D. A. Rees

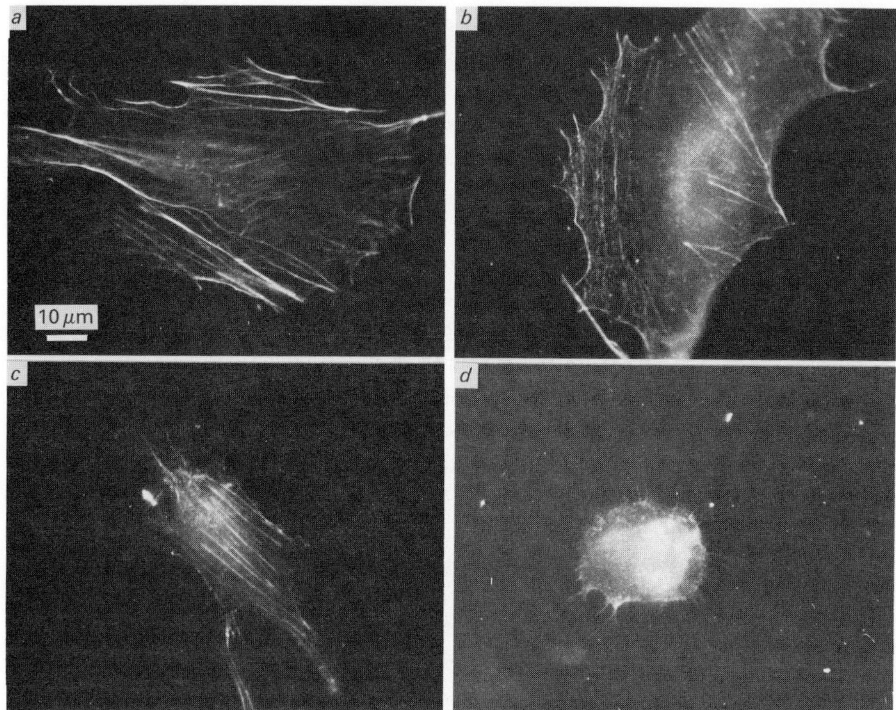

Fig. 11. The action of trypsin (25 μg ml^{-1} in HBSS for 10 min) on the actin distribution in BHK cells with and without preincubation with con A (20 μg ml^{-1} in HBSS for 20 min). Cells were treated with (*a*) HBSS; (*b*) trypsin: note breaking up of stress fibres; (*c*) con A followed by trypsin: note cell has contracted significantly but some stress fibres remain intact; (*d*) con A, trypsin, soy bean trypsin inhibitor (100 μg ml^{-1} in HBSS, 10 min) followed by methyl α-D-mannoside (αMM, 0.05 M in HBSS) for 15 min: note rounding of cell and loss of stress fibres. All incubations were at 37 °C with brief washing with HBSS between; reagents, fixation and immunofluorescent staining were as previously described (Badley *et al.*, 1978). Extreme care was necessary in processing of cells pretreated with con A and trypsin since this was found to render them very fragile.

& R. Safford, unpublished) suggests that it might intervene more effectively with the function of forms of actomyosin directed towards rounding. The mechanism is unlikely to involve direct physical constraints by actomyosin on all the receptors which patch, because biochemical (Flanagan & Koch, 1978) and other (Ash & Singer, 1976) evidence suggest that most linkages to actin do not occur until the patching event. Perhaps the nucleation threshold is determined by a sub-population of protein molecules that is permanently linked to

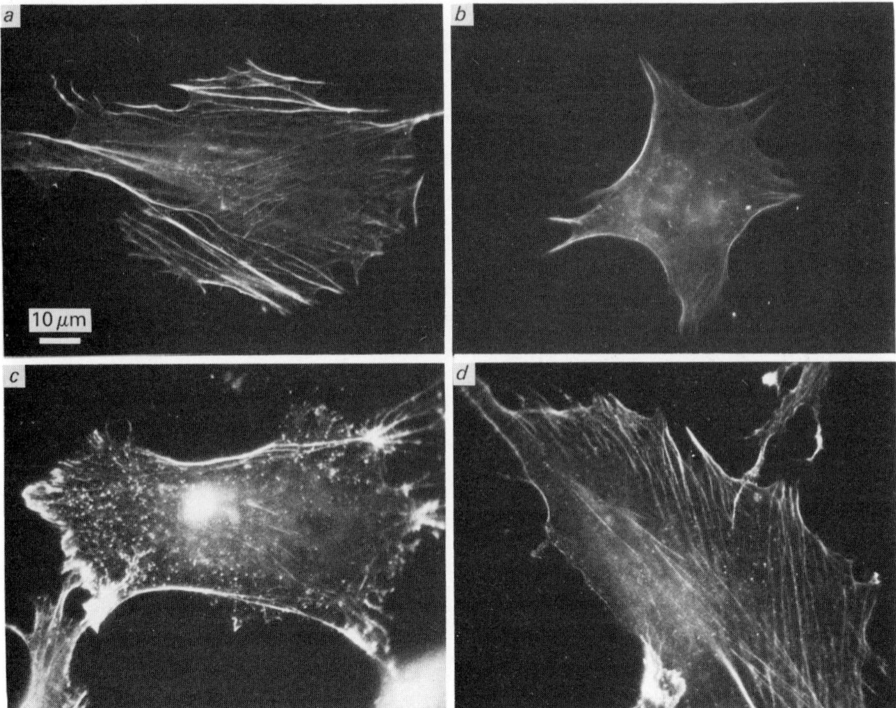

Fig. 12. The action of EGTA (0.05 % in Ca^{2+}, Mg^{2+}-free PBS for 10 min) on the actin distribution in BHK cells with and without preincubation with con A (20 μg ml^{-1} in PBS for 20 min). Cells were treated with (*a*) PBS; (*b*) EGTA: note break up of stress fibres; (*c*) con A followed by EGTA: note absence of stress fibres although cells remain spread; (*d*) con A, EGTA, αMM (0.05 M in HBSS for 15 min): note reappearance of stress fibres. Incubations were at 37 °C and immunofluorescent staining was performed as in Fig. 11.

actin and is also essential for patch formation. The orientation of these molecules in the membrane, and hence their availability for patching, would then change with the state of cellular actomyosin. One candidate for this sub-population would of course be the class of ricin receptors mentioned above.

RECEPTOR CONSTRAINTS ON ACTOMYOSIN

We have previously shown that the pretreatment of spread fibroblasts with con A can block their detachment by trypsin and EGTA and that, for trypsin, this protection includes the partial preservation of the cytoskeleton against dispersal (Rees, Lloyd & Thom, 1977).

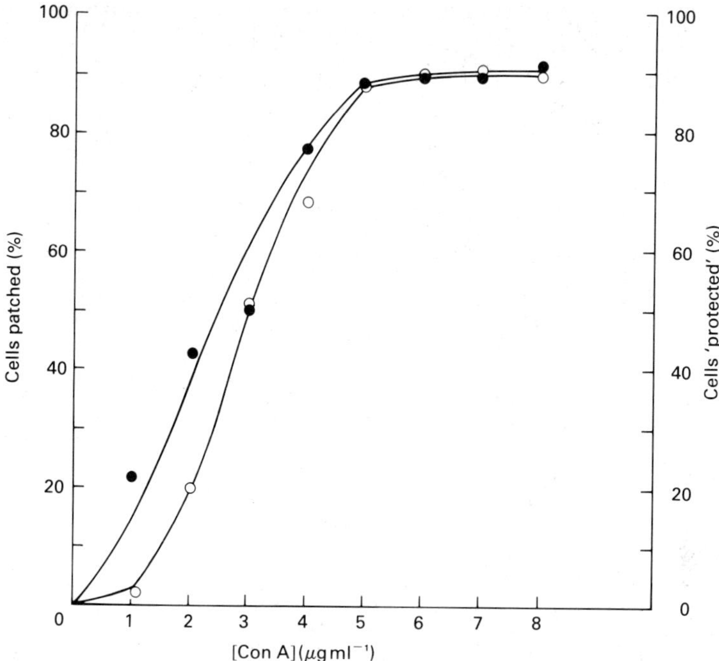

Fig. 13. The parallel between the extent of detachment (2) of rat dermal fibroblasts (16C) by EGTA (0.05 % in Ca^{2+} and Mg^{2+}-free Hank's balanced salt solution), after pretreatment with con A (concentrations ranging from 1–10 μg ml^{-1} for 20 min at 37 °C), and the extent of patching on cells incubated similarly with con A (○). Methods as for Fig. 6 and Fig. 10.

Knowing now that the dispersal of stress fibres is an early event in detachment by both agents, we have re-investigated the phenomenon by immunofluorescence microscopy and find some differences between the action of the two reagents.

The dispersal of stress fibres by trypsin is partially blocked by pretreatment of the cells with con A even though marked changes often occur in cell shape; the stress fibres that remain tend to be thinner than in untreated cells and are those which are originally at the lower cell surface. If the cells are subsequently washed with trypsin inhibitor and then treated with methyl α-D-mannopyranoside to displace the bound con A, dispersal of these remaining stress fibres occurs rapidly (Fig. 11; R. A. Badley, A. Woods & D. A. Rees, in preparation). Thus the action of con A was to block the expression of the tryptic lesion rather than to prevent the primary action of the enzyme. The rapid expression of the lesion with hapten would suggest that the primary events occur on the outer cell surface. The simplest interpretation of these results is therefore that a tryptic lesion on the outer cell surface (perhaps cleavage of fibronectin) normally leads to some receptor rearrangement to which is coupled the dispersal of stress fibres, whereas in con-A-treated cells the receptor movement is constrained so that stress fibres do not disperse until the lectin constraint is released. Whether the relevant receptors are at focal adhesions or elsewhere is not yet known.

In contrast, con A does not appear to protect stress fibres against dispersal by EGTA (Fig. 12; R. A. Badley, A. Woods & D. A. Rees, in preparation) and the cells remain spread and attached to an extent that closely parallels the extent of patching (Fig. 13; D. A. Rees and R. Safford, unpublished). This correlation with patching would suggest that the mechanism of protection against detachment may have something in common with other events which parallel the progress of patching, such as certain shape changes and the inhibition of phagocytosis, and which seem to be driven by patch-induced contraction of the actomyosin meshwork (Thom *et al.*, 1979). The con-A-treated cells do however retain preconditions for stress fibre assembly because the fibres reappear within minutes of returning to Hank's balanced salt solution in the presence of methyl α-D-mannopyranoside (Fig. 12; R. A. Badley, A. Woods & D. A. Rees, in preparation). The mechanism of action of EGTA therefore seems to bypass the relationships of cell surface molecules so that stress fibres disperse just as they do in the absence of con A. However, the cell surface relationships would seem to retain a 'memory' for stress fibre reassembly so that once again, if in a somewhat different way, this system also demonstrates the interdependence of stress fibre organisation and the distribution of external receptors.

CONCLUSIONS

We have described several relationships between features of the outer cell surface and the state of organisation of one form of actomyosin, the stress fibre, which is responsible for flattening the cell body to 'grip' the substratum. The stress fibre terminates at a specialised assembly having different composition and organisation on either side of the membrane but regular side-to-side structures having matching periodicities. Further away from these adhesion points, a class of receptors which is probably identical to those at the adhesions not only codistributes with the stress fibres but disperses in parallel with them. The patching of membrane receptors is subject to a threshold apparently having its origin in some barrier to receptor clustering imposed by the condensed form of actomyosin in spread cells. Conversely, the receptor distribution clearly constrains stress fibres because external cross-linkage can limit the dispersal of stress fibres by trypsin and, while not preventing their dispersal by EGTA, does preserve some associated features which allow very rapid reassembly.

All this evidence shows that the organisation of stress fibres is correlated with features of the organisation of external carbohydrate receptors, and that the state of each can be influenced by agents acting on the other. Stress fibres might represent only one of several forms of actin and associated proteins having this type of relationship with the cell membrane, and might be 'special' only in that the behaviour is conveniently visualised with the optical microscope.

Acknowledgements. We thank Dr D. Thom, Mr C. G. Smith, Mr R. Safford and Mrs D. S. Cox for allowing us to cite their unpublished results in this review, and for discussions.

REFERENCES

Ash, J. F. & Singer, S. J. (1976). Concanavalin A-induced transmembrane linkage of concanavalin A surface receptors to intracellular myosin-containing filaments. *Proceedings of the National Academy of Sciences, USA*, **73**, 4575–9.

Badley, R. A., Lloyd, C. W., Woods, A., Carruthers, L., Allcock, C. & Rees, D. A. (1978). Mechanisms of cellular adhesion. III. Preparation and preliminary characterisation of adhesions. *Experimental Cell Research*, **117**, 231–44.

Badley, R. A., Woods, A., Smith, C. G. & Rees, D. A. (1979). Actomyosin relationships with surface features in fibroblast adhesion. *Experimental Cell Research*. (in press.)

Bornens, M., Karsenti, E. & Avrameas, S. (1976). Cooperative binding of concanavalin A to thymocytes at 4 °C and micro-redistribution of concanavalin A receptors. *European Journal of Biochemistry*, **65**, 61–9.

Cuatracasas, P. (1973). Interaction of wheat germ agglutinin and concanavalin A with isolated fat cells. *Biochemistry*, **12**, 1312–23.

Flanagan, J. & Koch, G. L. E. (1978). Cross-linked surface Ig attaches to actin. *Nature, London*, **275**, 278–81.

Franzini-Armstrong, C. (1973). The structure of a simple Z-line. *Journal of Cell Biology*, **58**, 630–42.

Gachelin, G., Buc-Caron, M.-H., Lis, H. & Sharon, N. (1976). Saccharides on teratocarcinoma cell plasma membranes: their investigation with radioactively labelled lectins. *Biochimica et Biophysica Acta*, **436**, 825–32.

HEGGENESS, M. H., ASH, J. F. & SINGER, S. J. (1978). Transmembrane linkage of fibronectin to actin-containing filaments in cultured human fibroblasts. *Annals of the New York Academy of Sciences*, **312**, 414–17.

HILL, A. V. (1913). XLVII The combinations of haemoglobin with oxygen and with carbon monoxide. *Biochemical Journal*, **7**, 471–80.

HYNES, R. O. & DESTREE, A. T. (1978). Relationships between fibronectin (LETS protein) and actin. *Cell*, **15**, 875–86.

JOHNSON, G. S. & PASTAN, I. (1972). Cyclic AMP increases the adhesion of fibroblasts to substratum. *Nature, London*, **236**, 247–9.

KARSENTI, E., BORNENS, M. & AVRAMEAS, S. (1977). Control of density and micro-distribution of concanavalin A receptors in rat thymocytes at 4 °C. *European Journal of Biochemistry*, **75**, 251–6.

POLLACK, R. & RIFKIN, D. (1975). Actin-containing cables within anchorage-dependent rat embryo cells are dissociated by plasmin and trypsin. *Cell*, **6**, 495–506.

PRUJANSKY, A., RAVID, A. & SHARON, N. (1978). Cooperativity of lectin binding to lymphocytes and its relevance to mitogenic stimulation. *Biochimica et Biophysica Acta*, **308**, 137–46.

REES, D. A., BADLEY, R. A., LLOYD, C. W., THOM, D. & SMITH, C. G. (1978). Glycoproteins in the recognition of substratum by cultured fibroblasts. In *Cell–Cell Recognition* (32nd Symposium of the Society for Experimental Biology), ed. A. S. G. Curtis, pp. 241–60. Cambridge University Press.

REES, D. A., LLOYD, C. W. & THOM, D. (1977). Control of grip and stick in cell adhesion through lateral relationships of membrane glycoproteins. *Nature, London*, **267**, 124–8.

REISNER, Y., LIS, H. & SHARON, W. (1976). On the importance of the binding of lectins to cell surface receptors at low lectin concentrations. *Experimental Cell Research*, **97**, 446–8.

ROSEN, J. J. & CULP, L. A. (1977). Morphology and cellular origins of substrate-attached material from mouse fibroblasts. *Experimental Cell Research*, **107**, 139–49.

SCATCHARD, G. (1949). The attraction of proteins for small molecules and ions. *Annals of the New York Academy of Sciences*, **51**, 660–72.

SHIELDS, R. & POLLOCK, K. (1976). The adhesion of BHK and pyBHK cells to the substratum. *Cell*, **3**, 31–8.

THOM, D., COX, D. S., SAFFORD, R. & REES, D. A. (1979). Cooperativity of lectin binding to fibroblasts and its relation to cellular actomyosin. *Journal of Cell Science*, **39**, 117–36.

WEBER, K., RATHKE, P. C. & OSBORN, M. (1978). Cytoplasmic microtubular images in glutaraldehyde-fixed tissue culture cells by electron microscopy and by immunofluorescence microscopy. *Proceedings of the National Academy of Sciences, USA*, **75**, 1820–4.

Mechanisms of fibroblast locomotion

G. A. DUNN

Strangeways Research Laboratory, Cambridge CB1 4RN, UK

One of the most puzzling aspects of fibroblast motility is the continual activity often seen in the flattened lamellar area of a spread cultured fibroblast. This complex activity frequently accompanies locomotion, but it can also occur in certain circumstances when the fibroblast is stationary. It consists of the production of ruffles (Abercrombie, Heaysman & Pegrum, 1970*a*) or, occasionally, blebs or microspikes (Harris, 1973), usually at the leading edge of a locomoting fibroblast. If these protrusions persist, they are usually propagated rearwards towards the nucleus for a short distance before they collapse back into the dorsal surface of the leading lamella; the speed of ruffles has been estimated by Abercrombie *et al.* (1970*a*) to fall mainly in the range 2–3 μm min^{-1} relative to the substratum. Abercrombie *et al.* (1970*b*) observed that a series of indefinite shadows in the leading lamella may also participate in this steady movement towards the nucleus at approximately the same speed. Another phenomenon appears to be closely related to these activities: if the locomoting fibroblast encounters particles of debris on the substratum, these often adhere to the fibroblast and are transported towards the nucleus at the same speed of approximately 2–3 μm min^{-1}, usually on the dorsal surface of the leading lamella (Abercrombie *et al.*, 1970*b*), but occasionally on the ventral surface (Harris & Dunn, 1972). The particles on the dorsal surface become approximately stationary relative to the cell on reaching the region immediately surrounding the nucleus, and if there are many, they form a ring around the nucleus (Harris, 1973). This behaviour of particles has a striking resemblance to yet another class of phenomena: if antigens (Edidin & Weiss, 1972) or lectin receptors (Vasiliev *et al.*, 1976) of the fibroblast surface are induced to form a 'cap' by cross-linking with antibodies or lectins, this cap also forms a ring around the nucleus on the dorsal surface of the spread cell.

Abercrombie *et al.* (1970*b*, 1972) attempted to explain the relationship between these varied phenomena by proposing that the surface of the lamella of a locomoting fibroblast continuously moves towards the nuclear region, where it is disassembled and recirculated to the periphery via the cytoplasm. They further proposed that a reassembly of the surface components, close to the cell margin, accounts for the formation of ruffles and the closely related process of forward spreading of the cell margin. Harris (1973, 1976) sought to attribute a more general significance to this surface-flow hypothesis. He argued that if the cell has the machinery to transport particles towards the nucleus – perhaps cortical arrays of microfilaments which pull the whole lamellar surface towards the nucleus – then the same machinery could be responsible for exerting a force on the substratum. He also attributed protrusion to the membrane flow but disagreed with Abercrombie *et al.* about the actual mechanism, as I will discuss later. Thus two essential requirements for locomotion, the forward protrusion of the leading margin and the exertion of force on the substratum, could be derived from a single mechanism – the circulation of the lamellar surface. Unfortunately, the idea of the cell membrane or surface moving bodily towards the nucleus is not easily compatible with observations on the capping phenomenon (see discussion by Bretscher, 1976, and reply by Harris, 1976) and the surface-flow hypothesis has consequently attracted much criticism.

My colleagues, Michael Abercrombie and Julian Heath, and I believe that a much more satisfactory general hypothesis of fibroblast locomotion can now be constructed along the same lines as the surface-flow hypothesis but with an important change of emphasis from the lamellar surface to a cytoplasmic contractile meshwork which is postulated to contract continuously towards the nucleus during locomotion. An appropriate meshwork, pervading the entire cytoplasm, has recently been demonstrated in critical-point-dried fibroblasts; this is the trabecular meshwork of Wolosewick & Porter (1976). Webster, Henderson, Osborn & Weber (1978) have now shown that such meshworks contain actin. Furthermore, Isenberg *et al.* (1976) have shown that the cytoplasm of glycerinated models of spread adenocarcinoma cells can be induced to contract towards the nucleus under appropriate conditions. Such induced contractions are, however, discontinuous; once a glycerinated model has contracted no further contraction can be induced. In order for a continuous contraction to occur in the living cell it is necessary to further

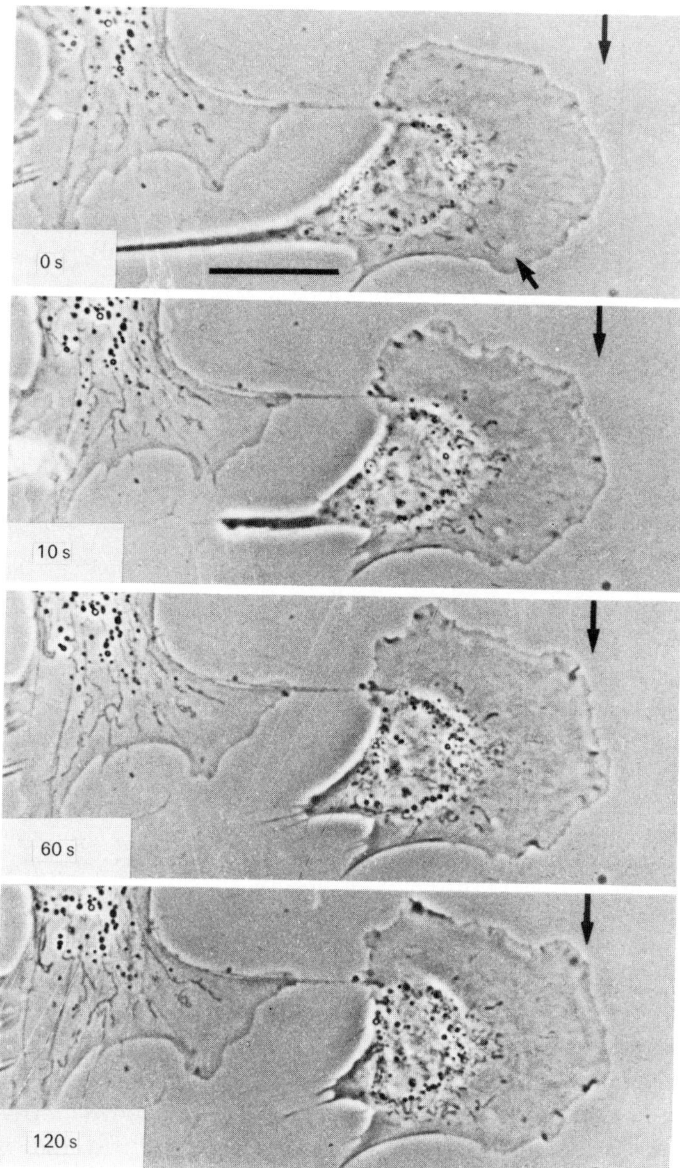

Fig. 1. A two-minute sequence of phase contrast micrographs shown retraction and protrusion activity after detaching the tail of a chick heart fibroblast using micromanipulation. Most of the tail retraction is complete by 10 s and increased ruffling activity starts at approximately this time. The mottled shadow pattern is not well defined in this cell, except that a single discrete pale area, which moves towards the nucleus, can be seen in the lower part of the leading lamella (arrowed). Scale bar = 20 μm. Vertical arrows mark a fixed reference point on the substratum.

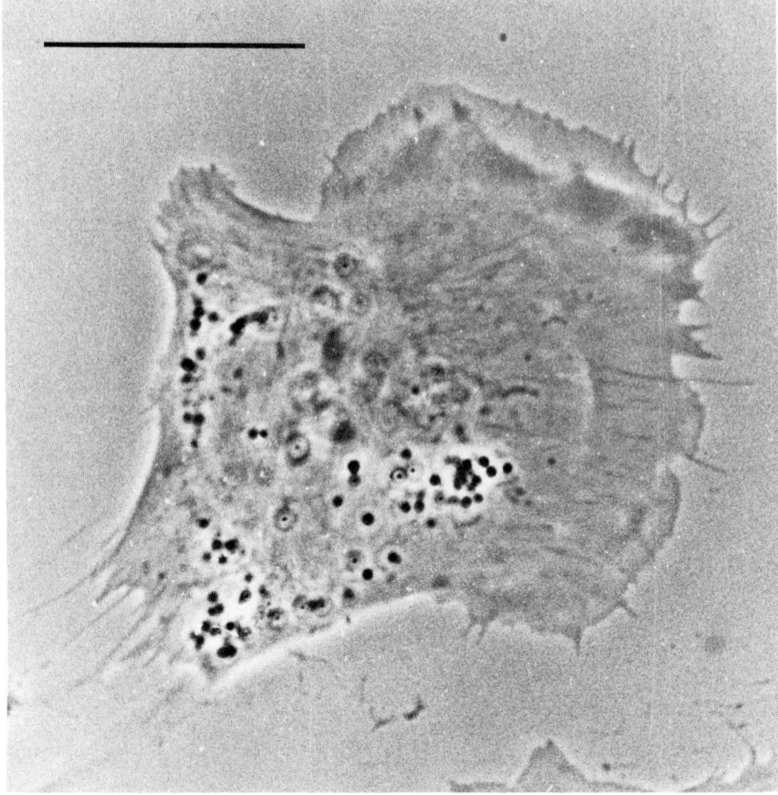

Fig. 2. Chick heart fibroblast in phase contrast 1 min after a spontaneous tail detachment. Microspikes can be seen at the leading edge. The large amorphous phase density in the upper part of the leading lamella is transitional with ruffles at its leftmost extremity. The arc-shaped structure, visible in the middle of the leading lamella, is not yet fully formed. Its mean speed towards the nucleus was measured as 2.6 μm min^{-1}. Scale bar = 20 μm.

postulate that the material of the contractile meshwork undergoes a cyclical activity, analagous to that proposed for the surface in the surface-flow hypothesis. The meshwork must continuously disassemble in the nuclear region and return to the marginal region to reassemble and participate again in the contraction.

The first advantage of this hypothesis is a simpler explanation of the movement towards the nucleus of ruffles and other protrusions. It is no longer necessary to attribute any particular mechanical properties to the cell surface, only to propose that the internal structure of the protrusions is mechanically integral with the meshwork. J. P. Heath and I have recently made some observations on cultured fibroblasts which support this (Dunn & Heath, in preparation). We exploited the increased protrusive activity shown in Fig. 1, which follows the artificial detachment of the tail of a spread fibroblast using a micromanipulator (Harris, 1971; Chen, 1978). Not only is protrusive activity increased, but it often changes from one form to another; thus ruffling may change to blebbing or to the production of microspikes, and intermediate forms of individual protrusions may occur. But we also observed three forms of activity which appeared not to consist of protrusions. The mottled shadow pattern, first described by Abercrombie *et al.* (1970*b*), moves continuously from the margin to the nuclear region and does not appear to change in intensity on tail detachment (Fig. 1). Large amorphous structures, much denser than shadows in phase contrast, often appear close to the margin after tail detachment (Fig. 2). These densities show transitional forms with the three types of protrusion and, like the protrusions, they usually disappear after moving only a short distance inwards from the margin. Curious arc-shaped structures which appear infrequently in the leading lamella were first observed by Harris and me. They appear more frequently after tail detachment and can arise at various places in the leading lamella, including the margin, and they move steadily towards the nucleus, sometimes persisting until they reach the nuclear region (Fig. 2). Occasionally they develop from the amorphous dense structures, but more usually they appear as a pale area which rapidly forms into a thin dense line with its concave face towards the nucleus. Their ultrastructure was described by Heath & Dunn (1978) as consisting mainly of microfilaments (Fig. 3). The phenomenon of movement towards the nucleus therefore appears to be quite general, not only amongst protrusions, but amongst structures which are internal in the lamella and yet are obviously related to protrusions.

The fact that all these structures move at speeds within the same narrow range, as nearly as can be determined, attests to the common origin of their movement and Heath and I chose the movement of one type, the arc-shaped structures, as representative of the supposed movement of the meshwork. These structures have the advantages that their position is clearly defined, that they are found to occur in all positions between margin and nuclear region, and that they show relatively long excursions. Their mean speed is 2.28 μm min^{-1} and we detected no significant difference in their speed whether they

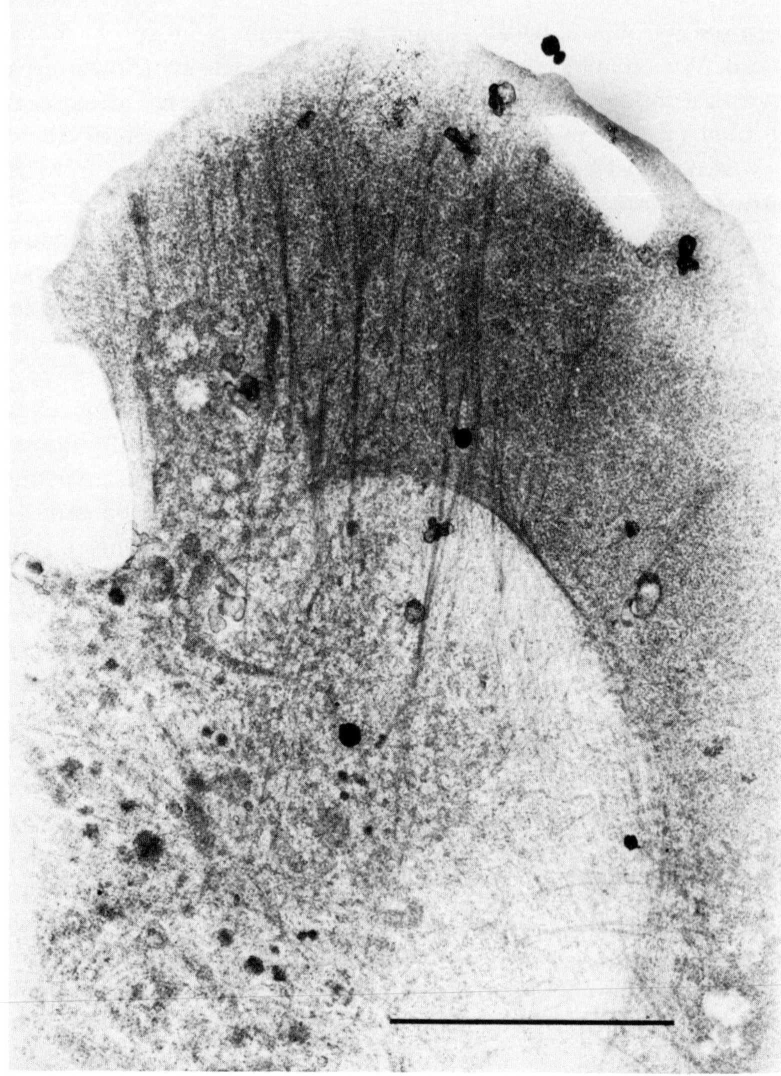

Fig. 3. High-voltage electron micrograph of thick horizontal section of chick heart fibroblast. The arc-shaped structure in the middle of the photograph consists of a band of microfilaments whose central region lies close to the dorsal surface of the lamella. The structure demarcates an anterior region of cytoplasm which contains an unusually high density of polysomes. Scale bar = 5μm.

occurred before tail detachment or within the 10 min period following detachment. But they decelerate steadily as they approach the nucleus, which indicates that the contraction is distributed throughout the lamella. Those which reach the rear of the lamella decelerate rapidly and come to rest, relative to the cell, in the nuclear region, approximately 10 μm from the nucleus. This suggests that the contraction is more concentrated at the rear of the lamella and this concords with the observation of Webster *et al.* (1978) that close to the nucleus there is a transition between the denser meshwork surrounding the nucleus and the less dense meshwork occupying the lamella.

The fibroblast tail detachment experiments also provide more direct support for the continuous contraction hypothesis. Chen (1977) has recently shown that the tail retraction which follows detachment is partially an active contraction and, since tail retraction invariably leads to a wave of increased protrusive activity, it may be speculated that the continual protrusive activity of normal locomotion results from a continuous active contraction of the cytoplasmic meshwork. Heath and I considered two possible explanations of the causal link between contraction and protrusion. The first is based on the proposal of Harris (1973) that an inward pull on the surface membrane causes an internal hydrostatic pressure which leads to herniation at weak points of the cell surface. This view was supported by his observation that blebs could be collapsed, or their formation prevented, by raising the external osmotic pressure. The rapid contraction which follows tail detachment might also be expected to raise the internal pressure and thus account for the increased protrusion. Harris's experiments certainly suggest that a pressure difference is required for the formation of blebs, but we think it unlikely that a pressure increase is the primary cause of blebs or other protrusions. If it were, then the course of tail retraction should exactly coincide with the course of increased protrusive activity since a compression wave should travel across the cell within a fraction of a microsecond. Yet we invariably observed a delay of 10–20 s before the crest of increased protrusive activity – even in the case of blebs. The second possibility is derived from the suggestion of Abercrombie *et al.* (1970*b*) that protrusion is caused by the reassembly of the 'full thickness of the surface' close to the leading margin. In the case of the continuous contraction hypothesis, we may assume that the surface plays a passive role in protrusion and consider the possibility

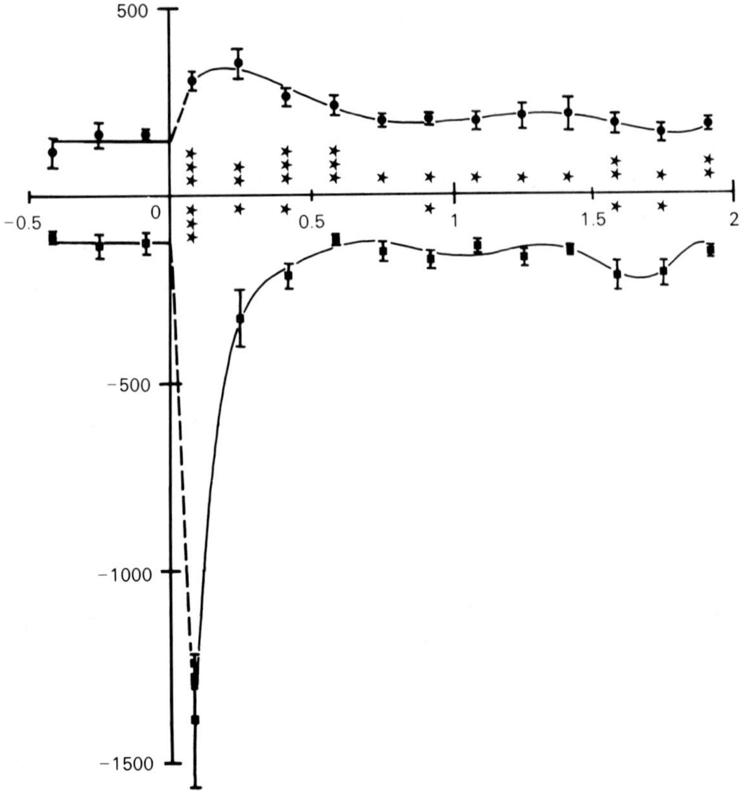

Fig. 4. Abscissa: time after tail detachment in min; ordinate: protrusion rate (positive values) and retraction rate (negative values) in $\mu m^2\ min^{-1}$. Circles and squares are respectively mean protrusion rate and mean retraction rate of ten fibroblasts over each 10 s time interval ± standard error of mean. The curves are respectively fifth- and eight-degree polynomials fitted to data by least-squares method. Each data point after time 0 was compared with the appropriate mean basal rate (before time 0) using paired 2-way t-tests. Significant differences are indicated by 3 stars ($p < 0.001$), 2 stars ($p < 0.01$), or 1 star ($p < 0.1$).

that protrusion is caused by the reassembly of the cytoplasmic meshwork at the cell margin. Harris (1973) objected to the proposal of Abercrombie *et al.* on the grounds that the surface membrane is probably too flexible to exert the necessary lateral pressure but this objection is obviously less valid if applied to this proposal of meshwork assembly. We therefore suggest that protrusion is part of the continuous cycle of meshwork contraction, disassembly, transport to the periphery, reassembly and contraction, and that the delay between a large induced contraction and the subsequent wave of increased protrusion is the time taken by the two intermediate stages of the cycle.

If this view of protrusion is correct, then not only is protrusion dependent on contraction but, since protrusion is the assembly of contractile material, contraction should be dependent on protrusion. A large and sudden disturbance of such an interdependent system might therefore be expected to result in oscillation. Heath and I will present evidence that such oscillations occur in fibroblasts after tail detachment (Dunn & Heath, in preparation). In these experiments we measured the protrusion rate as the rate at which the cell margin spreads over the substratum and simultaneously we measured the retraction rate as the rate at which the margin uncovers the substratum; the units were $\mu m^2\ min^{-1}$ in each case. The results of taking such measurements over 10 s intervals, and pooling them for ten fibroblasts, are shown in Fig. 4. The delay between peak retraction activity and peak protrusion activity is clearly shown, but the graph also shows a suggestion of an oscillation, in both activities, following the initial peaks. A two-way analysis of variance of the data in the interval 0.5–2 min showed that the variation in the retraction rate with time was significantly greater than could be attributed to random fluctuation. An analysis of protrusion rate over the same time interval did not, however, yield significant results, but we found a significant negative correlation between protrusion rate and retraction rate over this time interval – a result which would be expected if protrusion and retraction were both varying with time but out of phase with each other.

Consideration of the other important aspect of fibroblast locomotion, the exertion of a force on the substratum, leads again to the proposal of continuous contraction. Abercrombie *et al.* (1971) found oblique bundles of microfilaments, in electron micrographs of vertical sections of the leading lamella, which have their anterior origins in adhesion plaques and extend obliquely away from the substratum towards the central nuclear region. They proposed that the fibroblast exerts force on the substratum by the contraction of these cytoplasmic bundles, and this is now supported by three lines of evidence. Firstly, immunospecific labelling of similar bundles found in a variety of cells revealed that they contain an almost complete array of the muscle proteins: actin (Lazarides & Weber, 1974), myosin (Weber &

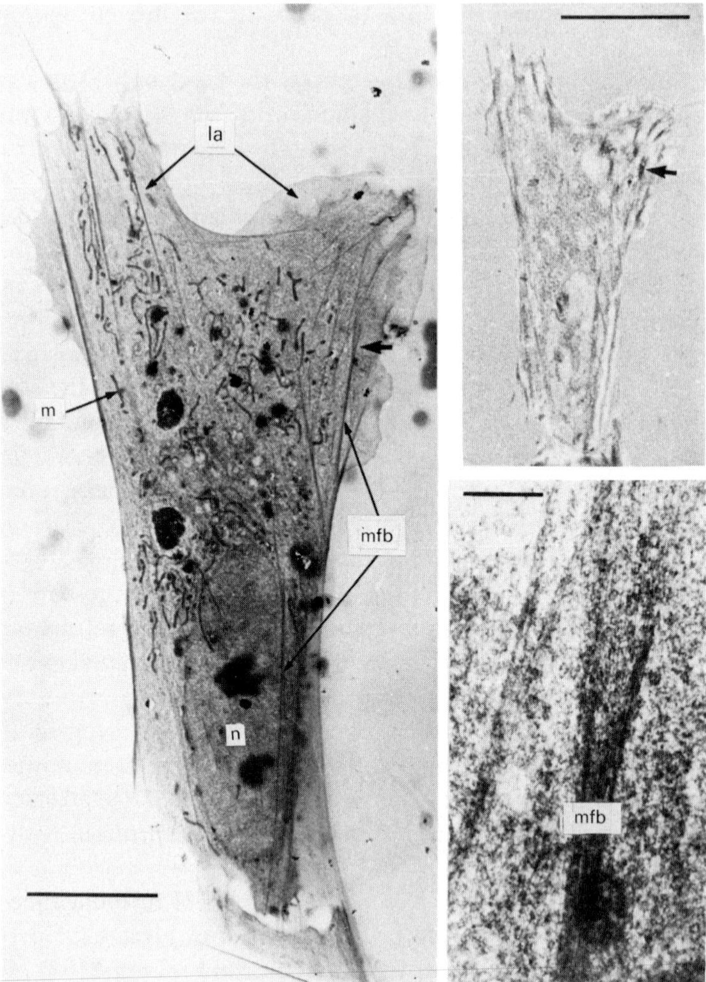

Fig. 5. *Left*: high-voltage electron micrographs of whole chick heart fibroblast showing microfilament bundles (mfb) within the leading lamella. Scale bar = 10 μm. *Upper right*: same cell in interference reflection. Note the association of focal contacts (black streaks) with the anterior terminations of the microfilament bundles in the high-voltage picture (example arrowed). Scale bar = 20 μm. *Lower right*: high-voltage electron micrograph of the arrowed focal contact. Scale bar = 0.5 μm. (Reproduced from Heath & Dunn, 1978).

Groeschel-Stewart, 1974), tropomyosin (Lazarides, 1975), α-actinin (Lazarides & Burridge, 1975) and M-protein (Schollmeyer *et al.*, 1976). Secondly, Isenberg *et al.* (1976) produced direct evidence of the contractility of similar bundles dissected from adenocarcinoma cells using a laser microbeam. And thirdly, Heath & Dunn (1978), by studying the dynamics of the adhesive relations between locomoting fibroblasts and the substratum, inferred that the oblique microfilament bundles of the leading lamella shorten as the cell advances. The discrete focal contacts, seen in the interference-reflection microscope to be distributed throughout the leading lamella and at the tail of moving fibroblasts, are known to be sites of adhesion to the substratum (Abercrombie & Dunn, 1975; Izzard & Lochner, 1976) and each is invariably associated with a microfilament bundle which usually extends obliquely towards the nuclear region (Heath & Dunn, 1978) as shown in Fig. 5. This invariable association, found in 200 cases of focal contacts near the leading margin, implies that the bundles form at the same time, and as rapidly, as the associated focal contacts. Fleischer & Wohlfarth-Bottermann (1975) have shown that the contractile cytoplasm of *Physarum* becomes strongly oriented into microfilament bundles during isometric contraction but not during isotonic contraction. Abercrombie, Dunn & Heath (1977) and Heath & Dunn (1978) therefore suggested that the rapid bundle development and persistence in fibroblasts might be similarly caused by the development of mechanical stress in a cytoplasmic microfilament meshwork. This is supported by the observation of Chen (1977) that, when the tension is suddenly released in a fibroblast tail by tail detachment, the microfilament bundles of the tail became rearranged into a meshwork. If the cytoplasmic meshwork were continuously contracting towards the nucleus in a moving fibroblast, then the formation of a focal contact, by merely connecting a small portion of the meshwork to the substratum and preventing it from moving towards the nucleus, would cause the rapid development of stress in the meshwork between the focal contact and the nuclear area. The expected stress pattern is therefore closely similar to the observed pattern of oblique microfilament bundles. Of course, bundle contraction is not strictly isometric in locomoting fibroblasts. The bundles of the leading lamella shorten at the same speed as the cell moves forward, approximately 1 μm min^{-1}, which is considerably slower than the speed of the postulated meshwork contraction, approximately 2–3 μm min^{-1}. A simplified view is therefore that the formation of a focal contact slows down the contraction of a portion of the meshwork which causes stress to build up which, in turn, leads to bundle formation. But the simultaneous disappearance of both bundles and their focal contacts at the rear of the leading lamella of locomoting fibroblasts suggests that the oblique bundle/focal contact complex can only exist as a unit and that when a focal contact reaches the non-contracting meshwork which surrounds the nucleus, the local stress subsides causing the disappearance of both the bundle and its focal contact.

Abercrombie *et al.* (1977) and Couchman & Rees (in preparation) now believe that the oblique microfilament bundle system is not the only mechanism for exerting tractive force on the substratum in fibroblasts. Some primary chick fibroblasts, especially those not in contact with neighbouring cells, show no focal contacts in interference reflection but only the large amorphous close contacts which often coexist with focal contacts and which Heath & Dunn (1978) have found to be seldom associated with microfilament bundles. Since these fibroblasts locomote perfectly well, they must exert force on the substratum via the close contacts. The proposed contracting meshwork is an ideal candidate for the exertion of this force, but the force must be too small to cause bundle formation. It is reasonable to suppose that the force exerted is directly related to the reduction in contraction speed caused by the adhesions, and one explanation for the lack of bundles is that the meshwork is not connected directly to the substratum at a close contact. It may be that the force is transmitted across an interface of shear, separating the meshwork from the substratum, by a process akin to frictional or viscous drag. There are several possibilities for the position of this interface, including within the surface membrane. The fluid-mosaic model for the cell surface (Singer & Nicolson, 1972) allows that components of the cell surface connected to the meshwork could interact by molecular collision with others connected to the substratum in the region of a close contact.

As Harris (1973) suggested, the movement of particles attached to the lamellar surface might well employ the same machinery as that which exerts the tractive force on the substratum. In the case of the continuous-contraction hypothesis, the components of the cell surface which directly connect the meshwork to the substratum, at a focal contact, could also directly connect the meshwork to a particle; but the freely moveable particle would not stress the meshwork and cause a bundle to form. Space is too limited here to enter into a discussion of whether the continuous-contraction hypothesis is consistent with the observations on capping in fibroblasts, but suffice it to say that Bretscher (personal communication) believes that a valid alternative to his lipid-flow hypothesis (Bretscher, 1976) is that two-dimensional precipitates within the surface are raked towards the nuclear region by actively moving surface components.

So far I have avoided discussing an essential feature of any general hypothesis of cell locomotion, the mechanism of polarity. If the locomoting machinery of the spread fibroblast were evenly distributed all round the nucleus, directed movement would be impossible. Protrusion must be localised in certain parts of the cell margin and the force exerted on the substratum must have a resultant approximately directed towards the region of maximum spreading over the substratum. In the case of the continuous-contraction hypothesis, both these requirements could be met if the return flow of the disassembled meshwork components were directed to certain parts of the margin, thus localising protrusion in parts of the margin and causing the local assembly of contractile meshwork. Microtubules have been shown to be involved in the maintenance of fibroblast polarity (Vasiliev *et al.*, 1970, Harris, 1971) and the treatment of fibroblasts with anti-tubulin drugs can result in an apparently disorganised contractile activity of the cytoplasm (Harris, 1971). The distribution of the cytoplasmic microtubules in a spread fibroblast (Osborn & Weber, 1976) is appropriate to attribute to them the role of directing the transport of disassembled meshwork elements, and thus spatially organising the continuous contraction.

Acknowledgements. Much of this paper is the result of discussions with my colleagues M. Abercrombie and J. P. Heath. The research was supported by an MRC programme grant to M. Abercrombie and an MRC project grant to G. A. Dunn.

REFERENCES

Abercrombie, M. & Dunn, G. A. (1975). Adhesions of fibroblasts to substratum during contact inhibition observed by interference reflexion microscopy. *Experimental Cell Research*, **92**, 57–62.

Abercrombie, M., Dunn, G. A. & Heath, J. P. (1977). The shape and movement of fibroblasts in culture. In *Cell and Tissue Interactions*, ed. J. W. Lash & M. M. Burger, pp. 57–70. New York: Raven Press.

Abercrombie, M., Heaysman, J. E. M. & Pegrum, S. M. (1970*a*). The locomotion of fibroblasts in culture. II. 'Ruffling'. *Experimental Cell Research*, **60**, 437–44.

Abercrombie, M., Heaysman, J. E. M. & Pegrum, S. M. (1970*b*). The locomotion of fibroblasts in culture. III. Movements of particles on the dorsal surface of the leading lamella. *Experimental Cell Research*, **62**, 389–98.

Abercrombie, M., Heaysman, J. E. M. & Pegrum, S. M. (1971). The locomotion of fibroblasts in culture. IV. Electron microscopy of the leading lamella. *Experimental Cell Research*, **67**, 359–67.

Abercrombie, M., Heaysman, J. E. M. & Pegrum, S. M. (1972). Locomotion of fibroblasts in culture. V. Surface marking with concanavalin A. *Experimental Cell Research*, **73**, 536–9.

Bretscher, M. S. (1976). Directed lipid flow in cell membranes. *Nature, London*, **260**, 21–3.

Chen, Wen-Tien (1977). Retraction of the trailing edge during fibroblast movement. *Journal of Cell Biology*, **75**, 416a.

Chen, Wen-Tien (1978). Induction of spreading during fibroblast movement. *Journal of Cell Biology*, **79**, 83a.

Edidin, M. & Weiss, A. (1972). Antigen cap formation in cultured fibroblasts: a reflection of membrane fluidity and cell motility. *Proceedings of the National Academy of Sciences, USA*, **69**, 2456–9.

Fleischer, M. & Wohlfarth-Bottermann, K. E. (1975). Correlation between tension force generation, fibrillogenesis and ultrastructure of cytoplasmic actomyosin during isometric and isotonic concentrations of protoplasmic strands. *Cytobiologie*,**10**, 339–65.

Harris, A. K. (1971). The role of adhesion and the cytoskeleton in fibroblast locomotion. Ph.D. thesis, Yale University.

Harris, A. K. (1973). Cell surface movements related to cell locomotion. In *Locomotion of Tissue Cells*, Ciba Foundation Symposium. **14**, 3–20.

Harris, A. K. (1976). Recycling of dissolved plasma membrane components as an explanation of the capping phenomenon. *Nature, London*, **263**, 781–3.

Harris, A. K. & Dunn, G. A. (1972). Centripetal transport of attached particles on both surfaces of moving fibroblasts. *Experimental Cell Research*, **73**, 519–23.

Heath, J. P. & Dunn, G. A. (1978). Cell to substratum contacts of chick fibroblasts and their relation to the microfilament system. A correlated interference-reflexion and high-voltage electron microscope study. *Journal of Cell Science*, **29**, 197–212.

Isenberg, G., Rathke, P. C., Hulsmann, N., Franke, N. W. & Wohlfarth-Botterman, K. E. (1976). Cytoplasmic actomyosin fibrils in tissue culture cells. Direct proof of contractility by visualization of ATP-induced contraction in fibrils isolated by laser microbeam dissection. *Cell and Tissue Research*, **166**, 427–43.

Izzard, C. S. & Lochner, L. (1976). Cell-to-substrate contacts in living fibroblasts: an interference-reflexion study with an evaluation of the technique. *Journal of Cell Science*, **21**, 129–59.

Lazarides, E. (1975). Tropomyosin antibody: the specific localisation of tropomyosin in non-muscle cells. *Journal of Cell Biology*, **65**, 549–61.

LAZARIDES, E. & BURRIDGE, K. (1975). α-Actinin: immunofluorescent localisation of a muscle structural protein in non-muscle cells. *Cell*, **6**, 289–98.

LAZARIDES, E. & WEBER, K. (1974). Actin antibody: the specific visualisation of actin filaments in non-muscle cells. *Journal of Cell Biology*, **68**, 202–19.

OSBORN, M. & WEBER, K. (1976). Cytoplasmic microtubules in tissue culture cells appear to grow from an organising structure towards the plasma membrane. *Proceedings of the National Academy of Sciences, USA*, **73**, 867–71.

SCHOLLMEYER, J. E., FURCHT, L. T., GOLL, D. E., ROBSON, R. U. & STROMER, M. H. (1976). In *Cell Motility, Localization of Contractile Proteins in Smooth Muscle Cells and in Normal and Transformed Fibroblasts*, ed. R. Goldman, T. Pollard & J. Rosenbaum, pp. 361–88. Cold Spring Harbor Conferences on Cell Proliferation.

SINGER, S. J. & NICOLSON, G. L. (1972). The fluid mosaic model of the structure of cell membranes. *Science*, **175**, 720–31.

VASILIEV, J. M., GELFAND, I. M., DOMNINA, L. V., DORFMANN, N. A. & PLETYUSHKINA, O. Y. (1976). Active cell edge and movement of concanavalin A receptors of surface of epithelial and fibroblastic cells. *Proceedings of the National Academy of Sciences, USA*, **73**, 4085–9.

VASILIEV, J. M., GELFAND, I. M., DOMNINA, L. V., IVANOVA, O. Y., KOMM, S. G. & OLSHEVSKAJA, L. V. (1970). Effect of colcemid on the locomotory behaviour of fibroblasts. *Journal of Embryology and Experimental Morphology*, **24**, 625–40.

WEBER, K. & GROESCHEL-STEWART, U. (1974). Antibody to myosin: the specific visualisation of myosin containing filaments in non-muscle cells. *Proceedings of the National Academy of Sciences, USA*, **71**, 4561–4.

WEBSTER, R. E., HENDERSON, D., OSBORN, M. & WEBER, K. (1978). Three-dimensional electron microscopical visualisation of the cytoskeleton of animal cells: Immunoferritin identification of actin and tubulin-containing structures. *Proceedings of the National Academy of Sciences, USA*, **75**, 5511–15.

WOLOSEWICK, J. J. & PORTER, K. R. (1976). Stereo high-voltage electron microscopy of whole cells of the human diploid line W1-38. *American Journal of Anatomy*, **147**, 303–24.

Microfilament–membrane interactions in the mechanism of capping

G. L. E. KOCH

MRC Laboratory of Molecular Biology, Hills Road, Cambridge CB2 2QH, UK

INTRODUCTION

It is now securely established that the actin-rich microfilaments are major constituents of most types of eukaryotic cells. In some cells, such as lymphocytes, actin can make up over 10% of the total cytoplasmic protein. This ubiquity and quantitative dominance of actin, coupled with the high degree of conservation of the protein's structure during evolution, has led to the view that microfilaments play a crucial role in a variety of cellular processes. Thus it is not surprising that they have been implicated in adhesion, locomotion, mitosis, phagocytosis, cytoplasmic streaming etc. An important feature of many of these processes is the fact that they appear to require some form of interaction between the microfilaments and the plasma membranes of the cells concerned. Thus in the adhesion sites of fibroblasts and the phagocytic sites of macrophages, local accumulations of microfilaments are readily detected at the appropriate sites. These foci of microfilaments are probably directly involved in the functions of these sites. Close associations between microfilaments and membranes are also very evident in the various microprojections of cells such as microvilli and filopodia, since these structures virtually consist of elongated sacs of plasma membrane with their lumens filled by bundles of microfilaments. In terms of general cellular morphology too, it appears that most of the patent microfilaments of cells are located in the submembranous or cortical region of the cell, either as a meshwork of filaments or as the organised bundles referred to as stress fibres, often appearing to make direct contacts with the plasma membrane. In cases such as adhesion plaques (Abercrombie, Heaysman & Pegrum, 1971) and sites of phagocytosis a rather specific interaction between the microfilaments and the membrane can be inferred from the fact that the accumulations are generally confined to the sites concerned, and some interaction with the membrane must occur to ensure that the events occurring outside the membranes are coordinated with those occurring within the cell. For example, phagocytosis is usually confined to the points of attachment of the particles to the cell surface. Thus it is essential that the foci of microfilaments should form at such sites and not others. Furthermore, it has been observed (Griffin & Silverstein, 1974) that the phagocytosis is confined to the sites of attachment of ingestible particles alone. It is difficult to visualise how these events can be coordinated without some form of interaction between the microfilaments and the membrane. A further implication of these studies is that in phenomena such as phagocytosis, which involve a highly segmental response to an extracellular stimulus, some form of transmembrane signalling mechanism must operate to coordinate the intra- and extracellular events. Other observations also suggest the need for specific transmembrane signalling between cell surface components and microfilaments. Thus proteolysis of surface components under circumstances where the proteolysis is almost certainly confined to the external surface of the cell can cause the disaggregation of the microfilament bundles, whilst certain cells which lack the capacity to adhere and spread can be restored to the latter phenotype by the addition of fibronectin. This also leads to the restoration of the microfilament bundles and other cytoskeletal elements.

Although the concepts of microfilament–membrane interaction and transmembrane signalling between surface receptors and microfilaments are widely accepted, the molecular mechanisms which operate during these processes remain elusive. Recently however studies on the 'capping' of surface receptors have shed new light on possible mechanisms, and in this article these observations will be considered and assessed.

THE NATURE OF CAPPING

Capping is the popular term for the ligand-induced polar redistribution of cell surface receptors (Taylor, Duffus, Raff & de Petris, 1971). The major requirement for the induction of the redistribution is the use of a multivalent ligand which can cross-link the surface receptors (Loor, Forni & Pernis, 1972; de Petris & Raff, 1973*a*). Consequent to the cross-linking, which manifests itself as a 'patchy' distribution of the receptor, there is a translocation of the patches to one pole of the cell. This is particularly striking in cells such as lymphocytes. The requirement for an active mechanism for the translocation of the patches can be inferred from the requirement for metabolic energy, since capping is inhibited by agents such as cyanide or azide. Thus the polar distribution of receptors cannot be the result of passive aggregation of the patched receptors.

The analogy between capping and cell motility was recognised from the earliest studies on the phenomenon (Taylor *et al.*, 1971; de Petris & Raff, 1973*b*). Thus some receptors, notably surface immunoglobulin on lymphocytes, cap at rates corresponding to the translation of the patched receptors at upto 5 μm min^{-1} (de Petris & Raff, 1973*b*). This compares favourably with the rate of particle movement, i.e. 3–4 μm min^{-1}, observed on the surfaces of fibroblasts during locomotion (Abercrombie, Heaysman & Pergrum, 1970). It has been estimated that the rate of lymphocyte movement over solid surfaces can be as high as 30 μm min^{-1} (Trowell, 1958). Therefore capping does not seem to occur as fast as lymphocyte movement on suitable substrata, but it probably compares with the rate of lymphocyte movement in suspension. That capping is related to lymphocyte movement is also suggested by the site of cap formation. During lymphocyte movement, the cells assume a characteristic hand-mirror shape, with the broad region in front and the narrow one trailing (Trowell, 1958). The latter is usually referred to as the uropod. During capping on lymphocytes, the cross-linked receptors move towards the uropod and accumulate around it (Schreiner & Unanue, 1976*b*). Thus the polarity of capping closely parallels the polarity of normal lymphocyte movement. An analogous situation exists with fibroblast capping, since the cross-linked receptors move backwards from the leading edges of the lamellipodia, although in this case they rarely pass the perinuclear region (Edidin & Weiss, 1972). The behaviour of the patched receptors is also strongly reminiscent of the movement of particles over the cell surface during fibroblast movement (Abercrombie *et al.*, 1970).

Another significant analogy between locomotion of cells and capping lies in their sensitivity to agents which disrupt cytoskeletal elements. The main agents used have been colchicine, which disrupts microtubules, and cytochalasin B, which probably operates by disrupting microfilaments (see below). The effects of colchicine are variable and depend on the particular system being analysed. Thus colchicine has no effect on Ig capping (Taylor *et al.*, 1971; Loor *et al.*, 1972), enhances the capping of Con A receptors (Edelman, Yahara & Wang, 1973) or complements the inhibitory effect of cytochalasin B on capping generally (de Petris, 1974; Unanue & Karnovsky, 1974) when applied to the same cell type, i.e. lymphocytes. In general, the effect of colchicine seems to be slightly enhancing towards capping. Thus it has an analogous effect to that on cell motility, since colchicine will stimulate cytoplasmic streaming in macrophages (Bhisey & Freed, 1971*b*) and enhance macrophage and lymphocyte movement (Bhisey & Freed, 1971*a*; Unanue & Karnovsky, 1974). The apparently analogous effects of colchicine on capping and movement are also evident in fibroblasts, except that in such cells the agent is inhibitory towards both phenomena (Gail & Boone, 1971; Edidin & Weiss, 1972).

In contrast with colchicine, cytochalasin B always has an inhibitory effect on capping, although the actual extent can be variable (Taylor *et al.*, 1971; de Petris, 1974). Interpretation of this observation has been complicated by doubts about the mechanism by which cytochalasin B acts on cells. The original assumption was that cytochalasin B acted on the the microfilaments of cells (Schroeder, 1970). However, the observation that cytochalasin B inhibits glucose transport across plasma membranes (Kletzien, Perdue & Springer, 1972) raised doubts about this view. More recently, studies *in vitro* have provided strong support for the original concept that cytochalasin B acts by disrupting cellular microfilaments. Spudich & Lin (1972) showed that cytochalasin B could reduce the viscosity of rabbit skeletal muscle actomyosin, ostensibly by acting on the actin filaments. These studies were questioned on the grounds that the concentrations of the agent employed were very high and that no parallel effect on striated muscle could be detected. More recently, two independent studies have confirmed the disruptive effects of cytochalasin B on actin filaments at concentrations which produce effects *in vivo*. Using microviscometric analyses it was found (Maclean, Griffith & Pollard, 1978) that micromolar concentrations of cytochalasin B reduced the interactions between actin filaments but not between subunits in the filaments. In a study using gels formed with actin filaments and macrophage actin-binding proteins (Hartwig & Stossel, 1978) it was found that the stability of such gels was markedly reduced by micromolar concentrations of cytochalasin B. In the latter study, the results were interpreted as showing that the agent acted by introducing breaks in the filaments. The explanation for the difference in interpretation is not obvious but, irrespective of this particular issue, it does appear that cytochalasin B can act directly on microfilaments. Therefore the parallel effect of microfilament disruption on cell motility and capping further emphasises the analogy between the two phenomena.

In summary, capping is the polarised translocation of cross-linked surface receptors by an active process involving the microfilament systems of cells.

MODELS OF CAPPING

Broadly speaking, two types of model have been proposed to account for the capping of cross-linked surface receptors. These are the anchorage model (Berlin, Oliver, Ukena & Yin, 1974; de Petris & Raff, 1973*b*; Edelman, 1976; Nicolson, 1976) and the flow models. The latter exist in two forms, membrane flow (Harris, 1976) and lipid flow (Bretscher, 1976). The basic difference between the models is that the anchorage model views the capping as resulting from the attachment of the surface receptors to cytoskeletal elements and their translocation as reflecting the polarised movement of the cytoskeleton, whilst the flow models do not demand any physical associations between receptors and the cytoplasm, since the translocation results from a more or less global flow of whole membrane from one end of the cell to the other. All the models accommodate the link between capping and motility. The flow models envisage the flow as a consequence of the need to generate new membrane at the leading edge of a moving cell. The anchorage model sees capping as reflecting the normal cytoskeletal movements associated with cell motility. Clearly, therefore, the crucial factor that distinguishes the two models is that the anchorage model demands some transmembrane association between cell surface receptors and microfilaments, whilst the flow models require no such association.

The membrane-flow model

The membrane-flow model for capping was proposed (Harris, 1976) as an extension of the studies on particle movement over the surfaces of fibroblastic cells during locomotion. Since the latter was rationalised in terms of the constant addition of new membrane at the leading edge of the moving cell, it was proposed that an analogous flow exists in all motile cells. This aspect of the model is clearly consistent with the fact that the polarity of capping and movement are invariably the same. According to this model, capping is observed because the receptors are rendered visible by the ligand and possibly because the binding of the ligand actually induces the events which lead to the directed membrane flow.

Direct experimental tests of this model do not seem to have been performed. However, it is possible to assess it in the light of numerous other studies carried out on capping cells. If capping reflects a general flow of membrane, then the capping of one receptor should automatically lead to the coordinate capping of all other receptors. This is clearly not the case. However, it is argued that the crucial experiment is the application of a multivalent ligand to one receptor and a univalent ligand to another, under which circumstances both receptors should cap simultaneously. Whilst this precise experiment has not been reported, analogous ones have been performed, i.e. the capping of one receptor followed by antibody or *Fab* fragment to a second (Neauport-Sautes, Lilley, Silvestre & Kourilsky, 1973). These experiments clearly confirm that only cross-linked receptors are capped. This observation might be reconciled with the membrane-flow model if it is argued that the continuous regeneration of membrane components restores the normal levels of the second receptor before the second antibody is applied. However, this does not explain the failure of the capped receptor to be restored simultaneously by the same mechanism (Taylor *et al.*, 1971; Unanue & Karnovsky, 1974). One observation which might be indicative of membrane flow is the spontaneous capping of surface immunoglobulin observed on lymphocytes under certain circumstances (Schreiner, Braun & Unanue, 1976). However, the fact that the spontaneous capping is selective for surface immunoglobulin does not support the idea of a generalised flow of membrane components.

The lipid-flow model

The basic premise of the lipid-flow model is that there is a flow of plasma membrane lipids caused by the addition of lipid at the leading end of a cell and a balanced removal at the trailing end (Bretscher, 1976). Since the flow is confined to lipid, other plasma membrane components, such as proteins, do not move with the flow unless they have diffusion constants below a critical level. In general, it is assumed that most plasma membrane proteins and receptors diffuse fast enough to overcome the flow. However, when cross-linked into patches, the rates of diffusion are decreased sufficiently to cause them to become caught up in the flow and swept along to one pole of the cell, i.e. to cap. The main advantage of the lipid-flow model over membrane-flow is that it provides a simple explanation for the need for cross-linking in order to detect capping. In fact, if lipid flow did exist it would certainly suffice to explain the capping phenomenon. Unfortunately, like the membrane-flow model, direct evidence for the model does not seem to have been obtained. Therefore the model can only be evaluated on its theoretical foundations. The first question that arises is: What is the need for lipid flow? In the case of membrane flow the need for recycling was rationalised on the basis of a need to produce new membrane at the leading edge of a moving cell at a rate considerably greater than at which new membrane could be synthesised. However, as pointed out previously (de Petris & Raff, 1973*b*), a forward displacement of membrane components, which is feasible with fluid membranes, could easily serve the same purpose. In the case of lipids the need for polarised recycling is even less important since lipid molecules which diffuse at very rapid rates can be introduced at any point in the plasma membrane and could rapidly equilibrate towards regions of new membrane synthesis. Therefore, whilst it is almost certain that lipid recycling does occur, there is no obvious reason for the process to be as highly polarised, as demanded by the lipid flow model. In the case of cells growing in serum even the requirement for recycling becomes less stringent, since the serum provides a substantial store of lipid for cellular growth and division (Cornell, Grove, Rothblat & Horwitz, 1977). It should be noted that if rapid polarised lipid recycling does occur it must be an energy-dependent process. Furthermore, if lipid flow is the explanation for receptor capping, the recycling must occur in the absence of movement, since capping itself can occur in non-motile cells (Schreiner & Unanue, 1976*a*). Finally, even if a flow does occur in moving cells it must result from a production of lipid at the leading edge of the cell grossly in excess of the requirements for movement. In general, therefore, if the polarised flow of lipid does exist it must do so for reasons which are rather obscure. This does not imply that the flow does not exist but rather that there is no obvious reason for its existence.

The second hypothetical requirement for polarised lipid flow is the molecular filter which ensures that recycling is confined to membrane lipids alone. Since the molecular filter is directly responsible for generating the polarised flow of lipid, its location can be fairly accurately predicted, i.e. it must occupy a site distal to the site at which the cap forms. In the case of lymphocytes capping surface immunoglobulin, it has been shown that the cap is highly localised at the tail of the uropod. Consequently, the lipid-filtering mechanism must be localised to that region of the uropod. Furthermore, to account for the rapid rate of Ig capping it is necessary to assume that more than 10% of the surface lipid is being re-absorbed at this site every minute. Therefore, the posterior region of the uropod would be expected to manifest at least some sign of this gross absorption process. However, numerous examinations of the uropods of normal and capped cells (de Petris & Raff, 1972; Unanue & Karnovsky, 1974; de Petris, 1975; Schreiner & Unranue, 1976*a*) have failed to show the local concentrations of internalised lipid predicted by the model. The model also speculates that coated vesicles (Pearse, 1975) might be the device by which lipid is selectively re-absorbed by cells. However, coated vesicles and pits do not show the polarised distribution predicted by the model. In fact it is now questionable whether coated vesicles could play the filtering role required by the model, since it is now established that receptors for low-density lipoproteins, insulin, epidermal growth factor and α2-macroglobulin accumulate preferentially in the coated pits and vesicles (Schlessinger, Schechter, Willingham & Pastan, 1978; Maxfield *et al.*, 1978; Willingham & Pastan, 1978). Thus the vesicles appear not to be lipid-bearing structures exclusively. In summary, evidence for the existence of the polarised selective lipid re-absorbing mechanism required by the lipid-flow model is still unavailable.

The other major theoretical premise of the lipid-flow model pertains to the rates of lateral diffusion of surface receptors. In order to account for the rapid rate of capping of receptors such as surface Ig, a rate of lipid flow greater than 3 μm min^{-1} must be assumed. Thus, receptors with lateral diffusion rates greater than the flow rate, i.e. with diffusion constants greater than 10^{-10} cm^2 s^{-1} would not cap spontaneously. However, by the same argument receptors diffusing slower than 10^{-10} cm^2 s^{-1} should be caught up in the flow and be capped continually. Thus the observation that the diffusion constants of many receptors which do not cap spontaneously, but do so when cross-linked, have diffusion constants orders of magnitude below

10^{-10} cm^2 s^{-1} (Edidin, 1977) argues against the existence of an oriented flow of lipid as the explanation for capping. As pointed out (Edidin, 1977) rhodopsin (Poo & Cone, 1974) is probably an inappropriate choice as the prototype for protein diffusion, since it appears to be the exception rather than the rule.

A further prediction of the lipid-flow model is that all patched receptors should cap at essentially the same rate and with comparable efficiency. However, it has been repeatedly observed that the capping of receptors on lymphocytes does not proceed at the same rate or with equal efficiency (Schreiner & Unanue, 1976*a*). Thus, the capping of surface Ig is usually complete in under 5 min, whilst other surface receptors, such as the histocompatibility antigens, θ antigen or concanavalin A receptors take over 30 min to cap, and the extent of capping is usually considerably less than total. This wide difference in capping rates and efficiencies are not easily reconciled with a flow mechanism, since the latter would be expected to sweep all receptors, once patched, at the same rate. Another aspect of the same issue is the fact that some surface receptors fail to cap at all even when extensively patched. Notable amongst these are the receptors for hormones such as insulin and epidermal growth factor, as well as serum components such as $\alpha 2$ macroglobulin (Maxfield *et al.*, 1978; Willingham & Pastan, 1978). The failure to cap does not reflect the type of cell used, since the 3T3 fibroblasts have been widely used to study the capping of other receptors. These patent differential effects on patching of different receptors are not easily reconciled with the existence of a continuous flow of lipid, which should not be able to distinguish between different types of patched receptors since they would all possess the reduced diffusion constant predicted by the model. These arguments also apply to the membrane flow model.

As mentioned before, both the membrane- and lipid-flow models suffer from lack of direct experimental support. One type of observation advanced as indirect support of the models, particularly the latter, is the capping of glycolipids. This was first observed in studies using the GM1 ganglioside (Révész & Greaves, 1975) and has also been shown to occur with bacterial glycolipids introduced into lymphocyte membranes (Brailovsky, Wigam & Pesant, 1977). Since glycolipids only penetrate one half of the lipid bilayer they cannot make direct contacts with the cytoplasm of the cell and consequently their capping must arise from some other mechanism than direct anchorage (see below). However, this view presupposes that glycolipids cannot interact with other membrane components, particularly transmembrane proteins. The feasibility of such interactions has been demonstrated, e.g. with ESR techniques to measure the mobility of glycolipids in synthetic and natural membranes. In one study (Shanon & Grant, 1978) it was suggested that the glycolipids bound preferentially to the glycoproteins in the membrane and an association between glycolipids and an integral membrane glycoprotein, glycophorin was observed. This ability of glycolipids to interact with membrane proteins is not surprising, since cell surfaces are known to have glycosyltransferase and lectin-like molecules which could bind to glycolipids through their carbohydrates. The possibility of carbohydrate–carbohydrate associations coupled with interactions between the lipid moiety and the intramembranous region of integral proteins further enhances the feasibility of such associations. Because of this capacity of interactions between lipid and protein in the plane of the membrane, the demonstration of even phospholipid capping does not argue against the anchorage model. It is noteworthy that in a careful study using DNP-labelled phospholipids incorporated into cell membranes (Schroit & Pagano, 1978) anti-DNP antibodies were found to patch phospholipids but not to cause any capping.

In summary, therefore, the flow models have not received direct support for their basic premise, i.e. the existence of a directed flow of either whole membrane or lipid. Furthermore, some of the implications and predictions of the models have not been confirmed. In the absence of experimental support for the models they remain interesting but intuitive explanations for the capping of surface receptors.

The anchorage model

The basic premise of the anchorage model is that surface receptors which have been cross-linked become attached to the cytoskeleton and are translocated to one pole of the cell as a consequence of this attachment. The polarity of capping is thought to reflect the polarity of cytoskeleton movements in the cells. The latter are probably related to locomotion especially in view of the identity in the polarity of uropod formation and lymphocyte movement. Therefore, in general terms the anchorage model envisages that polarised distributions of the cytoskeleton are necessary for the normal movements of cells, and capping of surface receptors results from the attachment of the surface receptors to the cytoskeleton following cross-linking, so that the receptors ultimately assume the distribution of the cytoskeleton. The predictions of the model are: (*a*) the capping of surface receptors should be inhibited by agents which interfere with the cytoskeleton; (*b*) the distribution of patched and capped surface receptors should closely parallel the distribution of cytoskeletal elements; (*c*) it should be possible, at least in some cases, to isolate the putative complexes between surface receptors and cytoskeletal elements, particularly those in which capping is both rapid and efficient, e.g. surface immunoglobulin.

Evidence that agents which disrupt cytoskeletal elements affect the capping of surface receptors has been reviewed extensively (de Petris & Raff, 1973*b*; Edelman, 1976; Nicolson, 1976; Schreiner & Unanue, 1976*a*). There is fairly general agreement that the cytoskeletal elements most likely to be directly involved in capping are the microfilaments, and that the microtubules probably play a secondary and possibly regulatory role. However, it is clear that the studies using the disruptive agents can only provide indirect evidence for the anchorage model. The advent of the antibodies to cytoskeletal proteins such as actin, myosin and tubulin, together with the development of techniques for the visualisation of the proteins in cells which have been rendered permeable, has permitted the distributions of surface receptors and cytoskeletal elements to be examined. Studies on capped cells (Schreiner *et al.*, 1976; Bourguignon & Singer, 1977; Toh & Hard, 1977) clearly showed that there was a coordinate redistribution of the surface receptors, actin and myosin into the cap regions of the cells examined. However, these analyses are open to the criticism that the cap region is a major distortion of the cell and that coordinate redistributions might occur independent of any associations between the capped receptors and the cytoskeletal elements. In contrast, these objections are less applicable to the observations that there is coordinate distribution of actin and myosin (Bourguignon & Singer, 1977) into the regions where surface receptors have been patched but not capped. Since the patches represent subtle redistributions of the surface receptors, the close similarity between the distributions of surface receptors and patches is very striking. The patching of cytoskeletal elements is confined to actin and myosin, and patching of tubulin does not seem to occur, suggesting that the association is confined to microfilaments and associated proteins. This is consistent with earlier studies, which indicated a secondary role for the microtubules in capping. These studies using the fluorescence microscope indicate that the surface receptors are not bound to the microfilaments until they have been cross-linked by the ligand concerned. Further evidence for this has been obtained from studies on the major histocompatibility antigen and Con A receptors of fibroblasts (Ash, Louvard & Singer, 1977). These studies showed that the receptors which were essentially randomly distributed over the cell surface became oriented over the stress fibres upon cross-linking with antibody or lectin. It should be noted that in the studies on lymphoid cells (Bourguignon & Singer, 1977) the formation of the patches seems to stabilise the microfilaments which are attached to the patches, since membrane-linked microfilaments could not be detected in cells without patches.

One area of disagreement in the studies in which the coordinate redistribution of patched receptors and cytoskeletal elements have been examined pertains to the redistribution of myosin. The studies of Bourguignon & Singer (1977) showed that myosin redistributed with several different surface receptors, e.g. the antigen and Con A receptors. In contrast, Braun, Fujiwara, Pollard & Unanue (1978*a*, *b*) observed that the distribution of myosin into patches and caps only occurred consistently when surface Ig was being cross-linked, and in the case of other surface receptors the distribution of myosin did not coincide with that of the redistributed receptors. One possible explanation for these differences is that the techniques used to study the distribution of myosin, i.e. frozen sections in the former and acetone extraction in the latter, are responsible for the differences. However, in view of the repeated observation that surface Ig capping is substantially more efficient than that of other surface receptors (de Petris & Raff, 1973*b*; Stackpole, Jacobson & Lardis, 1974; Schreiner & Unanue, 1976*a*) the possibility that quantitative or qualitative differences exist between different mechanisms of capping must be considered. In this connection it is interesting that analyses of Con A capping in *Dictyostelium* amoebae (Condeelis, 1978) show that cap formation increases the amount of both actin and myosin associated with isolated plasma membrane ghosts. Furthermore, actin and myosin are the major components of isolated caps, and electron micrographs reveal the presence of both thick myosin-like and thin actin filaments at the cytoplasmic face of the plasma membrane. In this system it appears that myosin does associate with the cap region. It will be interesting to see whether caps and patches from other systems can be isolated and subjected to a similar analysis in order to determine the detailed ultrastructural features of the associations.

The major limitation of the above evidence for the anchorage model is the inability to determine whether direct associations between the surface receptors and the cytoskeleton actually occur. This is an important point, since the anchorage model predicts that the association occurs through protein–protein interactions between receptors and cytoskeletal elements. Evaluating this question even with available high-resolution microscopy is still impossible. An alternative approach is the following. If patching and capping depend on the formation of stable protein–protein interactions between surface receptors and microfilaments, isolation of either should reveal the association with the other. One approach would be to use the conventional immunoprecipitation techniques to isolate the antigen–antibody complexes when antibodies are used for the capping and examine these for proteins such as actin and myosin. However, the problem of specificity does arise, since in molar terms the antigens are present in minute quantities relative to the actin in cells such as lymphocytes, and a small amount of contamination will give a false result. On the other hand, isolation of the microfilaments by a specific affinity technique would not suffer from this limitation, since the probability of extensive non-specific contamination by a minor component must be small. The principal requirement, however, is to obtain a specific affinity technique for microfilaments and an obvious possibility is the use of myosin. Myosin has two advantages in this respect: it has a high and specific affinity for actin filaments, and it forms insoluble filaments at low ionic strength. These insoluble filaments can be used directly as an affinity matrix for actin filaments without the need for attachment to another matrix such as agarose. Although attachment of the surface receptors to the myosin filaments would be consistent with their attachment to microfilaments, it is not impossible that it results from coprecipitation with the myosin or from direct attachment of the surface receptors to the myosin filaments. This requires an additional specificity control in which the myosin filaments are presaturated with actin. If the binding of the surface receptors is blocked by the presaturation, this would strongly suggest that the receptors were attached to the microfilaments.

The myosin-affinity technique has been applied to lysates of cells with surface immunoglobulin at different stages of the capping process (Flanagan & Koch, 1978). The study shows that uncrosslinked surface immunoglobulin marked with radioiodinated *Fab* fragments of anti-immunoglobulin is not bound to the microfilaments of both lymphocytes and myeloma cells. In contrast, a large proportion of the surface Ig is attached to the microfilaments when divalent antibody is used to cross-link the surface receptors. This attachment is complete at the patching stage of the process, showing that cross-linking is sufficient for the association to occur. Since these studies have been carried out on lysates of cells in which the lipid bilayer of the plasma membrane was dissolved away with detergent, the existence of protein–protein associations between the surface receptors and the microfilaments seems inescapable.

The major limitation of the experiments using the myosin-affinity technique relates to the difficulty in distinguishing between an association formed before cell lysis and one formed after. In direct tests on this possibility (Koch, unpublished observations) actin filaments were mixed with various protein aggregates and immune precipitates and analysed by the myosin-affinity technique. In general no attachment of the aggregates and complexes to the actin could be detected. In the light of these observations it is worth evaluating the widely held view that actin or actin-filaments are 'sticky' (Bray, 1975). The concept was based on the observation that several proteins and enzymes showed a capacity to bind strongly to actin. However, the actual number was rather small and it is not inconceivable that the associations are biologically significant. The other line of evidence suggesting that actin binds non-specifically to proteins is the observation that actin often appears as a contaminent in immunoprecipitates (Delovitch, Fegelman, Barber & Frelinger, 1979). However, it should be noted that this probably reflects the fact that cells generally contain large amounts of filamentous actin, some of which would have an obvious tendency either to coprecipitate during the immunoprecipitation or become trapped in the precipitate. The important point is that these data do not imply that actin is sticky but rather that it is a major constituent of cells. Thus even a small amount of cross-contamination would be patent. In fact the idea that actin and actin filaments are intrinsically sticky is difficult to reconcile with the universal occurrence and quantitative dominance of actin in cells, since any tendency of this protein, above all others, to form non-specific associations would paralyse the insides of cells. Therefore in the absence of direct evidence that actin is an especially adherent protein, the possibility that patches of surface Ig associate with microfilaments after cell lysis remains hypothetical.

In summary, several different types of evidence indicate that the anchorage model remains the most satisfactory explanation for the capping phenomenon.

VALENCY MODULATION AND TRANS-MEMBRANE ASSOCIATIONS

The studies on the capping of cell surface receptors have suggested some important principles in membrane–microfilament interactions. The observation that cross-linking can lead to the attachment of surface receptors to the microfilaments implies that the receptors, and by association the membrane itself, are in equilibrium with the microfilament network. In general the equilibrium is a weak one favouring the dissociated state, but upon cross-linking, as a consequence of a change in valency of one of the components, the equilibrium will shift towards association. The model is based on the simple thermodynamic principle that an equilibrium can be shifted by altering the entropy change associated with the reaction. Analogous phenomena are known to occur in other systems. Thus immunoglobulin which has been aggregated by heat or antibody treatments different enough to make a common conformational change unlikely will bind to the Fc receptors of cells, whereas unaggregated Ig has a very low affinity (Phillips-Quagliata, Levine, Quagliata & Uhr, 1971). Similarly, the activation of complement by antigen–antibody complexes (Müller-Eberhard, 1970) requires a multivalent state of the immunoglobulin molecule.

Evidence for the weak interactions discussed above is difficult to acquire since their very weakness prevents direct analysis. However, it is conceivable that the very low diffusion constants of many membrane components (Edidin, 1977) reflects the presence of these weak interactions. Another line of evidence for the weak interactions is the spontaneous capping observed with surface Ig on lymphocytes (Schreiner *et al.*, 1976). This capping only occurs on cells with highly deformed shapes, but as with induced capping the caps accumulated in the region of the uropod. In a separate study (de Petris, 1978) it was found that surface Ig could be concentrated into the microvilli of lymphocytes in cells subjected to ATP depletion. In both cases the effect was apparently specific for Ig excluding general membrane movements. These observations could be explained in terms of an equilibrium between the surface Ig and the microfilaments such that the former will tend to regions such as uropods and microprojections which contain high local concentrations of the microfilaments.

The major outstanding question relating to microfilament–membrane interactions during processes such as capping concerns the nature of the transmembrane link between the surface receptors and the microfilaments. If it is accepted that most surface receptors can form associations with microfilaments (Bourguigno & Singer, 1977), and that many surface receptors such as Ig and glycolipids are not transmembrane proteins, it becomes necessary to postulate the existence of some transducing protein to effect the connection between the receptor and the microfilaments. Since it is unlikely that each receptor has a separate transducing protein, it has been suggested that a general transducer 'X' (Bourguignon & Singer, 1977) effects this function. Thus 'X' could be a transmembrane protein which spontaneously binds to microfilaments and attaches to surface receptors when the latter have been cross-linked. Clearly, therefore, the ultimate test of the anchorage model is the isolation and characterisation of the hypothetical transducing protein or proteins. An interesting hypothetical question concerns the possible function of such a transducing protein. One possibility is that it participates in the packaging of plasma membrane components by virtue of its capacity to bind these on the extracellular face. The packaged components could then be transported to the cell surface as inverted vesicles which possess the capacity to bind to microfilaments via the transducing protein. Such an arrangement could be necessary to ensure passage of the new membrane through the cortical layer of microfilaments which appear to act as a barrier for the other cytoplasmic organelles. It must be emphasised that the above scheme is only proposed as an example of the possible functions of the hypothetical transducing protein.

CONCLUSIONS

Available analyses of the capping phenomenon strongly favour the anchorage model as the explanation for this phenomenon, particularly that of receptors such as surface immunoglobulins of lymphocytes. These transmembrane associations constitute a simple device for attaching microfilaments to membranes. Furthermore, the ability to modulate these associations by altering the state of cross-linking of the transmembrane components provides a mechanism for regulating the extent of the association between membranes and microfilaments. More detailed analysis will be required to determine whether or not a general transducing mechanism for the attachment of surface receptors to microfilaments exists and to establish the concept of membrane–microfilament associations on a biochemical basis.

REFERENCES

ABERCROMBIE, M., HEAYSMAN, J. E. M. & PERGRUM, S. M. (1970). The locomotion of fibroblasts in culture. III. Movements of particles on the dorsal surface of the leading lamella. *Experimental Cell Research*, **62**, 389–98.

ABERCROMBIE, M., HEAYSMAN, J. E. M. & PEGRUM, S. M. (1971). The locomotion of fibroblasts in culture. IV. Electron microscopy of the leading lamella. *Experimental Cell Research*, **67**, 359–67.

ASH, J. F., LOUVARD, D. & SINGER, S. J. (1977). Antibody-induced linkages of plasma membrane proteins to intracellular actomyosin-containing filaments in cultured fibroblasts. *Proceedings of the National Academy of Sciences, USA*, **74**, 5584–8.

BERLIN, R. D., OLIVER, J. M., UKENA, T. E. & YIN, H. H. (1974). Control of cell surface topography. *Nature, London*, **247**, 45–6.

BHISEY, A. M. & FREED, J. J. (1971*a*). Ameboid movement induced in cultured macrophages by colchicine or vinblastine. *Experimental Cell Research*, **64**, 419–29.

BHISEY, A. N. & FREED, J. J. (1971*b*). Altered movement of endosomes in colchicine-treated cultured macrophages. *Experimental Cell Research*, **64**, 430–8.

BOURGUIGNON, L. Y. W. & SINGER, S. J. (1977). Transmembrane interactions and the mechanism of capping of surface receptors by their specific ligands. *Proceedings of the National Academy of Sciences, USA*, **74**, 5031–5.

BRADLEY, R. H., IRELAND, M., RASMUSSEN, N. S. & MAISEL, H. (1978). Isolation and identification of cytoplasmic filaments in the vertebrate lens. *Journal of Cell Biology*, **79**, 263–3a (MI1716).

BRAILOVSKY, C. A., WIGAM, V. N. & PEASANT, S. (1977). Cell surface motility of a bacterial R-form glycolipid mR595, after binding to rat cells, and its effect on the motility of concanavalin A receptors. *Experimental Cell Research*, **109**, 389–95.

BRAUN, J., FUJIWARA, K., POLLARD, T. D. & UNANUE, E. R. (1978*a*). Two distinct mechanisms for redistribution of lymphoblast surface macromolecules. I. Relationship to cytoplasmic myosin. *Journal of Cell Biology*, **79**, 409–18.

BRAUN, J., FUJIWARA, K., POLLARD, T. D. & UNANUE, E. R. (1978*b*). Two distinct mechanisms for redistribution of lymphoblast surface macromolecules.

II. Contrasting effects of local anesthetics and a calcium ionophore. *Journal of Cell Biology*, **79**, 419–26.

Bray, D. (1975). Sticky actin. *Nature, London*, **256**, 616–17.

Bretscher, M. B. (1976). Directed lipid flow in cell membranes. *Nature, London*, **260**, 21–2.

Condeelis, J. (1978). A direct role for the actin cytoskeleton in the mobility of cell surface receptors. *Journal of Cell Biology*, **79**, 263–3a (MI1715).

Cornell, R., Grove, G. L., Rothblat, G. H. & Horwitz, A. F. (1977). Lipid requirement for cell cycling. The effect of selective inhibition of lipid synthesis. *Experimental Cell Research*, **109**, 299–307.

Delovitch, T. L., Fegelman, A., Barber, B. H. & Frelinger, J. A. (1979). Immunochemical characterisation of the *Ly*-8.2 murine lymphocyte alloantigen: possible relationship to actin. *Journal of Immunology*, **122**, 326–33.

de Petris, S. (1974). Inhibition and reversal of capping by cytochalasin B, vinoblastine and colchicine. *Nature, London*, **250**, 54–6.

de Petris, S. (1975). Concanavalin A receptors, immunoglobulins and θ antigen of the lymphocyte surface. *Journal of Cell Biology*, **65**, 123–46.

de Petris, S. (1978). Preferential distribution of surface immunoglobulins on microvilli. *Nature, London*, **272**, 66–8.

de Petris, S. & Raff, M. C. (1972). Distribution of immunoglobulin on the surface of mouse lymphoid cells as determined by immunoferritin electron microscopy. Antibody induced, temperature dependent redistribution and its implications for membrane structure. *European Journal of Immunology*, **2**, 523–35.

de Petris, S. & Raff, M. C. (1973*a*). Normal distribution, patching and capping of lymphocyte surface immunoglobulin studied by electron microscopy. *Nature, London*, **241**, 257–9.

de Petris, S. & Raff, M. C. (1973*b*). Fluidity of the plasma membrane and its implications for cell movement. In *Locomotion of Tissue Cells*, (CIBA Symposium 14) ed. M. Abercrombie, pp. 27–41. Amsterdam: Elsevier.

Edelman, G. M. (1976). Surface modulation in cell recognition and cell growth. *Science*, **192**, 218–26.

Edelman, G. M., Yahara, I. & Wang, J. L. (1973). Receptor motility and receptor-cytoplasmic interactions in lymphocytes. *Proceedings of the National Academy of Sciences, USA*, **70**, 1442–6.

Edidin, M. (1977). Lateral diffusion and the function of cell plasma membranes. In *Progress in Immunology*, III, ed. T. E. Mandel, pp. 17–22. Amsterdam: North-Holland.

Edidin, M. & Weiss, A. (1972). Antigen cap formation in cultured fibroblasts: a reflection of membrane fluidity and of cell motility. *Proceedings of the National Academy of Sciences, USA*, **69**, 2456–9.

Flanagan, J. & Koch, G. L. E. (1978). Cross-linked surface Ig attaches to actin. *Nature, London*, **269**, 697–8.

Gail, M. H. & Boone, C. W. (1971). Effect of colcemid on fibroblast motility. *Experimental Cell Research*, **65**, 221–7.

Gonatas, N. K., Kim, S. U., Stieber, A. & Avrameas, S. (1977). Internalisation of lectins in neuronal GERL. *Journal of Cell Biology*, **73**, 1–13.

GRIFFIN, F. M., Jr. & SILVERSTEIN, S. C. (1974). Segmental response of the macrophage plasma membrane to a phagocytic stimulus. *Journal of Experimental Medicine*, **139**, 323–36.

HARRIS, A. K. (1973). Cell surface movements related to cell locomotion. In *Locomotion of Tissue Cells* (CIBA Symposium 14), ed. M. Abercrombie, pp. 3–20. Amsterdam: Elsevier.

HARRIS, A. K. (1976). Recycling of dissolved plasma membrane components as an explanation of the capping phenomenon. *Nature, London*, **263**, 781–3.

HARTWIG, J. H. & STOSSEL, T. P. (1978). Cytochalasin B dissolves actin gels by breaking actin filaments. *Journal of Cell Biology*, **79**, 271–1a (MI1741).

KLETZIEN, R. F., PERDUE, J. F. & SPRINGER, A. (1972). Cytochalasin A and B. Inhibition of sugar uptake in cultured cells. *Journal of Biological Chemistry*, **247**, 2964–6.

LOOR, F., FORNI, L. & PERNIS, B. (1972). The dynamic state of the lymphocyte membrane. Factors affecting the distribution and turnover of surface immunoglobulins. *European Journal of Immunology*, **2**, 203–12.

MACLEAN, S., GRIFFITH, L. M. & POLLARD, T. D. (1978). A direct effect of cytochalasin B upon actin filaments. *Journal of Cell Biology*, **79**, 267–7a (MI1728).

MAXFIELD, R. M., SCHLESSINGER, J., SHECHTER, Y., PASTAN, I. & WILLINGHAM, M. C. (1978). Collection of insulin, EGF and α_2-macroglobulin in the same patches on the surface of cultured fibroblasts and common internalisation. *Cell*, **14**, 805–10.

MAXFIELD, R. M., WILLINGHAM, M. C., DAVIES, P. J. A. & PASTAN, I. (1979). Amines inhibit clustering of α_2-macroglobulin and EGF on the fibroblast cell surface. *Nature, London*, **277**, 661–2.

MÜLLER-EBERHARD, H. J. (1970). The molecular basis of the biological activities of complement. *Harvey Lectures*, **66**, 75–104.

NEAUPORT-SAUTES, C., LILLY, F., SILVESTRE, D. & KOURILSKY, F. M. (1973). Independence of *H-2K* and *H-2D* antigenic determinants on the surface of mouse lymphocytes. *Journal of Experimental Medicine*, **137**, 511–26.

NICOLSON, G. L. (1976). Transmembrane control of the receptors on normal and tumor cells. I. Cytoplasmic influence over cell surface components. *Biochimica et Biophysica Acta*, **457**, 57–108.

PEARSE, B. M. F. (1975). Coated vesicles from pig brain: purification and biochemical characterisation. *Journal of Molecular Biology*, **97**, 93–8.

PHILLIPS-QUAGLIATA, J. M., LEVINE, B. B., QUAGLIATA, F. & UHR, J. W. (1971). Mechanisms underlying binding of immune complexes to macrophages. *Journal of Experimental Medicine*, **133**, 589–601.

POO, M. M. & CONE, R. A. (1974). Lateral diffusion of rhodopsin in the photoreceptor membrane. *Nature, London*, **247**, 438–41.

RÉVÉSZ, T. & GREAVES, M. (1975). Ligand-induced redistribution of lymphocyte membrane ganglioside GM1. *Nature, London*, **257**, 103–6.

SCHLESSINGER, J., SHECHTER, G., WILLINGHAM, M. C. & PASTAN, I. (1978). Direct visualisation of insulin and epidermal growth factor on living fibroblastic cells. *Proceedings of the National Academy of Sciences, USA*, **75**, 2659–63.

SCHREINER, G. F., BRAUN, J. & UNANUE, E. R. (1976). Spontaneous redistribution of surface immunoglobulin in the motile B lymphocyte. *Journal of Experimental Medicine*, **144**, 1683–8.

SCHREINER, G. F., FUJIWARA, K., POLLARD, T. D. & UNANUE, E. R. (1977). Redistribution of myosin accompanying capping of surface Ig. *Journal of Experimental Medicine*, **145**, 1393–8.

SCHREINER, G. F. & UNANUE, E. R. (1976*a*). Membrane and cytoplasmic changes in B lymphocytes induced by ligand-surface immunoglobulin interaction. *Advances in Immunology*, **24**, 37–165.

SCHREINER, G. F. & UNANUE, E. R. (1976*b*). Calcium-sensitive modulation of Ig capping: evidence supporting a cytoplasmic control of ligand–receptor complexes. *Journal of Experimental Medicine*, **143**, 15–31.

SCHROEDER, T. E. (1970). The contractile ring. Fine structure of dividing mammalian (HeLa) cells and the effect of cytochalasin B. *Zeitschrift für Zellforschung und Mikroskopische Anatomie*, **109**, 431–49.

SCHROIT, A. J. & PAGANO, R. E. (1978). Introduction of antigenic phospholipids into the plasma membrane of mammalian cells: organisation and antibody-induced lipid redistribution. *Proceedings of the National Academy of Sciences, USA*, **75**, 5529–33.

SHANON, J. F. & GRANT, C. W. M. (1978). A model for ganglioside behaviour in cell membranes. *Biochimica et Biophysica Acta*, **507**, 280–93.

SILVERSTEIN, S. C., STEINMAN, R. M. & COHN, Z. A. (1977). Endocytosis. *Annual Reviews of Biochemistry*, **46**, 669–722.

SPUDICH, J. A. & LIN, S. (1972). Cytochalosin B, its interaction with actin and actomyosin from muscle. *Proceedings of the National Academy of Sciences, USA*, **69**, 442–6.

STACKPOLE, C. W., JACOBSON, J. B. & LARDIS, M. P. (1974). Two distinct types of capping of surface receptors on mouse lymphoid cells. *Nature, London*, **248**, 232–4.

TAYLOR, R. B., DUFFUS, W. P. H., RAFF, M. C. & DE PETRIS, S. (1971). Redistribution and pinocytosis of lymphocyte surface immunoglobulin molecules induced by anti-immunoglobulin antibody. *Nature, London*, **233**, 225–9.

TOH, B. H. & HARD, G. C. (1977). Actin co-caps with concanavalin A receptors. *Nature, London*, **269**, 695–7.

TROWELL, O. A. (1958). The lymphocyte. *International Review of Cytology*, **7**, 235–93.

UNANUE, E. R. & KARNOVSKY, M. J. (1974). Ligand-induced movement of lymphocyte membrane macromolecules. V. Capping, cell movement and microtubular function in normal and lectin-treated lymphocytes. *Journal of Experimental Medicine*, **140**, 1207–20.

VASILIEV, V. M., GELFAND, I. M., DOMNINA, L. V., DORFMAN, N. A. & PLETYNSHKINA, O. Y. (1976). Active cell edge and movements of concanavalin A receptors of the surface of epithelial and fibroblastic cells. *Proceedings of the National Academy of Sciences, USA*, **73**, 4085–9.

WILLINGHAM, M. C. & PASTAN, I. (1978). The visualisation of fluorescent proteins in living cells by video intensification microscopy (VIM). *Cell*, **13**, 501–7.

Desmosomes and filaments in mammalian epidermis

CHRISTINE J. SKERROW AND DAVID SKERROW

Department of Dermatology, University of Glasgow, Glasgow G11 6NU, UK

Desmosome–tonofilament complexes constitute an epidermal cytoskeleton which distributes mechanical stress throughout the tissue, minimising the possibilities of localised damage (Mercer, 1961; McNutt & Weinstein, 1973). Both epidermal desmosomes and the tonofilament protein, prekeratin, have been isolated and their molecular structures investigated (Matoltsy, 1964; Skerrow & Matoltsy, 1974*a*, *b*). In subsequent sections these structures will be discussed in relationship to normal and pathological epidermal differentiation.

DESMOSOMES

In epidermis, desmosomes are oval or circular areas of enhanced intercellular adhesion, ranging from 0.1 to 1.0 μm in diameter and occupying a high proportion of the surface of keratinocytes at the later stages of differentiation. The membrane at the junction is modified by an accumulation of material, seen within the intercellular space and the peripheral cytoplasm (Fig. 1) and by the presence of clusters of up to several thousand membrane-associated particles within the bilayer itself (Fig. 2; Farquhar & Palade, 1963; Kelly, 1966; Breathnach, 1975). Isolated desmosomes contain 75 % protein which can be separated into more than 20 polypeptide chains with molecular weights ranging from 15–230 K. Thus, in contrast to that of the gap junction, desmosomal protein has a level of complexity comparable with that of plasma membrane preparations (Skerrow & Matoltsy, 1974*b*).

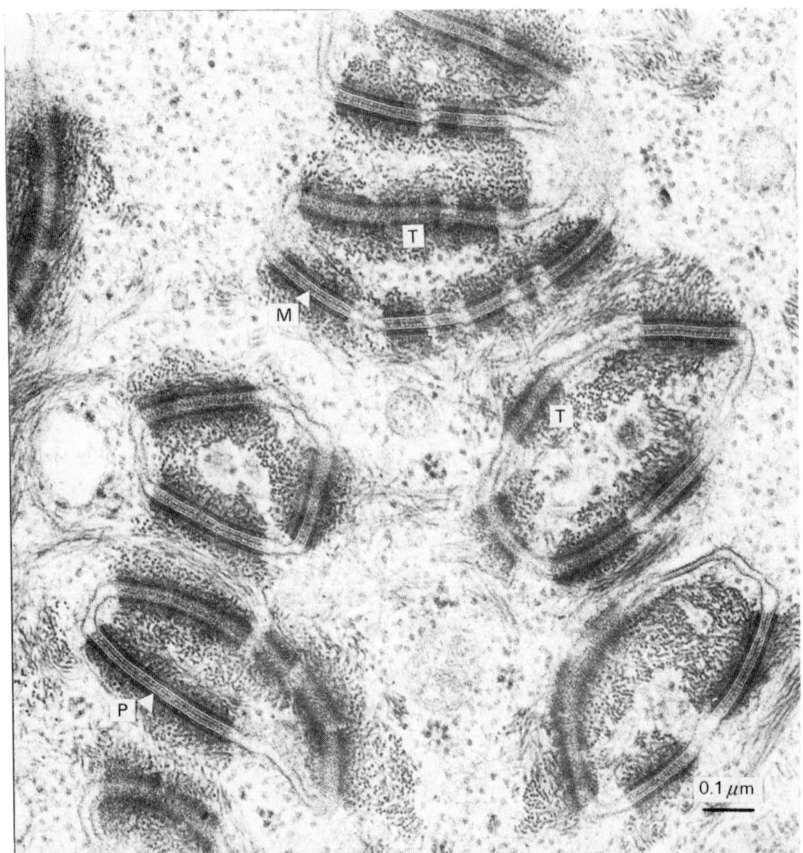

Fig. 1. Desmosome–tonofilament complexes in the spinous layer of human epidermis, with tonofilaments seen predominantly in transverse section. P, desmosomal plaque; M, midline; T, tonofilaments.

The desmosomal interspace: adhesive interactions

Intercellular material at the desmosome (the desmosomal interspace) is approximately 30–40 nm in width, moderately electron dense and bisected by a denser midline which marks the plane of intercellular contact (Fig. 1). It has been variously suggested that the appearance of the midline and the enhanced adhesion at the desmosome is due to the presence of calcium ions between acidic glycoproteins (Benedetti & Emmelot, 1968); additional protein ligands (Borysenko & Revel, 1973) or overlapping of filamentous loops extending from each cell surface (Kelly, 1966). There is evidence that the relative

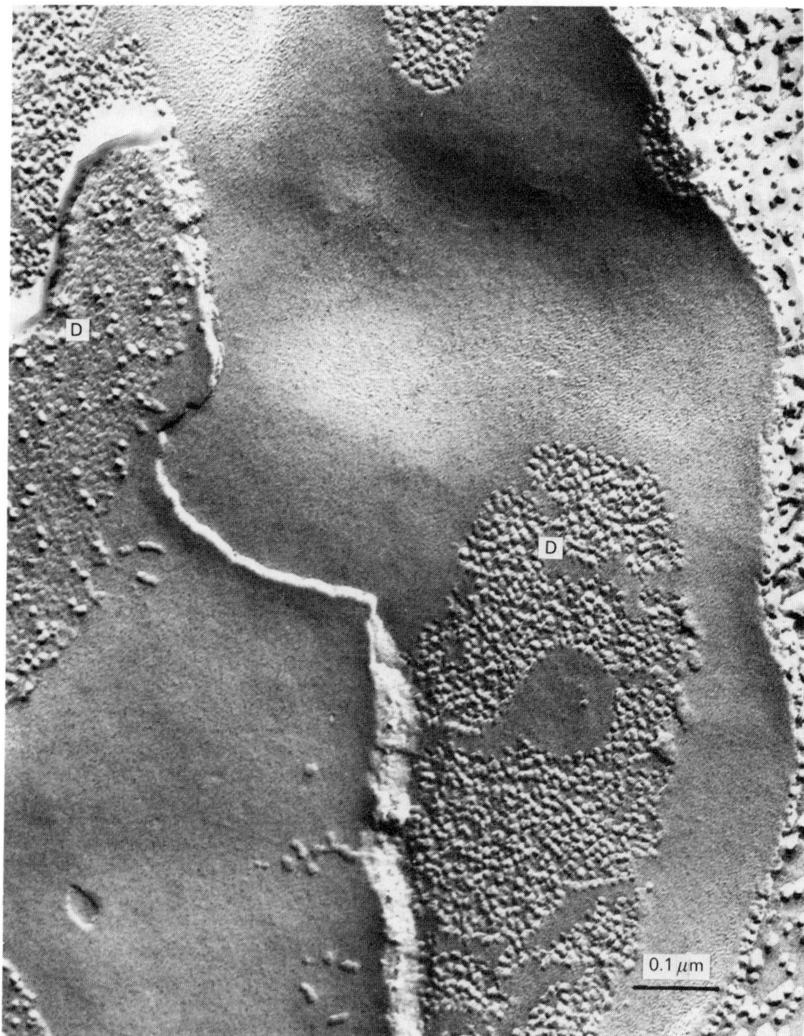

Fig. 2. Freeze-fractured desmosomes (D) in the horny layer of human epidermis consisting of clusters of membrane-associated particles. Photograph by A. S. Breathnach.

importance of interactions involving calcium and protein to desmosomal adhesion is not the same in all epithelia (Borysenko & Revel, 1973).

Citric acid-sodium citrate, pH 2.6 (CASC) -isolated desmosomes are structurally well preserved and show no tendency to split longitudinally. They therefore represent an area of contact between two cells isolated with its adhesive properties grossly intact. Isolated desmosomes contain several polypeptides, comprising over 20% of the total protein, which react with periodic acid–Schiff's reagent (PAS) and whose molecular weights are in the range 130–140 K (Skerrow & Matoltsy, 1974*b*). Desmosomal polysaccharide is located exclusively in the intercellular space (Mercer, Jahn & Maibach, 1968) which is, therefore, the presumptive location of these molecules. Some indication that they have an adhesive role is given by the observation that exhaustive extraction with ethylenediamine tetra-acetic acid (EDTA) removes approximately 50% of the PAS-reactive material, and that this is accompanied by extraction and widening of the desmosomal interspace. This widening occurs on either side of the midline which is swollen but not removed by the treatment (C. J. Skerrow, unpublished observations). These preliminary results support the view (Borysenko & Revel, 1973) that adhesion at epidermal desmosomes is not simply due to bridging by calcium ions situated in the midline. They also suggest that desmosomal glycoproteins consist of both peripheral proteins and also of integral membrane proteins firmly embedded in the bilayer (Nicolson & Poste, 1976).

The desmosomal plaque: membrane strengthening and filament attachment

In situ, the plaque is a dense, sharp-edged layer of material, approximately 15 nm in width, which fulfils the dual role of strengthening the junctional membrane and attaching bundles of tonofilaments. Electron microscopy of ultra-thin sections and of freeze-fractured desmosomes has revealed the presence in the plaque of filaments whose diameter of 4–5 nm is less than that (8 nm) of tonofilaments (Kelly, 1966; McNutt & Weinstein, 1973). It has been suggested that the plaque is a highly condensed 'filamentous mat' (Fawcett & McNutt, 1969; McNutt, 1970). In many cell types, a network of filaments with similar diameters is associated with the cytoplasmic surface of the plasma membrane, and is believed to function in strengthening and supporting the membrane (Juliano, 1973) and as part of the link between the membrane and the cytoskeleton (Nicolson, 1976). Both functions are obviously analogous to those of the desmosomal plaque. The relationship of the proteins which form these filaments to each other and to myosin is a matter of debate

(Hiller & Weber, 1977). However, they have molecular weights in the range of 200–250 K and can be selectively eluted from membrane preparations at low ionic strength (Tillack, Marchesi, Marchesi & Steers, 1970). It is, therefore, significant that isolated desmosomes contain a very high proportion (27% of their total weight) of two polypeptides with molecular weights 210 K and 230 K which are present in approximately equimolar amounts and can be completely and selectively removed by low ionic strength media (Skerrow, 1979*a*).

Apart from a reduction in the electron density and thickness of the plaque, removal of the 210 K and 230 K polypeptides leaves the desmosomal structure remarkably unchanged, and desmosomal adhesion is apparently unaffected (Skerrow, 1979*a*). These findings may partially explain the differences in protein composition between CASC-isolated desmosomes and those prepared by a recently reported method involving prolonged exposure to low ionic strength media of pH 9, 10 or 11 (Drochmans *et al.*, 1978). The structures obtained by the latter procedure resemble CASC-isolated desmosomes extracted at low ionic strength and are selectively depleted in high molecular weight proteins. Additional extraction might be expected to result from the use of pH values of up to 11, as it has been found that CASC-isolated desmosomes completely and rapidly dissolve at pH 12 (Skerrow & Matoltsy, 1974*a*).

Desmosome–filament attachment

In epidermis, tonofilaments do not enter the desmosomal plaque, but are separated from it by a relatively electron-lucent zone. This zone has been reported to contain a finely filamentous 'connecting component' which overlaps with the end of tonofilament bundles to form a dense band (Fig. 1: Kelly, 1966; Brody, 1968; Skerrow & Matoltsy, 1974*a*). In both low ionic strength and CASC-isolated desmosomes, the cytoplasmic surface of the plaque bears a layer of fine filaments extending perpendicularly to the plaque for approximately 70 nm. These filaments are clearly continuous with the plaque and not separated from it by an electron-lucent zone and there is biochemical evidence that, in CASC-isolated desmosomes, they are not composed of tonofilament proteins (Skerrow & Matoltsy, 1974*a*, *b*). It seems probable that the filamentous layer consists of the connecting filaments normally present in the electron-lucent zone, made visible by the dissolution of the attached tonofilaments in CASC.

The filamentous layer persists unchanged after low ionic strength extraction (Skerrow, 1979*a*) and is therefore not composed, as was first suggested, of the 210 K and 230 K polypeptides present in the plaque (Skerrow & Matoltsy, 1974*b*). It appears to be a biochemically distinct component which acts as an intermediary between the plaque and tonofilaments. There is an interesting possibility that connecting filaments contain α-actinin. This suggestion, which could readily be tested by labelling isolated desmosomes with an anti-α-actinin antibody is based on, firstly, the observation that anti-α-actinin antibodies bind to desmosomes in intestinal epithelium (Schollmeyer *et al.*, 1974; Craig & Pardo, 1979); secondly, on the presence in isolated desmosomes of a major polypeptide whose molecular weight of 90 K (Skerrow & Matoltsy, 1974*b*) is in the range for α-actinin; and thirdly on the analogous function of α-actinin as an intermediary between the plasma membrane and cytoplasmic filaments (Mooseker & Tilney, 1975). α-actinin is known to bind to microfilaments (Lazarides & Burridge, 1975) and interactions between desmosomes and microfilaments have been demonstrated in a number of non-epidermal tissues (Toselli & Pepe, 1968; Lentz & Trinkaus, 1971). However, in normal epidermis the desmosomal plaque, with the intervention of one or more connecting components, binds an α-type fibrous protein, prekeratin, which is not actin-like. The structure of prekeratin and its possible relationship to the proteins of intermediate filaments will be discussed below.

Conclusion: the nature of the desmosome

Studies on the nature, location and functions of the numerous polypeptides identified in isolated desmosomes are still at a preliminary stage. However, the ultrastructural, histochemical and biochemical data so far available suggest that the desmosome is a localised concentration of molecules concerned with adhesion, membrane strengthening and filament binding which are similar to those serving analogous functions in non-junctional areas of plasma membranes. If this is so, the characteristic properties of the desmosome may be ascribed to the presence of specific mechanisms for the accumulation and stabilisation of large clusters of these molecules rather than any unique properties of the molecules themselves. These mechanisms

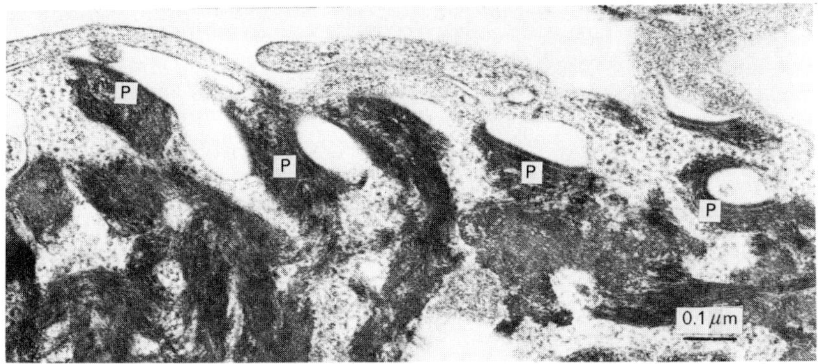

Fig. 3. Residual plaque–tonofilament complexes (P), still intact after trypsinisation of human epidermis for 4 h, in the process of ingestion of the cell.

remain unknown: however, some indication of plaque stability is given by the observation that, after disruption of epidermal intercellular adhesion by trypsin, the residual plaque remains intact, rigid and still binds tonofilaments. Residual plaque–tonofilament complexes persist unchanged after trypsinisation for 12–24 h, during which time they are engulfed by the cell (Fig. 3; Overton, 1975; Skerrow, 1979*b*).

TONOFILAMENTS

Tonofilaments, about 8 nm in diameter, are the major differentiation product of epidermal keratinocytes, comprising more than 60% of the mass of terminally differentiated cells. At near-neutral pH values, these filaments are insoluble in the absence of a denaturing agent such as urea or sodium dodecyl sulphate (SDS). However, the discovery that they can be solubilised from the living cells of bovine epidermis using CASC pH 2.6, led to the isolation of the tonofilament protein prekeratin (Matoltsy, 1964, 1965). The particular significance of prekeratin is that it is the only intact multichain fibrous protein to have been extracted from any keratinising tissue and, as such, it affords a unique opportunity to study the molecular architecture of tonofilaments.

The molecule has now been characterised in some detail (Skerrow, Matoltsy & Matoltsy, 1973; Skerrow, 1974; Skerrow, 1977; Steinert, 1978) and a current working hypothesis of its structure is shown in Fig. 4. It consists of a dimer of three-stranded subunits each of which

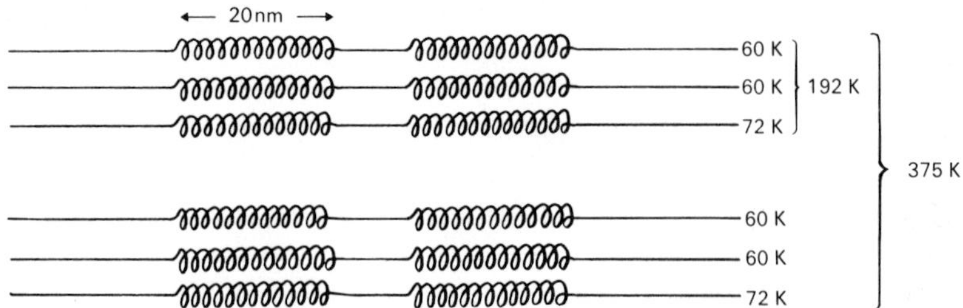

Fig. 4. Diagrammatic representation of the structure of the prekeratin molecule. The molecular weights of the individual chains and the whole molecule, shown on the right, are as determined for bovine prekeratin (Skerrow, 1974). Corresponding molecular weights of the chains of human prekeratin are 70 K, 63 K and 55 K (Skerrow, 1977). The three α-helices of each subunit are most probably wound around each other in a three-stranded coiled-coil and two 20 nm coiled-coils are separated by a relatively short length of non-α-helix. There are no interchain disulphide or other covalent bonds. The relatively small number of tryptic peptides produced by total digestion of prekeratin (Skerrow, 1972) suggests that both inter- and intra-chain sequence homologies could exist within the molecule.

Data for this model were taken from Skerrow, Matoltsy & Matoltsy (1973) and Skerrow (1974) which was confirmed and extended by Steinert (1978).

contains two α-helical regions about 20 nm long in which the helices are most probably arranged in the form of triple-stranded coiled-coils. The non-α-helical regions are of unknown conformation and are particularly rich in glycine and serine. Some workers have described more complex overall chain compositions for prekeratin, suggesting possible microheterogeneity, but have confirmed the presence of three-chain structural units within the molecule (Steinert, 1975; Lee & Baden, 1976; Steinert, 1978).

Tonofilaments have been classified as intermediate filaments on the grounds of their similarity of diameter (Fig. 5) solubility characteristics and possible function to filaments found in many eukaryotic cell types. The tubular appearance and α-helix content of intermediate filaments from BHK-21 cells are further points of similarity with tonofilaments (Steinert, Zimmerman, Starger & Goldman, 1978) but neither of these is a diagnostic characteristic.

Biochemical studies of the other types of intermediate filament protein have not, so far, been sufficiently detailed to allow a valid comparison with the model of prekeratin presented above. The polypeptide chain compositions of most of the intermediate filament proteins appear to be simpler than that of prekeratin, consisting of one or two major chains with molecular weights about 55 K. On SDS

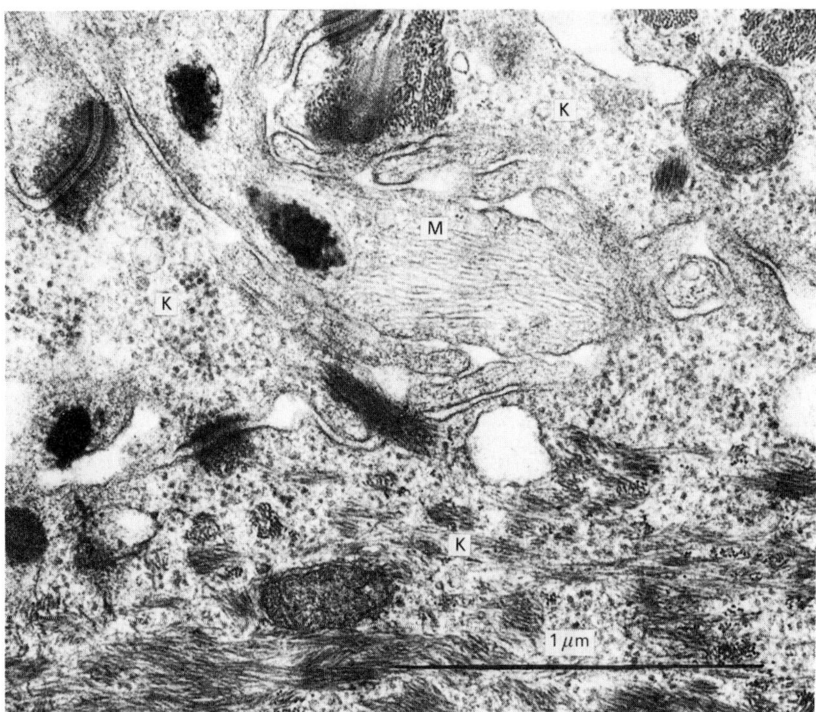

Fig. 5. Tonofilaments and intermediate filaments in human basal keratinocytes (K) and a process of a melanocyte (M) respectively. The diameters of the individual filaments are similar, but tonofilaments are more electron dense and show a greater tendency to aggregate.

gels, these may comigrate with one of the prekeratin chains, but other prekeratin chains are absent in the intermediate filament proteins (Freudenstein, Franke, Osborn & Weber, 1978; Franke, Schmid, Weber & Osborn, 1979). Peptide mapping of the 55 K chains from brain and smooth muscle intermediate filaments has shown them to be different (Davison, Hong & Cooke, 1977) but a direct comparison with prekeratin has not been made.

Recently, antibodies to prekeratin and other intermediate filament proteins have been used, in immunofluorescence microscopy, to distinguish between different classes of intermediate filaments. By this technique, the prekeratin-containing filaments of epithelium-derived cells appear to be immunologically distinct from other intermediate filaments, which can themselves be subclassified (Franke, Schmid, Osborn & Weber, 1978). Some epithelial cell lines such as PtK_2 and HeLa have two systems of intermediate filaments, one of which decorates with anti-prekeratin and the other which, after Colcemid capping, decorates with an antibody to mouse 3T3 cell intermediate filament protein (Franke *et al.*, 1978, 1979). In these and other similar studies the identity and purity of the antigens and the specificity of the antibody binding may be questioned (Bennett *et al.*, 1978). However, the techniques used clearly decorate systems of filaments which parallel the appearance and distribution of intermediate or tonofilaments in the various cell types studied.

It is possible that the inert structural proteins of tonofilaments and intermediate filaments do not have precise sequence requirements and consequently, accept a high mutation rate. This would lead to a relatively rapid divergence of sequences and hence characteristics such as peptide maps and antigenic determinants, which could obscure basic structural similarities. The determination of whether these proteins are descended from a common protein ancestor or have converged evolutionarily will, then, probably require a precise comparison of sequence data which is currently unavailable. A similar argument applies to the comparison of prekeratin and fibrinogen, which are two molecules known to possess considerable structural similarities (Skerrow, 1974, 1977) but which are immunologically distinct (D. Skerrow, unpublished observations).

TONOFILAMENT MODIFICATIONS DURING EPIDERMAL DIFFERENTIATION

Comparisons of the polypeptide chain composition of human epidermal living and terminally differentiated horny cells have shown differences which have been attributed to the modification or removal of one of the prekeratin chains, molecular weight *c.* 55 K, by differential gene activation, proteolysis or cross-linking by covalent bonds other than disulphide bonds (Skerrow & Hunter, 1978). If one of these is shown to be the case it will be the first molecular modification to this protein demonstrated to occur during epidermal differentiation. An abnormality occurring in the common skin disease psoriasis (Skerrow & Hunter, 1978; Thaler *et al.*, 1978) appears to be due to the defective synthesis of the *c.* 70 K chain of prekeratin and this could be an early and important factor in the pathogenesis of the condition (Skerrow & Hunter, 1978; I. Hunter and D. Skerrow, unpublished observations).

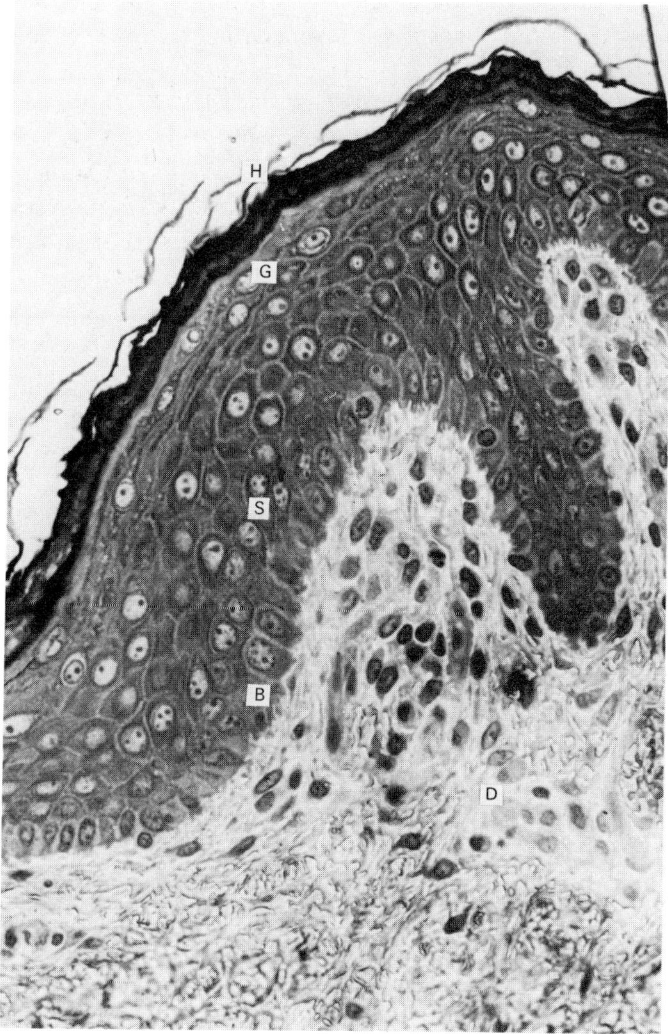

Fig. 6. Human epidermis, showing the sequence of changes in the shape of keratinocytes, which become progressively flattened towards the surface of the skin. D, dermis; B, basal layer; S, spinous layer; G, granular layer; H, horny layer.

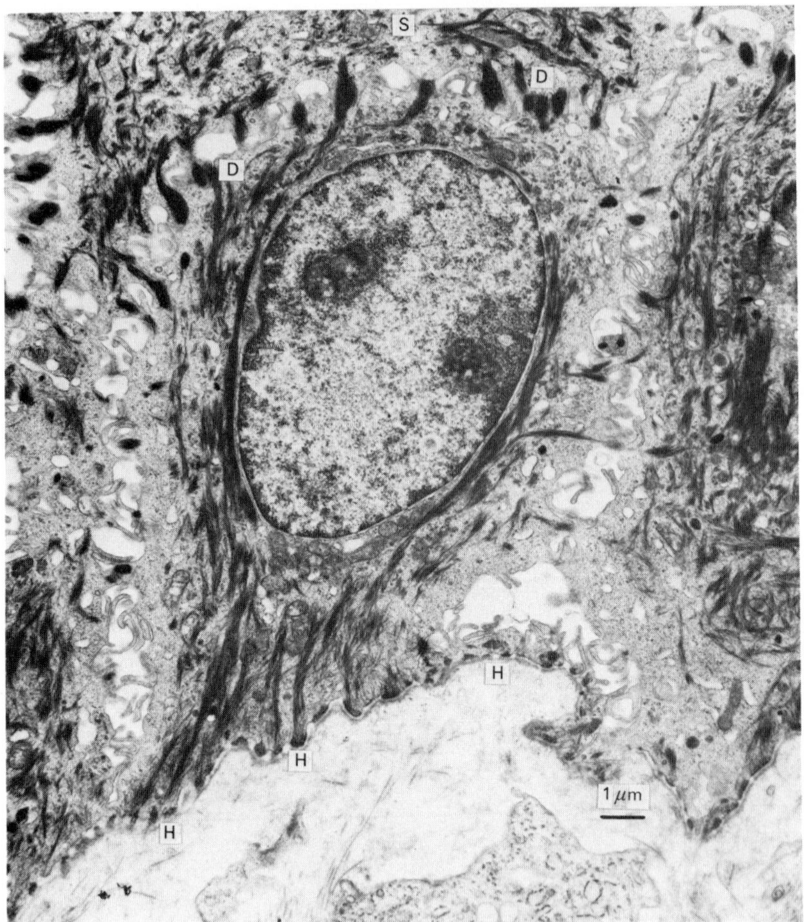

Fig. 7. Human basal keratinocytes attached to the dermis by hemidesmosomes (H) and to other keratinocytes by desmosomes (D) mainly located on the apical surface adjacent to the spinous layer (S).

DESMOSOME MODIFICATIONS DURING EPIDERMAL DIFFERENTIATION

In epidermis, desmosome–tonofilament complexes exist in a differentiating tissue in which a continuous flow of keratinocytes passes from the mitotically active basal layer to the dead, outermost horny layer, from which the cells are ultimately lost by desquamation (Fig. 6). The differentiation process, which is also termed keratinisation has been described in detail from an ultrastructural point of view

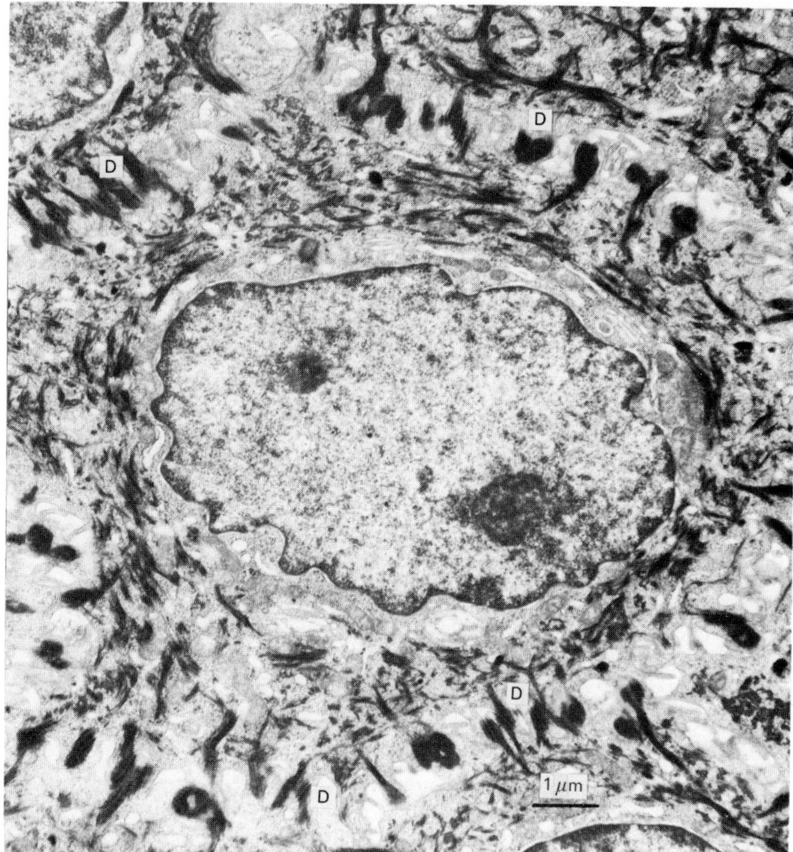

Fig. 8. Human spinous keratinocytes attached to each other by numerous desmosomes (D) formed between surface interdigitations.

(Breathnach, 1975; Matoltsy, 1975; Wolff & Wolff-Schreiner, 1975). The events of epidermal differentiation which involve desmosomes (reviewed by Skerrow, 1978) can be briefly summarised as follows. The dividing basal cell population forms relatively few desmosomes and its cell surfaces are highly folded (Fig. 7). Migration of individual postmitotic basal cells into the spinous layer is accompanied by a three-fold increase in desmosome formation (Fig. 8; Klein-Szanto, 1977). In the upper spinous layer, cells become increasingly flattened in a direction parallel with the surface of the skin (Fig. 6) and their surfaces begin to unfold. The onset of this process coincides with the appearance at the cell periphery of masses of fine filaments which tend

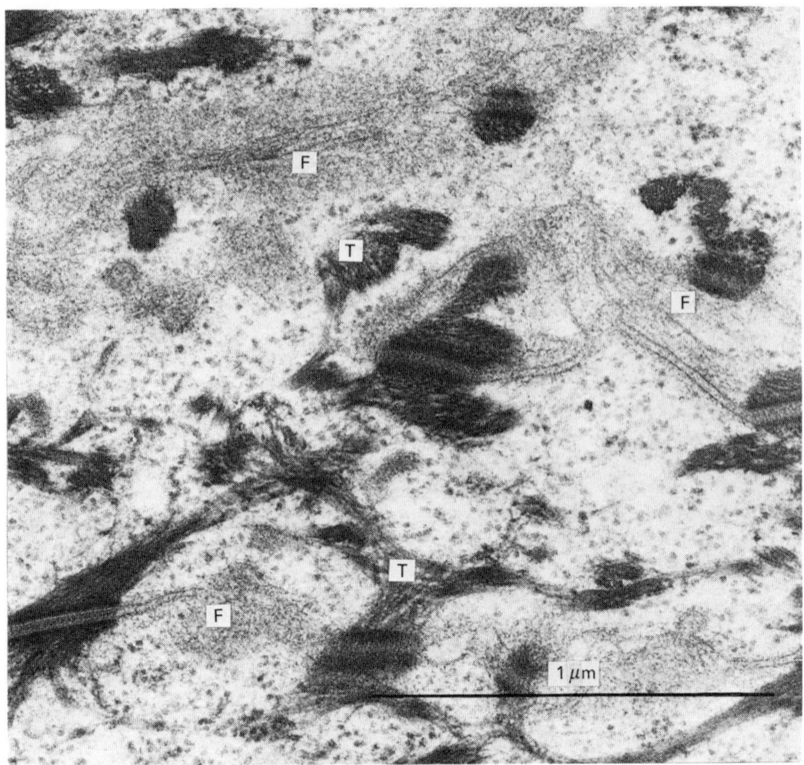

Fig. 9. Portions of human keratinocytes from the upper spinous layer, showing masses of microfilaments (F), 4–5 nm in diameter, in the peripheral cytoplasm. These filaments are clearly distinct from tonofilaments (T) and have a similar electron density to the plasma membrane, which they tend to obscure.

to obliterate the plasma membrane (Fig. 9). Occasionally, organised bundles of microfilaments are observed parallel with the direction of flattening and not, apparently, associated with desmosomes (Skerrow, 1978). The number of these bundles observed may be artificially low, as they are converted into disorganised networks by osmium tetroxide fixation (Pollard, 1976). The appearance of microfilament bundles at this level suggests that they are actively involved in cell surface unfolding in a manner analogous to that demonstrated in embryonic tissue (Szollosi, 1970). At the transition between the spinous and granular layers, the plasma membrane becomes more rigid (Takigawa, Imamuri & Ofuji, 1978), effectively freezing adhesive contacts, and the material in the desmosomal interspace becomes ultrastructurally modified and resistant to trypsin (Skerrow, 1979*b*). Upper granular

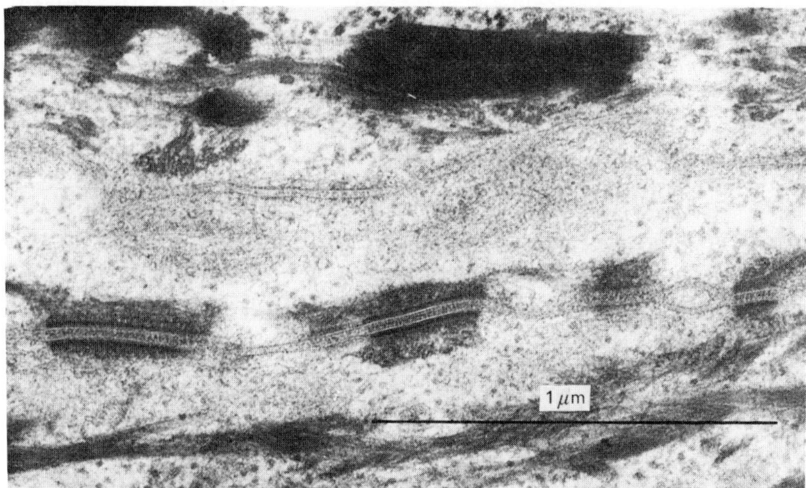

Fig. 10. Portions of human granular keratinocytes showing smooth, closely apposed, plasma membranes.

cell plasma membranes are smooth and closely apposed with an average outer leaflet separation of 15 nm (Fig. 10). Desmosomes between the dead, dehydrated horny cells undergo further ultrastructural changes (Raknerud, 1975) and, finally, adhesion between the outer layers of horny cells decays and individual cells desquamate.

The processes of cell division, migration of basal cells and flattening of spinous and granular cells are clearly incompatible with permanent attachment of cells by desmosomes. Other situations are known in which individual cells move within a tissue containing numerous desmosomes (Epstein, Conant & Krasnobrod, 1966). The lability of desmosomes has been demonstrated in culture (Klaus, Benson & Moellman, 1971). However, whilst structures which could be stages in either desmosome formation or dispersal are observed (Skerrow, 1978) there are no data available on the average life of the desmosome at various stages of differentiation in normal epidermis.

In many senses, adult epidermis can be said to resemble embryonic tissue (Cohen, 1969). The pattern of programmed changes in cell shape (Fig. 6), types of cell contacts (Skerrow, 1978) and cell surface histology (Mercer, Jahn & Maibach, 1968; Nieland, 1973) together with the presence of regulated cell movements has obvious parallels with observed events in embryonic development. However, few studies of normal or pathological epidermal differentiation, along the lines developed for other tissues, have been made, although this would appear to be a fruitful approach. A particularly interesting question, for example, is whether the relative cell movements observed in epidermis are driven and guided by differences in adhesive properties which arise as keratinocytes differentiate. Sorting out has been demonstrated to occur, with desmosome formation, in a mixture of epithelial cells taken from chick embryos at stages prior and subsequent to their acquisition of a greatly increased capacity to form desmosomes. These movements occur in the presence of cytochalasin B (Overton, 1975). This situation is apparently analogous to that during the basal–spinous cell transition in epidermal differentiation. A basal cell which has reached the stage of postmitotic maturity at which the three-fold increase in capacity to form desmosomes occurs might form junctions preferentially with the spinous cells in the layer above and be pulled out of the basal layer. Occasionally, pedunculated basal cells which appear to be spreading their contacts with the spinous cells above are observed (Christophers, 1971) and in many basal cells desmosomes are located preferentially on the apical surface, in contact with a spinous cell (Fig. 7). As yet, no experiments have been performed to demonstrate directly the existence of adhesive preferences, based on a differential capacity to form desmosomes, in mixed populations of keratinocytes.

In the acantholytic disorders, adhesion of keratinocytes to each other or to the dermis is lost at a specific level which is characteristic of the disease (Sneddon, 1972). In no case has the nature of the primary defect, the sequence of events involved in lesion production or the reason for the localised expression of the disease been established. It is not practicable to obtain sufficient quantities of pathological tissue for isolation of desmosomes and direct determination of molecular abnormalities as has been done for tonofilaments. However, ultrastructural studies on the effects on normal epidermis of various treatments which lead to cell detachment can provide models for acantholytic diseases (Skerrow, 1979*b*). Furthermore, the active agents identified in some of these diseases produce their specific effects on normal epidermis *in vitro* (Elias *et al.*, 1974; Michel & Ko, 1974; Barnett, 1978). Parallel studies of these systems will provide information both on the nature of adhesion within normal epidermis and the mechanisms of acantholytic disease.

Acknowledgement. This work was supported by grants from the Medical Research Council. Fig. 2 was kindly provided by Professor A. S. Breathnach.

REFERENCES

BARNETT, M. L. (1978). Effect of pemphigus antibodies on desmosomal structure in vitro. *Journal of Investigative Dermatology*, **70**, 141–2.

BENEDETTI, E. L. & EMMELOT, P. (1968). Structure and function of plasma membranes isolated from liver. In *The Membranes*, ed. A. J. Dalton & F. Hagenau, pp. 33–120. New York, London: Academic Press.

BENNETT, G. S., FELLINI, S. A., CROOP, J. M., OTTO, J. J., BRYAN, J. & HOLTZER, H. (1978). *Proceedings of the National Academy of Sciences, USA*, **75**, 4364–8.

BORYSENKO, J. Z. & REVEL, J. P. (1973). Experimental manipulation of desmosome structure. *American Journal of Anatomy*, **137**, 403–22.

BREATHNACH, A. S. (1975). Aspects of epidermal ultrastructure. *Journal of Investigative Dermatology*, **65**, 2–15.

BRODY, I. (1968). An electron microscope study of the junctional and regular desmosomes in normal human epidermis. *Acta dermato-Venereologica*, **48**, 290–302.

CHRISTOPHERS, E. (1971). Cellular architecture of the stratum corneum. *Journal of Investigative Dermatology*, **56**, 165–9.

COHEN, J. (1969). Interactions in the skin. *British Journal of Dermatology*, **81**, suppl. 3, 46–54.

CRAIG, S. W. & PARDO, J. V. (1979). Alpha-actinin localization in the junctional complex of intestinal epithelial cells. *Journal of Cell Biology*, **80**, 203–10.

DAVISON, P. F., HONG, B-S. & COOKE, P. (1977). Classes of distinguishable 10 nm cytoplasmic filaments. *Experimental Cell Research*, **109**, 471–4.

DROCHMANS, P., FREUDENSTEIN, C., WANSON, J-C., LAURENT, L., KEENAN, T. W., STADLER, J., LELOUP, R. & FRANKE, W. W. (1978). Structure and biochemical composition of desmosomes and tonofilaments isolated from calf muzzle epidemis. *Journal of Cell Biology*, **79**, 427–43.

ELIAS, P. M., FRITSCH, P., TAPPEINER, G., MITTERMAYER, H. & WOLFF, K. (1974). Experimental staphylococcal toxic epidermal necrolysis (TEN) in adult humans and mice. *Journal of Laboratory and Clinical Medicine*, **84**, 414–24.

EPSTEIN, W. L., CONANT, M. A. & KRASNOBROD, H. (1966). Molluscum contagiosum: normal and virus infected epidermal cell kinetics. *Journal of Investigative Dermatology*, **46**, 91–103.

FARQUHAR, M. G. & PALADE, G. E. (1963). Junctional complexes in various epithelia. *Journal of Cell Biology*, **17**, 375–412.

FAWCETT, D. W. & MCNUTT, N. S. (1969). The ultrastructure of the cat myocardium. I. Ventricular papillary muscle. *Journal of Cell Biology*, **42**, 1–45.

FRANKE, W. W., SCHMID, E., OSBORN, M. & WEBER, K. (1978). Different intermediate-sized filaments distinguished by immunofluorescence microscopy. *Proceedings of the National Academy of Sciences, USA*, **75**, 5034–38.

FRANKE, W. W., SCHMID, E., WEBER, K. & OSBORN, M. (1979). HeLa cells contain intermediate-sized filaments of the prekeratin type. *Experimental Cell Research*, **118**, 95–109.

FRANKE, W. W., WEBER, K., OSBORN, M., SCHMID, E. & FREUDENSTEIN, C. (1978). Antibody to prekeratin. *Experimental Cell Research*, **116**, 429–45.

FREUDENSTEIN, C., FRANKE, W. W., OSBORN, M. & WEBER, K. (1978). Reaction of tonofilament-like intermediate-sized filaments with antibodies raised against isolated defined polypeptides of bovine hoof prekeratin. *Cell Biology, International Reports*, **2**, 591–600.

HILLER, G. & WEBER, K. (1977). Spectrin is absent in various tissue culture cells. *Nature, London*, **266**, 181–3.

JULIANO, R. L. (1973). The proteins of the erythrocyte membrane. *Biochimica et Biophysica Acta*, **300**, 341–78.

KELLY, D. E. (1966). Fine structure of desmosomes, hemidesmosomes and an adepidermal globular layer in developing newt epidermis. *Journal of Cell Biology*, **28**, 51–72.

KLAUS, S. N., BRANSON, S. H. & MOELLMANN, G. E. (1971). Lability of desmosomes in cell culture. *Journal of Investigative Dermatology*, **56**, 402 (abstract).

KLEIN-SZANTO, A. J. P. (1977). Stereologic baseline data of normal human epidermis. *Journal of Investigative Dermatology*, **68**, 73–8.

LAZARIDES, E. & BURRIDGE, K. (1975). Alpha-actinin: immunofluorescence localization of a muscle structural protein in non-muscle cells. *Cell*, **6**, 289–98.

LEE, L. D. & BADEN, H. P. (1976). Organization of the polypeptide chains in mammalian keratin. *Nature, London*, **264**, 377–9.

LENTZ, T. L. & TRINKAUS, J. P. (1971). Differentiation of the junctional complex of surface cells in the developing Fundulus blastoderm. *Journal of Cell Biology*, **48**, 455–72.

MCNUTT, N. S. (1970). Ultrastructure of intercellular junctions in adult and developing cardiac muscle. *American Journal of Cardiology*, **25**, 169–83.

MCNUTT, N. S. & WEINSTEIN, R. S. (1973). Membrane ultrastructure at mammalian intercellular junctions. *Progress in Biophysics and Molecular Biology*, **26**, 45–101.

MATOLTSY, A. G. (1964). Prekeratin. *Nature, London*, **201**, 1130–1.

MATOLTSY, A. G. (1965). Soluble prekeratin. In *Biology of the Skin and Hair Growth*, ed. A. G. Lyne & B. F. Short, pp. 291–305. Sydney: Angus & Robertson.

MATOLTSY, A. G. (1975). Desmosomes, filaments and keratohyalin granules: their role in the stabilization & keratinization of the epidermis. *Journal of Investigative Dermatology*, **65**, 127–42.

MERCER, E. H. (1961). In *Keratin and Keratinization*. Oxford: Pergamon Press.

MERCER, E. H., JAHN, R. A. & MAIBACH, H. I. (1968). Surface coats containing polysaccharides on human epidermal cells. *Journal of Investigative Dermatology*, **51**, 204–14.

MICHEL, B. & KO, C. S. (1974). Effect of pemphigus or bullous pemphigoid sera and leukocytes on normal human skin in organ cultures. An in vitro model for the study of bullous diseases. *Journal of Investigative Dermatology*, **62**, 541–2.

MOOSEKER, M. S. & TILNEY, L. G. (1975). Organization of an actin filament-membrane complex. *Journal of Cell Biology*, **67**, 725–43.

Nieland, M. L. (1973). Epidermal intercellular staining with fluorescein-conjugated phytohemagglutinins. *Journal of Investigative Dermatology*, **60**, 61–6.

Nicolson, G. L. (1976). Trans-membrane control of the receptors on normal and tumor cells 1. Cytoplasmic influence over cell surface components. *Biochimica et Biophysica Acta*, **457**, 57–108.

Nicolson, G. L. & Poste, G. (1976). The cancer cell. *New England Journal of Medicine*, **295**, 197–203.

Overton, J. (1975). Experiments with junctions of the adhaerens type. *Current Topics in Developmental Biology*, **10**, 1–34.

Pollard, T. D. (1976). Cytoskeletal function of cytoplasmic contractile proteins. *Journal of Supramolecular Structure*, **5**, 317–34.

Raknerud, N. (1975). The ultrastructure of interfollicular epidermis of the hairless (hr/hr) mouse III. Desmosomal transformation during keratinization. *Journal of Ultrastructural Research*, **52**, 32–51.

Schollmeyer, J. V., Goll, D. E., Tilney, L., Mooseker, M., Robson, R. & Stromer, M. (1974). Localization of α-actinin in non-muscle material. *Journal of Cell Biology*, **63**, (3 part 2): 304a (abstract).

Skerrow, D. (1972). A repeating subunit of soluble prekeratin. *Biochimica et Biophysica Acta*, **257**, 398–403.

Skerrow, C. J. (1978). Intercellular adhesion and its role in epidermal differentiation. *Investigative and Cell Pathology*, **1**, 23–37.

Skerrow, C. J. (1979*a*). Selective extraction of desmosomal proteins by low ionic strength media. *Biochimica et Biophysica Acta*, **579**, 241–5.

Skerrow, C. J. (1979*b*). The experimental production of high level intraepidermal splits. *British Journal of Dermatology* (in press).

Skerrow, C. J. & Matoltsy, A. G. (1974*a*). Isolation of epidermal desmosomes. *Journal of Cell Biology*, **63**, 515–23.

Skerrow, C. J. & Matoltsy, A. G. (1974*b*). Chemical characterization of isolated epidermal desmosomes. *Journal of Cell Biology*, **63**, 524–30.

Skerrow, D. (1974). The structure of prekeratin. *Biochemical and Biophysical Research Communications*, **59**, 1311–16.

Skerrow, D. (1977). The isolation and preliminary characterisation of human prekeratin. *Biochimica et Biophysica Acta*, **494**, 447–51.

Skerrow, D. & Hunter, I. (1978). Protein modifications during the keratinization of normal and psoriatic human epidermis. *Biochimica et Biophysica Acta*, **537**, 474–84.

Skerrow, D., Matoltsy, A. G. & Matoltsy, M. N. (1973). Isolation and characterization of the α-helical regions of epidermal prekeratin. *Journal of Biological Chemistry*, **248**, 4820–26.

Sneddon, I. B. (1972). Bullous eruptions. In *Textbook of Dermatology*, ed. A. Rook, D. S. Wilkinson & F. J. E. Ebling, pp. 1296–1333. Oxford. Blackwell Scientific.

Steinert, P. M. (1975). The extractions and characterization of bovine epidermal α-keratin. *Biochemical Journal*, **149**, 39–48.

Steinert, P. M. (1978). Structure of the three-chain unit of the bovine epidermal keratin filament. *Journal of Molecular Biology*, **123**, 49–70.

Steinert, P. M., Zimmerman, S. B., Starger, J. M. & Goldman, R. D. (1978). 10 nm filaments of hamster BHK-21 cells and epidermal keratin filaments have similar structures. *Proceedings of the National Academy of Science, USA*, **75**, 6098–101.

Szollosi, D. (1970). Cortical cytoplasmic filaments of cleaving eggs: a structural element corresponding to the contractile ring. *Journal of Cell Biology*, **44**, 192–209.

Takigawa, M., Imamura, S. & Ofuji, S. (1978). Surface distribution of pemphigus antibody-binding substance(s) on isolated guinea pig epidermal cells. An immunoferritin electron microscopic study. *Journal of Investigative Dermatology*, **71**, 182–5.

Thaler, M. P., Fukuyama, K., Inoue, N., Cram, D. L. & Epstein, W. L. (1978). Two tris-urea-mercaptoethanol extractable polypeptides found uniquely in scales of patients with psoriasis. *Journal of Investigative Dermatology*, **70**, 38–41.

Tillack, T. W., Marchesi, S. L., Marchesi, V. T. & Steers, E. (1970). A comparative study of spectrin: a protein isolated from red blood cell membranes. *Biochimica et Biophysica Acta*, **200**, 125–31.

Toselli, P. A. & Pepe, F. A. (1968). The fine structure of the intersegmental abdominal muscle of the insect Rhodnius prolixus during the molting cycle 1. Muscle structure at molting. *Journal of Cell Biology*, **37**, 445–61.

Wolff, K. & Wolff-Schreiner, E. C. (1975). Trends in electron microscopy of skin. *Journal of Investigative Dermatology*, **65**, 39–57.

Molecular and supramolecular cell surface events during the process of normal and neoplastic cell adhesion *in vitro*

R. RAJARAMAN AND J. M. MACSWEEN

Departments of Medicine and Microbiology, Dalhousie University and Camp Hill Hospital, Halifax, NS, Canada

INTRODUCTION

Cell adhesion appears to be the central phenomenon related directly or indirectly to other processes such as locomotion (Gaile & Boone, 1972), morphogenesis (Wiseman, Steinberg & Phillips, 1972), intercellular communication (Gilula, Reeves & Steinbach, 1967), contact inhibition (Abercrombie, 1975) and regulation of cell division (Shin, Freedman, Risser & Pollack, 1975). A loss or reduction in cellular adhesive strength in neoplastic cells (Coman, 1944) is usually accompanied by unlimited division potential (Ponten, 1976), anchorage independence for growth (Stoker, O'Neill, Berryman & Waxman, 1968), altered membrane permeability (Hatanaka, 1974) and the presence of surface proteases (Burger, 1973). For the above reasons, the molecular aspects of cell adhesion have been of great interest in cell biology.

During cell adhesion and locomotion, several molecular and supramolecular events occur both on the cell surface and in the cytoplasm beneath the cell membrane. We have studied some aspects of cell surface events using normal and neoplastic cells *in vitro*.

Effect of serum on cell adhesion

Recent studies on the mechanism of cell adhesion *in vitro* have revealed that serum proteins influence cell deformation and spreading (Grinnell, 1976*a*, *b*, *c*; Rajaraman, Rounds, Yen & Rembaum, 1974, 1975; Rajaraman *et al.*, 1977; Rajaraman, MacSween & Fox, 1978*a*; Rajaraman, Westermark, Vaheri & Ponten, 1978*b*). Certain cells such as human conjunctiva cells (Taylor, 1961), human lung fibroblasts MRC5 (Witkowski & Brighton, 1971), WI38 (Rajaraman *et al.*, 1974, 1975), human glia cells (Rajaraman *et al.*, 1978*a*), human monocytes (Rajaraman *et al.*, 1977) and guinea pig macrophages (Fox, Fernandez & Rajaraman; 1977) can undergo rapid adhesion and spreading on the substratum in serum-free medium. Others, such as normal BHK21 and BHK21-C13 cells, BHK-py, L929, HeLa and CHO-A10 cells (Grinnell, 1976*a*, *b*, *c*) and WI-38VA13-2RA (2RA) cells (Rajaraman *et al.*, 1975) require serum proteins for deformation and spreading *in vitro*. Using different normal and neoplastic cells grown *in vitro* under standard culture conditions, we have studied the role of serum requirement by these cells for adhesion and spreading.

It has been shown earlier that cells adhere and progressively deform and spread on the substratum in sequential stages such as (i) initial attachment, (ii) filopodial growth, (iii) cytoplasmic webbing and (iv) eventual flattening of the nucleus (Rajaraman *et al.*, 1974). In the studies reported below, the extent of cell spreading was quantitated

---

Fig. 1. (*a*) Histogram showing per cent well-spread cells (ordinate) in medium containing 0 % serum (left column with broad stripes) and 20 % serum (right column, narrow stripes) for each cell type tested. MRC5, WI38, FKC (foetal kidney) and FSK (foreskin) are human normal cells; 2RA and VA4 are SV-40 transformed human cells; HeLa cells of human in-vivo neoplastic origin (carcinoma). The normal cells spread irrespective of the presence or absence of serum, while the neoplastic cells require serum proteins for deformation and spreading.

(*b*) Incubation of 2RA cells in suspension in 20 % foetal calf serum (thin stripes) or 10 % human serum (black column) after which the cells were rinsed twice in BSS and plated in serum-free medium. Ordinate: Per cent well-spread cells in 4 h.

(*c*) 2RA cells used to 'adsorb' the serum cell-spreading factor. Cell spreading activity of the supernatants after being 'adsorbed' with different numbers of 2RA cells is expressed as per cent well-spread cells. Duplicate aliquots of 2 ml of 20 % serum supernatant medium were incubated for 3 h with 2RA cells in suspension at room temperature. The supernatant media were then collected by pelleting the cells and the cell-deforming activity of these supernatants was assayed with freshly harvested 2RA cells. The cell-deforming activity of the supernatants was inversely proportional to the number of 2RA cells with which the aliquots of media were pre-incubated.

(*d*) Fibronectin is the active component in plasma involved in cell spreading. Cell-spreading assay was performed using 2RA cells. Ordinate: per cent well-spread cells after 3 h incubation at 37 °C. (a) cell spreading in serum-free medium; (b) cell spreading in medium supplemented with 25 μg ml^{-1} of fetal calf serum; (c) cell spreading in medium supplemented with 25 μg ml^{-1} human plasma; (d) cell spreading in medium supplemented with 25 μg ml^{-1} fibronectin-free human plasma and (e) cell spreading activity of 25 μg human plasma fibronectin alone.

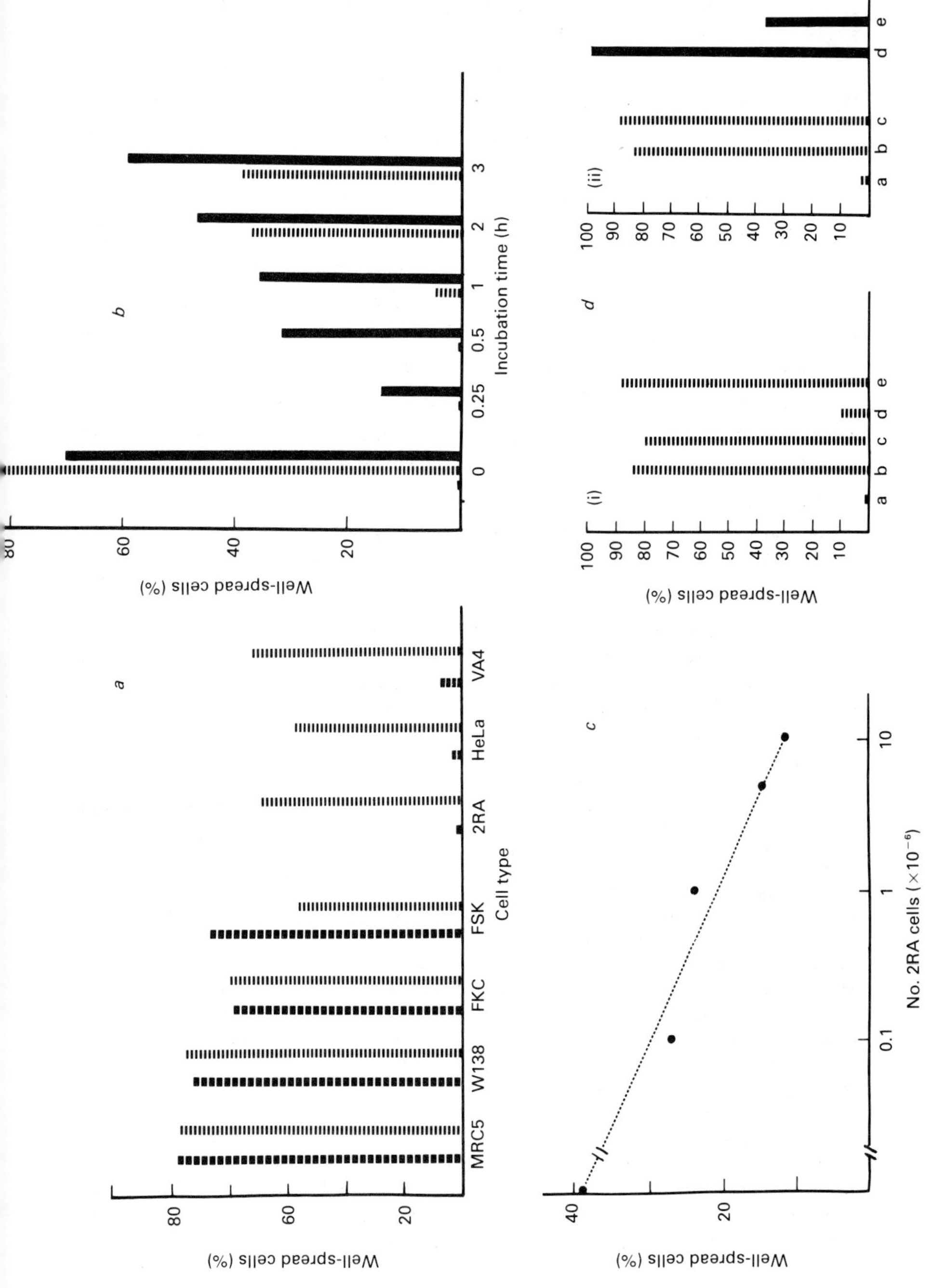
a
Well-spread cells (%)
80
60
20
MRC5
W138
FKC
FSK
2RA
HeLa
VA4
Cell type
b
Well-spread cells (%)
80
60
40
20
0
0.25
0.5
1
2
3
Incubation time (h)
c
Well-spread cells (%)
40
20
0.1
1
10
No. 2RA cells (×10⁻⁶)
d
Well-spread cells (%)
(i)
(ii)
100
90
80
70
60
50
40
30
20
10
a
b
c
d
e

by counting the proportion of well-spread cells (in stage iv: flattening of the nucleus with well-spread cytoplasm) using a Zeiss inverted phase contrast microscope at 400 × magnification. Cells were allowed to spread under appropriate conditions (described in the legends) for the required duration and fixed in 2% glutaraldehyde in 0.1 M phosphate buffer at pH 7.4 for scoring the well-spread cells (200–400 cells per treatment).

In the absence of serum in the medium, both normal WI38 cells and Sv40 transformed WI-38VA12-2RA (2RA) cells are able to attain initial attachment to the substratum within about 20 min after layering. However, normal WI38 cells spread rapidly and flatten following this initial attachment in the absence of serum proteins, while the transformed 2RA cells cannot do so unless serum proteins are present (Rajaraman *et al.*, 1974, 1975). This behaviour has been true for several normal and transformed cells studied so far (Fig. 1*a*). 2RA cells growing in medium containing 10% serum for 2, 3, 4, 5 and 7 days were resuspended and assayed for spreading over a 3 h period in the absence and presence of 20% serum. The cells from 2- and 3-day-old cultures showed 13% and 4% well-spread cells in the absence of serum, while the cells from 4 or more day-old cultures showed no spontaneous spreading and were totally dependent on serum proteins for spreading. Thus, the serum requirement for cell spreading by these transformed cells increased with the age of the subculture.

On serum-coated coverslips 2RA cells spread very well. That the transformed cells can recruit the required factors directly from the medium was suggested by the following two experiments: (*a*) 2RA cells were kept in suspension in 20% foetal calf serum with intermittent agitation for different times and layered on coverslips after rinsing. At least 1 h suspension was required for these cells to be able to spread subsequently in serum-free medium, while only 15 min pre-exposure to human AB serum was sufficient for subsequent spreading of 2RA cells in the absence of serum (Fig. 1*b*); (*b*) Increasing number of 2RA cells were kept in suspension for 3 h at room temperature in small test tubes containing equal aliquots (2 ml) of medium with 10% foetal calf serum, and the cell-spreading activity of the supernatant medium was assayed with fresh 2RA cells; the cell-spreading activity of the different aliquots of supernatants decreased with increasing numbers of cells incubated in the medium previously (Fig. 1*c*).

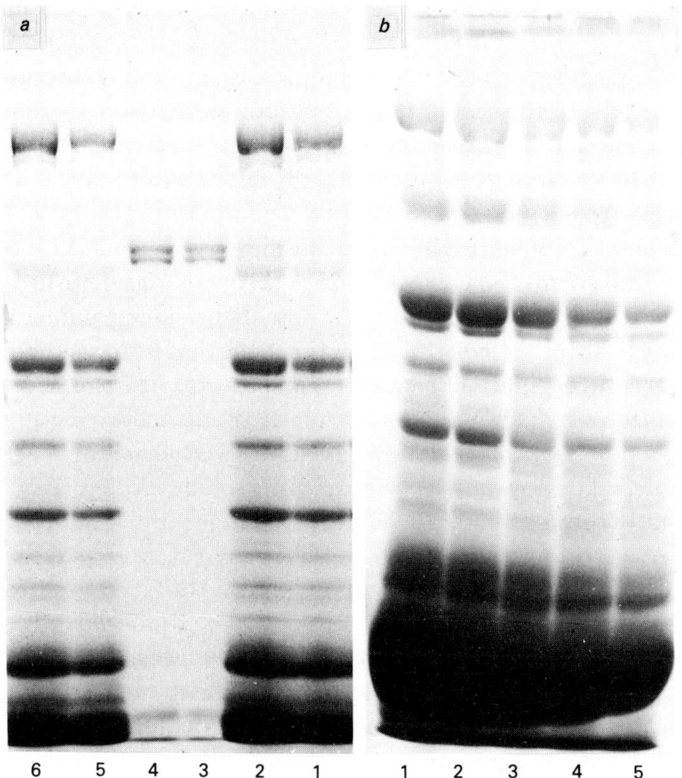

Fig. 2. (*a*) Sodium dodecyl sulphate–polyacrylamide gel (5 %) electrophoresis pattern of human plasma before removal of fibronectin (tracks 1 (218 μg protein) and 2 (109 μg protein)); human plasma fibronectin isolated by the gelatin–Sepharose affinity column run in the presence of mercaptoethanol showing two major bands in the 220–250 K region (tracks 3 (4.4 μg) and 4 (3.3 μg)); the minor band at the bottom appears to be breakdown product of fibronectin, since the antiserum produced by this preparation shows only one band when run against the plasma or purified fibronectin in the Ouchterloney immunodiffusion analysis. Tracks 5 (225 μg) and 6 (113 μg) show human plasma after the removal of fibronectin by the gelatin–Sepharose affinity column. (*b*) SDS–polyacrylamide gel (5 %) electrophoresis of 50 % human plasma in MEM incubated for 30 min at 25 °C with increasing number of 2RA cells in suspension. The cells were removed by centrifugation and supernatant samples were run in SDS–PAGE. Note the loss of intensity in the bands of several plasma proteins, the band representing fibronectin (arrow) being conspicuously reduced with the increase in the number of cells used to adsorb. Number of cells used to adsorb from tracks 1–5 : 0×10^6; 3×10^6; 6×10^6; 12×10^6 and 24×10^6 (0.125 ml of plasma per track).

Fibronectin as an active factor in fibroblastic cell adhesion

Fibronectin, a major cell surface glycoprotein is found on the surface of normal cells but is absent or present only in reduced amounts on transformed cells (Gahmberg & Hakamori, 1973; Hynes, 1973, 1976; Hogg, 1974; Vaheri & Ruoslahti, 1974; Robbins *et al.*, 1974; Yamada & Weston, 1974; Vaheri, Ruoslahti, Westermark & Ponten, 1976; Keski-Oja, 1977, Vaheri, 1977). Since a similar molecule called cold insoluble globulin (CIG) is also found in serum or plasma (Morrison, Edsall & Miller, 1948; Ruoslahti, Vaheri, Kuusela & Linder, 1973; Ruoslahti & Vaheri, 1974, 1975), it appeared that fibronectin might be the serum or plasma factor responsible for the spreading of transformed cells. Therefore, we isolated plasma fibronectin by a gelatin–Sepharose affinity column (Engvall & Ruoslahti, 1977) and studied the cell-spreading activity of fibronectin, whole plasma and fibronectin-free plasma using 2RA cells. These studies showed that the plasma fibronectin alone was sufficient for the spreading of transformed cells, while fibronectin-free plasma lost its cell-spreading activity (Figs. 1*d*(a) and 2*a*). The 2RA cells in suspension were able to recruit ^{125}I-labelled fibronectin from the medium (Fig. 1*d*(b)) and when suspended in plasma, the transformed cells depleted fibronectin along with other unknown proteins from plasma (Fig. 2*b*).

Distribution of surface fibronectin in the early stages of cell spreading

The above experiments indicated that fibronectin played a major role in cell adhesion. Therefore, immunofluorescence technique was used to study the dynamic changes in cell surface fibronectin distribution pattern during the process of normal and neoplastic cell spreading. Human glia (normal) and glioma (neoplastic) cells were used in these studies. Anti-human fibronectin (anti-FN) raised in rabbit was used in conjunction with fluorescein-conjugated goat anti-rabbit IgG to demonstrate indirect immunofluorescence staining of fibronectin on cells fixed with 2% paraformaldehyde. The non-spread glia cells showed varying amounts of fibronectin on their surface distributed in a diffuse fashion within 20–30 min after layering; 5–10% of cells in the initial stages of filopodial and lamellipodial formation showed a patchy distribution of fibronectin in a given preparation superimposed on the diffuse distribution of fibronectin (Fig. 3*a*, *b*). The patchy distribution of fibronectin disappeared with slightly increased

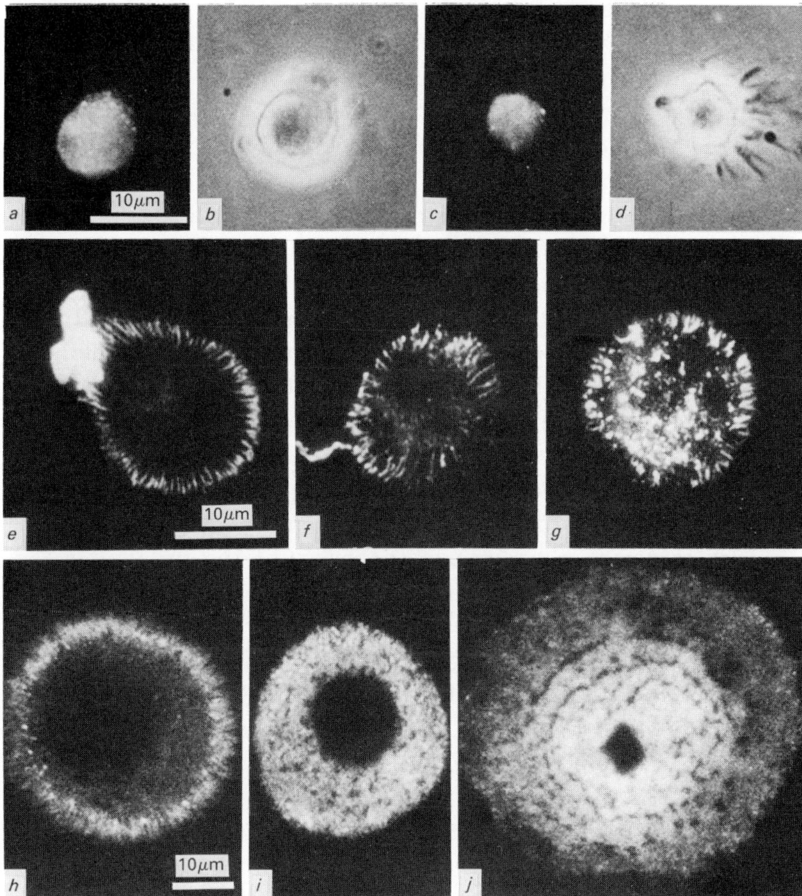

Fig. 3. (*a*, *b*) Glia cell with the beginnings of lamellipodia on one side shows diffuse surface distribution of fibronectin in (*a*). The patches of fibronectin are preferentially prominent near the protruding lamellipodia. (*c*, *d*). Partially spread glia cell with lamelipodia and filopodia in (*b*) shows only diffuse fibronectin on the cell surface; no patchy distribution is visible at this stage.

(*e–g*) Foot-prints of Triton X-100 extracted glia cells. The progressive stages of deposition of fibronectin in the foot-prints of glial cells spreading in the presence of serum. Note the discontinuous nature of the deposition.

(*h–j*) The progressive stages of deposition of fibronectin in the foot-prints of glial cells spreading in the absence of serum. Note the contiguous deposition of fibronectin with islands of fibronectin free areas. Also the central area which was the initial site of cell substratum contact is still free of fibronectin stain. (From Rajaraman *et al.*, 1978*a*.)

spreading (Fig. 3*c*, *d*). It has been suggested that the membrane flow and relocation of the cell surface initiated by the cell substratum contact may be preceded by reorganisation such as patching of membrane receptors (Shields & Pollock, 1974; Maroudas, 1975; Juliano & Gagalong, 1977; Rees, Lloyd & Thom, 1977). This patching of fibronectin is probably analogous to antibody or lectin-induced patch formation in lymphocytes and transformed cells (Taylor, Duffus, Raff & De Petris, 1971; Nicolson, 1973).

Since fixation of cells before exposure to ligands or antibody inhibits mobility of surface antigens (Rosenblith *et al.*, 1973), the patching of fibronectin observed by us is not due to cross-linking of fibronectin molecules by anti-FN; rather this patching has taken place before fixation in response to the cell–substratum contact at a specific stage in cell spreading and may, therefore, represent 'adhesion patches' rich in fibronectin. The low frequency of cells showing patchy distribution of fibronectin at a given time suggests that this is a very transient phenomenon (probably of the order of a few minutes) restricted to a short duration before the morphological alterations due to spreading reach an advanced stage.

In the late filopodial and cytoplasmic webbing stage, the diffuse distribution of fibronectin in the upper surface was less evident and finally disappeared completely from the upper surface of well-spread cells. Since the upper surface was usually devoid of fibronectin, a study for 'foot-print' pattern of fibronectin on the substratum was carried out by extracting glia cells in progressive stages of spreading with a 10 min exposure to 0.1 % Triton X-100 in phosphate-buffered saline before staining for fibronectin. Fibronectin deposition was in the form of radial filaments beneath the advancing filopodia at the cell peripheries (Fig. 3*e–g*). In the presence of serum, further deposition of fibronectin progressed toward the centre in a centripetal direction in the form of discontinuous patches (Fig. 3*f*, *g*), eventually giving rise to a mosaic distribution of fibronectin. In the absence of serum, glia cells deposited fibronectin in a contiguous mosaic pattern throughout the underside of the cell, leaving small fibronectin-free islands (Fig. 3*h–j*).

Non-spread glioma cells also showed diffuse distribution of fibronectin on their surface, but failed to develop a patchy distribution of fibronectin and did not spread in the absence of serum proteins (up to 3 h). When the glioma cells were allowed to spread in the presence of serum, no fibronectin was detectable in their foot-prints or on their upper surface.

DISCUSSION

These results indicate that normal diploid cells as a rule are capable of adhesion and spreading in the absence of serum or exogenous fibronectin. The initial cell–substratum contact stimulates a propagative phenomenon in normal cells in the form of global patching of adhesion sites rich in fibronectin (Fig. 3 *a*, *b*) much akin to the global effect observed in lymphocytes landing on con-A derivatised nylon fibres (Edelman, Wang & Yahara, 1976). Thus, normal cells can use glass or plastic surfaces as matrix for spreading even in the absence of serum proteins, since these cells synthesise and deploy abundant amounts of fibronectin.

In the absence of serum proteins, normal cell membrane is in close apposition to the substraum as shown by scanning and transmission electron microscopy (Witkowski & Brighton, 1972; Pegrum & Maroudas, 1975). Immunofluorescence studies (Fig. 3 *e–j*) also indicates a tighter binding between the cell surface and the substratum in the absence of serum. This would explain the trypsin and EDTA resistance of adhesive contacts in the absence of serum (Takeichi, 1971; Culp & Buniel, 1976; Rajaraman *et al.*, 1978*b*).

The sequence of fibronectin deposition during the progressive stages of cell spreading indicates the existence of at least two types of adhesive contacts between the cell and the substratum: (*a*) an initial fibronectin-poor (primary) adhesion and (*b*) a fibronectin-mediated secondary adhesion (spreading) triggered by the primary contact and executed by filopodia and lamellipodia. Even though fibronectin alone appears to be sufficient for cell spreading, eventually the cell may secrete microexudates containing several proteinaceous components, when the adhesive contacts are futher strengthened (Weiss, Poste, MacKearmin & Willett, 1975; Culp, 1976). Recent studies suggest that fibronectin has binding sites for other matrix components such as collagen hyaluronic acid and proteoglycans (Hahn & Yamada, 1979; Ruoslahti, Hayman & Engvall, 1979; Yamada, Hahn & Olden, 1979).

In contrast, neoplastic cells can adhere but are unable to spread in the absence of serum or fibronectin, and they are not capable of performing the global patching response as the normal cells do upon contact with the substratum. But they can spread on a preformed matrix of serum proteins or fibronectin and can also recruit fibronectin from the medium to form matrix for spreading. Different transformed cells synthesise varying amounts of fibronectin (Vaheri *et al.*, 1976; Wartiowaara, Linder, Ruoslahtib, Vaheri, 1976; Hynes, 1976; Yamada, Yamada & Pastan, 1977) but are unable to retain large amounts of fibronectin for efficient adhesion. Since transformed cells spread only in the presence of fibronectin, it appears that they deploy very small quantities of fibronectin for adhesion and are, therefore, weakly adherent. This would account for the lack of detectable fibronectin in the foot-prints of neoplastic cells by the immunofluorescence technique.

Metastasis *in vivo* depends on successful attachment of malignant cells at the new site (Fidler, 1975). The facts that transformed cells recruit fibronectin for efficient adhesion and can also spread on fibronectin-coated substrata are suggestive of similar molecular events during metastasis of malignant cells *in vivo*. Tumour cells can readily recruit fibronectin from the blood and they also can adhere to fibronectin on the surface of platelets. Tumour cells in circulation have often been observed in association with platelets (Chew, Josephron & Wallace, 1976). Since antibody-mediated cytoxity appears to be very efficient only on non-adherent cells, adhesion of tumour cells to platelet surfaces may be of significance in their survival against host defence mechanisms.

Acknowledgement. The work reported here was supported in part by the Medical Research Council of Canada, Grant No. MA-6462, the International Cancer Data Bank Programme of the National Cancer Institute under Contract No. 1-(0-65 34) with the International Union Against Cancer and Dalhousie University Internal Medical Research Foundation.

REFERENCES

Abercrombie, M. (1975). The contact behaviour of invading cells. In *Cellular Membranes and Tumor Cell Behaviour. 20th Annual Symposium on Fundamental Cancer Research*, p. 21. Baltimore, Md: The William & Wilkins Co.

Burger, M. M. (1973). Surface changes in transformed cells detected by lectins. *Federation Proceedings*, **32**, 91.

Chew, E. C., Josephron, R. L. & Wallace, A. R. (1976). Morphologic aspects of the arrest of circulating cancer cells. In *Fundamental Aspects of Metastasis*, ed. L. Weiss, p. 121. Amsterdam: North Holland.

Coman, D. E. (1944). Decreased mutual adhesiveness, a property of cells from squamous cell carcinomas. *Cancer Research*, **4**, 625.

Culp, L. A. (1976). Molecular composition and origin of substrate-attached material from normal and virus transformed cells. *Journal of Supramolecular Structure*, **5**, 239.

Culp, L. A. & Buniel, J. F. (1976). Substrate-attached serum and cell proteins in adhesion of mouse fibroblasts. *Journal of Cellular Physiology*, **88**, 89.

Edelman, G. M., Wang, J. L. & Yahara, I. (1976). Surface modulating assemblies in mammalian cells. In *Cell Motility, Book A*, ed. R. Goldman, T. Pollard and J. Rosenbaum, p. 305. Cold Spring Harbor Laboratory.

Engvall, E. & Ruoslahti, E. (1977). Binding of soluble form of fibroblast surface protein, fibronectin, to collagen, *International Journal of Cancer*, **20**, 1.

Fidler, I. J. (1975). Mechanisms of cancer invasion and metastasis. In *Cancer Research : Biology of Tumors : Surfaces, Immunology and Comparative Pathology*, ed. F. F. Becker, vol. 4, p. 101. New York: Plenum Publications Corpn.

Fox, R. A., Fernandez, L. A. & Rajaraman, R. (1977). Migration inhibition produced by sodium periodate oxidation of the macrophage membrane and reversal by sodium borohydride. *Scandinavian Journal of Immunology*, **6**, 1151.

Gahmberg, C. G. & Hakamori, S. (1973). Altered growth behavior of malignant cells associated with changes in externally labelled glycoprotein and glycolipid. *Proceedings of the National Academy of Sciences, USA*, **70**, 3329.

Gaile, M. H. & Boone, C. W. (1972). Cell-substrate adhesivity: a determinant of cell motility. *Experimental Cell Research*, **70**, 33.

Gilula, N. B., Reeves, D. R. & Steinbach, A. (1967). Metabolic coupling, ionic coupling and cell contacts. *Nature, London*, **235**, 262.

Grinnell, F. (1976*a*). The serum dependence of baby hamster kidney cell attachment to a substratum. *Experimental Cell Research*, **97**, 265.

Grinnell, F. (1976*b*). Cell spreading factor. Occurrence and specificity of action. *Experimental Cell Research*, **102**, 51.

Grinnell, F. (1976*c*). Biochemical analysis of cell adhesion to a substratum its possible relevance to cell metastasis. *Progress in Clinical and Biological Research*, **9**, 227.

Hahn, L.-H. E. & Yamada, K. M. (1979). Identification and isolation of a collagen binding fragment of the adhesive glycoprotein fibronectin. *Proceedings of the National Academy of Sciences, USA*, **76**, 1160–3.

Hatanaka, M. (1974). Transport of sugars in tumour cell membranes. *Biochemica et Biophysica Acta*, **355**, 77.

Hogg, N. M. (1974). A comparison of membrane proteins of normal and transformed cells by lactoperoxidase labelling. *Proceedings of the National Academy of Sciences, USA*, **71**, 489.

Hynes, R. O. (1973). Alteration of cell-surface proteins by viral transformation and by proteolysis. *Proceedings of the National Academy of Sciences, USA*, **70**, 3170.

Hynes, R. O. (1976). Cell surface proteins and malignant transformation. *Biochemica et Biophysica Acta*, **458**, 73.

Juliano, R. L. & Gagalong, E. (1977). The adhesion of Chinese Hamster cells. I. Effects of temperature, metabolic inhibitors and proteolytic dissection of cell surface macromolecules. *Journal of Cellular Physiology*, **92**, 209.

Keski-Oja, J. (1977). Fibronectin, a major cell surface-associated glycoprotein. Properties in cultured cells, polymerization, and loss in viral transformation. Ph.D. Thesis, Department of Virology, University of Helsinki, Helsinki, Finland.

Maroudas, N. G. (1975). Polymer exclusion, cell adhesion and membrane fusion. *Nature, London*, **254**, 695.

Morrison, P., Edsall, R. & Miller, S. G. (1948). Preparation and properties of serum and plasma proteins XVIII. The separation of purified fibrinogen from fraction I of human plasma. *Journal of the American Chemical Society*, **70**, 3103.

Nicolson, G. L. (1973). Temperature-dependent mobility of concanavalin A sites on tumor cell surfaces. *Nature, London*, **243**, 218.

Pegrum, S. M. & Maroudas, N. G. (1975). Early events in fibroblast adhesion to glass. An electron microscopic study. *Experimental Cell Research*, **96**, 416.

Ponten, J. (1976). The relationship between in vitro transformation and tumor formation in vivo. *Biochemica et Biophysica Acta*, **458**, 397.

Rajaraman, R., Fox, R. A., Vethamany, V. G., Fernandez, L. A. & MacSween, J. M. (1977). Adhesion and spreading behavior of human peripheral blood mononuclear cells (PBMC) in vitro. *Experimental Cell Research*, **107**, 179.

Rajaraman, R., Rounds, D. E., Yen, S. P. S. & Rembaum, A. (1974). A scanning electron microscope study of cell adhesion and spreading in vitro. *Experimental Cell Research*, **88**, 327.

Rajaraman, R., Rounds, D. E., Yen, S. P. S. & Rembaum, A. (1975). Effects of Ionenes on normal and transformed cells. In *Polyelectrolytes and their Applications*, ed. A. Rembaum and E. Selegny, p. 163. Dordrecht-Holland: D. Reidel Publishing Co.

Rajaraman, R., MacSween, J. M. & Fox, R. A. (1978*a*). Models of normal and transformed cell adhesion, and capping and locomotion in vitro. *Journal of Theoretical Biology*, **74**, 177.

Rajaraman, R., Westermark, B., Vaheri, A. & Ponten, J. (1978*b*). Immunofluorescence studies on fibronectin distribution patterns during adhesion, deformation and spreading of human glial and glioma cells. *Annals of the New York Academy of Sciences*, **312**, 444.

Rees, D. A., Lloyd, C. W. & Thom, D. (1977). Control of grip and stick in cell adhesion through lateral relationships of membrane glycoproteins. *Nature, London*, **267**, 124.

Robbins, P. W., Wickus, G. G., Branton, P. E., Gaffney, B. J., Hirschberg, C. B., Fuchs, P. & Blumberg, P. M. (1974). The chick fibroblast cell surface after transformation by Rous Sarcoma virus. *Cold Spring Harbor Symposium on Quantitative Biology*, **39**, 1173.

Rosenblith, J. Z., Ukena, I. E., Yin, H. H., Berlin, R. P. & Karnovsky, M. J. (1973). A comparative evaluation of the distribution of concanavalin A binding sites on the upper surfaces of normal, virally transformed, and protease-treated fibroblasts. *Proceedings of the National Academy of Sciences, USA*, **70**, 1625.

Ruoslahti, E., Hayman, E. G. & Engvall, E. (1979). Interaction of fibronectin with collagens, and its role in cell adhesion. *Journal of Supramolecular Structure*, (suppl.), **3**, 173.

RUOSLAHTI, E. & VAHERI, A. (1974). Novel human serum protein from fibroblast plasma membrane. *Nature, London*, **248**, 789.

RUOSLAHTI, E. & VAHERI, A. (1975). Interaction of soluble fibroblast surface antigen with fibrinogen and fibrin. Identity with cold insoluble globulin of human plasma. *Journal of Experimental Medicine*, **141**, 497.

RUOSLAHTI, E., VAHERI, A., KUUSELA, P. & LINDER, E. (1973). Fibroblast surface antigen: a new serum protein. *Biochemica et Biophysica Acta*, **322**, 352.

SHIELDS, R. & POLLOCK, K. (1974). The adhesion of BHK and PyBHK cells to the substratum. *Cell*, **3**, 31.

SHIN, S., FREEDMAN, V. H., RISSER, R. & POLLACK, R. (1975). Tumorigenicity of virus-transformed cells in nude mice is correlated specifically with anchorage independent growth in vitro. *Proceedings of the National Academy of Sciences, USA*, **72**, 4435.

STOKER, M., O'NEILL, C., BERRYMAN, S. & WAXMAN, V. (1968). Anchorage and growth regulation in normal and virus transformed cells. *International Journal of Cancer*, **3**, 683.

TAKEICHI, M. (1971). Changes in the properties of cell substrate adhesion during cultivation of chicken fibroblasts in vitro in a serum free medium. *Experimental Cell Research*, **68**, 88.

TAYLOR, A. C. (1961). Attachment and spreading of cells in culture. *Experimental Cell Research* (suppl.), **8**, 154.

TAYLOR, R., DUFFUS, W. P. H., RAFF, M. C. & DE PETRIS, S. (1971). Redistribution and pinocytosis of lymphocyte surface immunoglobulin molecules induced by anti-immunoglobulin antibody. *Nature, London*, **233**, 225.

VAHERI, A. (1977). Surface proteins of virus transformed cells. In *Virus Transformed Cell Membranes*, ed. C. Nicolson, p. 139. New York and London: Academic Press.

VAHERI, A. & RUOSLAHTI, E. (1974). Disappearance of a major cell type specific surface glycoprotein antigen (SF) after transformation of fibroblasts by Rous Sarcoma Virus. *International Journal of Cancer*, **13**, 519.

VAHERI, A., RUOSLAHTI, E., WESTERMARK, B. & PONTEN, J. (1976). A common cell-type specific surface antigen in cultured human glial cells and fibroblasts: loss in malignant cells. *Journal of Experimental Medicine*, **143**, 64.

WARTIOVAARA, J., LINDER, E., RUOSLAHTI, E. & VAHERI, A. (1976). Distribution of fibroblast surface antigen. Association with fibrillar structures of normal cells and loss upon viral transformation. *Journal of Experimental Medicine*, **140**, 1522.

WEISS, L., POSTE, G., MACKEARMIN, A. & WILLETT, K. (1975). Growth of mammalian cells on substrates coated with cellular microexudates I. Effect on cell growth at low population densities. *Journal of Cell Biology*, **64**, 135.

WISEMAN, L. L., STEINBERG, M. S. & PHILLIPS, H. M. (1972). Experimental modulation of intercellular cohesiveness: reversal of tissue assembly patterns. *Development Biology*, **28**, 498.

WITKOWSKI, J. A. & BRIGHTON, W. D. (1971). Stages of spreading of human diploid cells on glass surfaces. *Experimental Cell Research*, **68**, 372.

Witkowski, J. A. & Brighton, W. D. (1972). Influence of serum on attachment of tissue cells to glass surfaces. *Experimental Cell Research*, **70**, 41.

Yamada, K. M., Hahn, L.-H. E. & Olden, K. (1979). Structure and function of fibronectins. *Journal of Supramolecular Structure* (suppl.), **3**, 167.

Yamada, K. M. & Weston, J. A. (1974). Isolation of a major cell surface glycoprotein from fibroblasts. *Proceedings of the National Academy of Sciences, USA*, **71**, 3492.

Yamada, K. M., Yamada, S. S. & Pastan, I. (1977). Quantitation of a transformation sensitive adhesive cell surface glycoprotein. *Journal of Cell Biology*, **74**, 647.

INDEX

Page numbers in bold type refer to figures, legends to figures or tables.